Communications in Computer and Information Science 2729

Series Editors

Gang Li ⓘ, *School of Information Technology, Deakin University, Burwood, VIC, Australia*

Joaquim Filipe ⓘ, *Polytechnic Institute of Setúbal, Setúbal, Portugal*

Zhiwei Xu, *Chinese Academy of Sciences, Beijing, China*

Rationale

The CCIS series is devoted to the publication of proceedings of computer science conferences. Its aim is to efficiently disseminate original research results in informatics in printed and electronic form. While the focus is on publication of peer-reviewed full papers presenting mature work, inclusion of reviewed short papers reporting on work in progress is welcome, too. Besides globally relevant meetings with internationally representative program committees guaranteeing a strict peer-reviewing and paper selection process, conferences run by societies or of high regional or national relevance are also considered for publication.

Topics

The topical scope of CCIS spans the entire spectrum of informatics ranging from foundational topics in the theory of computing to information and communications science and technology and a broad variety of interdisciplinary application fields.

Information for Volume Editors and Authors

Publication in CCIS is free of charge. No royalties are paid, however, we offer registered conference participants temporary free access to the online version of the conference proceedings on SpringerLink (http://link.springer.com) by means of an http referrer from the conference website and/or a number of complimentary printed copies, as specified in the official acceptance email of the event.

CCIS proceedings can be published in time for distribution at conferences or as post-proceedings, and delivered in the form of printed books and/or electronically as USBs and/or e-content licenses for accessing proceedings at SpringerLink. Furthermore, CCIS proceedings are included in the CCIS electronic book series hosted in the SpringerLink digital library at http://link.springer.com/bookseries/7899. Conferences publishing in CCIS are allowed to use our online conference service (Meteor) for managing the whole proceedings lifecycle (from submission and reviewing to preparing for publication) free of charge.

Publication process

The language of publication is exclusively English. Authors publishing in CCIS have to sign the Springer CCIS copyright transfer form, however, they are free to use their material published in CCIS for substantially changed, more elaborate subsequent publications elsewhere. For the preparation of the camera-ready papers/files, authors have to strictly adhere to the Springer CCIS Authors' Instructions and are strongly encouraged to use the CCIS LaTeX style files or templates.

Abstracting/Indexing

CCIS is abstracted/indexed in DBLP, Google Scholar, EI-Compendex, Mathematical Reviews, SCImago, Scopus. CCIS volumes are also submitted for the inclusion in ISI Proceedings.

How to start

To start the evaluation of your proposal for inclusion in the CCIS series, please send an e-mail to ccis@springer.com

Vicente Garcia Diaz · I-Hsien Ting · Kai Wang
Editors

Multidisciplinary Social Networks Research

12th International Conference, MISNC 2025
Oviedo, Spain, September 3–5, 2025
Proceedings

Editors
Vicente Garcia Diaz ⓘ
University of Oviedo
Oviedo, Spain

I-Hsien Ting ⓘ
National University of Kaohsiung
Kaohsiung, Taiwan

Kai Wang ⓘ
National University of Kaohsiung
Kaohsiung, Taiwan

ISSN 1865-0929 ISSN 1865-0937 (electronic)
Communications in Computer and Information Science
ISBN 978-3-032-09944-0 ISBN 978-3-032-09945-7 (eBook)
https://doi.org/10.1007/978-3-032-09945-7

© The Editor(s) (if applicable) and The Author(s), under exclusive license to Springer Nature Switzerland AG 2026

This work is subject to copyright. All rights are solely and exclusively licensed by the Publisher, whether the whole or part of the material is concerned, specifically the rights of translation, reprinting, reuse of illustrations, recitation, broadcasting, reproduction on microfilms or in any other physical way, and transmission or information storage and retrieval, electronic adaptation, computer software, or by similar or dissimilar methodology now known or hereafter developed.
The use of general descriptive names, registered names, trademarks, service marks, etc. in this publication does not imply, even in the absence of a specific statement, that such names are exempt from the relevant protective laws and regulations and therefore free for general use.
The publisher, the authors and the editors are safe to assume that the advice and information in this book are believed to be true and accurate at the date of publication. Neither the publisher nor the authors or the editors give a warranty, expressed or implied, with respect to the material contained herein or for any errors or omissions that may have been made. The publisher remains neutral with regard to jurisdictional claims in published maps and institutional affiliations.

This Springer imprint is published by the registered company Springer Nature Switzerland AG
The registered company address is: Gewerbestrasse 11, 6330 Cham, Switzerland

If disposing of this product, please recycle the paper.

Preface

It is our great pleasure to welcome you to the *12th Multidisciplinary International Social Networks Conference (MISNC 2025)*, held from September 3–5, 2025, at the University of Oviedo, Spain. Since its inception in 2014 in Kaohsiung, Taiwan, MISNC has grown into an international forum for scholars, researchers, and practitioners to exchange ideas, present research findings, and explore innovations in the ever-evolving domain of social networks. Over the past decade, the conference has traveled across Asia, Europe, and North America—spanning cities such as Matsuyama, Japan; Bangkok, Thailand; Saint-Étienne, France; Wenzhou, China; Bergen, Norway; Phuket, Thailand; and Bali, Indonesia—continually expanding its reach and fostering a truly global research community.

MISNC is dedicated to advancing multidisciplinary research in social networking, a field that bridges technology, communication, sociology, business, and beyond. The conference serves as a platform for cross-disciplinary dialogue, enabling participants from academia and industry to share perspectives, methodologies, and real-world applications. Through such collaboration, MISNC not only strengthens academic inquiry but also addresses the needs and challenges faced by practitioners, thereby shaping the future directions of social network research and practice.

This year, MISNC 2025 received 89 submissions, out of which 42 papers were accepted for inclusion in the proceedings. Reviews were double blind, and submissions received three reviews each on average. The authors represent 14 countries and regions, including China, Colombia, Ecuador, Finland, France, Germany, Indonesia, Japan, Malaysia, Mexico, Spain, Taiwan, Thailand, and the UK. These contributions reflect the diversity and richness of the global research community engaged in advancing theories, methodologies, and applications in areas such as socionetwork strategies, intelligent informatics, education, health care, software engineering, information management, and many more.

We would like to extend our sincere gratitude to all authors for their valuable contributions, to the program committee members and reviewers for their diligent efforts in ensuring the quality of the proceedings, and to our keynote speakers and session chairs for enriching the conference with their expertise. Our heartfelt appreciation also goes to the organizing committee and supporting institutions whose dedication made MISNC 2025 possible.

We hope that MISNC 2025 in Oviedo, Spain, provided not only a stimulating academic exchange but also an opportunity for participants to build enduring collaborations and friendships. We hope you had a successful and memorable conference experience.

Program Co-chairs

Daniel Fernández Lanvin
Basit Shahzad
Nitin Singh
Kai Wang

Organization

Honorary Chair

Tzung-Pei Hong — National University of Kaohsiung, Taiwan

General Chairs

Juan Manuel Cueva Lovelle — Universidad de Oviedo, Spain
Vicente Garcia Diaz — University of Oviedo, Spain
Dario Liberona — Seinäjoki University of Applied Sciences, Finland
I-Hsien Ting — National University of Kaohsiung, Taiwan

Program Chairs

Daniel Fernández Lanvin — Universidad de Oviedo, Spain
Basit Shahzad — University of Prince Mugrin, Saudi Arabia
Nitin Singh — IIM Ranchi, India
Kai Wang — National University of Kaohsiung, Taiwan

Organizing Chair

Jordán Pascual Espada — University of Oviedo, Spain

Steering Committee Chairs

Cathy Hsing-Tzu Lin — National University of Kaohsiung, Taiwan
George Chang — Kean University, USA
Vincent Tseng — National Yang Ming Chiao-Tung University, Taiwan
Shiro Uesugi — Matsuyama University, Japan
Chih-Chien Wang — National Taipei University, Taiwan
Guandong Xu — University of Technology Sydney, Australia
Katsutoshi Yada — Kansai University, Japan

Publication Chairs

Chung-Hong Lee	National Kaohsiung University of Science and Technology, Taiwan
Hsin-Chang Yang	National University of Kaohsiung, Taiwan

Publicity Chairs

Chian-Hsueng Chao	National University of Kaohsiung, Taiwan
I-Wen Fang	Tatung University, Taiwan
Hiroshi Koga	Kansai University, Japan
Yen-Hsien Lee	National Chiayi University, Taiwan
Kazunori Minetaki	Kindai University, Japan
Charles Perez	Paris School of Business, France
Takashi Okamoto	Ehime University, Japan
Karina Sokolova	Paris School of Business, France
Didi Sundiman	Universitas Universal, Indonesia
Ming-Chia Hsieh	I-Shou University, Taiwan
Chuan-Chun Wu	I-Shou University, Taiwan

Special Issue Chairs

Guan-Lin Chen	Shu-Te University, Taiwan
Chien-Liang Lin	Ming Chuan University, Taiwan

Industry Chairs

Chia-Chi Chang	National University of Kaohsiung, Taiwan
Chia-Sung Yen	National Chin-Yi University, Taiwan

Special Session Chairs

Jerry Chun-Wei Lin	Silesian University of Technology, Poland
Yu-Lung Wu	I-Shou University, Taiwan

Program Committee Members

Dan Braha	New England Complex Systems Institute, USA
Teruyuki Bunno	Kindai University, Japan
Chao-Ching Chang	National Taipei University of Education, Taiwan
Chia-Chi Chang	National University of Kaohsiung, Taiwan
Chih-Hao Chang	National Kaohsiung University of Science and Technology, Taiwan
Nai-Wen Chang	Kun Shan University, Taiwan
Chian-Hsueng Chao	National University of Kaohsiung, Taiwan
Richard Chbeir	University of Pau & Pays Adour, France
Yunwei Chen	Chengdu Library of Chinese Academy of Sciences, China
Ping Chou	National University of Kaohsiung, Taiwan
João Cordeiro	University of Évora, Portugal
Tsai-Hsin Chu	National Chiayi University, Taiwan
Yu Cui	Kansai University, Japan
Min-Yuh Day	National Taipei University, Taiwan
Juan Carlos Figueroa-García	Universidad Distrital Francisco José de Caldas, Colombia
Hafiz Budi Firmansyah	University of Geneva, Switzerland
Begoña Cristina Pelayo García-Bustelo	University of Oviedo, Spain
Vicente Garcia Diaz	University of Oviedo, Spain
Jordán Pascual Espada	University of Oviedo, Spain
I-Wen Fang	Tatung University, Taiwan
Zhiwei Guo	Chongqing Technology and Business University, China
Tzung-Pei Hong	National University of Kaohsiung, Taiwan
Han-Fen Hu	University of Nevada, Las Vegas, USA
Ching-Yu (Austin) Huang	Kean University, USA
Ming-Chia Hsieh	I-Shou University, Taiwan
Carol Hsu	University of Sydney, Australia
Pin-Rui Hwang	National United University, Taiwan
Hiroki Idota	Kindai University, Japan
Jason Jung	Chung-Ang University, South Korea
Vinothini Kasinathan	Asia Pacific University of Technology & Innovation, Malaysia
Chutisant Kerdvibulvech	National Institute of Development Administration, Thailand
Hiroshi Koga	Kansai University, Japan
Ying-Feng Kuo	National University of Kaohsiung, Taiwan

Christine Largeron	Université Jean Monnet, France
Cheng-Yu Lai	Chung Yuan Christian University, Taiwan
Chia-Wen Lee	Xiamen Ocean Vocational College, China
Yen-Hsien Lee	National Chiayi University, Taiwan
Ching Li	National Taiwan Normal University, Taiwan
Chih-Chin Liang	National Formosa University, Taiwan
Paoling Liao	National Kaohsiung University of Science and Technology, Taiwan
Dario Liberona	Seinäjoki University of Applied Sciences, Finland
Cathy Hsing-Tzu Lin	National University of Kaohsiung, Taiwan
Chien-Liang Lin	Ming Chuan University, Taiwan
Jerry Chun-Wei Lin	Silesian University of Technology, Poland
Wen-Yang Lin	National University of Kaohsiung, Taiwan
Juan Manuel Cueva Lovelle	Universidad de Oviedo, Spain
Jaime Solís Martínez	University of Oviedo, Spain
Kazunori Minetaki	Kindai University, Japan
Dzulkifli Mukhtar	Universiti Malaysia Kelantan, Malaysia
Sanetake Nagayoshi	Shizuoka University, Japan
Roman Neruda	Institute of Computer Science, Academy of Sciences of the Czech Republic, Czechia
Edward Rolando Núñez-Valdez	University of Oviedo, Spain
Takashi Okamoto	Ehime University, Japan
Jo Hui-Chun Pan	Southern Taiwan University of Science and Technology, Taiwan
Charles Perez	Paris School of Business, France
Amer Rashee	University of South Denmark, Denmark
Suppanunta Romprasert	Srinakharinwirot University, Thailand
Hidenobu Sai	Hokkai-Gakuen University, Japan
Norihito Seki	Hokkai-Gakuen University, Japan
Muhammad Ikhsan Setiawan	Narotama University, Indonesia
Rong-An Shang	Soochow University, Taiwan
Jun Shen	University of Wollongong, Australia
Nitin Singh	IIM Ranchi, India
Karina Sokolova	Paris School of Business, France
Gautam Srivastava	Brandon University, Canada
Didi Sundiman	Universitas Universal, Indonesia
Sheikh Abu Taher	Jahangirnagar University, Bangladesh
Jeffrey Chi-Feng Tai	National Chiayi University, Taiwan
I-Hsien Ting	National University of Kaohsiung, Taiwan
Yu-Chuan Tsai	National University of Kaohsiung, Taiwan
Masatsugu Tsuji	Kobe International University, Japan
Masashi Ueda	Kyoto Sangyo University, Japan

Yasushi Ueki	Institute of Developing Economies, Japan
Shiro Uesugi	Matsuyama University, Japan
Chih-Chien Wang	National Taipei University, Taiwan
Kai Wang	National University of Kaohsiung, Taiwan
Li Weigang	University of Brasília, Brazil
Belfin Robinson	University of North Carolina at Chapel Hill, USA
Quyuan Wang	Chongqing Technology and Business University, China
Chienhsing Wu	National University of Kaohsiung, Taiwan
Chuan-Chun Wu	I-Shou University, Taiwan
Yu-Lung Wu	I-Shou University, Taiwan
Sachiko Yanagihara	University of Toyama, Japan
Chin-Sheng Tang	Yuan Ze University, Taiwan
Katsutoshi Yada	Kansai University, Japan
Hsin-Chang Yang	National University of Kaohsiung, Taiwan
Shu-Chen Yang	National University of Kaohsiung, Taiwan
Chia-Sung Yen	National Chin-Yi University of Technology, Taiwan
Yi Zuo	Kansai University, Japan
Keyi Zhong	Loughborough University, UK

Contents

Multidisciplinary Social Networks Research

How Visual Design Influences the Continuous Use of AI Chatbots: The Role of Cute Appearance and Rich Facial Expressions 3
    Jian-Ren Hou, Yu-Ching Chang, Ying-Feng Kuo, and Yi-Hsun Lin

An Empirical Study on User Profile Analysis and SEO Performance: A Case of Taiwan Cultural Memory Bank 2.0 16
    Mei-Yun Hsu, I-Hsien Ting, Yun-Hsiu Liu, and Kazunori Minetaki

Artificial Intelligence in Natural Science Education: A Case Study on Students' Environmental Literacy 30
    Chia-Sung Yen

Liquid Consumption Behavior in Online Purchasing: Preliminary Discussion of Online Survey Results 39
    Hiroshi KOGA

Organizational Memory Facilitates Effective Learning from Failure: Evidence from a Questionnaire Survey Conducted in 2024 in Japanese Companies 47
    S. Nagayoshi and J. Nakamura

Exploring the Relationship Between Local Election Results and Online Public Opinion in Taiwan: A Case Study of Taitung County 59
    I-Hsien Ting, Yen-Chih Chiu, Yun-Hsiu Liu, Kazunori Minetaki, and Chia-Sung Yen

Factors Influencing Value Co-creation in Live Streaming E-Commerce Ecosystems: The Microfoundations Theory Perspective 70
    Kai Wang, Yong-Yi Chang, and Jia-Chi Lin

Analyzing the Improvement of Students' Performance in Group Discussion: A Comparative Study of Online and Face-to-Face Collaborative Learning - A Case Study of Japanese Undergraduate Students . . . 81
    Laila Diana Khulyati and Sanetake Nagayoshi

Factors Influencing Continuance Intention of Watching in VTuber Live Streaming: A Co-creation Experience Perspective 94
    Kai Wang, Yen-Chan Lee, and Chih-Hsuan Yeh

Entropy-Based Anomaly Detection for Cybersecurity Threats in Network Traffic 106
    Han-Wei Hsiao and Yun-Zhen Lee

Designing an Agentic AI–Driven Framework for Decision-Making in Industrial Maintenance 118
    Miguel Alonso-González, Vicente García-Díaz, and Benjamín López Pérez

Comparing Human and Machine-Labeled Sentiment Analysis for Disaster Response 131
    Hafiz Budi Firmansyah and Aidil Afriansyah

Reconfiguring Knowledge Management Systems for the Web 3.0 Ecosystem 142
    Yu Cui, Hiroki Idota, and Masaharu Ota

A LLM-Driven Agent System for Automotive Fault Diagnosis with Integrated Reasoning and Expertise Sharing 156
    Tzu-Ting Weng, Jun-Teng Sun, Zhen-Xin Fu, and Chung-Hong Lee

Sales Predictive Model with Customer Segmentation Integration 167
    Alex Mejía and Priscila Valdiviezo-Diaz

Study on the Use of Generative AI by Japanese IT Freelance Engineers: Effectuation as a Theory of Entrepreneurial Behavior 178
    Kentaro Goto

Relational Taxonomy of Cyberattacks: A Model for Threat Classification and Connection in Digital Environments 192
    Mikel Ferrer Oliva, José Amelio Medina Merodio, José Javier Martínez Herraiz, and Carlos Cilleruelo Rodriguez

Taxonomy of Human Errors in Smart Energy Systems Cybersecurity 206
    Mikel Ferrer Oliva, José Amelio Medina Merodio, Alberto Larena Luengo, and José Javier Martínez Herraiz

Exploring New Business in Japanese SMEs from the Perspective of Employees' Engagement 217
    Kazunori Minetaki, Hiroki Idota, and Teruyuki Bunno

Facial Expression-Based Monitoring of Fatigue and Stress in Care Facility Workers 226
    Kazunori Minetaki, I-Hsien Ting, Teruyuki Bunno, and Hiroki Idota

Facial and Vocal Cues in Daily Life: A Multimodal Framework
for Behavioral Insight 238
   Kazunori Minetaki and I-Hsien Ting

Question-Answering Systems for Tourism: Development of a RAG-Based
Prototype for Ecuadorian Places 247
   Yahaira Benítez-Morejón, Janneth Chicaiza, and Priscila Valdiviezo-Diaz

Key Approaches to Neuroleadership: Perspective from the Public
Educational Institutions of Montería 263
   *Diana Patricia Eljach Hernández, Leonardo Antonio Díaz Pertuz,
   Helmer Muñoz Hernández, Kavir Ala Oviedo Prioló,
   and Yamid Fabian Hernández Julio*

Structural Analysis as a Tool to Establish the Influence of ICT
in the Educational Processes of Rural Institutions in San Antero, Cordoba,
Colombia 279
   *Leonardo Antonio Díaz Pertuz, Diana Patricia Eljach Hernández,
   Yamid Fabian Hernández Julio, Helmer Muñoz Hernández,
   and Kavir Ala Oviedo Prioló*

An Efficient Federated Utility-Mining Algorithm 298
   Tzung-Pei Hong, Jing-Chi Yang, Yu-Chuan Tsai, and Chun-Hao Chen

The Study of Value Creation and AI Agent in Social Network: Governance
and Ethical Risks Aspects 311
   Chian-Hsueng Chao, Pei-Chen Hsu, Xiu-Hua Wei, and Tian-Yi See

Enhancing Sustainability Education Through Social Media Platforms 324
   Dario Liberona

Exploring Awareness in Social Media Regarding Europe (EU) and Taiwan
Relations 338
   Marcel Rother and Dario Liberona

How to Highlight the Lowest Values: Adapting Bar Chart Features
to Emphasize Data Minima 354
   Laura Montenegro, Jordán Pascual Espada, and Juan Luis Carús Candás

Generative AI and the Future of Social Content: Opportunities and Challenges

Enhancing K-means Clustering in B2B Customer Segmentation: A Comparative and Hybrid Approach of Recursive Feature Elimination, Correlation Analysis, and Lasso Regularization 369
   Daisy Ipatzi Bello, Sabeen Tahir, and Stefania Paladini

Virtual or Human, Real or Rendered: Assessing the Persuasiveness of Synthetic Influencers 385
   Karina Sokolova

Do Autonomous Agents Exhibit Consistent Personality Traits in Open Social Environments? 396
   Charles Perez and Samir El Hassani

Utility Evaluation of Synthetic Data by Variational Autoencoder 408
   Natsuki Sano and Yejun Tao

Advanced Social Computing through AI-centric Multidisciplinary Fusion

A Determination Framework of Quantitative Element of Criminal Offenses Integrating Explainable Artificial Intelligence 419
   Quyuan Wang, Yuxin Liao, Xia Hu, Yuping Tu, Run Zeng, and Zhiwei Guo

Quantification of Fish Feeding Behavior with MC-YOLO and Image Texture Features in Recirculating Aquaculture Systems 431
   Jiachun Zhou, Junchao Yang, Xueni Pan, Haiyan Huang, Li Liu, Yanzheng Gao, and Yu Shen

Surfacing Fish Detection Based on MSSYOLO in Recirculating Aquaculture System 443
   Ruiwen Xiao, Junchao Yang, Haiyan Huang, Xueni Pan, Yu Shen, and Yanzheng Gao

Design of LoRA Tunning-Assisted Pretrained LLM Structure for Sentiment Analysis in Online E-Commerce 455
   Yuanyuan Cai, Zhiwei Guo, Han Zhao, and Bo Liu

BEDGKT: A Behavior-Enhanced Dynamic Graph Knowledge Tracing Model 467
   Rongkui Yu and Ying Wang

Network, Economic and Business Studies

Enhancing Energy Policy Planning in Thailand Through AI-Based
Forecasting of Crude Oil Imports 481
    Suppanunta Romprasert, Tanapat Romprasert, and Danai Tanamee

Digital Transformation as an Enabler of Organizational Green Learning
Orientation and Ambidextrous Green Innovation: Organizational
Information Processing Perspectives 498
    Yi-Chun Huang, Min-Li Yang, and Miao-Hui Chiu

Consumer Attitude Toward QR Code Payments with Smartphone in Japan:
Satisfaction and Dissatisfaction 515
    S. Nagayoshi and K. C. Abisikha

Feasibility Analysis on Green Business Plan from Biomass 527
    Danai Tanamee and Suppanunta Romprasert

Author Index 539

Multidisciplinary Social Networks Research

How Visual Design Influences the Continuous Use of AI Chatbots: The Role of Cute Appearance and Rich Facial Expressions

Jian-Ren Hou[1], Yu-Ching Chang[1], Ying-Feng Kuo[2(✉)], and Yi-Hsun Lin[2]

[1] Institute of Information Management, National Cheng-Kung University, Tainan, Taiwan
[2] Department of Information Management, National University of Kaohsiung, Kaohsiung, Taiwan
fredkuo@nuk.edu.tw

Abstract. This study, grounded in the UTAUT2 model, examines the effects of cute appearance and rich facial expressions of AI chatbots on performance expectancy, effort expectations, social influence, and hedonic motivation, as well as how these factors influence users' intentions to continue using the chatbots. The results indicate that, compared to a non-cute appearance and non-rich facial expressions, a cute appearance and rich facial expressions significantly enhance users' performance expectancy and hedonic motivation. However, no significant effects were found for effort expectancy and social influence. Additionally, path analysis reveals that performance expectancy, social influence, and hedonic motivation have a significant positive impact on users' intention to continue using AI chatbots, while effort expectancy shows no significant effect.

Keywords: AI · chatbots · Extended Unified Theory of Acceptance and Use of Technology (UTAUT2) · cute appearance · facial expressions

1 Introduction

Chatbots are computer programs that simulate human language (written or spoken) using natural language processing (NLP) and interact with users through digital interfaces. In recent years, they have gained widespread attention in the industry, serving as the digital frontlines for many businesses. Traditional chatbots rely on predefined dialogue flows to match user inputs and generate natural language responses. More recently, AI-powered chatbots have been equipped with advanced speech recognition and NLP technologies, enabling them to comprehend more complex conversations and provide precise responses to user needs [57]. Additionally, chatbots are not influenced by emotions, allowing them to maintain consistently friendly interactions while handling large volumes of user inquiries, thereby enhancing their commercial value.

As AI chatbots increasingly simulate human-like conversations and engage in more intimate dialogues, their perceived identity as machines becomes a critical factor. Prior research suggests that whether users perceive their conversational partner as a robot or a human significantly influences communication styles and interaction outcomes [24, 36]. People often assume that chatbots are less capable of processing emotions compared to humans [38], which can reduce trust and negatively impact the overall interaction experience [19].

Cuteness, defined as the ability to attract attention in an endearing or charming manner, is often associated with baby-like features [9]. Exposure to baby-like stimuli has been shown to elicit positive emotions [9, 28, 43]. Incorporating cute elements into product design can evoke a stronger sense of comfort compared to identical products without such features [15]. In nonverbal communication, facial expressions play a more significant role in conveying emotions than other communication cues [39]. In interpersonal communication, facial expressions serve as key informational cues that reflect emotions [22, 31]. Since humans are highly skilled at interpreting behaviors through facial expressions, incorporating expressive facial features into chatbot design can enhance the naturalness and stability of human-computer interactions [21]. Therefore, this study posits that endowing AI chatbots with a cute appearance or expressive facial features can increase their appeal and improve user experience.

With the expansion of market size and the continuous introduction of new products, competition among businesses has intensified. Understanding whether consumers continue to use a technology remains a central concern for both academia and industry. UTAUT2 (Unified Theory of Acceptance and Use of Technology) [55] is one of the most widely used models for explaining technology adoption. Therefore, this study first examines the effects of cute appearance and facial expressiveness on key UTAUT2 constructs and further explores their impact on users' intention to continue using AI chatbots.

This study offers several important theoretical contributions. By extending the UTAUT2 framework to incorporate design-related affective cues, specifically the cuteness of chatbot appearance and the richness of facial expressions, it deepens our understanding of how non-functional and emotionally engaging design elements influence users' perceptions and their intentions to continue using AI-powered chatbots. While the UTAUT2 model has traditionally emphasized cognitive and utilitarian factors such as performance expectancy and effort expectancy, this study introduces aesthetic and emotional dimensions, showing that a cute appearance and expressive facial features significantly enhance both performance expectancy and hedonic motivation. These findings enrich the existing literature by highlighting the importance of visual and emotional design in shaping technology acceptance, particularly in human-computer interaction contexts where emotional engagement is essential.

From a practical standpoint, the results offer valuable insights for designers and businesses seeking to increase user engagement with AI chatbots. As competition intensifies in digital customer service and conversational AI markets, this study suggests that enhancing the visual and expressive design of chatbots, such as incorporating baby-like features or animated facial expressions, can meaningfully improve users' perceptions of usefulness and enjoyment. This improvement in user experience can, in turn, encourage continued usage. Furthermore, by identifying which UTAUT2 constructs are significantly influenced by these design features, practitioners can make more informed decisions about chatbot development. Focusing on elements that enhance performance expectancy and hedonic motivation may ultimately lead to higher user retention.

2 Literature Review and Hypothesis Development

2.1 Utaut2

Venkatesh et al. [54] proposed the Unified Theory of Acceptance and Use of Technology (UTAUT) to provide a comprehensive framework for explaining and predicting user acceptance of technology. Later, Venkatesh et al. [55] extended this model to UTAUT2, enhancing its explanatory power in the context of consumer technology adoption. UTAUT2 comprises seven key constructs: performance expectancy, effort expectancy, social influence, facilitating conditions, hedonic motivation, price value, and habit.

Numerous studies have confirmed that performance expectancy and effort expectancy positively influence users' intentions to adopt information systems (e.g., [20, 53, 56]). Additionally, social influence has been recognized as a critical factor in promoting technology adoption [20, 35, 56]. In consumer contexts, hedonic motivation plays a key role in technology acceptance and usage behavior [10, 14].

However, in this study's context, facilitating conditions, price value, and habit are less applicable and are therefore excluded from the research model. Facilitating conditions refer to the availability of resources and support necessary for users to perform a given behavior [55]. As this study examines AI chatbot usage in an era where computers and the internet are widely accessible, most users already possess the necessary technical resources, making facilitating conditions less relevant to continued usage behavior. Price value represents the trade-off between the perceived benefits and monetary costs of an application [55]. Since many AI chatbots currently offer free versions, price considerations have minimal influence on usage intentions. Habit, which reflects users' past experiences and behavioral tendencies in technology use [55], is also excluded, as AI chatbots remain in the developmental stage, and users' accumulated experience varies significantly.

Based on these considerations, this study focuses on four UTAUT2 constructs—performance expectancy, effort expectancy, social influence, and hedonic motivation—to examine their effects on users' intentions to continue using AI chatbots.

2.2 Cute Appearance and Facial Expressions

Cuteness is generally defined as a characteristic that evokes positive emotions in an endearing manner [9, 28, 43]. In academic literature, cuteness is often associated with "baby schema," which includes infantile features and behaviors such as a large forehead, big eyes, chubby cheeks, plump limbs, and clumsy yet curious movements [25].

In product design, cuteness is a subjective perception primarily shaped by visual features [43]. Consumers tend to be more receptive to cute products or brands and are more forgiving of their flaws [7, 29, 49]. When interacting with cute products, consumers generally exhibit more positive and patient attitudes and may even be more willing to learn complex operations [49]. Additionally, cuteness enhances attentiveness, encouraging users to perform tasks more carefully while reducing frustration from operational errors [43, 47].

In recent years, cuteness has become a prominent trend in AI application design, particularly in its influence on user experience through non-functional features [11, 34, 37]. A cute appearance increases a product's appeal and consumer engagement, thereby influencing product evaluation, purchase intentions, and word-of-mouth recommendations [33, 42, 46, 50]. Furthermore, cute elements can evoke positive emotions, enhance consumer emotional engagement, and provide psychological benefits such as relaxation and stress relief [15, 28, 43].

Facial expressions are one of the most important means of information exchange in interpersonal interactions [40]. In addition to conveying biological information, such as identity, gender, and age through facial features, facial expressions also offer insights into an individual's emotional state and even social traits, including personality [22, 31]. Since people naturally rely on facial expressions to interpret behaviors, animated avatars that simulate speech, gestures, and facial expressions can significantly enhance the naturalness and stability of human-computer interactions [21].

2.3 Hypothesis Development

Performance expectancy refers to the belief that using a particular technology will enhance job performance [54]. Cuteness has been shown to improve focus, increasing user attentiveness during task execution and reducing frustration caused by operational errors [43, 47]. Additionally, cuteness can enhance trust in the product [58]. Moreover, rich facial expressions enable AI chatbots to deliver more vivid responses, facilitating easier understanding and interaction [8], while also fostering greater trust [45], thereby enhancing performance expectancy. Based on these insights, we propose the following hypotheses:

H1a: Compared to a non-cute appearance, a cute appearance is more effective in enhancing expectancy.
H1b: Compared to non-rich facial expressions, rich facial expressions is more effective in enhancing performance expectancy.

Effort expectancy refers to the perceived ease of learning and using a particular technology [54]. When a product incorporates cute elements, users tend to display more positive and patient attitudes, even being willing to learn more complex operations [49]. Additionally, rich facial expressions can improve non-verbal communication, making a robot's instructions and responses more intuitive and reducing cognitive load [16]. Based on this, the following hypotheses are proposed:

H2a: Compared to a non-cute appearance, a cute appearance is more effective in enhancing effort expectancy.
H2b: Compared to non-rich facial expressions, rich facial expressions are more effective in enhancing effort expectancy.

Social influence refers to the user's perception that significant others (e.g., family and friends) believe they should use the technology [54]. A cute appearance enhances product interest and consumer attention, which, in turn, influences product evaluation, purchase intention, and word-of-mouth recommendations [33, 42, 46, 50]. Rich facial expressions can also enhance the interactivity of AI chatbots, making them more socially engaging [45, 52]. Therefore, we propose the following hypotheses:

H3a: Compared to a non-cute appearance, a cute appearance is more effective in enhancing social influence.
H3b: Compared to non-rich facial expressions, rich facial expressions are more effective in enhancing social influence.

Hedonic motivation refers to the enjoyment and pleasure derived from using the technology itself [10]. A cute appearance tends to evoke human affection and fondness [33, 42, 46, 50], and it also provides stronger psychological recovery and therapeutic effects [15, 28, 43], leading users to form an emotional connection with AI chatbots, thus enhancing their enjoyment during use. Rich facial expressions can alleviate feelings of boredom [32], making interactions more vivid and engaging, further enhancing users' hedonic experiences [45]. Accordingly, we propose the following hypotheses:

H4a: Compared to a non-cute appearance, a cute appearance is more effective in enhancing hedonic motivation.
H4b: Compared to non-rich facial expressions, rich facial expressions are more effective in enhancing hedonic motivation.

Performance expectancy refers to the degree to which users believe adopting the technology will improve their job performance [54]. If users believe that a technology can provide effective information and assistance, thereby increasing efficiency, they are more likely to choose and continue using it [12]. Previous research has demonstrated that performance expectancy positively influences technology adoption behavior [2, 6, 59]. When users believe that an AI chatbot can quickly provide accurate information and reduce search time, they are more likely to continue using the technology. Therefore, we propose the following hypothesis:

H5: Performance expectancy will enhance users' intention to continue using the AI chatbot.

Effort expectancy refers to the perceived ease of use of a particular technology [54]. If users find a technology easy to operate and low in learning cost, they are more likely to use it [30]. Conversely, if users believe that a technology requires too much learning and adaptation, they may abandon it, even if it has potential value [17]. Previous research has indicated that the ease of use of a technology is positively related to continued usage behavior [13]. Therefore, if users find using an AI chatbot easy, it will increase their intention to continue using it. Based on this, we propose the following hypothesis:

H6: Effort expectancy will enhance users' intention to continue using the AI chatbot.

Social influence refers to the user's perception that significant others (e.g., family and friends) believe they should adopt the technology [54]. Social influence has a particularly strong effect on user behavior intentions when the technology is still in the early stages of diffusion [1]. Additionally, social influence increases users' willingness to adopt chatbots [5, 48, 51]. Therefore, if users perceive that their social network widely accepts and recommends AI chatbots, they will be more likely to continue using them. Based on this, we propose the following hypothesis:

H7: Social influence will enhance users' intention to continue using the AI chatbot.

Hedonic motivation refers to the enjoyment and pleasure gained from using the technology [10]. In information technology research, hedonic motivation has been identified as a key factor influencing usage intention [3, 4, 44]. When users find interactions with AI chatbots enjoyable and entertaining, they are more likely to continue using them. Therefore, we propose the following hypothesis:

H8: Hedonic motivation will enhance users' intention to continue using the AI chatbot.

3 Research Methodology

3.1 Participants

This study collected data through an online questionnaire distributed via Facebook in Taiwan, resulting in 156 valid responses. Among the respondents, 44.2% identified as male and 55.8% as female. The largest age groups were 19 to 24 years old (36.5%) and 25 to 34 years old (21.2%). In terms of education, nearly half of the participants held a bachelor's degree (48.7%), followed by those with a master's degree (30.1%). Regarding occupation, the most common group was students, accounting for 51.3% of the sample. Furthermore, a large majority of participants (91.7%) reported having previously used a commercially available AI chatbot.

3.2 Research Design

This study employed a 2 (Appearance: cute vs. non-cute) × 2 (Facial expressions: rich vs. non-rich) between-subjects design, resulting in four experimental conditions. Participants were randomly assigned to one of the conditions. Furthermore, a standardized conversation with the AI chatbot was designed, with the content remaining consistent across all conditions. Prior to the experiment, a pretest was conducted to ensure the appropriateness of the manipulation.

The scales used in this study—performance expectancy, effort expectancy, social influence, hedonic motivation, and intention to continue use—were adapted from the UTAUT2 model [55], with modifications made to fit the research context. A seven-point Likert scale was used to measure the responses, ranging from strongly disagree (1) to strongly agree (7).

4 Results

4.1 Manipulation Check

Independent samples t-tests were conducted to verify the success of the manipulation of appearance and facial expressions. The mean for the cute appearance (M = 4.870, SD = 1.324) was significantly higher than that for the non-cute appearance (M = 3.290, SD

= 1.145) (t = 8.001, p < .001). Similarly, the mean for the rich facial expression (M = 4.970, SD = 1.258) was significantly higher than that for the non-rich facial expression (M = 2.830, SD = 1.333) (t = 10.314, p < .001), indicating that the manipulations were successful.

4.2 Hypothesis Test for H1a-H4b

MANOVA was employed to test hypotheses H1a-H4b. Given that the interaction effect between appearance and facial expression did not significantly influence performance expectancy (F = 0.035, p = .851), effort expectancy (F = 2.583, p = .110), social influence (F = 0.831, p = .364), and hedonic motivation (F = 0.612, p = .435), we proceeded to test the main effects of appearance and facial expression separately.

In terms of performance expectancy, the cute appearance (M = 4.848, SD = 1.011) was significantly higher than the non-cute appearance (M = 4.450, SD = 1.030) (F = 6.413, p < .05), thus supporting H1a. Additionally, the rich facial expression (M = 4.949, SD = 1.002) was significantly higher than the non-rich facial expression (M = 4.355, SD = 0.990) (F = 13.873, p < .001), supporting H1b.

Regarding effort expectancy, no significant difference was found between the cute appearance (M = 5.617, SD = 0.968) and the non-cute appearance (M = 5.591, SD = 0.807) (F = 0.032, p = .840), thus failing to support H2a. Similarly, no significant difference was observed between the rich facial expression (M = 5.705, SD = 0.847) and the non-rich facial expression (M = 5.503, SD = 0.925) (F = 2.023, p = .161), failing to support H2b.

For social influence, no significant difference was found between the cute appearance (M = 4.536, SD = 1.299) and the non-cute appearance (M = 4.178, SD = 1.090) (F = 5.121, p = .060), thus failing to support H3a. Similarly, no significant difference was found between the rich facial expression (M = 4.513, SD = 1.270) and the non-rich facial expression (M = 4.205, SD = 1.135) (F = 2.546, p = .102), failing to support H3b.

For hedonic motivation, the cute appearance (M = 5.177, SD = 1.003) was significantly higher than the non-cute appearance (M = 4.606, SD = 1.223) (F = 13.081, p < .001), supporting H4a. Similarly, the rich facial expression (M = 5.205, SD = 1.045) was significantly higher than the non-rich facial expression (M = 4.586, SD = 1.172) (F = 12.146, p < .001), supporting H4b.

4.3 Hypothesis Test for H5-H8

We first conducted confirmatory factor analysis (CFA) to assess the reliability, convergent validity, and discriminant validity of the scales. Table 1 presents the factor loadings, composite reliability, and average variance extracted (AVE) for each construct, while Table 2 provides the inter-variable correlations and square roots of AVE.

Regarding reliability, the composite reliability (CR) ranged from 0.874 to 0.931, exceeding the threshold of 0.7 [26], indicating good reliability. For convergent validity, the factor loadings for all constructs exceeded the threshold of 0.70, and the average variance extracted (AVE) for each construct was above the threshold of 0.5 [23], confirming the scales' good convergent validity. Additionally, the square roots of the AVE for each construct were greater than the correlation between that construct and others [23], demonstrating good discriminant validity for the scales employed in this study.

This study examined the variance inflation factor (VIF) to assess the presence of multicollinearity. Regression analysis was performed with continuous usage intention as the dependent variable and the other constructs as independent variables. The VIF values for performance expectancy (VIF = 2.815), effort expectancy (VIF = 1.213), social influence (VIF = 2.025), and hedonic motivation (VIF = 2.464) were all below the threshold value of 3.3 [18], indicating no issues with multicollinearity.

Subsequently, structural equation modeling (SEM) was conducted using Smart-PLS 4 to test hypotheses H5–H8. The path analysis results revealed that performance expectancy had a significant positive effect on continuous usage intention ($\beta = 0.288$, $p < .001$), supporting H5. Effort expectancy had no significant effect on continuous usage intention ($\beta = 0.020$, $p = 0.622$), so H6 was not supported. Social influence had a significant positive effect on continuous usage intention ($\beta = 0.177$, $p < .005$), supporting H7. Hedonic motivation also had a significant positive effect on continuous usage intention ($\beta = 0.508$, $p < .001$), supporting H8. The combined explanatory power (R^2) of performance expectancy, effort expectancy, social influence, and hedonic motivation for continuous usage intention was 76.5%, demonstrating strong explanatory power [27].

Table 1. Factor loadings, CR, and AVE for each construct

Construct	Item	Factor loading	CR	AVE
Performance expectancy (PE)	PE1	0.866	0.874	0.797
	PE2	0.924		
	PE3	0.887		
Effort expectancy (EE)	EE1	0.866	0.923	0.757
	EE2	0.899		
	EE3	0.859		
	EE4	0.856		
Social influence (SI)	SI1	0.928	0.903	0.838
	SI2	0.935		
	SI3	0.883		
Hedonic motivation (HM)	HM1	0.902	0.895	0.820
	HM2	0.896		
	HM3	0.919		
Continuous usage intention (CU)	CU1	0.909	0.931	0.825
	CU2	0.935		
	CU3	0.913		
	CU4	0.876		

Table 2. Inter-variable correlations and square roots of AVE

	PE	EE	SI	HM	CU
PE	**0.893**				
EE	0.417	**0.870**			
SI	0.684	0.270	**0.916**		
HM	0.752	0.385	0.647	**0.906**	
CU	0.642	0.343	0.698	0.831	**0.908**

Note: Diagonal elements represent the square root of AVE by that construct.

5 Conclusions and Discussions

5.1 Conclusions

The study, based on the UTAUT2 model, investigates how a cute appearance and rich facial expressions in AI chatbots influence performance expectancy, effort expectancy, social influence, and hedonic motivation, as well as how these factors shape users' intentions to continue using the chatbots. The findings reveal that, compared to a non-cute appearance and non-rich facial expressions, a cute appearance and rich facial expressions significantly enhance users' performance expectancy and hedonic motivation. However, they have no significant impact on effort expectancy and social influence. Furthermore, path analysis indicates that performance expectancy, social influence, and hedonic motivation enhance users' intentions to continue using AI chatbots, whereas effort expectancy do not show a significant effect.

5.2 Theoretical Implications

This study provides new insights into the application and extension of the UTAUT2 model. First, the findings confirm that a cute appearance and rich facial expressions are more effective in enhancing performance expectancy and hedonic motivation than a non-cute appearance and non-rich facial expressions, supporting previous research on cuteness and emotional attraction [9, 15]. Second, the study reveals that effort expectancy does not significantly influence continuous usage intention, which contrasts with the predictions of the traditional UTAUT2 model. This may reflect the intuitive nature of AI chatbots, which do not require substantial learning effort, thereby reducing the impact of effort expectancy in this context. Furthermore, the study confirms the positive role of social influence on continuous usage intention, highlighting the importance of social factors in human-computer interaction. This contribution expands existing literature by addressing the limited attention given to the impact of visual design on social influence [24, 36].

5.3 Practical Implications

In terms of practical implications, the findings offer valuable guidance for the visual design and application of AI chatbots. First, companies developing chatbots can improve user experience by incorporating cute appearance and rich facial expressions, which can enhance performance expectancy and hedonic motivation, leading to greater continuous usage intention. Second, as performance expectancy and hedonic motivation have a stronger influence on continuous usage intention, companies should focus on improving the chatbot's utility and entertainment value, such as by providing more accurate responses and incorporating interactive expression features. Finally, the positive effect of social influence on continuous usage intention suggests that companies can emphasize the social interaction value of AI chatbots in their marketing strategies, such as encouraging users to share their experience to further reinforce continuous usage behavior.

5.4 Limitations and Future Research

While this study offers important insights into AI chatbot design, it does have several limitations. First, the study participants may have been limited by personal backgrounds and preferences, which could result in a sample that does not fully represent all potential user groups. Second, the focus of the study was primarily on cute appearance and facial expression richness, without considering other potential visual or interactive design elements, such as voice style, animation effects, or the degree of anthropomorphism, which could also impact usage intention.

Future research could explore the following directions: (1) Comparison across different usage contexts: Future studies could explore how cute appearance and facial expression richness affect users in various application contexts, such as customer service, healthcare, or educational assistance. (2) Additional visual and interactive design factors: Future research could include other design elements that influence user experience, such as anthropomorphism, emotional voice expressions, and animation smoothness, to further refine the behavioral model of AI chatbot usage. (3) Impact of individual differences: Research could examine how individual characteristics, such as technology acceptance, cultural background, and gender, moderate the effects of cute appearance and facial expressions on usage intention, offering more refined market segmentation strategies.

References

1. Adapa, A., Nah, F.F.-H., Hall, R.H., Siau, K., Smith, S.N.: Factors influencing the adoption of smart wearable devices. Int. J. Human-computer Interact. **34**(5), 399–409 (2018)
2. Ahmed, M.H., et al.: Intention to use electronic medical record and its predictors among health care providers at referral hospitals, north-West Ethiopia, 2019: Using unified theory of acceptance and use technology 2 (UTAUT2) model. BMC Med. Inform. Decis. Mak. **20**(1), 207 (2020)
3. Alalwan, A.A., Dwivedi, Y.K., Rana, N.P.: Factors influencing adoption of mobile banking by Jordanian bank customers: Extending UTAUT2 with trust. Int. J. Inf. Manage. **37**(3), 99–110 (2017)

4. Al-Azawei, A., Alowayr, A.: Predicting the intention to use and hedonic motivation for mobile learning: A comparative study in two Middle Eastern countries. Technol. Soc. **62**, 101325 (2020)
5. Al-Emran, M., Al-Qudah, A.A., Abbasi, G.A., Al-Sharafi, M.A., Iranmanesh, M.: Determinants of Using AI-Based Chatbots for Knowledge Sharing: Evidence From PLS-SEM and Fuzzy Sets (fsQCA). IEEE Trans. Eng. Manage. **71**, 4985–4999 (2023)
6. Anthony, B., Kamaludin, A., Romli, A.: Predicting academic staffs behaviour intention and actual use of blended learning in higher education: Model development and validation. Technol. Knowl. Learn. **28**, 1223–1269 (2021)
7. Beverland, M.B., Farrelly, F.J.: The quest for authenticity in consumption: Consumers' purposive choice of authentic cues to shape experienced outcomes. J. Consum. Res. **36**(5), 838–856 (2010)
8. Breazeal, C: Emotion and sociable humanoid robots. International Journal of Human-Computer Studies, **59**(1–2), 119–155 (2003)
9. Brosch, T., Sander, D., Scherer, K.R.: That baby caught my eye... attention capture by infant faces. Emotion **7**(3), 685–689 (2007)
10. Brown, S.A., Venkatesh, V.: Model of adoption of technology in the household: A baseline model test and extension incorporating household life cycle. MIS Q. **29**(4), 399–426 (2005)
11. Caudwell, C., Lacey, C.: What do home robots want? The ambivalent power of cuteness in robotic relationships. Convergence **26**(4), 956–968 (2020)
12. Chakava, M.H., Mberia, H.K., Gatero, G.: Relationship between performance expectancy and use of new media in scholarly communication by academic staff in public universities in Kenya. IOSR-JHSS **23**(6), 49–59 (2018)
13. Chatterjee, S., Bhattacharjee, K.K.: Adoption of artificial intelligence in higher education: A quantitative analysis using structural equation modelling. Educ. Inf. Technol. **25**(5), 3443–3463 (2020)
14. Childers, T.L., Carr, C.L., Peck, J., Carson, S.: Hedonic and utilitarian motivations for online retail shopping behavior. J. Retail. **77**(4), 511–535 (2001)
15. Chou, H.Y., Chu, X.Y., Chen, T.C.: The healing effect of cute elements. J. Consum. Aff. **56**(2), 565–596 (2022)
16. Darling, K.: Extending legal protection to social robots: The efect of anthropomorphism, empathy, and violent behavior towards robotic objects. In R. Calo, A. M. Froomkin, I. Kerr (Eds.), Robot Law, pp. 213–234. Edward Elgar, Cheltenham (2016)
17. Davis, F.D.: Perceived usefulness, perceived ease of use, and user acceptance of information technology. MIS Q. **13**(3), 319–340 (1989)
18. Diamantopoulos, A., Siguaw, J.A.: Formative versus reflective indicators in 42 organizational measure development: A comparison and empirical illustration. Br. J. Manag. **17**(4), 263–282 (2006)
19. Dietvorst, B.J., Simmons, J.P., Massey, C.: Overcoming algorithm aversion: People will use imperfect algorithms if they can (even slightly) modify them. Manage. Sci. **64**(3), 1155–1170 (2018)
20. Eckhardt, A., Laumer, S., Weitzel, T.: Who influences whom? Analyzing workplace referents' social influence on its adoption and non-adoption. J. Inf. Technol. **24**(1), 11–24 (2009)
21. Edlund, J., Beskow, J.: Pushy versus meek—using avatars to influence turn-taking behaviour. Proceedings of Interspeech-2007, 682–685 (2007)
22. Elfenbein, H.A., Ambady, N.: On the universality and cultural specificity of emotion recognition: A meta-analysis. Psychol. Bull. **128**(2), 203–235 (2002)
23. Fornell, C., Larcker, D.F.: Evaluating structural equation models with unobservable variables and measurement error. J. Mark. Res. **18**(1), 39–50 (1981)

24. Fox, J., Ahn, S.J., Janssen, J.H., Yeykelis, L., Segovia, K.Y., Bailenson, J.N.: Avatars versus agents: A meta-analysis quantifying the effect of agency on social influence. Human-Computer Interaction **30**(5), 401–432 (2015)
25. Golle, J., Lisibach, S., Mast, F.W., Lobmaier, J.S.: Sweet puppies and cute babies: Perceptual adaptation to babyfacedness transfers across species. PLoS ONE **8**(3), e58248 (2013)
26. Hair, J.F., Anderson, R.E., Tatham, R.L., Black, W.C.: Multivariate Data Analysis, 5th edn. Prentice-Hall, Upper Saddle River, NJ (1998)
27. Hair, J.F., Ringle, C.M., Sarstedt, M.: PLS-SEM: Indeed a silver bullet. J. Mark. Theory Pract. **19**(2), 139–152 (2011)
28. Hellén, K., Sääksjärvi, M.: Development of a scale measuring childlike anthropomorphism in products. J. Mark. Manag. **29**(1–2), 141–157 (2013)
29. Holbrook, M.B., Woodside, A.G.: Animal companions, consumption experiences, and the marketing of pets: Transcending boundaries in the animal human distinction. J. Bus. Res. **61**(5), 377–381 (2008)
30. Huang, D.H., Chueh, H.E.: Behavioral intention to continuously use learning apps: A comparative study from Taiwan universities. Technol. Forecast. Soc. Chang. **177**, 121531 (2022)
31. Izard, C.E.: Basic emotions, natural kinds, emotional schemas, and a new paradigm. Perspect. Psychol. Sci. **2**(3), 260–280 (2007)
32. Kumar, J.A.: Facial animacy in anthropomorphised designs: Insights from leveraging self-report and facial expression analysis for multimedia learning. Comput. Educ. **223**, 105150 (2024)
33. Lee, H.C., Chang, C.T., Chen, Y.H., Huang, Y.S.: The spell of cuteness in food consumption? It depends on food type and consumption motivation. Food Qual. Prefer. **65**, 110–117 (2018)
34. Lin, Y.T., Doong, H.S., Eisingerich, A.B.: Avatar design of virtual salespeople: Mitigation of recommendation conflicts. J. Serv. Res. **24**(1), 141–159 (2021)
35. Lu, J., Yao, J.E., Yu, C.S.: Personal innovativeness, social influences and adoption of wireless internet services via mobile technology. J. Strat. Inf. Syst. **14**, 245–268 (2005)
36. Lucas, G.M., Gratch, J., King, A., Morency, L.P.: It's only a computer: Virtual humans increase willingness to disclose. Comput. Hum. Behav. **37**, 94–100 (2014)
37. Lv, X.Y., Liu, Y., Luo, J.J., Liu, Y.Q., Li, C.X.: Does a cute artificial intelligence assistant soften the blow? The impact of cuteness on customer tolerance of assistant service failure. Ann. Tour. Res. **87**, 103114 (2021)
38. Madhavan, P., Wiegmann, D.A., Lacson, F.C.: Automation failures on tasks easily performed by operators undermine trust in automated aids. Hum. Factors **48**(2), 241–256 (2006)
39. Mandler, G.: Mind and body: Psychology of Emotion and Stress. Norton, New York (1984)
40. Mehrabian, A.: Communication without words. Psychol. Today **2**(4), 53–56 (1968)
41. Moffett, J.W., Folse, J.A.G., Palmatier, R.W.: A theory of multiformat communication: Mechanisms, dynamics, and strategies. J. Acad. Mark. Sci. **49**, 441–461 (2021)
42. Morreall, J.: Cuteness. Br. J. Aesthet. **31**(1), 39–47 (1991)
43. Nenkov, G.Y., Scott, M.L.: So cute I could eat it up: Priming effects of cute products on indulgent consumption. J. Consum. Res. **41**(2), 326–341 (2014)
44. Nikolopoulou, K., Gialamas, V., Lavidas, K.: Habit, hedonic motivation, performance expectancy and technological pedagogical knowledge affect teachers' intention to use mobile internet. Comput. Educ. Open **2**, 100041 (2021)
45. Qiu, L., Benbasat, I.: Evaluating anthropomorphic product recommendation agents: A social relationship perspective to designing information systems. J. Manag. Inf. Syst. **25**(4), 145–182 (2009)
46. Sanders, J.T.: On "cuteness." Br. J. Aesthet. **32**(2), 162–165 (1992)
47. Sherman, G.D., Haidt, J., Coan, J.A.: Viewing cute images increases behavioral carefulness. Emotion **9**(2), 282–286 (2009)

48. Sing, C.C., Teo, T., Huang, F., Chiu, T.K.F., Xing Wei, W.: Secondary school students' intentions to learn AI: Testing moderation effects of readiness, social good and optimism. Education Tech. Research Dev. **70**(4), 765–782 (2022)
49. Sprengelmeyer, R., et al.: The cutest little baby face: A hormonal link to sensitivity to cuteness in infant faces. Psychol. Sci. **20**(2), 149–154 (2009)
50. Sun, J., Nazlan, N.H., Leung, X.Y., Bai, B.: A cute surprise: Examining the influence of meeting giveaways on word-of-mouth intention. J. Hosp. Tour. Manag. **45**, 456–463 (2020)
51. Terblanche, N., Kidd, M.: Adoption factors and moderating effects of age and gender that influence the intention to use a nondirective reflective coaching chatbot. SAGE Open **12**(2), 1–16 (2022)
52. Thomaz, F., Salge, C., El, K., Hulland, J.: Learning from the dark web: Leveraging conversational agents in the era of hyper-privacy to enhance marketing. J. Acad. Mark. Sci. **48**(1), 43–63 (2020)
53. Van Raaij, E.M., Schepers, J.J.L.: The acceptance and use of a virtual learning environment in China. Comput. Educ. **50**(3), 838–852 (2008)
54. Venkatesh, V., Morris, M.G., Davis, G.B., Davis, F.D.: User acceptance information technology: toward a unified view. MIS Q. **27**(3), 425–478 (2003)
55. Venkatesh, V., Thong, J.Y.L., Xu, X.: Consumer acceptance and use of information technology: Extending the unified theory of acceptance and use of technology. MIS Q. **36**(1), 157–178 (2012)
56. Wang, Y.S., Wu, M.C., Wang, H.Y.: Investigating the determinants and age and gender differences in the acceptance of mobile learning. Br. J. Edu. Technol. **40**(1), 92–118 (2009)
57. Wilson, H.J., Daugherty, P.R., Morini-Bianzino, N.: The jobs that artificial intelligence will create. MIT Sloan Manag. Rev. **58**(4), 14–17 (2017)
58. Yim, A., Cui, A.P., Walsh, M.: The role of cuteness on consumer attachment to artificial intelligence agents. J. Res. Interact. Mark. **18**(1), 127–141 (2024)
59. Zhou, L.L., Owusu-Marfo, J., Asante Antwi, H., Antwi, M.O., Kachie, A.D.T., Ampon-Wireko, S.: Assessment of the social influence and facilitating conditions that support nurses' adoption of hospital electronic information management systems (HEIMS) in Ghana using the unified theory of acceptance and use of technology (UTAUT) model. BMC Med. Inform. Decis. Mak. **19**(1), 230 (2019)

An Empirical Study on User Profile Analysis and SEO Performance: A Case of Taiwan Cultural Memory Bank 2.0

Mei-Yun Hsu[1], I-Hsien Ting[2(✉)], Yun-Hsiu Liu[2], and Kazunori Minetaki[3]

[1] National Museum of Taiwan, History, Tainan City, Taiwan
myhsu@nmth.gov.tw
[2] National University of Kaohsiung, Kaohsiung City, Taiwan
iting@nuk.edu.tw
[3] Kindai University, Higashiosaka, Japan
kminetaki@bus.kindai.ac.jp

Abstract. Taiwan Cultural Memory Bank 2.0 is an online curation platform that invites the public to become curators, fostering diverse perspectives on Taiwan's society, humanities, natural landscapes, and daily life. Built on a material bank concept, the platform encourages users to co-create and curate their own works using shared resources or self-uploaded materials. At its core, the system follows a "collect, store, access, and reuse" model, supporting dynamic engagement with over three million cultural memory items from Taiwan. Users can search, browse, explore stories, and engage in creative applications and collaborative productions. Understanding user profiles is crucial for enhancing website service quality, particularly within the framework of the Visitor Relationship Management (VRM) model. This study conducts an empirical analysis of user profiles on the platform, examining demographic characteristics, browsing behaviors, and engagement patterns. Additionally, the research evaluates the platform's SEO performance, search visibility, and organic traffic effectiveness. Based on the findings, this study provides strategic recommendations for optimizing website management, improving user experience, and leveraging social media for enhanced digital outreach. The insights gained contribute to the broader discussion on digital cultural platforms and their role in audience engagement, online visibility, and networked communication.

Keywords: User Profile · SEO · Taiwan Cultural Memory Bank · Social Media · Google Analytics

1 Introduction

Taiwan Cultural Memory Bank 2.0 https://tcmb.culture.tw/ is a digital platform dedicated to preserving and revitalizing Taiwan's rich cultural heritage. It serves as an online curation environment where users are empowered to become curators themselves, offering personal and collective perspectives on Taiwan's society, humanities, natural landscapes, and everyday life. The platform is structured around a "material bank" concept, encouraging not only the use of existing resources but also the contribution of new materials. Through this, users can engage in co-creation, crafting their own curated works from a combination of shared and self-uploaded content, transforming the site into a participatory and evolving cultural archive (Fig. 1).

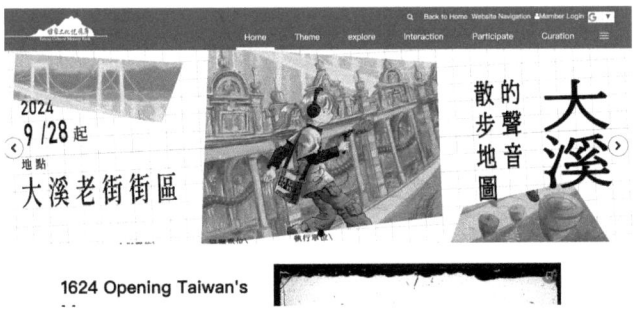

Fig. 1. The Website of Taiwan Cultural Memory Bank 2.0

From a system design perspective, the platform operates along four core functions: collect, store, access, and reuse. This structure ensures that cultural materials are not only preserved but also readily available for reinterpretation and creative deployment. Whether users are searching for specific themes, browsing through multimedia resources, exploring contextualized cultural stories, or assembling personalized exhibitions, the website facilitates an interactive and user-driven approach to cultural engagement. The system's flexibility and openness aim to bridge professional cultural institutions and the public, fostering a sense of shared stewardship over Taiwan's cultural memory.

Currently, the platform houses more than three million cultural memory items from across Taiwan, making it one of the most extensive repositories of its kind. These items include images, documents, videos, oral histories, and more—each contributing to a diverse and layered representation of Taiwan's cultural landscape. Users are invited to search, explore, and engage with these materials through story-based browsing and thematic exploration. In addition to individual use, the platform promotes creative applications and collaborative projects, supporting educational, artistic, and community-driven endeavors [4].

To better understand and enhance the effectiveness of Taiwan Cultural Memory Bank 2.0, this study focuses on analyzing user profiles and platform performance [10]. Using the Visitor Relationship Management (VRM) framework, the research investigates demographic characteristics, browsing patterns, and engagement behaviors [13]. It also evaluates the site's SEO performance, organic traffic, and search visibility. Furthermore, the study incorporates social media analytics and social network analysis (SNA) to assess how content is shared and circulated across platforms [9,17]. These insights aim to inform strategic recommendations for improving user experience, optimizing content delivery, and strengthening digital outreach through networked communication.

The remainder of this paper is organized as follows. Section 2 reviews the relevant literature on Visitor Relationship Management to establish the theoretical foundation of the study. As user profiling and search engine optimization are the two primary research focuses, related studies in these areas are also discussed in this section. Section 3 presents the findings on the online user profiles of Taiwan Cultural Memory Bank 2.0, while Sect. 4 details the results of the SEO performance analysis. Finally, Sect. 5 concludes the paper and offers recommendations for the future development of the Taiwan Cultural Memory Bank, as well as suggestions for further research.

2 Literature Review

2.1 Visitor Relationship Management

Visitor Relationship Management (VRM) in museums increasingly integrates concepts from customer relationship management (CRM) and knowledge management (KM) to enhance visitor engagement and institutional effectiveness [2,8]. By adopting CRM principles, museums aim to better understand and respond to the needs, interests, and behaviors of their visitors [5]. This approach is supported by a knowledge management framework, which provides a systematic model for managing and leveraging visitor data [11]. As shown in Fig. 2, the knowledge management cycle—comprising knowledge acquisition, storage, distribution, and use—serves as a foundation for informed decision-making across marketing, service, and content development within the museum. VRM thus becomes a strategic tool for building long-term visitor relationships and delivering more personalized cultural experiences [6].

The integration of data collection and analysis is central to this model, enabling museums to refine their marketing strategies, enhance service quality, and develop more targeted and relevant content [1]. Through the cyclical flow of knowledge—beginning with acquisition from visitor interactions, followed by structured storage, internal distribution, and practical application—museums can adapt to visitor needs in a more agile and data-informed manner [18]. At the heart of this model lies "visitor focus," which aligns marketing, content, and service in a triadic relationship designed to maximize engagement and satisfaction [14]. This knowledge-driven approach not only improves operational efficiency but also fosters deeper, more meaningful connections between museums and their audiences.

2.2 User Profile

A user profile is a structured collection of information that represents an individual user, typically encompassing both demographic data (such as age, gender,

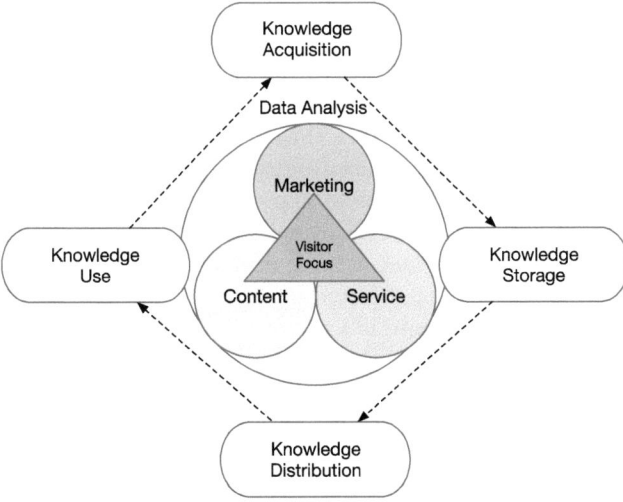

Fig. 2. The Visitor Relationship Management Model [6]

location) and behavioral data (such as preferences, interests, and interaction history). These profiles are foundational in digital systems where personalization, recommendation, and user-centric design are critical. In online environments, user profiles serve as a core element for tailoring content, improving user experience, and facilitating targeted communication. As digital platforms grow increasingly complex and user-driven, the role of user profiling becomes more significant in supporting data-informed strategies [7].

Understanding online user behavior is one of the most important applications of user profiling. This typically involves the analysis of user interactions captured through browsing histories, system logs, clickstreams, and other digital footprints. By mining and interpreting such data, researchers and practitioners can uncover patterns in user navigation, content consumption, and decision-making processes. These insights are particularly valuable for improving interface design, content structuring, and marketing effectiveness. Log data, in particular, provides a rich and often underutilized source for behavioral analysis, enabling dynamic and real-time understanding of user preferences [13].

A wide range of analytical methods and tools are available to support the creation and analysis of user profiles. Demographic analysis helps segment users by background characteristics, while log analysis focuses on user interactions over time. Social network analysis can reveal relationships and influence patterns within digital communities, and data visualization facilitates intuitive interpretation of complex behavioral trends [20]. Tools such as Gephi [16], Google Analytics, and platform-specific insights tools from Facebook and YouTube allow researchers and marketers to efficiently gather, process, and present user data [3]. These methods and tools collectively enhance the capacity to build comprehensive, actionable user profiles in various digital contexts.

2.3 SEO

Search Engine Optimization (SEO) refers to the strategic process of enhancing a website's visibility and ranking within search engine results by understanding and aligning with the algorithms that govern search engine behavior. By optimizing various elements of a website—such as content structure, metadata, keywords, and backlinks—SEO aims to improve the site's relevance and authority in the eyes of search engines. The ultimate goal is to increase organic traffic, improve search performance, and attract new users who are actively seeking related information. In addition, SEO practices often include diagnostic insights and recommendations for website improvement, thereby contributing to overall web usability and content quality [19].

However, the growing importance of SEO has also led to the emergence of automated techniques, including the use of bots and scripts designed to manipulate search engine rankings. These artificial methods can mimic user interactions or generate fake backlinks to boost visibility, raising concerns about the integrity of search results. As a result, recent research has focused on detecting and mitigating the impact of such bots to preserve fair competition and maintain the credibility of search engine ecosystems. The challenge lies in distinguishing between legitimate optimization efforts and deceptive practices, which continues to be an evolving area of interest in both academic and industry settings [12].

3 Online User Profile of Taiwan Cultural Memory Bank 2.0

To better understand online user behavior, we use Google Analytics as our primary tool. In addition to data from the Taiwan Cultural Memory Bank 2.0, we also collect data from the Taiwan Cultural Memory Bank 1.0. The analysis covers the period from November 29, 2022, to November 2023.

Since its launch on October 17, 2020, the Memory Bank 1.0 portal has been in operation for nearly two years and has accumulated a substantial user base. Consequently, its webpage views are significantly higher compared to those of the Memory Bank 2.0 platform. Although the Memory Bank 2.0 thematic platform initially exhibited lower traffic, the project team began publishing monthly feature articles starting in May 2023. These curated topics, tailored to each month, aim to attract public interest and are promoted through various channels associated with the Memory Bank. As a result, a noticeable upward trend in page views has been observed, as illustrated in Fig. 3.

Furthermore, in terms of basic user browsing behavior, the Taiwan Cultural Memory Bank 2.0 thematic platform outperforms the 1.0 portal website across all key engagement metrics, including average time on site, average pages viewed, and bounce rate, as shown in Table 1. These differences suggest that the higher bounce rate and lower engagement on the 1.0 portal may be attributed to system performance issues, which likely limited the depth of user interaction. In contrast, when the 2.0 platform was launched in 2023, it introduced enhancements

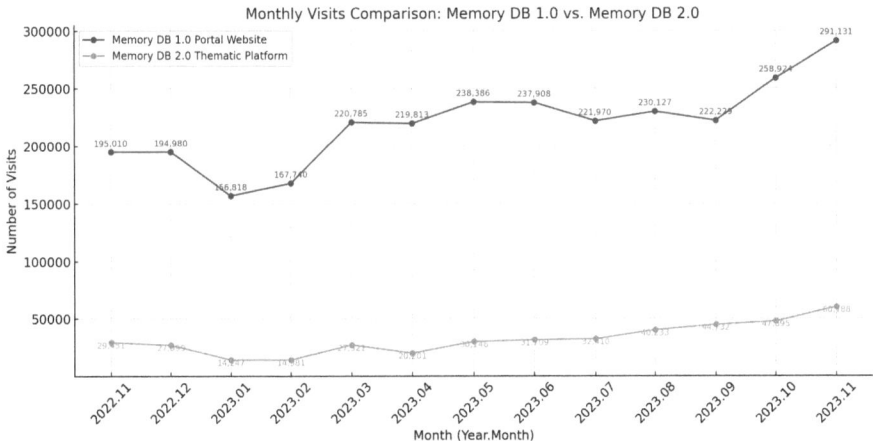

Fig. 3. Monthly Visits Comparison: Memory Database 1.0 vs. Memory Database 2.0

Table 1. User Engagement Metrics for Memory Bank Platforms

Metric	Memory Bank 1.0	Memory Bank 2.0
Average Time on Site	1 min 9 s	1 min 56 s
Average Pages Viewed	1.9 pages	7.02 pages
Bounce Rate	66%	35.2%

in search functionality, more structured and thematic content, and improved UI/UX design. These optimizations have contributed to greater user engagement and increased platform stickiness. Table 1 shows the details of user engagement metrics for memory bank platforms.

The Memory Bank 1.0 portal website began to show a notable upward trend in user growth starting in September 2023. This increase is primarily attributed to recent government-funded initiatives, such as community development and Living Aesthetics Center subsidy programs, which required participating teams to upload one to three pieces of content to the "Go Create" platform as a condition for project completion. As a result, many grant recipients registered for member accounts in order to create and upload content or stories, leading to a significant rise in new users.

Similarly, the Memory Bank 2.0 thematic platform has demonstrated a steady monthly growth trend. Notably, there was a substantial increase in new users in September, with 9,023 newly registered users—an evident jump compared to the previous month. Further investigation revealed that a large portion of these users originated from Singapore, as illustrated in Fig. 4 and Table 2, with the highest spike occurring on September 19. This suggests the possibility that a specific course or activity in Singapore was promoting the National Cultural Memory Bank. Additionally, November also saw a significant increase in new

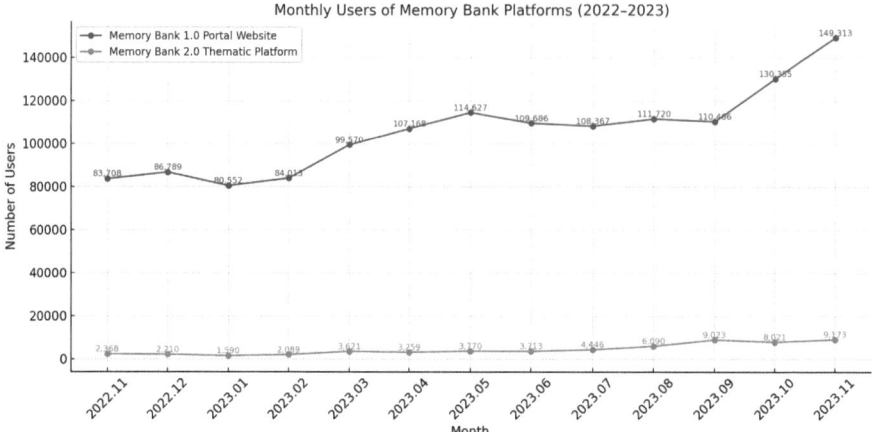

Fig. 4. Growth trend of new users on memory bank-related platforms

Table 2. Top 10 Countries by Number of New Users on Memory Bank 2.0 in September, 2023

Rank	Country	Number of New Users	Rank	Country	Number of New Users
1	Taiwan	7,815	6	China	29
2	Singapore	842	7	Malaysia	20
3	United States	103	8	Canada	14
4	Hong Kong	52	9	South Korea	11
5	Japan	49	10	Sweden	9

users, coinciding with the Memory Bank 2.0 "Island Translation Station" series of exchange events, which likely contributed to the surge in new user engagement.

Figures 5 and 6 present the age and gender distributions within the user profiles. Notably, the age distribution of users on both the Memory Bank 1.0 portal and the Memory Bank 2.0 thematic platform demonstrates a convergent trend, predominated by users in the 35 to 54 age bracket, representing the prime working-age demographic. This age group typically corresponds to individuals in mid-career stages, exhibiting a higher demand for digital resources, which may reflect the platform's significance in supporting professional development and lifelong learning. Concurrently, the gender distribution across both platforms also reveals a consistent pattern, characterized by a majority of female users. This gender skew may be attributed to the specific nature of the content or services offered by the platforms; for instance, if the platforms focus on areas such as health, education, or social networking, these domains generally attract greater female engagement.

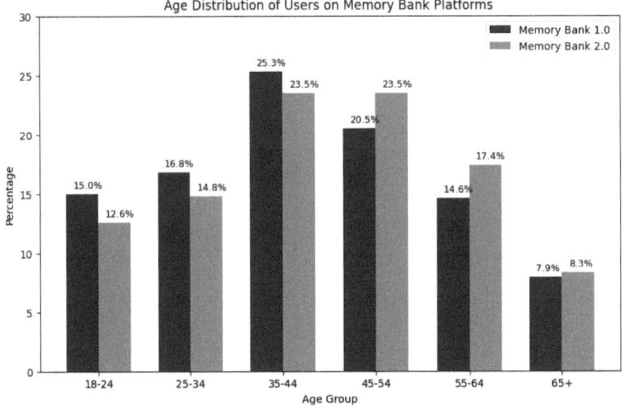

Fig. 5. Age Distribution of Users on Memory Bank Platforms

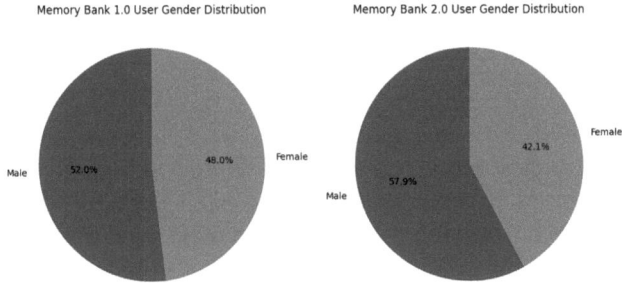

Fig. 6. Memory Bank 1.0 and 2.0 User Gender Distribution

4 SEO Performance of Taiwan Cultural Memory Bank 2.0

After gaining a comprehensive understanding of the user profiles, we implemented a series of SEO strategies tailored for the Taiwan Culture Memory Bank 2.0 website. These efforts included optimizing the site structure, annotating relevant keywords, and applying various SEO treatments to enhance the website's visibility and accessibility. Following these modifications, we collected user interaction and traffic data spanning from February 2023 to November 2024. This dataset was then analyzed using Google Analytics to evaluate user behavior, traffic sources, and overall site performance, providing valuable insights for further optimization and strategic planning.

As shown in Fig. 7, the Taiwan Culture Memory Bank 2.0 platform experienced a significant increase in page views beginning in January 2024, reaching a peak in May. During this period, the platform recorded an increase of approximately 1.48 million page views. This growth can be primarily attributed to the effectiveness of the implemented SEO strategies, which successfully enhanced

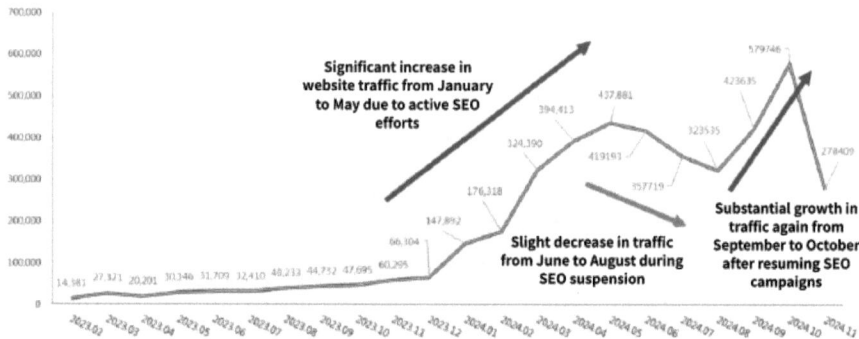

Fig. 7. Platform Traffic Trends and SEO Performance

the platform's visibility and attracted a substantial number of users. However, from June to August, a transition period in SEO deployment led to a temporary decline in website exposure, resulting in a decrease in both new visitors and overall page views. Compared to the average monthly page views of approximately 410,000 in April and May, the numbers dropped to around 360,000 per month during the transitional phase. Subsequently, in September and October, the reimplementation of SEO strategies contributed to a significant rebound in traffic, with page views rising to approximately 500,000. Furthermore, the decline observed in November, as shown in Fig. 7, is due to data being collected only up to November 15. Based on current trends, it is estimated that the total page views for November would exceed 500,000.

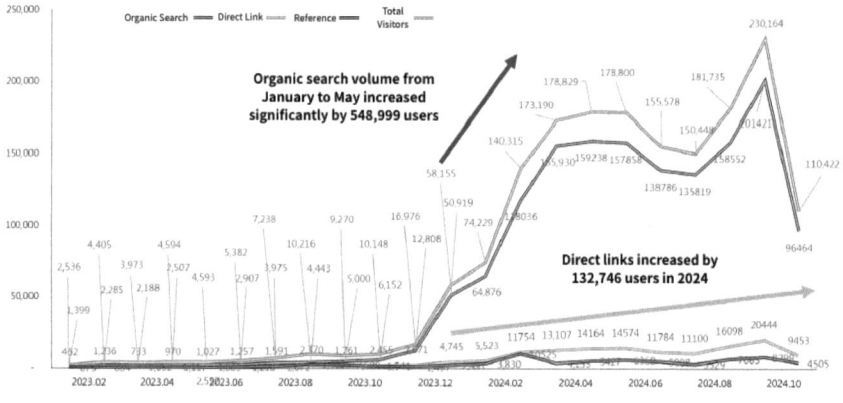

Fig. 8. SEO Performance of Visitor Source Analysis

As shown in Fig. 8, the majority of user traffic during the observed period originated from organic search. Following the implementation of SEO strategies, there was a noticeable and steady increase in organic search traffic starting in January 2024, with growth accelerating over time. However, a temporary suspension of SEO efforts between June and August led to a decline in organic search visits. In addition to boosting organic search traffic, SEO also contributed to an increase in direct traffic to the platform. Specifically, the average number of direct visitors rose from approximately 1,500 in 2023 to around 11,428 in 2024. These findings indicate that SEO optimization not only played a significant role in attracting new visitors but also effectively encouraged return visits from existing users. This suggests that SEO has a dual impact: it drives new traffic while also enhancing user engagement and loyalty, thereby increasing the number of dedicated platform users.

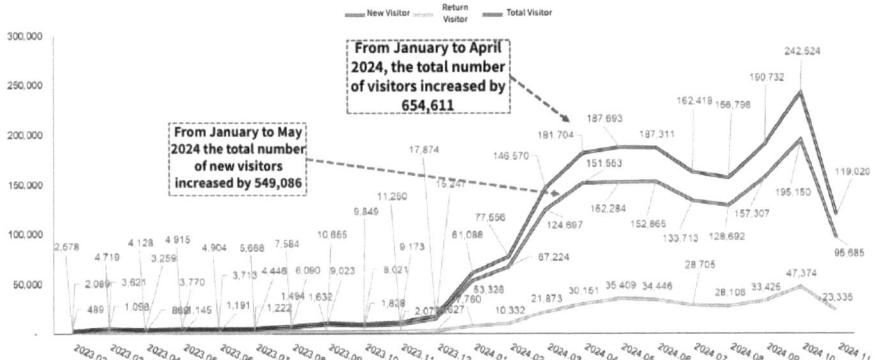

Fig. 9. SEO Performance of New and Return Visitors

As illustrated in Fig. 9, the number of visitors to the platform increased significantly from January 2024, reaching a peak in May. During this period, the total number of visitors grew by approximately 650,000, with the majority of this growth driven by an increase in new visitors—an estimated 540,000. This trend highlights the strong effectiveness of the SEO strategies in attracting new users to the platform. However, between June and August 2024, a transitional phase in SEO deployment resulted in a noticeable decline in new visitor traffic. Specifically, the average number of new visitors per month decreased from approximately 150,000 in April and May to around 130,000. Following the reimplementation of SEO efforts in September and October, the number of new visitors rebounded significantly, reaching approximately 190,000.

In addition to the analysis conducted using Google Analytics, we also examined the performance of the Taiwan Culture Memory Bank 2.0 Facebook fan page. The primary objective was to assess whether the SEO efforts, beyond enhancing the visibility of the official website, also had a measurable impact on the growth and engagement of the social media presence. As illustrated in Fig. 10, the number of followers on the Facebook fan page demonstrated steady growth throughout 2024. By November 24, 2024, the page had successfully reached its target of 33,000 followers, representing an increase of nearly 8,000 followers over

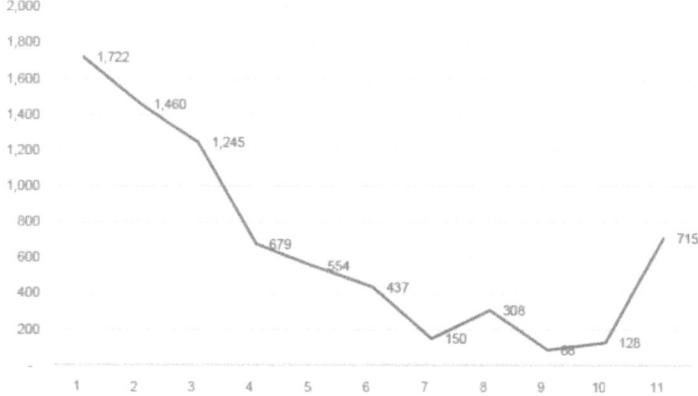

Fig. 10. The Increasing of Facebook Page Followers

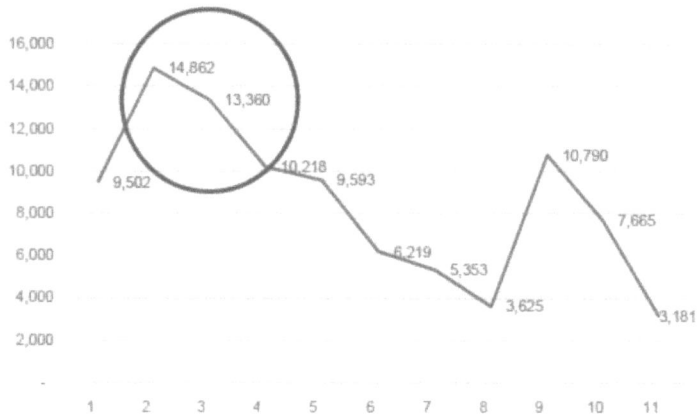

Fig. 11. The Number of Facebook Page Interactions

the course of the year. This suggests that SEO may have had a positive spillover effect on the platform's broader digital ecosystem.

Figure 11 presents the growth in user interactions on the Taiwan Culture Memory Bank 2.0 Facebook fan page, including metrics such as post likes, comments, and shares. The period from February to March 2024 saw the highest level of engagement throughout the year, driven by content aligned with seasonal events such as the Lantern Festival and International Women's Day. These thematic posts successfully captured audience interest and encouraged active participation. The notable increase in interactions during this period suggests that the SEO efforts not only enhanced website traffic but also contributed to improved visibility and engagement on social media platforms, demonstrating the broader effectiveness of the overall digital strategy.

5 Conclusion and Suggestions

In conclusion, this study highlights the critical interplay between user profile analysis, SEO optimization, and Visitor Relationship Management (VRM) in enhancing the performance of the Taiwan Cultural Memory Bank 2.0. By leveraging Google Analytics and strategic SEO interventions—such as restructuring the website, refining keyword annotations, and monitoring user behaviors—we observed substantial improvements in organic traffic, user engagement, and retention rates. These outcomes align with the core principles of VRM, which emphasize understanding and responding to visitor needs through data-informed decision-making. The platform's improved usability and visibility reflect how VRM, when integrated with SEO strategies, can foster more personalized and engaging user experiences that support both discovery and long-term interaction.

Moreover, the study demonstrates that the impact of SEO extends beyond the official website to the platform's social media presence. The observed growth in Facebook followers and interactions, particularly during culturally relevant periods, suggests a strong synergy between content strategy and digital outreach. These results underscore the importance of adopting a comprehensive VRM framework—one that not only captures demographic and behavioral data but also utilizes it to shape marketing efforts, optimize user engagement pathways, and strengthen audience loyalty. The findings provide a robust model for how cultural institutions can enhance their digital platforms by combining user-centered design, SEO performance, and relationship-driven strategies.

About future suggestion, Taiwan Cultural Memory Bank 2.0 is still in the stage of cultivating user engagement. To support this, SEO campaigns should be strategically scheduled to ensure continuity, helping to build user habits and loyalty. Given the clear impact of SEO on traffic growth, consistent investment and regular updates to keyword strategies based on market trends are essential. For seasonal and high-interest topics (e.g., Qixi Festival, Ghost Month), early content planning and SEO optimization, combined with social media activities, can further maximize visibility during peak periods.

Future efforts should also focus on engaging different user segments—such as lifestyle explorers, educational promoters, and creative storytellers—by offering targeted thematic content. While returning visitor rates are stable, deeper engagement can be encouraged through membership programs, personalized content, and timely updates. Tools like email notifications with tailored article suggestions can help turn new users into loyal ones and build a sustainable, active user community.

References

1. Anitha, J., Ting, I.H., Agnes, S.A., Pandian, S.I.A., Belfin, R.V.: Social media data analytics using feature engineering. In: Peter, J.D., Fernandes, S.L. (eds.) Advances in Ubiquitous Sensing Applications for Healthcare, Systems Simulation and Modeling for Cloud Computing and Big Data Applications, pp. 29–59. Academic Press (2020). https://doi.org/10.1016/B978-0-12-819779-0.00003-4
2. Bitgood, S., Shettel, H.: An overview of visitor studies. J. Museum Educ. **21**(3), 6–10 (1996)
3. Cutroni, J.: Google Analytics: Understanding Visitor Behavior. O'Reilly Media, Inc. (2010)
4. Falk, J.H.: A framework for diversifying museum audience. Museum News **77**(5), 36–39, 61 (1998)
5. Hooper-Greenhill, E.: Museums and Their Visitors, 1st edn. Routledge (1994). https://doi.org/10.4324/9780203415160
6. Hsu, M.Y., Ting, I.H.: Applying a combination model of knowledge management and visitor relationship management in the study of the visitors of historical museum. In: Uden, L., Ting, I.H. (eds.) Knowledge Management in Organisations. KMO 2023. Communications in Computer and Information Science, vol. 1825. Springer, Cham (2023). https://doi.org/10.1007/978-3-031-34045-1_23
7. Raad, E., Chbeir, R., Dipanda, A.: User profile matching in social networks. In: 2010 13th International Conference on Network-Based Information Systems, Takayama, Japan, pp. 297–304 (2010). https://doi.org/10.1109/NBiS.2010.35
8. Sanchez, R.: Strategic Learning and Knowledge Management. Wiley, Chichester (1996)
9. Scott, J.: Social Network Analysis: A Handbook. SAGE Publication (2000)
10. Sheng, C.W., Chen, M.C.: A study of experience expectations of museum visitors. Tour. Manage. **33**(1), 53–60 (2012). https://doi.org/10.1016/j.tourman.2011.01.023
11. Siu, N.Y.M., Zhang, T.J.F., Dong, P., Kwan, H.Y.: New service bonds and customer value in customer relationship management: the case of museum visitors. Tour. Manage. **36**(2013), 293–303 (2013). https://doi.org/10.1016/j.tourman.2012.12.001
12. Ting, I.-H., Minetaki, K., Hsu, M.Y., Yen, C.-S.: Applying social network embedding and word embedding for socialbots detection. In: Proceedings of the 2023 IEEE/ACM International Conference on Advances in Social Networks Analysis and Mining (ASONAM 2023), pp. 712–718. Association for Computing Machinery, New York (2024). https://doi.org/10.1145/3625007.3627306
13. Ting, I.H.: Special Issue: Applications and management aspects of social networks research. Rev. Socionetwork Strat. **16**, 571–572 (2022). https://doi.org/10.1007/s12626-022-00130-y
14. Ting, I.H., Zhang, Z., Wang, L.S.L.: A study of the effect of spiral of silence between different social networking platforms. In: Proceedings of the The 3rd Multidisciplinary International Social Networks Conference on SocialInformatics 2016, Data Science 2016 (MISNC, SI, DS 2016), Article 55, pp. 1–5. Association for Computing Machinery, New York (2016). https://doi.org/10.1145/2955129.2955183
15. Ting, I.-H., Yen, C.-S.: Opinion groups identification in blogsphere based on the techniques of web and social networks analysis. In: The 4th International Conference on Machine Learning and Computing, Hong Kong, China (2012)

16. Wajahat, A., Nazir, A., Akhtar, F., Qureshi, S., Razaque, F., Shakeel, A.: Interactively visualize and analyze social network Gephi. In: 2020 3rd International Conference on Computing, Mathematics and Engineering Technologies (iCoMET), pp. 1–9. IEEE (2020)
17. Wasserman, B., Faust, K.: Social Network Analysis: Methods and Applications. Cambridge University Press (1994)
18. Wielinga, B., Sandberg, J., Schreiber, G.: Methods and techniques for knowledge management: what has knowledge engineering to offer? Expert Syst. Appl. **13**(1), 73–84 (1997)
19. Yalçın, N., Köse, U.: What is search engine optimization: SEO? Procedia. Soc. Behav. Sci. **9**, 487–493 (2010)
20. Ziemkiewicz, C., Ottley, A., Crouser, R.J., Chauncey, K., Su, S.L., Chang, R.: Understanding visualization by understanding individual users. IEEE Comput. Graph. Appl. **32**(6), 88–94 (2012). https://doi.org/10.1109/MCG.2012.120

Artificial Intelligence in Natural Science Education: A Case Study on Students' Environmental Literacy

Chia-Sung Yen(✉)

Department of Cultural and Creative Industry, National Chin-Yi University of Technology, Taichung City, Taiwan
jsn1003@gmail.com

Abstract. This study examines how AI-driven plant identification applications can enhance students' environmental literacy, self-directed learning motivation, and awareness of carbon reduction initiatives. Using a qualitative research approach, this study also examines the role of AI tools in fostering learning motivation and ecological actions. The findings suggest that mobile applications enhance students' interest in and knowledge of plants. Future curriculum designs should consider integrating AI technology more effectively with environmental education to promote active learning and sustainability awareness.

Keywords: Artificial Intelligence · Plant Identification Applications · Environmental Literacy · Carbon Reduction Awareness · Digital Learning Tools

1 Introduction

Traditional environmental education primarily relies on classroom lectures, field observations, and textbook learning. However, research indicates that such static learning methods may not effectively enhance students' environmental awareness and pro-environmental behavior (Tilbury, 1995). In the digital age, students increasingly prefer interactive learning tools to acquire knowledge, making integrating environmental education with emerging technologies a pressing issue in education (Jickling & Wals, 2008).

Recent advancements in Artificial Intelligence (AI) have introduced new opportunities for environmental education. AI technology has been widely applied in biodiversity and environmental sciences, including plant identification applications. Despite numerous studies investigating AI's applications in education, research on the impact of AI-based plant identification applications on environmental literacy and carbon reduction awareness remains limited. Most existing studies focus on the effectiveness of these applications in biology education rather than their implications for environmental education (Silvertown, 2009). Thus, this study aims to bridge this research gap by employing qualitative methods to examine how AI-based plant identification applications influence high school students' environmental literacy and carbon reduction awareness, as well as their engagement in environmental actions.

This study aims to explore the application of AI-based plant identification tools in environmental education, with a particular focus on their impact on high school students' environmental literacy. Specifically, this study seeks to analyze whether students using these applications can develop an understanding of carbon sequestration and subsequently enhance their environmental action awareness. Additionally, this study investigates whether AI technology can boost students' self-directed learning motivation, encouraging them to explore environmental issues proactively and reinforce their environmental responsibility. The specific research objectives mainly focused on analyzing whether high school students using AI-based plant identification applications naturally associate plants with carbon sequestration and develop environmental literacy.

Through this analysis, this study aims to provide theoretical and empirical support for integrating AI technology with environmental education and offer practical recommendations for future curriculum design and educational technology applications. This study is guided by the following core research question to Do high school students using plant identification applications develop cognitive knowledge of plants?

2 Literature Discussion

Advances in artificial intelligence (AI) technology have led to the rapid development of personalized learning, Intelligent Tutoring Systems (ITS), and adaptive learning in the field of education. Personalized learning uses AI technology to automatically adjust learning content based on students' learning behaviors, progress, and needs so that students can learn in the way that works best for them. The intelligent tutor system mimics the guidance style of a human teacher and provides appropriate learning suggestions and feedback based on the real-time responses of learners, to improve learning effectiveness and engagement (Heffernan et al., 2022). Adaptive learning techniques allow the learning system to adjust the difficulty in real-time based on the student's performance, making the learning process more dynamic and tailored to individual needs.

Studies have shown that personalized learning technology can significantly improve students' interest and comprehension and analyze learners' behavior patterns through deep learning technology to provide more accurate learning plans (Siemens & Baker, 2012). In addition, the development of intelligent tutor systems has gradually broken through the limitations of traditional education, and studies have shown that AI tutors can not only provide guidance equivalent to that of human teachers but even better adapt to student's learning styles and needs some situations (Graesser et al., 2018). Adaptive learning techniques further optimize the learning experience, enabling students to achieve better learning outcomes in contexts that are adapted to their abilities and avoid excessive challenge or burnout (Adeyinka et al., 2025).

However, the application of AI technology in the field of education still faces many challenges. First, data privacy and security issues remain a major obstacle to AI adoption, especially in the protection of students' data, and better policies and technical support are still needed. Secondly, since adaptive learning techniques are highly dependent on big data and machine learning algorithms, their effectiveness is affected by the quality of the learning data, which may lead to inaccurate learning results if the training data is insufficient or biased (Basile & Tani, 2025). In addition, the acceptance of AI technology by teachers and students also affects the speed of its practical application, especially in an environment where the traditional education model is more stable, and it still takes longer to adapt and adjust (Vermeiren et al., 2025). Future research should focus on how to improve the explainability of AI technologies so that learners and educators can understand how AI systems work, and further improve the trust and learning experience.

The development of AI technology has shown great potential to improve learning motivation and learning effectiveness. Learning motivation is a key factor influencing learners' participation in learning activities and continuous engagement, and AI technology can effectively strengthen students' intrinsic motivation through gamification, personalized feedback mechanisms, and adaptive learning strategies (Kim et al., 2020). For example, studies have found that gamified learning systems combined with AI technology can enhance students' interest in learning, make the learning process more challenging and interactive, and make students more willing to participate in learning activities (Hamari et al., 2016). In addition, the real-time feedback and learning analysis provided by AI enable students to adjust their learning strategies according to their own learning pace, thereby improving learning efficiency and self-confidence.

In terms of learning outcomes, the application of AI technology can effectively improve students' knowledge acquisition and application ability. Studies have shown that AI-driven adaptive learning techniques can improve student performance by about 20%–30% (Baker & Siemens, 2014). In addition, AI technology can also provide real-time suggestions for students' learning bottlenecks, help students overcome learning difficulties, and improve their ability to internalize and apply knowledge (VanLehn, 2011). Recent studies have also shown that through AI-based learning analytics, teachers can better grasp students' learning status and further optimize teaching strategies to enable learners to learn in a more effective environment (Rodrigues et al., 2025).

However, there are still some challenges in the application of AI technology in terms of learning motivation and learning effectiveness. First, some studies have pointed out that if AI systems are not interactive enough, they may lead to a decrease in students' interest in learning, especially after long-term use, which may lead to "technology fatigue" (Vermeiren et al., 2025). In addition, the design of AI learning systems needs to consider the needs of different learners to ensure that the technology truly matches the learning styles and goals of learners. Future research should focus on how to integrate AI technology with traditional education models so that it can become an effective tool to support learning, rather than completely replacing the role of human teachers.

Environmental literacy is a comprehensive embodiment of an individual's awareness, attitude, and ability to act toward the environment, which can influence the choice and decision-making of his or her environmental behavior. In education, the improvement of environmental literacy is seen as a key factor in promoting pro-environmental behavior. In recent years, many scholars have studied the impact of environmental literacy on individual behavior and found that individuals with high environmental literacy are more inclined to take environmental actions, such as energy conservation, waste separation, and carbon emission reduction (McBeth & Volk, 2022). Ajaps and McLellan (2019) argue that the development of environmental literacy should encompass knowledge, skills, and behaviors rather than just imparting environmental knowledge because the transfer of knowledge alone is not enough to lead to action change.

Environmental Education is seen as one of the most effective ways to improve environmental literacy. Sterling (2021) proposes that interactive learning, such as problem-based learning and action research, can help students understand environmental problems more deeply and apply what they have learned in real-world situations to further promote environmental behavior change. In addition, Clayton and Myers (2019) highlight the correlation between environmental literacy and environmental behavior, and their research shows that by designing immersive learning experiences, students can develop a deeper empathy for environmental issues, which in turn increases the likelihood of actively participating in environmental actions.

Therefore, the future of environmental education should emphasize multi-level collaboration, including the joint participation of schools, communities, governments, and enterprises, to effectively promote students' sense of ecological responsibility and action.

3 Methodology

This study uses qualitative research methods to explore the learning process and environmental awareness changes of high school students after using artificial intelligence (AI) plant recognition applications through in-depth interviews. Qualitative research is suitable for exploring the deep meaning and individual experience of phenomena, especially in highly subjective topics such as learning attitudes, behavior change, and environmental literacy, which can provide rich data support (Creswell & Poth, 2018). Since the goal of this study is not only to confirm the impact of the app on learning outcomes but more importantly to understand how learners build environmental awareness during interactions, a qualitative approach is more likely to capture learners' real feelings and action shifts.

In this study, semi-structured interviews will be used as the main data collection method, and open-ended questions will be used to guide respondents to express their learning experiences and environmental attitudes. The interview will cover the impact of AI plant recognition applications on students' environmental literacy and the the use of AI technology motivates them to be more willing to participate in environmental action. Interviews will be conducted in person, face-to-face, or via video conference, each interview will be approximately 30–45 min long and will be recorded and transcribed to ensure data integrity. Thematic analysis will be used to classify and summarize the interview data to identify the key factors influencing the change in learners' environmental literacy (Braun & Clarke, 2021).

In this study, 20 high school students will be selected as the research subjects and the respondents will be ensured to be from different schools to improve the diversity and representativeness of the sample. The students will be divided into two groups: the primary study group and the control group, to compare the impact of AI plant identification applications.

The main research team will select students who have at least one month of experience in using the AI plant recognition application, and who can recall their learning process and environmental attitude changes. This group was selected to explore how

AI technology impacts learners' environmental literacy development and to analyze the potential impact of applications in promoting awareness of carbon reduction. The control group consisted of environmentally conscious students who did not use an AI plant identification app. This group was designed to compare the similarities and differences in environmental attitudes and behaviors between the two groups of students to determine whether the application had a significant impact on the learners' environmental actions and cognitive patterns.

In this study, in-depth interviews were used as the main data collection method to explore the learning process, environmental literacy development, and carbon reduction awareness formation process of high school students after using AI plant identification applications.

The interview process was designed as semi-structured interviews and focused on three core topics: students' experience of the plant identification app, whether students developed environmental literacy and carbon reduction awareness, and how students changed their learning behaviors and attitudes. The interview process will follow academic ethics to ensure the informed consent and privacy protection of the subjects, and the triangulation method will be used to improve the reliability of the data.

4 Results and Discussion

Based on the results of the interviews, this section explores the impact of AI plant identification applications on high school students' environmental literacy and carbon reduction awareness. Through Thematic Analysis, this study summarizes the main patterns of the learning experience, behavioral change, and environmental awareness development of the interviewed students, and further explains them with academic theories.

A total of 20 high school students were interviewed in this study, of which 10 were from the main study group (using an AI plant identification app) and 10 were from the control group (who did not use the app but were environmentally conscious). During the interview, all respondents expressed their basic concerns about environmental issues and provided their personal learning experiences and attitude-changing journeys. The results of the interview are as follows:

Respondent A (Principal Research Group, Year 2 of high school, using *PlantNet* for 3 months):

"I find the app very useful, especially when I am at school or outdoors, I can take photos and identify plants right away. Originally, I didn't have much interest in plants, but because this tool makes me feel like I'm solving puzzles and every time I discover a new plant, it's very rewarding. I will also share this information with my family and friends."

Respondent B (Principal Research Group, Year 1 of high school, using *Seek by iNaturalist* for 6 weeks):

"At first, I just thought it was a novelty, but after a long time, I started to pay attention to the changes in the plants on campus and even inquire about their role in the ecosystem. I didn't pay much attention to plants before, but now I try to

remember the names of different plants and even start to be more interested in biology."

Respondent C (control group, second year of high school, no experience with AI plant recognition applications):

"Our school has an ecological education course, and the teacher took us to do outdoor observation and taught us to identify different types of trees. I think it's a good way to learn because it's a hands-on experience, but sometimes I forget what the teacher taught, and I don't have a tool to help keep track of it."

Interviewee A (sophomore in high school, using *PlantNet* for 3 months):

"I wasn't interested in plants, but this app makes me feel like I'm doing an outdoor scavenger hunt. I take photos at school, in parks, and even in green spaces near my home to identify plants, and the fact that some plants can fix a lot of carbon dioxide makes me rethink the impact of plants on the environment."

Respondent B (freshman year of high school, using *Seek* for 6 weeks):

"I didn't pay much attention to the plants on campus before, but since using the app, I've started to wonder about the names of these plants, what they're used for, and even if they help reduce carbon emissions. This has given me a new perspective on environmental issues and I have started to take the initiative to care about the greening of the school."

Respondent C (control group, 2nd year of high school, no app):

"Our school's ecology class will take us to do outdoor observations, and the teacher will teach us to identify different plants. But sometimes I forget it after the course, I don't have the tools to keep track of it, and I feel like what I've learned isn't long-lasting."

Respondent D (3rd year of high school, using *PlantNet* for 5 months):

When I identified certain plants with the app, I started looking up their functions, such as which trees can absorb the most carbon dioxide. I never thought that planting certain plants could help reduce greenhouse gases, which made me more interested in carbon reduction actions."

Respondents' learning behaviors and attitudes can be divided into high-engagement learners and low-engagement learners. High-engagement learners generally showed a strong interest in the learning process of AI plant recognition apps, not only using the app for plant identification regularly but also interacting with other users through social features, such as sharing observations and participating in environmental discussions on the iNaturalist platform. In addition, high-engagement learners were more likely to seek out relevant environmental education resources, such as reading environmental articles or participating in environmental protection activities in schools, showing continuity of learning motivation.

In contrast, low-engagement learners are more passive in their learning behaviors, often using apps only in school lessons or in specific contexts, rather than incorporating them into their daily learning tools. In addition, some low-engagement learners said that their interest in AI applications was mainly due to the novelty of the technology, rather than a deep concern for environmental issues. This pattern of learning behavior is consistent with the Self-Determination Theory (Deci & Ryan, 1985), which states that learners are more likely to maintain long-term engagement when learning motivation is primarily derived from external incentives (e.g., teacher guidance or classwork) rather than intrinsic motivation.

Overall, this study confirms that AI plant identification applications have the potential to promote environmental education, but they still need to be combined with more holistic learning strategies to ensure that students can translate what they have learned into long-term environmental actions. The results of this study will serve as the basis for follow-up discussions and policy recommendations to further optimize the application of AI technology in environmental education.

5 Conclusion and Recommendations

This study explores the impact of AI plant recognition applications on high school students' environmental literacy and carbon reduction awareness. It analyzes students' learning journeys and behavioral changes through qualitative interviews. The results show that AI applications can improve students' environmental literacy and learning motivation, but their impact on environmental action still needs to be matched with a more complete educational strategy. This chapter synthesizes the findings and proposes conclusions, academic contributions, and future research directions.

This study used qualitative interviews to explore the impact of AI plant identification on high school students' environmental literacy and carbon reduction awareness and analyzed the changes in students' learning behaviors. The results show that AI plant recognition applications can effectively improve students' environmental knowledge, learning motivation, and identification ability. However, the impact on environmental protection actions is still limited by personal learning motivation and environmental factors. The specific conclusions of the study are as follows:

This study found that students who used AI plant recognition applications generally exhibited high levels of environmental literacy. Through real-time image recognition and interactive learning, they were able to raise their awareness of plant species and ecosystems and to associate the connection between plants and carbon reduction. These students are more likely to actively explore the natural environment and develop a deeper interest in environmental issues. This is in line with constructivist learning theory (Piaget, 1972), as AI applications provide an environment for self-directed learning that enables students to deepen their learning through exploration and practice.

In conclusion, this study confirms that the application of AI plant recognition has a positive effect on improving students' environmental literacy, but its impact on environmental protection actions still needs to be further optimized. This study responds to the research question of "Can AI technology improve environmental literacy and action" and provides enlightenment for the future development of environmental education.

This study provides the following academic contributions to the development of environmental education and AI education technology: Expanding the Application of AI in Environmental Education While existing research focuses on the application of AI in STEM education, this study focuses on how AI can improve students' environmental literacy and carbon reduction awareness. The results show that AI applications can enhance learning motivation and discernment, which provides a theoretical basis for the integration of AI and environmental education in the future.

References

Adeyinka, A.M., Yahaya, R.A., Oreka, S.J.V., Nagadau, M.A.: The role of artificial intelligence in enhancing adult education in Nigeria. *ResearchGate* (2025).https://www.researchgate.net/publication/388707917

Ajaps, S., McLellan, R.: We don't know enough": Environmental literacy and sustainable development. Environ. Educ. Res. **25**(8), 1177–1195 (2019). https://doi.org/10.1080/13504622.2019.1635266

Baker, R.S., Siemens, G.: Educational data mining and learning analytics. Cambridge Handbook of the Learning Sciences. **253–272** (2014). https://doi.org/10.1017/CBO9781139519526.015

Basile, S., Tani, M.: AI-powered adaptive learning systems and their impact on student performance. Educ. Technol. Soc. **28**(2), 198–215 (2025). https://doi.org/10.1109/EDUCON.2025.123456

Braun, V., Clarke, V.: Thematic analysis: A practical guide. *SAGE Publications* (2021)

Clayton, S., Myers, G.: Conservation psychology: Understanding and promoting human care for nature. Cambridge University Press (2019)

Creswell, J.W., Poth, C.N.: Qualitative inquiry and research design: Choosing among five approaches (4th ed.). SAGE Publications (2018)

Deci, E.L., Ryan, R.M.: Intrinsic motivation and self-determination in human behavior. Springer Science & Business Media (1985)

Graesser, A.C., Cai, Z., Morgan, B., Wang, L.: Intelligent tutoring systems. Cambridge Handbook of Instructional Feedback **188–208** (2018). https://doi.org/10.1016/B978-0-12-815866-4.00012-8

Hamari, J., Shernoff, D.J., Rowe, E., Coller, B., Asbell-Clarke, J., Edwards, T.: Challenging games help students learn: An empirical study on engagement, flow and immersion in game-based learning. Comput. Hum. Behav. **54**, 170–179 (2016). https://doi.org/10.1016/j.chb.2015.07.045

Heffernan, N.T., Heffernan, C.L., Lindquist, M., Van Inwegen, E.: The future of intelligent tutoring systems. Int. J. Artif. Intell. Educ. **32**(2), 345–365 (2022). https://doi.org/10.1007/s40593-022-00321-4

Jickling, B., Wals, A.E.: Globalization and environmental education: Looking beyond sustainable development. J. Curric. Stud. **40**(1), 1–21 (2008)

Kim, Y., Rothrock, L., Munson, M.: Enhancing motivation through AI-driven feedback in e-learning environments. Journal of Learning Analytics **7**(1), 45–62 (2020). https://doi.org/10.18608/jla.2020.71.4

McBeth, W., Volk, T.L.: Environmental literacy in secondary school education. J. Environ. Educ. **53**(2), 235–252 (2022). https://doi.org/10.1080/00958964.2021.1932210

Piaget, J.: The principles of genetic epistemology. Routledge & Kegan Paul (1972)

Rodrigues, L., Aranha, R.V., Costa, N.T.: Towards automating the personalization of gamified learning's aesthetics. *ResearchGate* (2025). https://www.researchgate.net/publication/388707917

Siemens, G., Baker, R.S.: Learning analytics and educational data mining: Towards communication and collaboration. Proceedings of the International Conference on Learning Analytics & Knowledge **252–254** (2012). https://doi.org/10.1145/2330601.2330661

Silvertown, J.: A new dawn for citizen science. Trends Ecol. Evol. **24**(9), 467–471 (2009)

Sterling, S.: Educating for the future: Rethinking education and sustainability. Sustain. Sci. **16**(1), 243–258 (2021). https://doi.org/10.1007/s11625-020-00842-y

Tilbury, D.: Environmental education for sustainability: Defining the new focus of environmental education in the 1990s. Environ. Educ. Res. **1**(2), 195–212 (1995)

VanLehn, K.: The relative effectiveness of human tutoring, intelligent tutoring systems, and other tutoring systems. Educational Psychologist **46**(4), 197–221 (2011). https://doi.org/10.1080/00461520.2011.611369

Vermeiren, H., Kruis, J., Bolsinova, M., Kloos, C.D.: Psychometrics of an Elo-based large-scale online learning system. Elsevier Artificial Intelligence (2025). https://doi.org/10.1016/j.artint.2025.102354

Liquid Consumption Behavior in Online Purchasing: Preliminary Discussion of Online Survey Results

Hiroshi KOGA

Kansai University, Osaka, Japan
hiroshi@kansai-u.ac.jp

Abstract. This paper explores "liquid consumption" in online shopping, focusing on experiences over ownership and impulsive buying. It examines how trends like fast purchasing, reliance on reviews, and non-ownership models impact consumer behavior. A survey in Japan revealed that liquid consumption influences online shopping habits, particularly through smartphone use and reviews. The study concludes that further research is needed to analyze these trends, especially in the age of social media.

Keywords: Liquid Consumption · Word-of Mouth · Online Purchasing

1 Introduction

The purpose of this study is to elucidate the relationship between two key concepts associated with emerging consumer behavior patterns that have accompanied the proliferation of social networking services (SNS).

The first concept is word-of-mouth (WOM). Traditionally, consumer behavior has been understood as a process in which individuals are stimulated by corporate advertising and promotion, memorize product or brand names, and proceed to make purchasing decisions. However, recent observations suggest a transformation in this behavioral pattern. Consumers now tend to search for WOM or user-generated experiences related to products of interest, develop purchase intentions when such reviews are favorable, and subsequently share their own experiences as WOM after purchase [1]. As a result, it is argued that the perceived value of products has shifted from physical functionality or brand image to experiential value. Such experiential value is disseminated via SNS platforms as consumer-generated media (CGM). Accordingly, considerable scholarly attention has been directed toward examining the current state and challenges of CGM in the context of experiential consumption [2].

The second concept is liquid consumption [4]. It is widely asserted that consumption has been shifting from goods-centered to experience-centered [3]. Traditional modes of consumption emphasized ownership, characterized as enduring, ownership-based, and material in nature [4]. In contrast, contemporary consumption trends reflect a weakening of attachment to ownership. Instead, there is a growing emphasis on access—rather than acquisition—through models such as subscriptions and sharing platforms. Consumers are increasingly willing to purchase products they find appealing via SNS, often without strong attachment, in pursuit of enviable experiences. The notion of liquid consumption has been proposed to conceptualize this new mode of consumption, and it is said to be inspired by Bauman's theory of liquid modernity [5].

Liquid consumption is typified by three main characteristics: its ephemeral nature (the transient relationship between consumers and products), its access-based logic (monetary value is placed on temporary usage rather than ownership), and its dematerialized orientation (emphasis on digital rather than physical goods). Among these, the access-based nature of liquid consumption appears to be closely intertwined with WOM. Moreover, the ephemeral nature of consumption may also be partially associated with WOM, insofar as consumers rely on transient, situation-dependent reviews in their decision-making. However, the relationship between dematerialization and WOM remains unclear. Perhaps as a result, little research to date has directly examined the referential connection between liquid consumption and WOM. This study therefore seeks to contribute to the literature by pursuing an integrative understanding of these two emerging paradigms through an interdisciplinary approach.

2 Previous Studies

2.1 Word of Mouth (WOM)

Word of mouth (WOM) serves as a crucial cue for consumers in evaluating the experiential value of products they are considering for purchase. Indeed, many products have gained widespread popularity as a result of positive WOM [6]. However, concerns have frequently been raised regarding the credibility of WOM, particularly in light of emerging issues such as fake reviews generated by artificial intelligence.

Despite these concerns, WOM continues to function as a critical decision-making resource in consumer behavior and remains an essential means of expressing product experiences. Consequently, research on WOM has expanded significantly in recent years. WOM is now recognized as a valuable subject for understanding the mechanisms of information sharing and diffusion in the era of social media. The fact that WOM can be quantitatively analyzed—through post volume, content, and user interaction on SNS— has further accelerated scholarly interest. Recent studies have increasingly focused on the relationship between WOM and brand engagement, as well as the connection between WOM and ethical consumption, highlighting the influence of WOM on consumers' values and worldviews.

2.2 Liquid Consumption: A Keyword in Experience-Oriented Consumption Research

Within the broader body of research emphasizing experiential value, a concept that has garnered increasing academic attention in recent years is liquid consumption. As previously noted, this concept draws inspiration from sociologist Zygmunt Bauman's [5] theory of Liquid Modernity. Building on Bauman's assertion that society has transitioned from a "solid" to a "liquid" state, liquid consumption has been proposed as a new mode of consumer behavior that contrasts with ownership-oriented consumption. This emerging mode prioritizes experience and immediate gratification over long-term possession.

The concept was formalized by Bardhi and Eckhardt [4], who identified four defining characteristics of liquid consumption; (1) instrumental rationality, (2) individualization, (3) risk and uncertainty, and (4) fragmentation of life and identity.

According to their argument, the driving force behind the rise of liquid consumption is the advancement of digital technologies. The proliferation of online platforms has not only facilitated the distribution and collection of information about goods and services, but also made it easier to purchase them via digital channels. As a result, consumer choices have expanded dramatically. This expansion of choice has, in turn, rendered consumer behavior more fluid and unstable, reflecting a departure from the relative stability of prior consumption models.

To clarify their position, the principal components of their argument are outlined below. One of the most fundamental ways in which liquid consumption differs from traditional consumption lies in the short-term orientation of consumer desire. Rather than emphasizing long-term ownership, contemporary consumers tend to prioritize instantaneous use or experiential engagement. First, the lifespan of products has become significantly shorter. Items with a limited duration of relevance—such as those associated with fast fashion—have attracted particular attention, contributing to a broader trend toward short-term ownership or disposability.

Second, there has been a notable shift from physical goods to experiences and services. Consumers increasingly favor temporary, event-driven experiences over the acquisition of tangible objects. This can be observed in the preference for events over merchandise and attractions over exhibitions.

Third, speed and immediacy have become central features of purchasing behavior. With the widespread availability of the internet and the evolution of mobile technologies, consumers have come to place greater value on immediate access to desired products and services.

Finally, there is the matter of brand and product fluidity. Contemporary consumers tend to exhibit low brand loyalty, with interests that are highly dynamic and constantly shifting. Rather than continuing to use well-established or familiar products, they often gravitate toward novel and seemingly more attractive alternatives. This behavioral shift reflects a broader cultural move away from enduring attachments and toward consumption defined by novelty and perceived experiential enhancement.

3 Method

3.1 Survey Outline

In this study, an online questionnaire survey was conducted with general consumers. The survey was carried out using the self-service survey tool (Freeasy) provided by i-Bridge Inc.

The survey participants were selected from registered monitors on the Freeasy platform. The sample consisted of men and women in their 20s to 60s who were employed in various forms of labor, including full-time employment, contract work, temporary dispatch work, public service, self-employment, freelance occupations, and part-time or hourly jobs.

The survey was conducted on a single day—March 13, 2025. On that day, responses were collected from 900 individuals, with 90 respondents from each gender and age group combination. The average age of respondents was 45.05 years, with a standard deviation of 13.60.

3.2 Measurement Items for Word-Of-Mouth

In this survey, the following questionnaire items were used to assess respondents' attitudes and behaviors regarding word-of-mouth (WOM). A four-point Likert scale was employed as the response format.

(M-1) I post positive word-of-mouth (WOM) after using a product or service.
(M-2) I post negative WOM after using a product or service.
(M-3) I post photos of purchased products on social media.
(M-4) I often talk to acquaintances about my shopping experiences.
(M-5) I trust products that have a large number of customer reviews.
(M-6) Seeing a high number of negative reviews makes me reconsider my purchase.
(M-7) A high number of positive reviews influences my decision to purchase.
(M-8) I refer to reviews by trusted celebrities or influencers.
(M-9) I trust reviews when many people express the same opinion.
(M-10) I trust reviews that are specific and easy to understand.
(M-11) I find reviews more trustworthy when they mention not only recommendations but also drawbacks.
(M-12) I do not trust reviews written in unnatural or awkward Japanese.
(M-13) I regard WOM as just one of many reference points when making purchase decisions.
(M-14) I often check WOM information when shopping or dining out.

3.3 Measurement Items for Liquid Consumption

In Japan, a scale to measure such liquid consumption tendencies has been developed (Kubota [6]). This measurement scale consists of five dimensions and their respective sub-items. Therefore, this study adopted the scale developed by Kubota [6].

Volatility

(1-1) Even after buying something, I often want to have something else immediately.
(1-2) Even if I still have usable items, I often want to replace them with new ones.
(1-3) Even if I buy something I think is good, I often find a better product shortly thereafter.

Non-ownership Consumption

(2-1) Borrowing is easier and better than buying.
(2-2) If the same result can be obtained, I prefer renting or sharing rather than buying.
(2-3) Most things do not need to be bought; it is fine to borrow them when needed.

Preference for Used and Reuse Goods

(3-1) If affordable reused or second-hand goods are available, I often buy them.
(3-2) If the product I am looking for is only available as a reused or second-hand item, I tend to buy it.

Experience-Oriented

- (4-1) I prefer to spend money on experiences rather than on buying things.
- (4-2) I want to experience things more than own things.
- (4-3) I want to spend money on experiences and memories rather than material goods.

Effort-Saving Orientation

- (5-1) When shopping, I do not want to spend much time choosing products.
- (5-2) I do not want to spend much effort or trouble when shopping.
- (5-3) I want to minimize the effort involved in shopping as much as possible.

4 Results

Word-of-mouth (WOM) has a significant influence not only on traditional offline (or in-person) shopping, but also on online shopping behaviors. At the same time, liquid consumption is deeply intertwined with online shopping. Therefore, examining the relationship between WOM and liquid consumption is expected to yield meaningful insights into the role and significance of WOM in the context of online consumer behavior.

It has been argued that the spread of online shopping accelerates tendencies toward liquid consumption. Moreover, marketing strategies specific to online platforms—such as personalized recommendations and mechanisms that encourage impulse purchases—are also believed to reinforce the characteristics of liquid consumption.

This raises an important question: How are these marketing techniques related to consumer-generated media (CGM), including online reviews and WOM? Clarifying these relationships constitutes a central objective of the present study. The following section provides an overview of the findings from the questionnaire survey.

4.1 General Trends in Liquid Consumption

To assess tendencies related to liquid consumption, a four-point Likert scale was employed using the measurement instrument described above. Table 1 presents the proportion of respondents who selected "Strongly Agree" or "Agree" (calculated as number of responses divided by sample size), the proportion of female respondents endorsing each item (number of female responses divided by the female subsample, n = 450), as well as the mean and standard deviation for each item.

Overall, the Effort-Saving Orientation and Experience Orientation dimensions showed relatively high endorsement rates, both exceeding 50%.

A statistically significant gender difference was observed in two items:

"I don't want to spend effort on shopping" (Effort-Saving Orientation), and
"I'd rather spend money on experiences than things" (Experience Orientation).

Table 1. Simple Aggregated Results of Liquid Consumption Tendencies

	%	Female %	Mean	S.D.		
I want to minimize shopping hassle.	61.4%	62.4%	2.68	0.77		
I don't want to spend effort on shopping.	57.9%	59.8%	2.62	0.78	F=4.119	p=0.041
I'd rather spend money on experiences than things.	57.3%	61.1%	2.62	0.74	F=9.485	p=0.002
I prefer spending on experiences over products.	55.4%	58.4%	2.59	0.73		
I prefer experiences to owning things.	52.4%	56.2%	2.54	0.74		
I don't want to spend too much time choosing products.	52.4%	51.1%	2.53	0.74		
I often buy secondhand if that's all that's available.	45.4%	43.6%	2.38	0.88		
I often buy secondhand if it's affordable.	42.1%	40.9%	2.32	0.86		
I prefer renting or sharing over buying if the result is the same.	34.4%	33.3%	2.15	0.84		
I often find a better product soon after buying one.	33.8%	32.9%	2.22	0.80		
I think most things can be borrowed instead of bought.	32.1%	30.7%	2.16	0.80		
I often want to buy something else right after buying something.	28.6%	29.6%	2.06	0.85		
I often want to replace things even if they're still usable.	27.3%	28.2%	2.05	0.82		
Renting is easier and better than buying.	25.7%	23.3%	2.00	0.82		

For all other items, no statistically significant differences between male and female respondents were identified.

In contrast, the endorsement rates for Non-ownership Consumption and Volatility items were generally lower, with less than 40% of respondents agreeing or strongly agreeing with those statements.

4.2 Factor Analysis Results for Liquid Consumption

Next, a factor analysis was conducted on the measurement items described above to examine whether the same five dimensions identified in Kubota's original study—the developer of the scale—could be extracted. However, in the present survey, the factors of "Non-ownership Consumption" and "Preference for Used and Reuse Goods" were not clearly distinguished and instead loaded onto a single factor. As a result, a four-factor structure was extracted (see Table 2).

Table 2. Factor Analysis Results of Liquid Consumption Tendencies

	Factor	after factor extraction			
	1st Factor	2nd Factor	3rd Factor	4th Factor	communalities
I prefer renting or sharing over buying if the result is the same.	0.784	0.141	0.103	0.159	0.670
I think most things can be borrowed instead of bought.	0.735	0.163	0.052	0.162	0.596
Renting is easier and better than buying.	0.662	0.079	0.106	0.294	0.542
I often buy secondhand if it's affordable.	0.625	0.028	0.072	0.146	0.418
I often buy secondhand if that's all that's available.	0.578	0.041	0.042	0.083	0.344
I prefer experiences to owning things.	0.109	0.825	0.088	0.031	0.701
I prefer spending on experiences over products.	0.104	0.818	0.115	0.097	0.702
I'd rather spend money on experiences than things.	0.119	0.786	0.167	0.058	0.663
I don't want to spend effort on shopping.	0.097	0.098	0.869	0.056	0.777
I want to minimize shopping hassle.	0.106	0.125	0.742	0.042	0.579
I don't want to spend too much time choosing products.	0.069	0.121	0.740	0.026	0.567
I often want to replace things even if they're still usable.	0.204	0.052	0.047	0.791	0.673
I often want to buy something else right after buying something.	0.202	0.050	0.040	0.778	0.650
I often find a better product soon after buying one.	0.226	0.076	0.037	0.678	0.518

Maximum likelihood with varimax rotation

4.3 Relationship Between Liquid Consumption and Technological Environment

Then, the author examined the relationship between trends in liquid consumption and the smartphone usage environment. For this analysis, we used the five dimensions of liquid consumption proposed by Kubota (volatility, non-ownership consumption, secondhand/reuse orientation, experience orientation, and effort reduction orientation) as the basis. We conducted an analysis using the total score of each construct scale, performed a variance analysis, and tested the differences in mean values.

First, when respondents had a "data plan with unlimited usage without worrying about data volume," no statistically significant relationship was observed except for non-ownership consumption.

Furthermore, individuals who answered, "I often use my smartphone during commute times such as on the train or bus," tended to show higher liquid consumption characteristics.

Additionally, those who stated, "I use paid services to avoid interruptions by ads because I want to focus on content such as videos, music, or games," showed a higher average for the secondhand/reuse orientation.

4.4 Relationship Between Liquid Consumption and Word-Of-Mouth (WOM)

Then, we explored the relationship between one of the characteristics of online purchases, word-of-mouth (WOM), and the traits of liquid consumption.

Table 3 presents the response rate for each of the above items, calculated by adding the percentages of those who "strongly agree" and "agree," along with the mean values and standard deviations for the 4-point scale, and the results of the variance analysis (testing of mean differences) for the five characteristics of liquid consumption.

Table 3. Relationship between Liquid Consumption Trends and Word-of-Mouth

	%	mean	S.D.	volatility	non-ownership	secondhand/reuse	experience	effort reduction
I don't trust reviews written in unnatural Japanese.	78.1%	3.15	0.85	$p<0.001$	$p<0.001$	$p<0.001$	$p<0.001$	$p<0.001$
I trust reviews that mention both pros and cons.	70.6%	2.80	0.77	$p=0.007$	$p<0.001$	$p<0.001$	$p<0.001$	$p=0.009$
I trust reviews that are specific and clear.	70.0%	2.80	0.75	$p=0.006$	$p<0.001$	$p<0.001$	$p<0.001$	$p<0.001$
I view reviews as just one source of information when making purchases.	68.4%	2.76	0.74	$p=0.087$	$p=0.002$	$p=0.013$	$p<0.001$	$p<0.001$
I reconsider my purchase after seeing many negative reviews.	64.6%	2.69	0.73	$p<0.001$	$p<0.001$	$p<0.001$	$p<0.001$	$p=0.002$
I trust reviews with many ratings.	60.7%	2.62	0.72	$p<0.001$	$p<0.001$	$p<0.001$	$p<0.001$	$p=0.005$
I trust reviews when many people share the same opinion.	59.6%	2.59	0.74	$p<0.001$	$p<0.001$	$p<0.001$	$p<0.001$	$p=0.005$
I consider purchasing if there are many positive reviews.	59.3%	2.60	0.74	$p<0.001$	$p<0.001$	$p<0.001$	$p<0.001$	$p<0.001$
I often check reviews before shopping or dining.	56.7%	2.59	0.78	$p<0.001$	$p<0.001$	$p<0.001$	$p<0.001$	$p=0.574$
I often share my shopping experiences with others.	32.0%	2.13	0.81	$p<0.001$	$p<0.001$	$p<0.001$	$p<0.001$	$p<0.001$
I rely on reviews from trusted celebrities or influencers.	31.2%	2.11	0.83	$p<0.001$	$p<0.001$	$p<0.001$	$p<0.001$	$p<0.001$
I write a positive review after using a product or service.	26.3%	1.96	0.84	$p<0.001$	$p<0.001$	$p<0.001$	$p<0.001$	$p<0.001$
I post photos of purchased items on social media.	24.8%	1.82	0.90	$p<0.001$	$p<0.001$	$p<0.001$	$p<0.001$	$p<0.001$
I write a negative review after using a product or service.	22.1%	1.85	0.85	$p<0.001$	$p<0.001$	$p<0.001$	$p<0.001$	$p<0.001$

Results showed that in many cases, there was a statistically significant relationship between word-of-mouth factors in online purchases and liquid consumption. The exceptions were for the statement, "I view reviews as just one source of information when making purchases" in relation to volatility, and "I often check reviews before shopping or dining," where no statistically significant relationship was observed. Regarding volatility, which involves purchasing more products and services at a faster pace, reviews may indeed be viewed merely as one source of reference. In the case of effort reduction orientation, checking reviews may be seen as a task separate from any sense of inconvenience.

5 Conclusion

In conclusion, this paper conducted an online survey to examine trends in liquid consumption in Japan and provided an overview of the situation. However, this paper is only a preliminary exploration. Further analysis and interpretation of the data are needed, and future work will focus on analyzing consumption trends in the social network era.

Acknowledgments. This work was supported by JSPS KAKENHI Grant Number JP23K01620 and the Kansai University Fund for Domestic and Overseas Research Support Fund, 2023.

Disclosure of Interests. The authors have no competing interests to declare that are relevant to the content of this article.

References

1. Setiawan, I., Kartajaya, H., Kotler, P.: Marketing 4.0: moving from Traditional to Digital. Wiley, Hoboken (2016)
2. Daugherty, T., Eastin, M.S., Bright, L.: Exploring consumer motivations for creating user-generated content. J. interactive advertising **8**(2), 16–25 (2008)
3. Pine, B. J., Gilmore, J. H.: The experience economy, 2nd ednition. Harvard Business Press, Boston, Massachusetts (2011)
4. Bardhi, F., Eckhardt, G.M.: Liquid consumption. J. Consum. Res. **44**(3), 582–597 (2017)
5. Bauman, Z.: Liquid Modernity. Polity Press (2000)
6. Kubota, Y.: "Development of a Scale to Measure the Liquidity of Consumption." Aoyama J. Bus. **56**(4), 109–129 (in Japanese) (2020)

Organizational Memory Facilitates Effective Learning from Failure: Evidence from a Questionnaire Survey Conducted in 2024 in Japanese Companies

S. Nagayoshi[1(✉)] and J. Nakamura[2]

[1] Shizuoka University, 3-5-1 Johoku, Chuo-ku, Hamamatsu, Shizuoka, Japan
`nagayoshi@inf.shizuoka.ac.jp`
[2] Chuo University, 742-1 Higashinakano, Hachioji, Tokyo, Japan
`jyulis.77f@g.chuo-u.ac.jp`

Abstract. Learning from failure is crucial for organizations to prevent the recurrence of failures and enhance their performance. However, numerous organizations encounter difficulties in effectively learning from failures, potentially because of the ineffective functioning of organizational memory. This study investigated the efficacy of the organizational memory mechanism in facilitating learning from organizational failure by extending a previously developed model of the organizational memory mechanism in learning from failure. A hypothetical model was constructed and analyzed using quantitative data collected through a questionnaire survey administered to employees of various Japanese companies in March 2024. Covariance structure analysis was employed to validate the model. The results indicate that organizational memory in the context of learning from failure exerts a significant influence on the effectiveness of organizational learning, with the extended hypothetical model demonstrating a high degree of fit. These findings suggest that a well-functioning organizational memory is associated with an increased likelihood of successful organizational learning from failure. This study contributes to the literature by extending the previous organizational memory model for organizational learning and knowledge management, and provides practical implications for organizations experiencing challenges with learning from failure. However, the limitations include the need for further investigation into the reproducibility of the model's goodness of fit and its applicability to organizations that successfully implement learning from failure.

Keywords: Organizational memory · Learning from failure · Successful Organizational learning

1 Introduction

The significance of learning from failure is widely acknowledged. Through the process of learning from failure, individuals can enhance their performance when confronting similar future challenges.

According to Hatamura [1], who examines failures and their countermeasures from an engineering perspective, failure is defined as "a human action that does not reach a defined goal" or "an undesirable and unexpected result of a human action." Two categories of failures are identified: "worthwhile" and "worthless." A worthwhile failure is defined as "a failure that cannot be avoided even with great care" and represents a step into the unknown. Conversely, "worthless failures" are defined as "failures other than worthwhile failures."

Iske [2] refers to failures as "brilliant failures," which he conceptualizes as organizational attempts to create value that fails to produce the originally intended results. These "brilliant failures" are classified into two categories: "serendipity," which involves the accidental discovery of important and unplanned outcomes that may lead to another desirable result or valuable learning experience.

Edmondson [3] delineated three typical examples of failures: avoidable, complex, and smart failures. According to the author, avoidable failures deviate from a known process and result in an unwanted outcome attributable to a lack of action, skill, or attention. Complex failures occur when an unprecedented and unusual combination of events and actions leads to unwanted outcomes caused by the addition of complexity, variety, and unprecedented factors to familiar situations. A smart failure is one in which a new endeavor is initiated and an unwanted result occurs, which is attributed to uncertainty, attempts, and risk-taking.

Organizations can learn more effectively from failures than from successes [4]. In the case of individual failure, through reflection, the individual can recall the cause of the failure and implement preventive measures to avoid recurrence. However, in the context of organizational failure, even if an individual who experienced failure remembers its cause and preventive measures through reflection, unless other members of the organization learn from the failure, they are likely to encounter the same failure. Therefore, it is crucial for organizations to retain the lessons learned from failure.

However, organizations face significant challenges in learning from their failures [5, 6], and many organizations demonstrate inadequate proficiency in this area. Consequently, instances arise where organizations repeatedly commit similar errors, indicating suboptimal capacity for failure-based learning. The factors contributing to an organization's ineffectiveness in learning from failures include the absence of psychological safety [3]; psychological barriers such as embarrassment, guilt, and excessive psychological burden associated with failure disclosure; and concerns related to maintaining organizational status [7]. Additionally, it is hypothesized that a lack of well-functioning organizational memory within an organization may be a contributing factor.

Through a literature review, Huber [8] defined organizational learning processes as knowledge acquisition, information distribution, information interpretation, and organizational memory. Organizational memory refers to the mechanism through which knowledge is stored for future use [8]. Organizational memory can be categorized into "data storage" and "data retrieval" [9]. There are two additional potential mechanisms for "data storage." One involves storing data in the cognitive faculties of the individuals comprising the organization, whereas the other entails recording data using documents, computers, and other tangible objects and instruments. The stored data were subsequently "retrieved" when necessary. Cognition and other factors shared by organizational members are hypothesized to play a crucial role in effective data retrieval [9–12], and Huber [8] emphasized the significance of computer utilization in Organizational Memory. Aggestam and Svensson [13] showed how digital applications facilitate knowledge sharing, based on a qualitative study of knowledge sharing among different healthcare providers and health professionals, as well as the need for complementary collaborative sessions.

This study is predicated on the premise that few organizations excel in organizational learning from failure, and that organizational failures are recurrent. The authors posit that one reason for this recurrence may be the ineffective functioning of organizational memory in learning. The objective of this study was to investigate the efficacy of the organizational memory mechanism in learning from organizational failure. Specifically, the authors constructed a hypothetical model that extends the organizational memory mechanism model of organizational learning from failure, as presented in a previous study, and analyzed this hypothetical model using quantitative data collected from a questionnaire through structural covariance analysis.

2 Research Methodology

This study employed a hypothesis testing quantitative methodology. This study will involve the formulation of a hypothesis that will be developed by extending previous research conducted by the authors. Specifically, the hypothesized model for this study was derived from an extension of the organizational memory mechanism model of organizational learning from failure, which was previously developed by the authors through case studies and subsequent generalization research. Subsequently, quantitative data were collected via an internet survey to test the hypothesized model. The collected data were subjected to covariance structure analysis using statistical analysis software. The results of the analysis are then discussed, and the hypothetical model constructed in this study is evaluated through comparison with the conventional model.

3 Constructing Hypothetical Model with Previous Study

3.1 Previous Study

Qualitative Study with Case Company. The authors conducted a case study of a Japanese company that successfully implemented organizational learning from failure using a computer-based knowledge-sharing database and developed a preliminary hypothetical model [14].

Quantitative Study for Generalization. Subsequently, to enhance the credibility of the preliminary hypothetical model that had been constructed, a questionnaire survey was administered to the employees of the case firm. The collected quantitative data were analyzed using covariance structure analysis to increase the credibility of the hypotheses [15].

However, because the hypothetical model was solely based on organizational learning practices from failures in specific companies, generalization is required for broader applicability.

Therefore, with the cooperation of a Japanese Internet survey company, the authors conducted a questionnaire survey among the company's monitored members who were employed by the organization. Utilizing quantitative data collected through this questionnaire survey, the authors tested the hypothesized model and determined its potential for generalization [16]. The results are shown in Fig. 1.

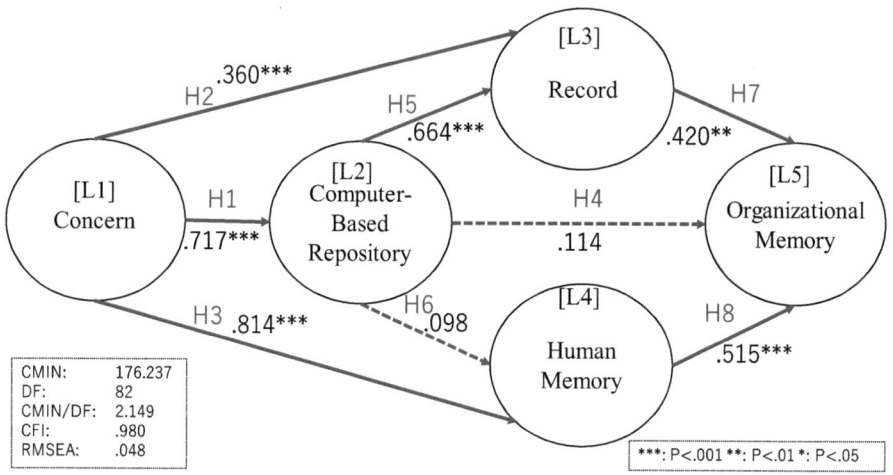

Fig. 1. Previous Analysis Results of Anonymity [16]

Nevertheless, this research does not elucidate whether the organizational memory mechanism is efficacious in organizational learning from failure, which constitutes the fundamental inquiry of this study. This study addresses this essential question by expanding the existing model.

3.2 Hypothetical Model

Huber [8] defined organizational learning processes as Information Acquisition, Knowledge Distribution, Information Interpretation, and Organizational Memory. Based on this definition, it is reasonable to infer that organizational memory is conducive to learning. Consequently, in this study, the authors expand the model [16] by incorporating the latent variable "Effective Organizational Learning from Failure." The extended hypothetical model is illustrated in Fig. 1. It is important to note that [L6: Effective Organizational Learning from Failure] in Fig. 2 represents the latent variable newly introduced in this study, building on the model [16] (Fig. 1).

The hypotheses that constitute this theoretical model are designated as Hypothesis1(H1) through Hypothesis9 (H9). H1 through H8 are the same as those in a previous study [16]. Thus, H9 represents a novel addition to this study.

Hypothesis 1 (H1): A matter of interest relevant to the work strengthens the use of computer-based knowledge repositories.
Hypothesis 2 (H2): A matter of interest relevant to work strengthens recording activity.

Hypothesis 3 (H3): A matter of interest that is relevant to work strengthens human memory.
Hypothesis 4 (H4): A computer-based knowledge repository directly strengthens organizational memory.
Hypothesis 5 (H5): A computer-based knowledge repository strengthens recording activity.
Hypothesis 6 (H6): Computer-based knowledge repositories strengthen human memory.
Hypothesis 7 (H7): Recording activities strengthen organizational memory.
Hypothesis 8 (H8): Human memory strengthens organizational memory.
Hypothesis 9 (H9): Organizational memory strengthens effective organizational learning from failures.

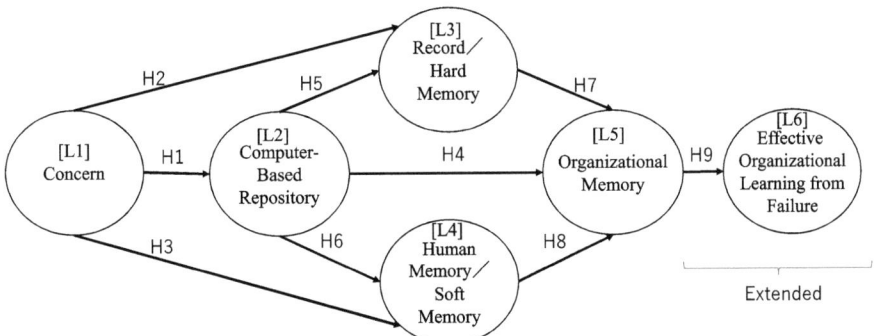

Fig. 2. Extended Model (Hypothetical Model in this study)

4 Data Collection and Analysis

4.1 Data Collection

The organization has 13 million members, representing the largest number of monitors in Japan. The questionnaire comprised of 50 questions, identical to the questionnaire for the case company described above. This is also consistent with the questionnaire that the authors administered on March 13, 2023, in their previous study. The questionnaire was developed using a survey platform called Freeasy, owned by the company conducting the survey, and was administered via the internet to monitor the members of the company. As the survey was designed for employees of organizations and companies, occupational attributes were restricted to company employees (regular employees), company employees (contract and temporary employees), managers and executives, public officials (excluding faculty members), and doctors and medical personnel. For the same reason, the age criterion was set at 20 years or older. The survey was administered on March 1, 2024, and firms were requested to provide 600 respondents. The survey was conducted on March 1, 2024, and the company was asked to provide a sample of 600 respondents with the option to utilize the survey provider's AI to eliminate fraudulent responses. According to the company, this service collects 120% of the target sample size and eliminates short- and straight-lining responses as fraudulent. However, it should be noted that if the number of fraudulent responses exceeds 20%, not all fraudulent responses can be eliminated. A sample size of approximately 385 participants was required to obtain data tolerance of 5%. When the authors removed the potentially fraudulent data from the survey data collected in a previous survey of the company's monitor members, they found that even with the addition of the AI fraudulent response elimination service provided by the company, they were unable to eliminate all fraudulent responses; therefore, the data for the analysis were not considered to be fraudulent. This reduced the data analyzed to 60–80% of the data provided by the company. Considering this, the authors requested that the firm provide a sample size of 600.

The authors examined 600 samples of data received from the company, meticulously eliminating data that could be considered fraudulent responses, and 502 samples were included in the analysis.

4.2 Data Analysis

Definition of Latent Variables and Data Consistency Investigation.
Cronbach's alpha was used to assess the reliability of the data. The calculations were performed using the Bell Curve for Excel (Social Survey Research Information Co., Ltd.). Generally, a Cronbach's alpha coefficient of 0.90 or higher is considered to indicate high internal consistency, 0.80 or higher is considered to demonstrate consistency, and 0.70 or higher is deemed an acceptable level of consistency.

Furthermore, a covariance structure analysis was conducted to evaluate the hypothesized model. In this analysis, the latent variables of the hypothetical model presented in Fig. 1 were defined by designating the survey items from the questionnaire as observed variables for the latent variables. This method of defining latent variables was also applied to the analysis of the questionnaire data collected from employees of multiple unspecified companies, as described in the subsequent sections.

The latent variable [L1:Concern] is defined by the following three observed variables: [M4: I can utilize my organization's failure cause investigation reports and failure recurrence prevention measures in my own work]; [M5: I can learn from my organization's failure cause analysis and failure recurrence prevention measures through its activity reports]; and [M6: Activity reports, such as failure cause analysis and failure recurrence prevention measures in my organization, contain information that may be relevant to me. M6: Activity reports such as analysis of the causes of failure and planning of measures to prevent recurrence of failure in your organization contain information that may be relevant to you.] The Cronbach's alpha coefficient for the latent variable [L1: Concern] was 75.56%, indicating an acceptable level of internal consistency.

The latent variable [L2: Computer-Based Repository] was defined by the following three observed variables: [M19: Your organization uses computers to search for failure prevention measures], [M20: Your organization stores failure prevention measures in a computer (system)], and [M21: Your organization shares failure prevention measures using a computer (system)]. Cronbach's alpha for the latent variable [L2 (Computer-Based Repository] also demonstrated internal consistency at 85.53%.

The latent variable [L3: Record] was defined by the following three observed variables: [M13: Your organization has a documented method to prevent failures], [M14:

Your organization has a manual in place to prevent failures], and [M15: Your organization has established rules on how to prevent failures]. The Cronbach's alpha coefficient for the latent variable [L3: Record] was 83.87%, demonstrating internal consistency.

The latent variable [L4: Human Memory] is defined by the following three observed variables: [M16: Members of the organization (supervisors, subordinates, and colleagues) remember how to prevent failures], [M17: Members of the organization (supervisors, subordinates, and colleagues) understand how to prevent failures], and [M18: Members of the organization (supervisors, subordinates, and colleagues) naturally implement methods to prevent failures]. Cronbach's alpha coefficient for the latent variable [L4: Human Memory] was 82.20%, indicating internal consistency.

The latent variable [L5: Organizational Memory] is defined by the following three observed variables: [M10: the organization has accumulated failure prevention measures], [M11: the organization can retrieve failure prevention measures when necessary], and [M12: the organization can recall failure prevention measures when necessary]. Cronbach's alpha for the latent variable [L5: Organizational Memory] was 81.60%, demonstrating internal consistency.

The latent variable [L6: Effective Learning from Failure] is defined by the following three observed variables: [M1: When a problem or failure occurs, the organization analyzes the causes and plans countermeasures to prevent recurrence], [M2: Your organization effectively learns from failure], and [M3: Your organization effectively analyzes failure and plans countermeasures]. The Cronbach's alpha for the latent variable [L5 Organizational Memory] was 80.41%, demonstrating internal consistency.

Consequently, the authors concluded that the data for all latent variables demonstrated consistency or exceeded an acceptable threshold, indicating that the collected data were suitable for hypothesis testing.

Covariance Structure Analysis. Covariance structure analysis was conducted using IBM SPSS AMOS29 software for statistical analysis to validate the hypothesized model. The results of covariance structure analysis of the hypothetical model are shown in Fig. 3.

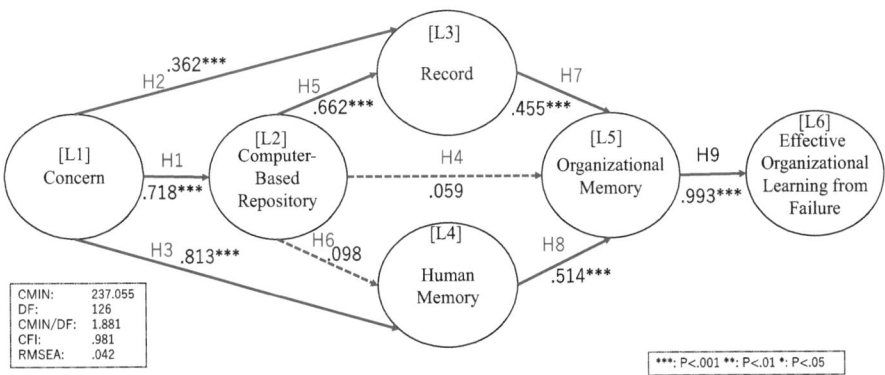

Fig. 3. Analysis Result

5 Discussion

5.1 Model Evaluation

Examination of the Comparative Fit Index (CFI), which indicates model fit, revealed a value of .981. Similarly, the Root Mean Square Error of Approximation (RMSEA) demonstrated a value of .042. Based on these results, the authors conclude that the hypothetical model exhibits a good fit.

The CFI for the pre-extension model [15] depicted in Fig. 1 was .980, whereas the CFI for the extension model in this study was .981, indicating a marginal improvement in fit. The RMSEA of the extended model was .048, whereas the RMSEA of the extended model in this study was .042. Based on these findings, the authors postulate that the goodness of fit of the extended model demonstrates an enhancement over that of the preextended model.

5.2 Hypothesis Verification

The hypothesis was tested based on the results of the analysis of the extended model.

Examining H1.[L1] → [L2] revealed a significant positive correlation at the 0.1% level. Therefore, H1 is supported, indicating that employees are more likely to use computers if it is a matter of interest to them within the organization. The analysis based on the pre-extension model yielded similar results.

Examining H2.[L1] → [L3] revealed a significant positive correlation at the 0.1% level. Therefore, it is plausible that recording would occur if it were a matter of interest to employees within the organization, and H2 was supported. The analysis based on the pre-extension model yielded similar results.

Examining H3.[L1] → [L4] revealed a significant positive correlation at the 0.1% level. Thus, if it is of interest to employees in the organization, they are more likely to remember it, and H3 is supported. The analysis based on the pre-extension model yielded similar results.

Examining H4.[L2] → [L5] showed no significant relationship. Therefore, it is unlikely that computer use directly leads to organizational memory; thus, H4 is not supported. Notably, the analysis based on the pre-extension model yielded similar results.

Examining H5.[L2] → [L3] demonstrated a significant positive correlation at the 0.1% level. Therefore, H5 is supported, suggesting that the use of computers facilitates recording. The analysis based on the pre-extension model yielded similar results.

Examining H6.[L2] → [L4] did not show a significant relationship. Therefore, it is unlikely that the use of computers facilitates human memory and H6 is not supported. Notably, the analysis based on the pre-extension model yielded similar results.

Examining H7.[L3] → [L5] showed a significant positive correlation at the 0.1% level. Therefore, H7 is supported, indicating that the recordings appear to reinforce organizational memory. The analysis based on the pre-extension model also showed a positive correlation, supporting H7. Although the significance level was 1% versus 0.1% in the extended model in the present study, the significance level was improved to 0.1%.

Examining H8.[L4] → [L5] showed a significant positive correlation at the 0.1% level. It appears that human memory reinforces organizational memory, and that H8 is supported. The analysis based on the pre-extension model yielded similar results.

Examining H9.[L5] → [L6], a hypothesis added to the extended model, demonstrated a significant positive correlation at the 0.1% level. The correlation coefficient was .993, indicating a strong correlation. Therefore, H9 is supported. In other words, organizational memory of organizational learning from failure strongly influences the effectiveness of organizational learning. This finding suggests that organizational learning from failure is more likely to be successful when organizational memory effectively functions.

6 Conclusion

The authors posited that one of the factors contributing to the recurrence of failures in organizations may be the ineffective functioning of organizational memory in organizational learning, given that few organizations demonstrate proficiency in learning from failure. Consequently, this study aimed to investigate the efficacy of the organizational memory mechanism in facilitating learning from organizational failures. The authors developed a hypothetical model that expands upon the organizational memory mechanism model of learning from failure, as presented in a previous study. They subsequently analyzed this hypothetical model using quantitative data collected through a questionnaire and employed structural covariance analysis to validate the model. These findings indicate that organizational memory in the context of learning from failure exerts a significant influence on the effectiveness of organizational learning, in addition to demonstrating a high degree of fit for the extended hypothetical model constructed in this study. These results suggest that a well-functioning organizational memory is associated with an increased likelihood of successful organizational learning from failure.

In previous studies, the authors constructed a model that demonstrates the formation of organizational memory in organizational learning from failure. Initially, the authors demonstrated that organizational memory results in successful organizational learning from failure. In this study, the authors not only re-examined the formation of organizational memory in organizational learning from failure but also demonstrated that organizational memory is effective in organizational learning from failure.

This research not only extends the previous organizational memory model for organizational learning and knowledge management but also has significant practical implications, given that many organizations struggle with organizational learning from failure, suggesting that organizational memory is critical to their success.

The model presented in this study, validated with data collected from unspecified firms, has the potential for broad generalizability. However, certain issues require attention for generalization. The analysis in this study was conducted using data collected through a questionnaire administered in March 2024 using a questionnaire similar to that conducted in March 2023. Owing to limitations in paper length, detailed descriptions were omitted; nevertheless, the results of the covariance structure analysis using data collected from the earlier questionnaire are presented in Fig. 4.

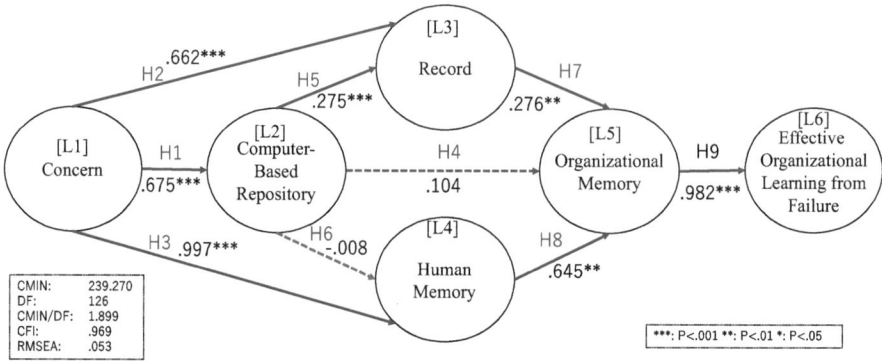

Fig. 4. Analysis Result base on 2023 Questionnaire

The CFI for the pre-extension model [16] shown in Fig. 1, which represents the goodness of fit of the pre-extension model conducted in a previous study, was .980. As previously stated, the CFI for the extended model based on data analysis collected in March 2024 was .981, marginally improving the goodness of fit, whereas that for the extended model based on data collected in March 2023 was .969, decreasing the goodness of fit. RMSEA for the pre-expansion model was .048. As noted above, the RMSEA for the extended model based on data analysis collected in March 2024 was .042, indicating a slightly better fit, whereas the RMSEA for the extended model based on data collected in March 2023 was .053, demonstrating a relatively better fit but a lower overall fit. Based on these results, it cannot be concluded that the model is necessarily reproducible with respect to goodness of fit. Thus, compared to the pre-expansion model [16], the model fit for the analysis of data collected in March 2023 decreased, whereas the model fit for the analysis of data collected in March 2024 improved. Therefore, it cannot be asserted that the expanded model is superior to the preexpansion model. The factors leading to this difference are unknown and require further investigation.

It is also not possible to verify whether this is applicable to unique organizations that have successfully implemented learning from failure, which the authors have previously studied.

The model constructed in this study failed to consider the relationship between the recordings and human memory. From a learning science perspective, it has been noted that recording facilitates a person's memory [17]; however, this aspect has not been considered.

The collected data included numerous fraudulent responses. The Internet survey company collected 1.2 times as much data as the target sample size and utilized the AI functionality provided by the company to eliminate fraudulent data and obtain data for the target sample size. However, this does not necessarily ensure that all fraudulent responses have been eliminated.

As noted in the study limitations above, compared to the pre-expansion model [16], the model fit of the analysis of data collected in March 2023 decreased, whereas the model fit of the analysis of data collected in March 2024 improved. The reasons for this discrepancy remain unknown and require further investigation by using additional questionnaires.

The authors posit that the model presented in this study requires validation to determine its applicability to organizations that successfully implement organizational learning from failure.

Furthermore, a comprehensive review of the literature on recording and memory in learning sciences will be conducted. The model developed in this study will be modified and analyzed using statistical methods to incorporate a path from the latent variable "Record" to "Human Memory." The authors also review the literature on records and memory related to science. A comparative analysis and discussion of the results obtained using the model constructed in this study are also necessary.

Finally, the questionnaire is revised to enhance the reliability of the collected data. Specifically, in this study, priority was given to minimizing the number of questions to reduce the burden on respondents. However, to improve the reliability of the data to be analyzed through the elimination of fraudulent response data, the authors will incorporate additional questions and control items to ensure consistency.

Acknowledgments. This research was supported by a Grant-in-Aid for Scientific Research (C) (Grant #22K01627) and sponsored by the Japan Society for the Promotion of Science (JSPS).

Disclosure of Interests. The authors have no competing interests to declare relevant to the content of this article.

References

1. Hatamura, Y.: Shippaigaku no Susume (Guideline to Failure). Kodansha Bunko (2005) (in Japanese)
2. Iske, P. L.: Institute of Brilliant Failures: Make Room to Experiment, Innovate, and Learn (Translated by Bovelander, M.). Business Contact Publishers (2019)
3. Edmondson, A. C.: The Fearless Organization – Creating Psychological Safety in the Workplace for Leaning, Innovation, and Growth. Wiley (2019)
4. Madsen, P.M., Desai, V.: Failing to learn? The effects of failure and success on Organizational Learning in the Global Orbital Launch Vehicle Industry. Acad. Manag. J. **5**(3), 451–476 (2010)
5. Cannon, M.D., Edmondson, A.C.: Confronting Failure: Antecedents and Consequences of Shared Beliefs about Failure in Organizational Work Groups. J. Organ. Behav. **22**, 167–177 (2001)
6. Cannon, M.D., Edmondson, A.C.: Failing to Learn and Learning to Fail (Intelligently): How Great Organizations Put Failure to Work to Innovate and Improve. Long Range Plan. **38**, 299–319 (2005)
7. Nagayoshi, S., Nakamura, J.: What Subdues Shame in Learning from Failure? Empirical Study on a Company in Japan. In: PACIS 2016 Proceedings, Paper 154 (2016) http://aisel.aisnet.org/pacis2016/154
8. Huber, G.P.: Organizational learning: The Contributing Processes and the Literatures. Organization Sciences **2**(1), 88–115 (1991)
9. Walsh, J.P., Ungson, G.R.: Organizational Memory. Acad. Manag. Rev. **16**(1), 57–91 (1991)
10. Mathieu, J.M., Maynardand, M. T., Rapp, T., Gilson, L.: Team Effectiveness 1997–2007: A Review of Recent Advancements and a Glimpse Into the Future. J. Manag. **34**(3), (2008)
11. Holyoak, K.J., Junn, E.N., Billman, D.O.: Development of analogical problem-solving skill. Child Dev. **55**(6), 2042–2055 (1984)

12. Austin, J.: Transactive Memory in Organizational Groups: The Effects of Content, Consensus, Specialization, and Accuracy on Group Performance. J. Appl. Psychol. **88**(5), 866–878 (2003)
13. Aggestam, L., Svensson, A.: How Digital Applications Can Facilitate Knowledge Sharing in Healthcare. Learn. Organ. **32**(1), 58–74 (2024). https://doi.org/10.1108/TLO-01-2024-0002
14. Nagayoshi, S., Nakamura, J.: How Computer Usage Affects Organizational Memory: A Case Study of a Japanese Company Implementing Organizational Learning from Failure. In: Proceedings of the 10th Multidisciplinary International Social Networks Conference, September 4–6, Phuket, Thailand. USA (2023) https://doi.org/10.1145/3624875.3624877
15. Nagayoshi, S., Nakamura, J.: Impact of Computer Usage on Organizational Memory and Learning from Failure: A Case Study of a Japanese Company. The Review of Socionetwork Strategies **18**, 349–371 (2024). https://doi.org/10.1007/s12626-024-00163-5
16. Nagayoshi, S., Nakamura, J.: Enhancing Organizational Memory through Computerized Recording and Facilitating Human Memory in Japanese Companies. In: Proceedings of Asia Pacific Conference on Information Management 2024(APCIM2024), October 31- November 2, 2024, Kyung Hee University, Seoul, Republic of Korea, pp.369–372 (2024)
17. Aiken, E.G., Thomas, G.S., Shennum, W.A.: Memory for a Lecture: Effects of Notes, Lecture Rate, and Informational Density. J. Educ. Psychol. **67**, 439–444 (1975)

Exploring the Relationship Between Local Election Results and Online Public Opinion in Taiwan: A Case Study of Taitung County

I-Hsien Ting[1,3](✉) [image], Yen-Chih Chiu[1,3], Yun-Hsiu Liu[1,3] [image], Kazunori Minetaki[2,3] [image], and Chia-Sung Yen[1,3] [image]

[1] National University of Kaohsiung, Kaohsiung City, Taiwan
iting@nuk.edu.tw
[2] Kindai University, Higashiosaka, Japan
kminetaki@bus.kindai.ac.jp
[3] National Chinyi University of Technology, Taichung City, Taiwan
csyen@ncut.edu.tw

Abstract. This study examines the relationship between online buzz and local election outcomes in Taiwan, with a focus on Taitung County. As social media becomes a major channel for public discourse, online buzz is increasingly seen as a factor influencing elections. However, its impact on local elections in Taiwan remains underexplored. This research addresses that gap through a comparative analysis of social media data and actual vote shares during the election period. A review of existing literature establishes the study's framework and highlights the need for empirical investigation in this area.

The findings aim to reveal whether online discussions align with electoral results and to what extent digital sentiment reflects voter behavior. The study also discusses methodological and data limitations that may affect interpretation. Beyond its academic value, the research offers practical insights into how online buzz can inform campaign strategies and enhance election predictions. By analyzing the Taitung County case, this study contributes to a deeper understanding of the role of online discourse in Taiwan's local elections and offers a foundation for future research in the field.

Keywords: Online Buzz · Online Public Opinion · Local Elections · Social Media Analysis · Taitung County

1 Introduction

In recent years, with the continuous advancement and widespread adoption of internet technologies, the internet has become one of the primary channels for political discourse and information dissemination. Within this virtual realm, online volume—referring to the quantity of discussions surrounding specific political events or candidates—has gradually emerged as a significant indicator in political analysis and electoral studies. The discourse and opinions reflected in online volume have the potential to influence voters' political attitudes and voting behavior. As such, understanding the relationship between online volume and electoral outcomes has become a focal point of interest in both academic and political spheres.

Taiwan, characterized by its political pluralism and high level of information accessibility, offers a unique context in which local elections constitute a vital component of the democratic system. These elections not only attract considerable attention from domestic and international scholars but also generate widespread engagement and discussion in online spaces. However, existing research on the correlation between online volume and local election outcomes in Taiwan remains limited, particularly in terms of studies employing data comparison methodologies for in-depth analysis. This study seeks to address this research gap by investigating the relationship between online volume and the results of Taiwan's local elections, utilizing comparative data analysis to provide a more comprehensive understanding of their interconnection.

The primary objective of this study is to investigate the relationship between online volume and the outcomes of local elections in Taiwan. Specifically, the research aims to: (1) analyze the correlation between the number of votes received by local election candidates and the volume of online discussions related to these elections during the corresponding periods; (2) examine the extent to which online volume influences electoral outcomes, as well as the underlying mechanisms through which such influence may occur; and (3) compare variations in the relationship between online volume and electoral results across different regions, time periods, and political contexts. Through these analytical dimensions, the study seeks to provide a more nuanced understanding of the interaction between digital discourse and electoral behavior in Taiwan, thereby contributing novel insights to the fields of political science, communication studies, and the broader social sciences.

This study carries significant theoretical and practical implications. From a theoretical perspective, an in-depth exploration of the relationship between online volume and local election outcomes enhances our understanding of the internet's role in shaping political behavior and contributes to the expansion of theoretical frameworks in political communication and digital media research. Practically, the findings of this study offer valuable guidance for political candidates and campaign strategists, enabling them to more effectively leverage online platforms in the planning and execution of election campaigns. By shedding light on the dynamics between digital engagement and electoral success, this research supports efforts to improve campaign strategies and increase competitiveness within Taiwan's democratic processes.

2 Literature Review

2.1 Social Networks and Social Metrics

Social Network Analysis (SNA) has emerged as a valuable methodological framework for examining public opinion in electoral contexts, particularly through the lens of social interactions and influence patterns within digital platforms. Originally rooted in sociology, SNA provides tools for observing and analyzing the structure and dynamics of social relationships, allowing researchers to uncover the roles and influence of individual actors within a network [9]. Its conceptual foundation and analytical techniques have been widely adopted across multiple disciplines, including information technology, management, and political science [3,7,8]. The fundamental metrics of SNA—such as centrality, betweenness, closeness, clustering coefficient, and structural holes—enable scholars to identify opinion leaders, track information flow, and assess the influence of network position on public discourse. In the context of elections, these metrics help elucidate how political opinions are disseminated, amplified, or silenced across social media, thereby offering critical insights into voter behavior and the formation of public sentiment [5].

With the exponential growth of social networking platforms, vast amounts of user-generated data have become available for analysis. This surge in data has attracted the attention of researchers across disciplines, particularly in the fields of information technology and communication. However, the sheer volume of data presents significant challenges for efficient analysis. Opinion mining—also known as sentiment analysis—has emerged as a pivotal solution, enabling the automated extraction of user sentiments from unstructured text. This technique focuses not only on identifying sentiment-laden expressions but also on transforming qualitative data into structured insights. Methods such as corpus-based and thesaurus-based opinion mining allow researchers to classify emotional tone, detect spam, and even forecast political trends based on social media discourse. Despite the widespread use of sentiment analysis, critical sociological dynamics, such as the Spiral of Silence (SOS) effect, have often been overlooked. Ting (2016) addressed this gap by proposing a framework to detect the SOS effect on social media, demonstrating that individuals may refrain from expressing dissenting opinions when perceiving their views as minority-held, thus impacting the reliability of sentiment-based predictions [4].

In parallel, the study of social network structures and metrics has deepened our understanding of user behavior within digital communities. Oerez and Ting (2022) examined how structural holes—gaps between non-redundant contacts in a network—can confer strategic advantages to individuals positioned within them [6]. Their findings revealed that the sustainability of such advantageous positions is strongly influenced by neighborhood similarity, offering nuanced insights into how social capital is formed and maintained in online environments. These structural characteristics are crucial in interpreting opinion dynamics, as users in structurally favorable positions may exert disproportionate influence on discourse, shaping visible sentiment trends. The integration of network theory with sentiment analysis thus provides a more comprehensive framework for understanding opinion formation and diffusion in social media contexts. Together, these studies underscore the necessity of incorporating both network metrics and sociopsychological effects, such as the SOS, to enhance the accuracy and depth of social media analytics.

2.2 Elections and Online Public Opinion

The intersection of social media and political behavior has become a critical area of scholarly inquiry, particularly in understanding how digital platforms influence electoral dynamics. Gil de Zúñiga and Valenzuela (2011) conducted a comprehensive cross-national study that demonstrated a significant positive correlation between social media use, social capital, civic engagement, and political participation. Their findings underscore the role of social media not only as a channel for news consumption but also as a catalyst for democratic participation across various cultural contexts. This research provides compelling evidence that online platforms can mobilize users toward political action, thereby altering traditional models of political engagement. The implications of their work are particularly relevant in the context of local elections, where community-based interactions and digital engagement can meaningfully influence voter sentiment and turnout [10].

Conversely, other scholars have urged caution regarding the use of social media data in electoral prediction. Jungherr, Jürgens, and Schoen (2012) offered a critical response to prior studies suggesting that Twitter content could reliably forecast election outcomes. Their critique highlighted methodological shortcomings, such as the representativeness of social media users and the overemphasis on volume-based sentiment analysis. They emphasized that while online discourse reflects certain aspects of public opinion, it is insufficient as a standalone predictive tool without considering broader contextual and demographic factors. Together, these contrasting perspectives illustrate both the potential and the limitations of using online public opinion in electoral studies. They provide a nuanced foundation for examining how digital platforms mediate political expression and voter behavior, informing the current study's investigation into the relationship between online volume and local election results in Taiwan [1,2].

3 Research Design and Data Collection

3.1 Research Method

This study adopts the Data Matching Method as its primary research approach to explore the relationship between local election outcomes and online public opinion in Taiwan. As a quantitative research method, data matching involves comparing datasets from different sources to identify correlations or associations between them. In this context, the study matches official data on local election results with corresponding data on online volume—specifically, the amount of discourse related to candidates or political events on digital platforms. Through this comparison, the research aims to quantitatively assess the extent to which fluctuations in online public opinion are linked to electoral performance, providing empirical evidence to support or challenge assumptions about the predictive value of online discourse.

3.2 Data Collection

This study utilizes two primary sources of data to investigate the relationship between online public opinion and the outcomes of Taiwan's 2024 Taitung County legislative election. First, electoral data were collected from the Central Election Commission (CEC) of Taiwan, which provides official statistics on vote counts, vote shares, and the election status of each candidate. Specifically, the study focuses on the three candidates—Liu Chao-hao (Candidate A), Huang Chien-pin (Candidate B), and Lai Kun-cheng (Candidate C) who contested in Taitung County. The data cover election results by township and include variables such as total votes received and winning margins. After collection, the electoral data underwent a thorough cleaning process to ensure completeness, consistency, and accuracy, which involved validating entries, correcting discrepancies, and preparing the dataset for analysis.

Second, online public opinion data were gathered using OpView (https://www.opview.com.tw/), a web-based social listening platform. This tool was employed to monitor and analyze the online visibility of the three candidates from December 2022 to December 2023. The dataset includes weekly and monthly volume trends, sentiment scores, platform-specific engagement rankings, and demographic breakdowns by gender and user groups. The sources analyzed include major social media platforms (e.g., Facebook, Twitter, Instagram), online forums (e.g., PTT https://www.ptt.cc/bbs/, Dcard https://www.dcard.tw/), and news websites. After collection, the data were processed to remove duplicates, filter out irrelevant discussions, and apply sentiment and topic analysis techniques. This preprocessing ensured that the data accurately reflected election-related public discourse, enabling meaningful comparisons with official electoral outcomes.

4 Data Analysis and Comparison

4.1 Online Public Opinion Analysis

An analysis of online supporters and commentators across various themes reveals distinct gender distributions for each of the three candidates. Candidate A demonstrates a male support rate of 76% and a female support rate of 23%. Candidate B shows a relatively balanced distribution, with 46% male and 53% female supporters. Candidate C has a male support rate of 79% and a female support rate of 20%.

In terms of thematic online discussions, the most frequently occurring topics for Candidate A include news, politics and society, lifestyle sharing, finance and investment, and travel and accommodation. For Candidate B, the dominant themes are politics and society, news, lifestyle sharing, travel and accommodation, and entertainment sharing. Candidate C's online discourse centers around politics and society, news, finance and investment, meteorology and geography, sports and fitness, arts and culture, as well as travel and accommodation. Notably, Candidate B's online support base is characterized by a more balanced gender distribution, whereas the commentary and engagement for Candidates A and C are predominantly male. The gender distribution of online supporters for each candidate is illustrated in Fig. 1.

Fig. 1. Gender Distribution of The Three Candidates

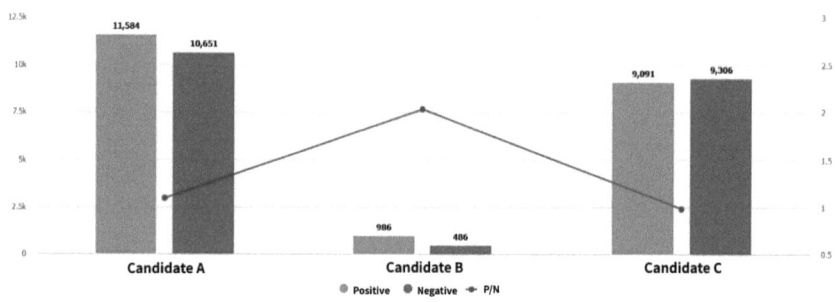

Fig. 2. Sentimental Analysis of Online Public Opinion

Figure 2 presents a sentiment analysis of online discussions across various platforms for the three candidates. Candidate A has a sentiment score of 1.09, with 11,584 positive mentions and 10,651 negative mentions. Candidate B shows a higher sentiment score of 2.03, based on 986 positive mentions and 486 negative mentions. Candidate C has a sentiment score of 0.98, with 9,091 positive mentions and 9,300 negative mentions. According to the sentiment analysis model, Candidate B has the highest sentiment score, indicating a more favorable public perception. Candidate A's sentiment results are relatively balanced, with a slightly higher number of positive mentions. In contrast, Candidate C has more negative mentions than positive ones, reflecting a less favorable online sentiment (Fig. 3).

An analysis of online platform distribution reveals significant differences in the digital visibility of the three candidates. Candidate A exhibited the highest overall presence, with 76.20% of mentions originating from social media, 19.02%

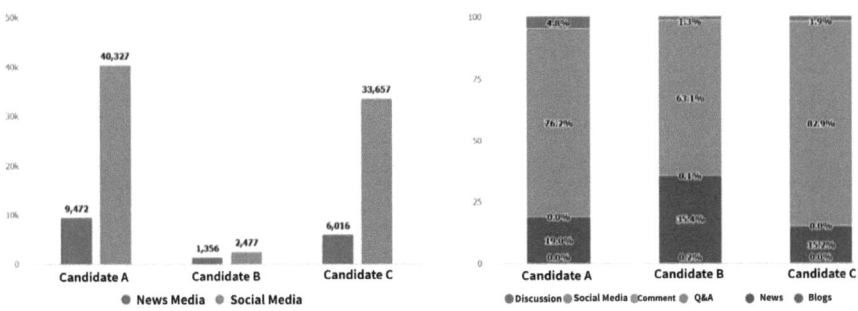

Fig. 3. The Comparison of Online Public Opinion in Different Platforms

from news outlets, 4.75% from forums, and a minimal 0.03% from blogs. Candidate C followed a similar pattern, with 82.89% of mentions on social media, 15.16% from news, and limited engagement across other platforms. In contrast, Candidate B had the lowest overall volume, with 63.08% of mentions from social media, 35.38% from news, and negligible figures in other categories. This suggests that Candidates A and C maintained stronger engagement across major platforms, particularly in social and news media, whereas Candidate B lagged behind significantly in both visibility and reach.

Further examination of the top 20 websites and channels by topic concentration underscores the differences in communication effectiveness. Candidate A ranked first, generating the highest volume of mentions (30,166) on Facebook fan pages. Candidate C ranked second with 22,070 mentions from Facebook key opinion leaders, indicating substantial influence through KOL networks. In comparison, Candidate B's most engaged channel—a Facebook fan page with 1,616 mentions—ranked only tenth, further reflecting limited digital traction. These findings indicate that Candidates A and C achieved broader and more impactful digital engagement, while Candidate B's presence remained comparatively weak across all measured dimensions.

The analysis of multi-topic volume trends across weekly (Fig. 4), monthly (Fig. 5), and overall (Fig. 7) timelines provides a clear view of how public attention fluctuated over different periods for each candidate. In the weekly analysis, Candidate A experienced peak daily volume on November 6, 2023, with 5,933 mentions, followed by a secondary peak on April 17, 2023, with 2,951 mentions. Candidate B reached their highest single-day volume on December 4, 2023, with 318 mentions, and the second highest on July 24, 2023, with 241 mentions. Candidate C recorded the highest daily volume on April 17, 2023, with 3,336 mentions, and the second highest on December 25, 2023, with 2,361 mentions. These peaks indicate that each candidate experienced surges in public attention at different times, likely reflecting distinct campaign events or media coverage.

In terms of monthly trends, the months of April and November 2023 saw heightened levels of discussion across all candidates, suggesting periods of increased political engagement or significant campaign milestones. The over-

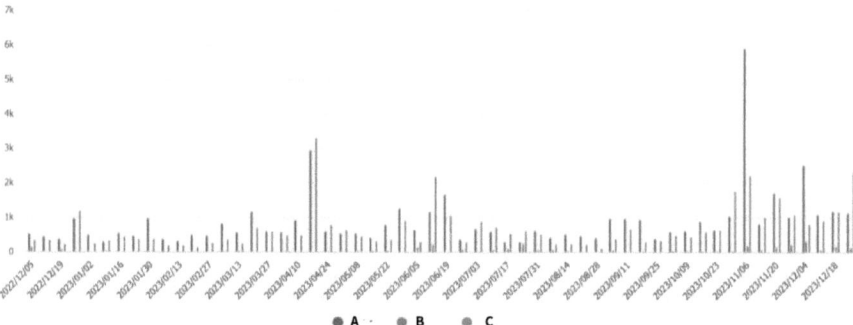

Fig. 4. Weekly Volume of Online Public Opinion

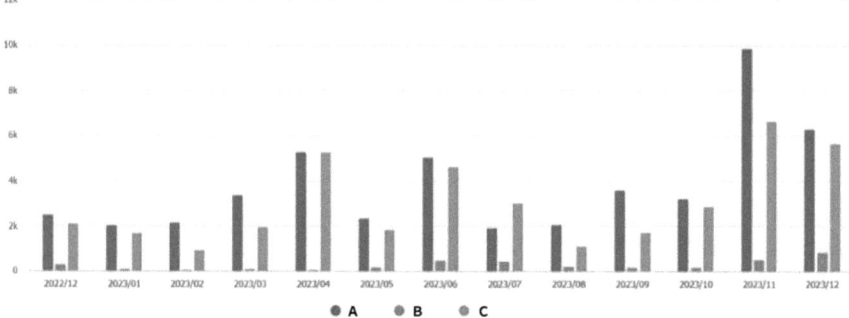

Fig. 5. Monthly Volume of Online Public Opinion

all volume analysis further underscores the disparity among candidates, with Candidate A receiving the highest total number of mentions at 49,799, followed by Candidate C with 39,673, and Candidate B with only 3,833. This data highlights not only the variation in peak engagement days among the candidates but also emphasizes Candidate A's consistently stronger presence and Candidate B's relatively limited visibility throughout the campaign period (Fig. 6).

4.2 The Comparison of Actual Vote Rate and Online Public Opinion

According to data provided by the Central Election Commission (CEC), the official results for the 11th Legislative Yuan election in Taitung County's Constituency 01 are as follows: Candidate B received 25,778 votes (34.79%) and was confirmed as the elected legislator. Candidate C followed with 23,420 votes (31.60%), and Candidate A received 18,744 votes (25.29%). The total number of registered voters in Taitung County was 115,336, with a voter turnout of 75,221, representing 65.22%. Of the total ballots cast, 74,103 (98.51%) were valid votes, while 1,118 (1.49%) were invalid.

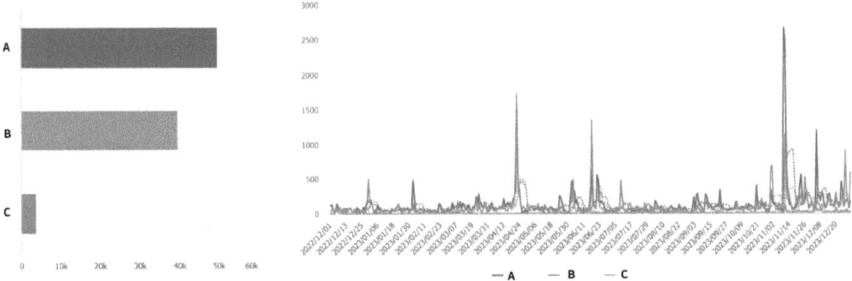

Fig. 6. Total Volume of Online Public Opinion

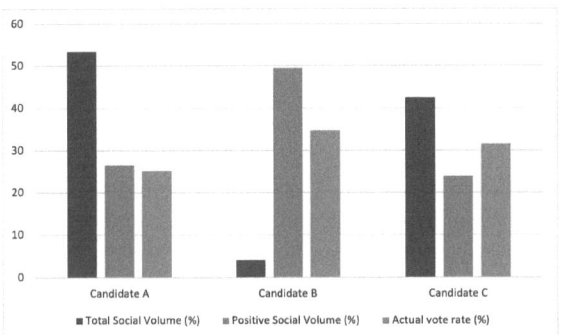

Fig. 7. The Comparison of Actual Vote Rate and Online Public Opinion

When comparing online presence and sentiment to election outcomes, notable discrepancies emerge. Candidate A garnered the highest volume of online mentions (49,799), yet received the lowest vote share (25.29%). Their sentiment score was 1.09, ranking second among the three candidates. In contrast, Candidate B had the lowest online visibility with only 3,833 mentions but achieved the highest sentiment score at 2.03, and ultimately won the election with the highest vote share (34.79%). Candidate C ranked second in both online mentions (39,673) and vote share (31.60%), but had the lowest sentiment score at 0.98, indicating a higher proportion of negative sentiment. These findings suggest that online volume alone is not a reliable predictor of electoral success; rather, sentiment quality and real-world campaign effectiveness play a more significant role.

5 Conclusion and Suggestions

This study employed data triangulation methods to investigate the relationship between online discourse volume and electoral outcomes in local elections in Taiwan. Contrary to initial expectations, the findings revealed an inverse relationship between online attention and actual voting results. While it was hypothesized that greater online presence would correlate positively with electoral success, the analysis demonstrated that high online visibility does not necessarily translate into more votes—particularly in small, localized constituencies. Instead, it is the quality and sentiment of online engagement, rather than the quantity, that appears to play a more decisive role in influencing voter behavior.

These findings highlight the need to reassess certain assumptions often made in political communication research, particularly those related to the impact of social media. The results underscore the importance of considering other influencing factors—such as grassroots campaigning, traditional media exposure, and candidate credibility within the local context—that may outweigh digital visibility. This unexpected outcome opens up valuable opportunities for further inquiry into the nuanced dynamics between online popularity and offline electoral performance.

Future research should adopt a more holistic approach by incorporating social, cultural, and political variables to better understand the complex interplay between digital presence and election outcomes. Comparative studies across different regions and electoral cycles would also shed light on temporal and spatial variations in these dynamics. Moreover, interdisciplinary collaboration—integrating perspectives from political science, communication studies, and sociology—could offer a more comprehensive framework for analyzing electoral behavior in the digital age. While our findings challenge conventional expectations, they pave the way for deeper exploration into the evolving relationship between the internet and democratic processes, with the ultimate goal of enhancing electoral fairness, transparency, and public engagement.

References

1. Jungherr, A., Jürgens, P., Schoen, H.: Why the pirate party won the german election of 2009 or the trouble with predictions: a response to Tumasjan, A., Sprenger, T. O., Sander, P. G., Welpe, I. M. "predicting elections with Twitter: what 140 characters reveal about political sentiment. Soc. Sci. Comput. Rev. **30**, 229–234 (2012). https://doi.org/10.1177/0894439311404119
2. Kwak, J.A., Cho, S.K.: Analyzing public opinion with social media data during election periods: a selective literature review. Asian J. Public Opin. Res. **5**(4), 285–301 (2018)
3. Lee, T.-L., Ting, I.-H., Liou, M.-M.: A study on product diffusion of social network sites - takes HTC as an example. In: The 2012 International Conference on Internet Studies, Bangkok, Thailand (2012)
4. Perez, C., Ting, I.-H.: Can you hold an advantageous network position? The role of neighborhood similarity in the sustainability of structural holes in social networks. Decis. Support Syst. **158** (2022). ISSN 0167-9236. https://doi.org/10.1016/j.dss.2022.113783
5. Scott, J.: Social Network Analysis: A Handbook. SAGE Publication, London (2000)
6. Ting, I.-H.: Detection of the spiral of silence effect in social media. J. Univ. Comput. Sci. **22**(3), 438–452 (2016)
7. Ting, I.: Towarding the effect of spiral of silence detection based on social network analysis and opinion mining techniques. In: CIST 2015: International Conference on Computer and Information Science and Technology, Ottawa, pp. 1–2 (2015)

8. Ting, I., Yen, C.: Opinion groups identification in blogsphere based on the techniques of web and social networks analysis. In: The 4th International Conference on Machine Learning and Computing, Hong Kong, China, pp. 76–80 (2012). Ting, I.-H., Chang, P.S., Wang, S.-L.: Understanding microblog users for social recommendation based on social networks analysis. J. Univ. Comput. Sci. **18**(4), 554–576 (2012). https://doi.org/10.3217/jucs-018-04-0554
9. Wasserman, B., Faust, K.: Social Network Analysis: Methods and Applications. Cambridge University Press, New York (1994)
10. Gil, H., de Zúñiga, N., Jung, S Valenzuela: Social media use for news and individuals' social capital, civic engagement and political participation. J. Comput.-Mediat. Commun. **17**(3), 319–336 (2012). https://doi.org/10.1111/j.1083-6101.2012.01574.x

Factors Influencing Value Co-creation in Live Streaming E-Commerce Ecosystems: The Microfoundations Theory Perspective

Kai Wang, Yong-Yi Chang, and Jia-Chi Lin(✉)

Department of Information Management, National University of Kaohsiung, Kaohsiung, Taiwan
`kwang@nuk.edu.tw, {m1123302,m1133311}@mail.nuk.edu.tw`

Abstract. As technology advances, the business model of live-streaming e-commerce has gained popularity, creating new ways of interaction and shopping channels for consumers. Today's consumers actively seek information about company reviews, product quality, and services before deciding to engage, making consumer participation crucial for businesses. Live streaming's success is no longer measured solely by customer numbers. Instead, consumer recognition and loyalty are key to retaining viewers and encouraging recommendations. Live-streaming e-commerce fosters emotional engagement between viewers and streamers for the purpose of facilitating purchases. The macro-level value co-creation in this context is influenced by microfoundations, including participants' perceived conditions, motivations, and interactions. This study employs microfoundations theories to examine consumer behavioral motivations, interactive factors, and their effects on value co-creation in live-streaming e-commerce.

Keywords: Live streaming e-commerce · Microfoundations · Value co-creation · Perceived benefits · Participant interaction

1 Introduction

With technological advancements, businesses increasingly employ live streaming platforms to enhance consumer experiences and create new business models. According to McKinsey & Company [1], global live streaming e-commerce sales will account for 10%–20% of all e-commerce by 2026. Live streaming e-commerce, characterized by immediacy, authenticity, and interactivity [2], addresses uncertainty in traditional online shopping and boosts purchase intentions [4]. For example, Sheri Hensley and her team sold 50 fashion items in two hours, generating US$25,000 in revenue [5].

The live-streaming ecosystem includes not only streamers and viewers but also platforms, brand companies, and suppliers. These actors drive value creation through resource integration and collaboration [6, 7]. Value co-creation emphasizes joint efforts among ecosystem members rather than a pure customer-company models [8]. Without the input of ecosystem members, resource integration and value co-creation will not occur [9]. While value co-creation is challenging to observe directly, micro-level engagement and resource integration are measurable [9]. Because innovation depends on the collaborative efforts among ecosystem participants [10], this study explores how micro-level factors such as personal characteristics (e.g., attention to detail, creativity, openness) and knowledge-sharing behaviors (e.g., motivation, control, ability, engagement) impact macro-level corporate innovation.

With the rapid growth of live streaming e-commerce, value co-creation among ecosystem participants has become crucial. In these ecosystems, value is co-created and evolves through continuous participant interactions. Real-time interactivity, a core feature of live streaming e-commerce, reshapes consumer behavior and value co-creation mechanisms.

Academically, this study draws on the microfoundations perspective to explore how factors at the micro level within the live streaming e-commerce ecosystem influence participants' engagement in value co-creation. By analyzing interaction patterns and perceived benefits, this study provides insights into value co-creation processes in the live streaming e-commerce context. Practically, the findings aim to help businesses and platforms enhance interaction mechanisms, emotional connections, word-of-mouth dissemination, and brand loyalty. Understanding these dynamics enables businesses to design competitive business models in the live streaming e-commerce market.

This study serves the following two research objectives:

1. To clarify how the interaction relationships between streamers and viewers in the live streaming e-commerce ecosystem influence value co-creation.
2. To clarify how viewers' perceived benefits in the live streaming ecommerce ecosystem affect value co-creation.

2 Theoreticalframework and Literaturereview

2.1 Live Streaming e-commerce

With technological advancements and changes in consumer behavior, live streaming e-commerce has emerged as a popular marketing opportunity that offers real-time, interactive, and authentic shopping experiences [2]. Live streamers showcase products through real-time video and interact with consumers in ways such as responding to questions via on-screen comments and direct responses [4], and addressing issues of communication delays and product uncertainty associated with traditional online and TV shopping [11, 12]. Live streaming e-commerce also incorporates entertainment and social engagement, allowing consumers to connect with streamers and other viewers through gifting, likes, and comments [13, 14]. Previous research has explored various aspects of live streaming e-commerce, which includes purchase intention [15, 16], impulse buying [17, 18], gifting intention [15, 19], and viewing intention [20]. These works demonstrate the unique value of live streaming e-commerce in driving consumer behavior. This study focuses on interaction and communication within the live streaming e-commerce context and examines how value co-creation helps businesses stand out in competitive markets.

2.2 Value Co-Creation in the Live Streaming Ecosystem

Value co-creation is one of the core concepts of service-dominant logic and refers to the practices with which customers and companies jointly create unique experiences and value through interaction and communication [21]. Previous studies on value co-creation draw from various perspectives that include management [22], marketing [23–27], and service logic [26, 28]. In a service ecosystem, value co-creation requires participants within the ecosystem to achieve unique experiences through interactive communications. In the context of live streaming e-commerce, enhancing consumer participation is crucial for businesses to establish connections and foster consumer loyalty [3]. Furthermore, the unique value of live streaming e-commerce lies in its ability to create distinctive experiences for consumers through interaction and communication. Consumer engagement in live interactions can enhance their experience and gain additional product and interaction value through engagement with streamers and businesses [30].

2.3 The Microfoundations Theory

Microfoundations theory provides a framework for understanding macro-level phenomena and resource allocation, emphasizing how micro-level factors influence organizations decisions and behaviors. Also, it assists the understanding of how individual interactions lead to macro-level outcomes and how organizational decisions are affected by individual behaviors [31]. Understanding macro-level phenomena requires insights into the micro-level elements that include individuals and their social interactions [9]. Coleman's model, known as the "Coleman's Boat" or "Coleman's Bathtub," illustrates the relationship between macro-level phenomena and micro-level factors [32–34].

At the macro-to-macro level, social facts directly influence social outcomes. At the macro-to-micro level, macro-level facts influence the conditions of individual actions through "situational mechanisms" [35]. Situational mechanisms refer to how macro-level conditions or contexts affect actor dispositions [9, 35]. The micro-to-micro level depicts how individual actors transform situational conditions for personal actions into actions through "action-formation mechanisms." Action-formation mechanisms refer to how individual actors incorporate situational conditions into their actions [9, 35]. The micro-to-macro level demonstrates how individual actions lead to macro-level outcomes through "transformational mechanisms" [35]. That is, transformational mechanisms describe how ecosystem members generate macro-level outcomes through their actions and interactions [35].

This research explores value co-creation in the live streaming e-commerce ecosystem. The transition from social facts to conditions of individual actions involve the investigation into the resource and institutional environment of the platform, which affect participants' action conditions, motivators, and drivers. The shift from conditions of individual action to individual actions illustrates how live streaming e-commerce participants' cognition, motivation, and opportunities may change in different situations and further influence their behavior [36]. The transition from individual action to social outcomes results in value co-creation through resource integration. This study focuses on the behavioral conditions, motivations, and interactions of viewer participation to clarify the operations within the live streaming ecosystem and develop more suitable strategies.

2.4 Interaction Among Ecosystem Participants

In the live streaming e-commerce ecosystem, participant interaction drives value co-creation among streamers, viewers, suppliers, brand companies, and platforms. Value co-creation is the integration of resources by businesses and stakeholders to create and share value [23]. Viewer interactions and parasocial interactions with streamers significantly enhance engagement and loyalty, and consequently foster value co-creation. From a microfoundations perspective, the real-time interactivity of live streaming enhances viewers' authentic experiences and promotes parasocial interaction [37]. This study examines how interactions with streamers influence value co-creation.

Online customer interactions build community harmony and influence customer cognition [38]. In live streaming e-commerce, viewer communication evolves into real-time, bidirectional interactions [14], with bullet comments fostering emotional connection and resonance [39]. Parasocial interaction describes the relationships between media figures and users [40, 41]. Through real-time video and bullet comments, streamers provide personalized Q&A, create the illusion of direct communication, and enhance emotional attachment and identification [37].

Viewers' perceived benefits play a critical role in the value co-creation process of live streaming e-commerce. From the perspective of microfoundations theory, the environment and institutional framework of the live streaming e-commerce ecosystem influence participants' action conditions and motivations. Understanding participants' perceived benefits in the context of live commerce helps uncovering their participation motivators and behavioral patterns. Perceived benefits refer to consumers' perceptions of values brought by products or services [43]. Providing benefits that meet consumer expectations is crucial for achieving positive outcomes.

This study categorizes perceived benefits in live streaming e-commerce into social benefits, self-esteem benefits, hedonic benefits, learning benefits, and convenience benefits. Social benefits refer to the degree to which social relationships improve through interaction with other community members [37]. Consumers' satisfaction with salespeople, loyalty, and peer recognition drive shopping intention and positive emotions [44–46]. Live streaming e-commerce facilitates easy communication among users, enhances interpersonal relationships, and provides a sense of fulfillment. Self-esteem benefits refer to the extent of achieving higher prestige, status, and self-efficacy [47]. Viewers express their opinions via features such as live-stream bullet screens. When these opinions are read aloud or discussed, they feel noticed and recognized, contributing to a sense of personal value. Hedonic benefits refer to the enjoyment and entertainment experiences users gain from live streaming, because these experiences motivate them to spend more leisure time watching [37]. Live streamers host events to attract viewers, enhance the atmosphere in the live room, and make viewers feel pleasure and satisfaction [49]. Learning benefits refer to the extent of acquiring knowledge related to products, thereby gaining better understanding of the products and their usage [47]. Viewers gain unique knowledge and professional insights about products by observing product information and reviews shared by live streamers and other viewers, and this information improves their skills and capabilities. The enhancement of understanding and knowledge will further increase viewer satisfaction [63]. Convenience benefits refer to the minimum effort and time consumers believe are needed to obtain services [50]. In the context of live streaming e-commerce, viewers can quickly access information and complete purchases, thus reducing time costs and enhancing the experience.

3 Research Model and Hypothesis Inference

Live streaming e-commerce has rapidly emerged as a key sales channel. This study focuses on live streaming e-commerce, aiming to examine factors leading to value co-creation in the live streaming ecosystem. As shown in the model in Fig. 1, we identify conditions and motivations influencing viewers' participation in value co-creation from a microfoundations perspective. Activities of ecosystem participants are categorized into viewer-viewer and parasocial interactions, and perceived benefits include social, self-esteem, hedonic, learning, and convenience benefits. This study explores how these factors shape viewers participation in value co-creation within the live streaming ecosystem.

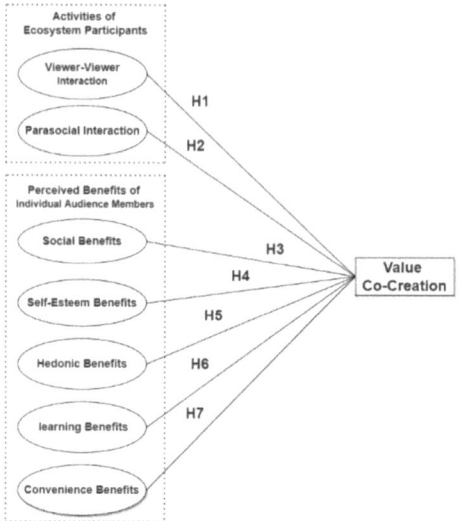

Fig. 1. Research Model

From the microfoundations perspective, participant interactions shape the atmosphere and arena of live streaming e-commerce. Viewer-Viewer interaction refers to the exchange of shopping advice and product information between viewers in live streaming e-commerce [51]. Real-time bullet comments allow viewers to share experiences and interact with each other. The interactions foster a positive atmosphere, assist information acquisition, and reduce psychological distances. Frequent interactions may lead viewers to perceive others as close friends and cause increasing participation and engagement. Perceived interactivity enhances self-efficacy and willingness to co-create value [52]. We therefore propose H1 as follows:

H1: Consumer-consumer interaction positively impacts value co-creation.

Live streaming e-commerce offers a rich, real-time interactive space, making parasocial interactions more likely [37]. Parasocial interaction reflects the perceived relationship between viewers and streamers and creates an illusion of real interpersonal connection [53]. Viewers often form emotional attachments to streamers who provide richer parasocial experiences [40, 42, 54], and the formation of the attachment increases interaction and purchase intention [55]. We thus propose H2:

H2: Parasocial interaction positively impacts value co-creation.

Viewer motivations in live streaming e-commerce relate closely to perceived benefits. Perceived benefits, defined as consumers' evaluations of their subjective experiences and advantages before taking further actions, are primary drivers of continued user engagement [47]. Perceived social benefits, which refer to improvements in social relationships, are key indicators of marketing performance, such as satisfaction, loyalty, and sales [56, 57]. In the context of live streaming e-commerce, viewers engage in real-time interactions with hosts and other viewers and discuss product information and daily experiences. These interactions enable quick responses, expand social networks, and strengthen relationships, thereby enhancing perceived social benefits. Improved customer experiences and positive attitudes increase the likelihood of consumers engaging in value co-creation behaviors [23]. We therefore propose H3 below:

H3: Viewers' perceived social benefits positively impact value co-creation.

Self-esteem benefit refers to enhanced prestige, status, and self-efficacy [47]. Interactions in live streams, such as suggestions, questions, and shared advice, allow viewers to feel noticed and recognized [37] and enhances self-worth and status [47, 58]. Recognition fosters positive emotions and attitudes and encourages value co-creation behaviors [23]. H4 is proposed as follows:

H4: Viewers' perceived self-esteem benefits positively impact value co-creation.

Hedonic motivation drives media use and technology adoption [59, 60]. Hedonic benefit refers to pleasure and entertainment from live streaming activities and motivate viewers to spend more time watching [47, 48]. Activities such as discounts, giveaways, and lotteries enhance enjoyment and customer experiences [61]. Greater intrinsic enjoyment leads to higher enthusiasm and performance [62].
H5 is thus proposed:

H5: Viewers' perceived hedonic benefits positively impact value co-creation.
Learning benefit refers to gaining product knowledge and usage understanding [47].

Real-time demonstrations and bullet comments provide professional insights and enhance satisfaction [63]. Clearer information increases perceived value and fosters emotional connections, commitment, and loyalty [64], further driving value co-creation [65]. We thus propose H6:

H6: Viewers' perceived learning benefits positively impact value co-creation.Convenience benefit reflects minimal effort and time in obtaining services [50, 66, 67].

Convenience of services positively influences word-of-mouth behaviors [68]. Viewers can use bullet comments for direct purchases to complete orders efficiently in live streaming e-commerce. Personalized guidance reduces time costs, enhances convenience, and fosters positive behaviors [23]. We therefore propose H7 below:

H7: Viewers' perceived convenience benefits positively impact value co-creation.

4 Research Method

4.1 Data Collection

This study aims to explore factors influencing viewer participation in value co-creation in live streaming e-commerce, with data collected through an online survey. The survey employs a snowball sampling method to solicit more respondents. The questionnaire is designed in Chinese and will be distributed via social media platforms including Facebook and Instagram.

To ensure the applicability and validity of the questionnaire items, expert reviews and a small-scale pilot test will be conducted before formal distribution. Based on the feedback of the expert review and pilot test, we will evaluate and revise inappropriate items, remove content with poor reliability and validity, and ensure the smooth progress of the survey. The research subjects are those who have experiences watching or participating in live streaming e-commerce purchases. To ensure respondents' focus while answering, the questionnaire includes two attention check questions [69]. Only those who pass the tests are included in subsequent analyses.

4.2 Variable Definition and Operationalization

This study, based on prior literature, provides operational definitions for viewer-viewer interaction, parasocial interaction, perceived social benefits, perceived self-esteem benefits, perceived hedonic benefits, perceived learning benefits, perceived convenience benefits, and value co-creation. All items are measured using a 7-point Likert scale, with 1 representing strongly disagree and 7 representing strongly agree. Measurement for viewer-viewer interaction is adapted from Cheung et al. [51] and reflects the extent to which viewers exchange advice or share product information with other viewers in live-streaming e-commerce. Definition for parasocial interaction is based on Sokolova & Kefi [53] and refers to the interactive relationship between viewers and live streamers that creates the illusion of a real interpersonal relationship by the viewers. Measurement items for perceived social benefits are adapted from Nambisan & Baron [47] and refer to the degree to which viewers experience improved interpersonal relationships in live streaming, manifested through larger and closer social circles and a stronger sense of belonging. Items for perceived self-esteem benefits, also adapted from Nambisan & Baron [47], reflect the extent to which viewers, through interactions with live streamers and other viewers, gain higher social prestige and status and achieve a sense of self-efficacy in live streaming e-commerce. Perceived hedonic benefits are defined as the degree to which viewers obtain pleasures and entertainment experiences from live streaming. The items are adapted from Jiang et al. [37]. Perceived learning benefits refer to the extent to which users acquire information and learn new knowledge through watching and participating in live streaming e-commerce. The measurement of perceived learning benefits is adapted from Nambisan & Baron [47]. The items for perceived convenience benefits, adapted from Benoit et al. [50], reflect the degree to which users perceive they spend least effort and time obtaining information, participating in interactions, and completing purchases during live-streaming sessions. Value co-creation refers to the process through which businesses, viewers, and other participants in the live streaming e-commerce ecosystem collaboratively achieve value creation process through ongoing interactions and exchanges. The measurement is adapted from Vargo and Lusch [70], Lusch and Vargo [29], and Vargo and Lusch [7].

5 Expected Contributions

This study delves into the interaction and perceived experiences among participants in the live streaming ecosystem from a microfoundations perspective. Results will reveal how the interaction fosters value co-creation. Academically, this research fills the gap between live streaming e-commerce and traditional sales models, providing a new perspective to understand the mechanisms of value co-creation. Practically, the findings will help businesses and platforms better understand how to promote value co-creation in the live streaming e-commerce ecosystem to enhance consumer engagement and satisfaction by optimizing interaction mechanisms.

Future research could further explore interaction patterns in live streaming ecosystems under different cultural contexts and how these patterns influence value co-creation. Additionally, studies can extend to examine technological innovations (e.g., artificial intelligence and virtual reality) in live streaming e-commerce and how these innovations further enhance interactivity and the potential for value co-creation. These directions will enable brand companies to design more effective business models, allowing them to stand out in highly competitive markets.

References

1. McKinsey & Company: It's showtime! How live commerce is transforming the shopping experience. https://www.mckinsey.com/capabilities/mckinsey-digital/our-insights/its-showtime-how-live-commerce-is-transforming-the-shopping-experience. Accessed 11 Dec 2023
2. Guo, Y., Zhang, K., Wang, C.: Way to success: understanding top streamer's popularity and influence from the perspective of source characteristics. J. Retail. Consum. Serv. **64**, 102786 (2022)
3. Hu, M., Chaudhry, S.S.: Enhancing consumer engagement in e-commerce live streaming via relational bonds. Internet Res. **30**(3), 1019–1041 (2020)
4. Wongkitrungrueng, A., Assarut, N.: The role of live streaming in building consumer trust and engagement with social commerce sellers. J. Bus. Res. **117**, 543–556 (2020)
5. Services Group.: 為什麼國外沒有直播帶貨?美國電商直播平台和發展現狀. https://www.eservicesgroup.com.cn/news/41171.html. Accessed 13 Dec 2023
6. Lusch, R.F., Vargo, S.L., Tanniru, M.: Service, value networks and learning. J. Acad. Mark. Sci. **38**, 19–31 (2010)
7. Vargo, S.L., Lusch, R.F.: Service-dominant logic: continuing the evolution. J. Acad. Mark. Sci. **36**, 1–10 (2008)
8. Shirazi, F., Wu, Y., Hajli, A., Zadeh, A.H., Hajli, N., Lin, X.: Value co-creation in online healthcare communities. Technol. Forecast. Soc. Chang. **167**, 120665 (2021)
9. Storbacka, K., Brodie, R.J., Böhmann, T., Maglio, P.P., Nenonen, S.: Actor engagement as a microfoundation for value co-creation. J. Bus. Res. **69**(8), 3008–3017 (2016)
10. Mazzucchelli, A., Chierici, R., Abbate, T., Fontana, S.: Exploring the microfoundations of innovation capabilities: Evidence from a cross-border R&D partnership. Technol. Forecast. Soc. Chang. **146**, 242–252 (2019)

11. Chen, A., Lu, Y., Wang, B.: Customers' purchase decision-making process in social commerce: A social learning perspective. Int. J. Inf. Manage. **37**(6), 627–638 (2017)
12. Zhou, L., Wang, W., Xu, J.D., Liu, T., Gu, J.: Perceived information transparency in B2C e-commerce: An empirical investigation. Information & Management **55**(7), 912–927 (2018)
13. Xin, M., Liu, W., Jian, L.: Live streaming product display or social interaction: How do they influence consumer intention and behavior? A heuristic-systematic perspective. Electron. Commer. Res. Appl. **67**, 101437 (2024)
14. Deng, Z., Benckendorff, P., Wang, J.: From interaction to relationship: Rethinking parasocial phenomena in travel live streaming. Tour. Manage. **93**, 104583 (2022)
15. Lu, Y., He, Y., Ke, Y.: The influence of e-commerce live streaming affordance on consumer's gift-giving and purchase intention. Data Science and Management **6**(1), 13–20 (2023)
16. Sun, Y., Shao, X., Li, X., Guo, Y., Nie, K.: How live streaming influences purchase intentions in social commerce: An IT affordance perspective. Electron. Commer. Res. Appl. **37**, 100886 (2019)
17. Xiang, L., Zheng, X., Lee, M.K., Zhao, D.: Exploring consumers' impulse buying behavior on social commerce platform: The role of parasocial interaction. Int. J. Inf. Manage. **36**(3), 333–347 (2016)
18. Lo, P.S., Dwivedi, Y.K., Tan, G.W.H., Ooi, K.B., Aw, E.C.X., Metri, B.: Why do consumers buy impulsively during live streaming? A deep learning-based dual-stage SEM-ANN analysis. J. Bus. Res. **147**, 325–337 (2022)
19. Zhou, J., Zhou, J., Ding, Y., Wang, H.: The magic of danmaku: A social interaction perspective of gift sending on live streaming platforms. Electron. Commer. Res. Appl. **34**, 100815 (2019)
20. Liu, Z., Li, J., Wang, X., Guo, Y.: How search and evaluation cues influence consumers' continuous watching and purchase intentions: An investigation of live-stream shopping from an information foraging perspective. J. Bus. Res. **168**, 114233 (2023)
21. Prahalad, C.K., Ramaswamy, V.: Co-creation experiences: The next practice in value creation. J. Interact. Mark. **18**(3), 5–14 (2004)
22. Payne, A.F., Storbacka, K., Frow, P.: Managing the co-creation of value. J. Acad. Mark. Sci. **36**, 83–96 (2008)
23. Wang, L., Zhang, R.S., Zhang, C.X.: Live streaming E-commerce platform characteristics: Influencing consumer value co-creation and co-destruction behavior. Acta Physiol (Oxf.) **243**, 104163 (2024)
24. Kohtamäki, M., Rajala, R.: Theory and practice of value co-creation in B2B systems. Ind. Mark. Manage. **56**, 4–13 (2016)
25. Terblanche, N.S.: Some theoretical perspectives of co-creation and co-production of value by customers. Acta Commercii **14**(2), 1–8 (2014)
26. Grönroos, C., Voima, P.: Critical service logic: making sense of value creation and co-creation. J. Acad. Mark. Sci. **41**, 133–150 (2013)
27. Wikström, S.: Value creation by company-consumer interaction. J. Mark. Manag. **12**(5), 359–374 (1996)
28. Wikström Grönroos, C.: Value co-creation in service logic: A critical analysis. Mark. Theory **11**(3), 279–301 (2011)
29. Lusch, R.F., Vargo, S.L.: Service-dominant logic: reactions, reflections and refinements. Mark. Theory **6**(3), 281–288 (2006)
30. Zhang, Y., Xu, Q.: Consumer engagement in live streaming commerce: Value co-creation and incentive mechanisms. J. Retail. Consum. Serv. **81**, 103987 (2024)
31. Felin, T., Foss, N.J., Ployhart, R.E.: The microfoundations movement in strategy and organization theory. Acad. Manag. Ann. **9**(1), 575–632 (2015)
32. Coleman, J.S.: Foundations of social theory. Harvard University Press (1990)
33. Abell, P., Felin, T., Foss, N.: Building micro-foundations for the routines, capabilities, and performance links. Manag. Decis. Econ. **29**(6), 489–502 (2008)

34. Mäkelä, K., Sumelius, J., Höglund, M., Ahlvik, C.: Determinants of strategic HR capabilities in MNC subsidiaries. J. Manage. Stud. **49**(8), 1459–1483 (2012)
35. Swedberg, R., Hedström, R.: Social mechanisms. Acta Sociologica **39**(3), 281–305 (1998)
36. Contractor, F., Foss, N.J., Kundu, S., Lahiri, S.: Viewing global strategy through a microfoundations lens. Glob. Strateg. J. **9**(1), 3–18 (2019)
37. Jiang, C., He, L., Xu, S.: Relationships among para-social interaction, perceived benefits, community commitment, and customer citizenship behavior: Evidence from a social live-streaming platform. Acta Physiol (Oxf.) **250**, 104534 (2024)
38. Luo, N., Zhang, M., Hu, M., Wang, Y.: How community interactions contribute to harmonious community relationships and customers' identification in online brand community. Int. J. Inf. Manage. **36**(5), 673–685 (2016)
39. Kang, K., Lu, J., Guo, L., Li, W.: The dynamic effect of interactivity on customer engagement behavior through tie strength: Evidence from live streaming commerce platforms. Int. J. Inf. Manage. **56**, 102251 (2021)
40. Frederick, E.L., Lim, C.H., Clavio, G., Walsh, P.: Why we follow: An examination of parasocial interaction and fan motivations for following athlete archetypes on Twitter. Int. J. Sport Commun. **5**(4), 481–502 (2012)
41. Horton, D., Wohl, R.: Mass communication and para-social interaction: Observations on intimacy at a distance. Psychiatry **19**(3), 215–229 (1956)
42. Hu, M., Zhang, M., Wang, Y.: Why do audiences choose to keep watching on live video streaming platforms? An explanation of dual identification framework. Comput. Hum. Behav. **75**, 594–606 (2017)
43. Keller, K.L.: Conceptualizing, measuring, and managing customer-based brand equity. J. Mark. **57**(1), 1–22 (1993)
44. Xie, J., Yoon, N., Choo, H.J.: How online shopping festival atmosphere promotes consumer participation in China. Fashion and Textiles **10**(1), 5 (2023)
45. Erjavec, J., Manfreda, A.: Online shopping adoption during COVID-19 and social isolation: Extending the UTAUT model with herd behavior. J. Retail. Consum. Serv. **65**, 102867 (2022)
46. Choi, Y.H., Choo, H.J.: Effects of Chinese consumers' relationship benefits and satisfaction on attitudes toward foreign fashion brands: The moderating role of country of salesperson. J. Retail. Consum. Serv. **28**, 99–106 (2016)
47. Nambisan, S., Baron, R.A.: Virtual customer environments: testing a model of voluntary participation in value co-creation activities. J. Prod. Innov. Manag. **26**(4), 388–407 (2009)
48. Han, M., Wu, J., Wang, Y., Hong, M.: A model and empirical study on the user's continuance intention in online China brand communities based on customer-perceived benefits. Journal of Open Innovation: Technology, Market, and Complexity **4**(4), 46 (2018)
49. Shen, Y.C.: What do people perceive in watching video game streaming? Eliciting spectators' value structures. Telematics Inform. **59**, 101557 (2021)
50. Benoit, S., Klose, S., Ettinger, A.: Linking service convenience to satisfaction: Dimensions and key moderators. J. Serv. Mark. **31**(6), 527–538 (2017)
51. Cheung, M.L., Pires, G.D., Rosenberger, P.J., Leung, W.K., Sharipudin, M.N.S.: The role of consumer-consumer interaction and consumer-brand interaction in driving consumer-brand engagement and behavioral intentions. J. Retail. Consum. Serv. **61**, 102574 (2021)
52. See-To, E.W., Ho, K.K.: Value co-creation and purchase intention in social network sites: The role of electronic Word-of-Mouth and trust–A theoretical analysis. Comput. Hum. Behav. **31**, 182–192 (2014)
53. Sokolova, K., Kefi, H.: Instagram and YouTube bloggers promote it, why should I buy? How credibility and parasocial interaction influence purchase intentions. J. Retail. Consum. Serv. **53**, 101742 (2020)
54. Brown, W.J., Basil, M.D.: Parasocial interaction and identification: Social change processes for effective health interventions. Health Commun. **25**(6–7), 601–602 (2010)

55. Li, Y., Li, X., Cai, J.: How attachment affects user stickiness on live streaming platforms: A socio-technical approach perspective. J. Retail. Consum. Serv. **60**, 102478 (2021)
56. Otto, A.S., Szymanski, D.M., Varadarajan, R.: Customer satisfaction and firm performance: Insights from over a quarter century of empirical research. J. Acad. Mark. Sci. **48**(3), 543–564 (2020)
57. Prebensen, N.K., Xie, J.: Efficacy of co-creation and mastering on perceived value and satisfaction in tourists' consumption. Tour. Manage. **60**, 166–176 (2017)
58. Yen, H.R., Hsu, S.H.Y., Huang, C.Y.: Good soldiers on the Web: Understanding the drivers of participation in online communities of consumption. Int. J. Electron. Commer. **15**(4), 89–120 (2011)
59. Venkatesh, V., Thong, J. Y., Xu, X.: Consumer acceptance and use of information technology: Extending the unified theory of acceptance and use of technology. MIS Quarterly, **157–178** (2012)
60. Yang, H.L., Lin, S.L.: The reasons why elderly mobile users adopt ubiquitous mobile social service. Comput. Hum. Behav. **93**, 62–75 (2019)
61. Hilken, T., De Ruyter, K., Chylinski, M., Mahr, D., Keeling, D.I.: Augmenting the eye of the beholder: Exploring the strategic potential of augmented reality to enhance online service experiences. J. Acad. Mark. Sci. **45**, 884–905 (2017)
62. Martocchio, J.J., Webster, J.: Effects of feedback and cognitive playfulness on performance in microcomputer software training. Pers. Psychol. **45**(3), 553–578 (1992)
63. Pang, H., Ruan, Y.: Disentangling composite influences of social connectivity and system interactivity on continuance intention in mobile short video applications: The pivotal moderation of user-perceived benefits. J. Retail. Consum. Serv. **80**, 103923 (2024)
64. Rezaei, S., Valaei, N.: Branding in a multichannel retail environment: Online stores vs app stores and the effect of product type. Inf. Technol. People **30**(4), 853–886 (2017)
65. Cui, X., Xie, Q., Zhu, J., Shareef, M.A., Goraya, M.A.S., Akram, M.S.: Understanding the omnichannel customer journey: The effect of online and offline channel interactivity on consumer value co-creation behavior. J. Retail. Consum. Serv. **65**, 102869 (2022)
66. Kim, W.B., Xie, J., Choo, H.J.: Role of perceived benefits of online shopping festival in Vietnam: Differences between millennials and generation Z. J. Retail. Consum. Serv. **75**, 103530 (2023)
67. Mombeuil, C., Uhde, H.: Relative convenience, relative advantage, perceived security, perceived privacy, and continuous use intention of China's WeChat Pay: A mixed-method two-phase design study. J. Retail. Consum. Serv. **59**, 102384 (2021)
68. Roy, S.K., Shekhar, V., Lassar, W.M., Chen, T.: Customer engagement behaviors: The role of service convenience, fairness and quality. J. Retail. Consum. Serv. **44**, 293–304 (2018)
69. Schleider, J.L., Weisz, J.R.: Using Mechanical Turk to study family processes and youth mental health: A test of feasibility. J. Child Fam. Stud. **24**, 3235–3246 (2015)
70. Vargo, S.L., Lusch, R.F.: Evolving to a new dominant logic for marketing. J. Mark. **68**(1), 1–17 (2004)

Analyzing the Improvement of Students' Performance in Group Discussion: A Comparative Study of Online and Face-to-Face Collaborative Learning - A Case Study of Japanese Undergraduate Students

Laila Diana Khulyati and Sanetake Nagayoshi(✉)

Department of Informatics, Shizuoka University, 3-5-1, Johoku, Chuo-ku, Hamamatsu, Shizuoka, Japan
`laila.diana.khulyati.21@shizuoka.ac.jp,`
`nagayoshi@inf.shizuoka.ac.jp`

Abstract. Despite their importance and benefits, collaborative learning faces many challenges and stereotypes. It may lead to unequal participation, unproductive, and a lack of engagement. However, the impact of the pandemic in the educational setting is not predetermined. This study found a method to improve students' satisfaction and learning experience by experimenting with both online and face-to-face (FTF) group discussions. The data was collected from undergraduate lectures using a questionnaire survey at a Japanese university. Present findings showed the importance of three key success factors: preparation (pre-assignment), contribution, and feedback. Although it was found that feedback was considered to affect collaborative learning as a mediator, to have no significant relationship with satisfaction and learning experience, these require further investigation. Furthermore, our results showed a positive relationship between satisfaction and learning experience in a group given an assignment before the classes. Our study underscores the importance of incorporating assignments (preparation) and effective collaborative learning strategies into the design of group discussions to improve students' performance, which affects the learning experience as well as their satisfaction. In addition, this study also provides an evidence-based analysis of the link between assignment availability, satisfaction, and learning performance.

Keywords: Online collaborative learning · Face-to-face collaborative learning · Group Discussion · Learning Performance

1 Introduction

Collaborative learning serves as a vital mechanism for fostering knowledge exchange, teamwork, and synergy among participants. It provides a platform for decision-making, and critical discussions—key skills needed in a globalized workforce. However, despite their importance, collaborative learning faces challenges such as unequal participation, cultural misunderstandings, and differing expectations of engagement, which can hinder learning outcomes. Judd et al. [7] explore that some students may prioritize personal goals over collaborative efforts, which can diminish overall productivity.

On the other hand, the Covid-19 pandemic has significantly impacted the mobility of human activities, including activities in education. Consequently, the use of online-based communication or video-conferencing systems has surged alongside the increase in students who study from home. In the new advanced era, Internet-based technology has brought both challenges and opportunities in the fields of education and training, particularly through online instruction. Online instruction, a form of distance education delivered over the Internet, is becoming increasingly popular. However, it is not without its criticism. Despite this, there remains a lack of comprehensive research to accurately assess the advantages and disadvantages of online instruction, particularly when compared to traditional FTF learning environments. Researchers and educators continue to be uncertain about how students' experiences in online learning differ from those in FTF learning environments.

The present study focuses on comparing both online and FTF collaborative learning conducted within a student group discussion setting to analyze the improvement of their performance. This study aims to gain a deeper understanding of how learning performance through learning experience and satisfaction can be improved, specifically within the context of group discussions. By examining the impact of both online and face-to-face group discussions on satisfaction and learning performance, the study suggests identifying best practices and strategies to optimize the outcomes of group discussions in both settings.

In particular, this study focuses on analyzing the behavior of students in group discussions, and examining how it can impact their understanding towards the topic they have discussed, as well as their contribution and writing assignments as a result. Group discussions are considered as an essential method of evaluating students' critical thinking ability, as well as their performances in sharing ideas and thoughts about specific issues. More importantly, group discussions have the potential to improve the quality of students' contributions, which can be practically valuable and useful in the future applications.

2 Literature Review

Collaborative learning is essential for academic and professional success. They play a crucial role in education, helping students develop critical thinking, communication, and teamwork skills. Consider the word "collaboration" itself: it evokes the idea of a method of interaction, sharing responsibility, and learning from one another. Sotto [17] stated that collaborative learning is "a case, where such a case includes the following main aspects: first, two or more students learn or try to learn something together; second, 'two or more may be explained as a pair, a small group (3–5 subjects) or a class (20–30 subjects)'; third, 'learn something' may be explained as following a course or perform learning activities such as problem-solving. Finally, 'together' may be explained as many forms of interaction which may be face-to-face or computer-mediated" [17].

Collaborative learning, particularly through group discussions, has proven to be an effective strategy for improving student learning. A study by Linton et al. [13] found that students working in groups performed better in a biology course compared to those studying alone. Similarly, Chen et al. [5] suggest that structured collaborative learning settings improve critical thinking and reasoning skills among students. Collaborative learning fosters active engagement by requiring students to take a proactive approach in managing and monitoring their teams and the knowledge generated during discussions (Le et al., [10]).

However, in recent years, digital technology has also played an influential role in collaborative learning practices. This shift has prompted a move toward a more "learner-centered" approach to course design, enabling students to assume a more active role in shaping their own learning processes. Awang-Hashim et al. [2] demonstrate that collaborative learning serves as a complete mediator in the relationship between student-faculty interaction, teaching quality, and relatedness, influencing reflective and integrative learning as well as higher-order thinking. Zhang et al. [20] explore how familiarity among group members affects learning outcomes, including teamwork satisfaction, engagement, and perceived knowledge construction, by comparing face-to-face and online collaborative learning environments.

Additionally, Kulkarni [9] studied massive online classes and discovered that the more geographically diversified the discussion groups, the higher the students' performance. Salter et al. [16] found that face-to-face discussions effectively generated interest and commitment among participants, while online discussions allowed for a more in-depth exploration of topics, leading to coherent conclusions.

Another key factor in collaborative learning is the role of instructional materials. According to León et al. [12], the authors found that providing slides before classes has a negative effect on students' performance. While other studies found that allowing students to download PowerPoint slides before attending a lecture improves their learning outcomes. This ongoing debate shows how different teaching (Chen et al. [4]).

However, despite its benefits, the research on how collaborative learning plays out in both online and FTF group discussions remains limited. This study aims to fill the gap by exploring whether collaboration enhances student satisfaction and learning experience in different discussion settings. As a result, the main research question that we seek to address in the current study is whether collaborative learning mediates the relationship between the factors of *satisfaction* and learning *experience, preparation (pre-assignment), contribution,* and *feedback*.

To address this lack of pertinent research, the following research questions are addressed in this study:

RQ1. How does preparation (pre-assignment) impact student participation in online and face-to-face group discussions?

RQ2. What is the difference in the impact of collaborative learning as the mediator in online and face-to-face group discussions?

3 Hypotheses

To evaluate the role of group discussion through the independent variables of preparation, contribution, and feedback, we used collaborative learning as a mediating variable and assessed learning outcomes with two cognitive learning dependent variables: satisfaction and learning experience, respectively. The following hypotheses are proposed (Fig. 1):

Hypothesis 1 (H1) The first set of hypotheses explores how preparation, cont9kribution, and feedback are positively correlated with each other.

> **H1-1** Preparation is positively correlated with Contribution. Preparation for learning activities enables students to actively contribute in group discussions.
> **H1-2** Preparation is positively correlated with Feedback. Preparation for learning activities also motivates students to engage in or receive feedback from group discussions.
> **H1-3** Contribution is positively correlated with Feedback. Contribution fosters students' engagement in the feedback process in group discussion.

Hypothesis 2 (H2) Preparation facilitates the capability to engage effectively in collaborative learning.

Hypothesis 3 (H3) Contribution strengthens the collaborative learning process by exchanging ideas and opinions among members.

Hypothesis 4 (H4) Feedback clarifies understanding and improves the group learning outcomes in collaborative learning.

Collaborative learning mediates the relationship between independent variables and dependent variables. Preparation, contribution, and feedback influence collaborative learning, which in turn promotes a better learning experience and higher satisfaction.

Hypothesis 5 (H5) Collaborative learning positively influences engagement and better learning experience.

Hypothesis 6 (H6) Collaborative learning positively influences participation among group and satisfying discussion experience.

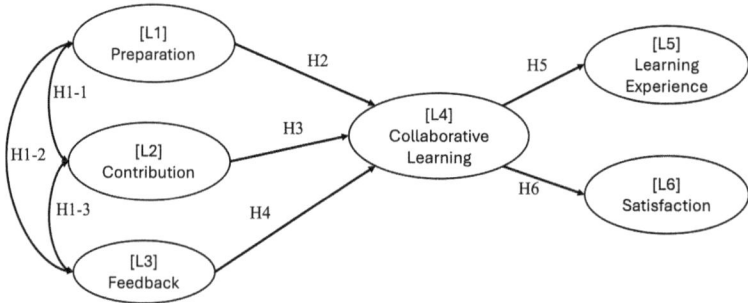

Fig. 1. Hypothesized Model.

4 Method

4.1 Case Study

This research utilizes both online and FTF lecture methods and considers the assignment submission structure as an attendance of the class. Students are required to participate in the lecture by discussing the main topic of the class. Within a week, students are asked to discuss certain topics in a group—particularly online through Zoom's *break-out room*. After the class is over, they need to answer the survey questionnaire anonymously and submit the assignment as an attendance of the class. The survey will be done to understand the development and implementation based on students' creative thinking by sharing the questionnaire after the lecture is over via Shizuoka University's Academic System Affair, where data is collected.

In this case study, both online and FTF group discussions with the students will be held twice a week. This study was conducted in a Management and Economics lecture at Shizuoka University, involving two groups. The first group consists of undergraduate students attending the Tuesday lecture. The second group attended the Wednesday lecture.

4.2 Design and Method

The survey aims to collect data on student productivity and satisfaction through a structured questionnaire. Responses are measured using a five-point Likert scale, (1 = very satisfied to 5 = very dissatisfied), to evaluate satisfaction levels across the four success factors identified in our study.

Responses were gathered with a total of 1,133 valid questionnaire data, both online (562) and FTF (571) over a month. Overall, 284 participants from four classes, 141 students in the online and 143 students in the FTF class at Shizuoka University, Japan, participated in this study. The demographic information for all participants is shown in Table 1 below.

Table 1. Students' demographic information.

	Online	Face-to-face
n (average)	141 students	143 students
- Team-A (Tuesday class)	63 students	43 students
- Team-B (Wednesday class)	78 students	100 students
Faculty		
- Engineering	78%	86%
- Informatics	22%	14%
Department of Engineering		
- Mechanical Engineering	23%	30%
- Electrical and Electronic Engineering	18%	19%

(*continued*)

Table 1. (*continued*)

	Online	Face-to-face
- Electronic Materials Science	21%	16%
- Chemistry and Biotechnology	10%	14%
- Mathematical System Engineering	7%	7%
Department of Informatics		
- Computer Science	7%	2%
- Social Informatics	8%	8%
- Behavourial Informatics	7%	3%
School Year		
- Second	46%	70%
- Third	5%	27%
- Fourth	4%	3%

The survey was conducted weekly during April and May, comparing two discussion methods:

1) No Topic Given (NONE): Students receive the discussion topic at the start of the lecture, requiring them to engage without prior preparation.
2) Topic Given (GIVEN): Students are informed of the discussion topic beforehand, allowing them to prepare and process the information provided.

Class sizes varied depending on the format. Online discussions range from 60 to 86 students, while FTF vary from 43 to 100 students. The study lasted four weeks, testing the two discussion methods across eight lectures on different topics, allowing for a comparison of how preparation impacts student engagement in discussions.

5 Results

5.1 Students' Performance on the Availability of Preparation (Pre-Assignment) on Group Discussion

The study examined the impact of preparation (pre-assignment) on students' performance in group discussion through 1,133 total survey responses (562 from online and 571 from FTF).

Analysis of preparation time revealed that most students, regardless of whether they received materials (pre-assignments), indicated 10 to 30 min students prepared for discussions. In online classes, Team-A ($M = 3.01$, $SD = 1.52$) and Team-B ($M = 2.72$, $SD = 1.51$). In FTF classes, Team-A ($M = 2.56$, $SD = 1.62$) and Team-B ($M = 2.67$, $SD = 1.64$). These findings suggest that most students, regardless of whether they received prior materials, did, in fact, prepare for it within a certain period.

5.2 Measurement Model

First, CFA was conducted to verify our hypotheses and evaluate the overall model fit using IBM SPSS AMOS 28.0. The independent variables (preparation, contribution, and feedback), while learning experience and satisfaction served as the dependent variables.

We define the latent variable of "Preparation" as composed of three observed variables, which are "Pre-assignment," "Topic," and "Content". Then, the latent variable of "Contribution" is composed of two observed variables, which are "Communication", and "Individual Contribution". We also define the latent variable of "Feedback" with two observed variables of "Conclusion" and "Convinceness". Next, the latent variable of "Collaborative Learning" is composed of three observed variables of "Group-member Contribution," "Start Smoothly", and "End Smoothly." Furthermore, the latent variable of "Learning Experience" with two observed variables of "Easiness to speak up", and "Comfortability". Finally, the latent variable of "Satisfaction" is composed of two observed variables of "Participation", and "Satisfying".

Satisfaction scores were obtained by scaling item responses according to standardized factor loading from CFA. The model fit for online measurement indicated CMIN = 273.2078, df = 70, CMIN/df = 3.901, CFI = 0.950, and RMSEA = 0.072. The FTF measurement indicated CMIN = 262.205, df = 70, CMIN/df = 3.746, CFI = 0.955, and RMSEA = 0.069. Both models indicate reasonable fit, with high-reliability scores (*Cronbach's alpha* = 0.9). *T-test* yielded *t-values* of 7.788 (online) and 2.24 (FTF), supporting model validity.

A structural equation model was performed to test our Hypothesis (Fig. 2 und 3). The results showed the correlation between independent variables of [L1], [L2], and [L3] was 0.81, 1.00, 0.89 (online), and 0.77, 1.03, 0.73 (FTF) with significance levels of less than 0.1%, respectively, which supports H1. A correlation between [L1] and [L4] was (0.63, online; 0,43, FTF), with a significance level of less than 0.1%, which H2 is supported. Then, the correlation between [L2] and [L4] was (0.60, online; 0.68, FTF), with a significance level of less than 0.1%, which supports H3.

However, it was observed that certain factors hypothesized to mediate satisfaction through collaborative learning exhibited a weak association. Notably, feedback [L3] to [L4] was not a significant moderator of this effect ($\beta = -0.29$, online; $\beta = -0.16$, FTF) and shows that there is no significant relationship to satisfaction, respectively. Therefore, H4 is not supported. Conversely, the correlation between [L4] and [L5] was (0.92, online; 0.97, FTF), with significance level of less than 0.1% which H5 is supported. Finally, the correlation between [L4] and [L6] was (0.97, online; 0.94, FTF), with a significance level of less than 0.1% which H6 is supported.

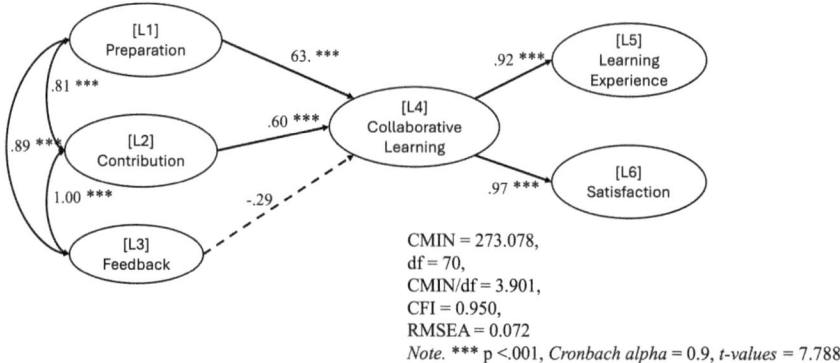

Fig. 2. Standardized estimates Model Analysis Result for Online.

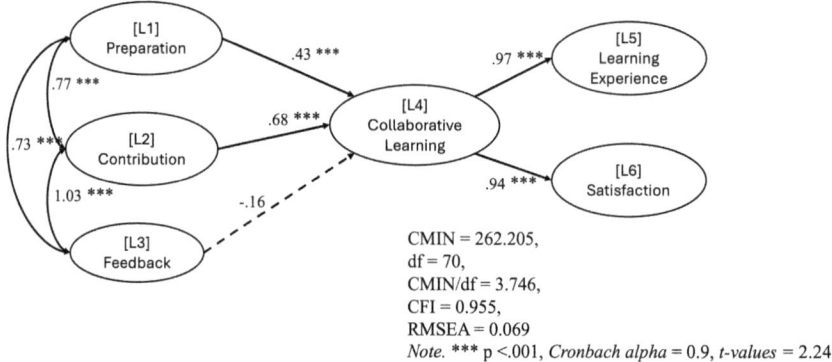

Fig. 3. Standardized estimates Model Analysis Result for FTF.

Further analysis presented in Tables 2 und 3 indicated that feedback (−.251, −.123), which is students' convinced level, did not significantly impact the satisfaction or learning experience through collaborative learning. This finding suggests that students accepted the discussion outcomes within the designated timeframe, regardless of the strength of their convincing level.

Moreover, contribution emerged as a crucial determinant of collaborative learning outcomes (1.519, 1.337). Providing preparation before discussion facilitated a smoother experience, allowing students to contribute effectively, which has a positive impact on the satisfaction and learning experience.

Table 2. The results of the Hypothesis testing for Online.

Variables		Estimate	p-value	Result
L1 Preparation	↔ L2 Contribution	1.003	***	Supported
L3 Feedback	↔ L2 Contribution	.806	***	Supported
L3 Feedback	↔ L1 Preparation	.888	***	Supported
L4 Collaborative Learning	→ L1 Preparation	1.054	***	Supported
L4 Collaborative Learning	→ L2 Contribution	1.519	***	Supported
L4 Collaborative Learning	→ L3 Feedback	−.251	.508	Rejected
L5 Learning Experience	→ L4 Collaborative Learning	1.113	***	Supported
L6 Satisfaction	→ L4 Collaborative Learning	1.132	***	Supported
O1 Pre-assignment	→ L1 Preparation	.331	.087	Supported
O2 Topic	→ L1 Preparation	1.043	***	Supported
O3 Content	→ L1 Preparation	1.000		
O4 Communication	→ L2 Contribution	1.450	***	Supported
O5 Individual-contribution	→ L2 Contribution	1.000		
O6 Conclusion	→ L3 Feedback	1.000		
O7 Convinceness	→ L3 Feedback	1.051	***	Supported
O9 Start smoothly	→ L4 Collaborative Learning	1.000		
O8 Member-contribution	→ L4 Collaborative Learning	1.060	***	Supported
O10 End smoothly	→ L4 Collaborative Learning	.865	***	Supported
O11 Easiness	→ L5 Learning Experience	1.000		
O12 Comfortability	→ L5 Learning Experience	1.179	***	Supported
O13 Participation	→ L6 Satisfaction	1.000		
O14 Satisfying	→ L6 Satisfaction	1.108	***	Supported

Notes. *** p < .001

Online. $df = 71$, p = .000, TLI = .926.

Table 3. The results of the Hypothesis testing for FTF.

Variables		Estimate	p-value	Result
L1 Preparation	↔ L2 Contribution	1.025	***	Supported
L3 Feedback	↔ L2 Contribution	.765	***	Supported
L3 Feedback	↔ L1 Preparation	.729	***	Supported
L4 Collaborative Learning	→ L1 Preparation	1.134	***	Supported
L4 Collaborative Learning	→ L2 Contribution	1.337	***	Supported
L4 Collaborative Learning	→ L3 Feedback	-.123	.065	Rejected

(*continued*)

Table 3. (*continued*)

Variables		Estimate	p-value	Result
L1 Preparation	↔ L2 Contribution	1.025	***	Supported
L5 Learning Experience	→ L4 Collaborative Learning	1.082	***	Supported
L6 Satisfaction	→ L4 Collaborative Learning	.994	***	Supported
O1 Pre-assignment	→ L1 Preparation	1.000		
O2 Topic	→ L1 Preparation	.963	***	Supported
O3 Content	→ L1 Preparation	.105	.565	Supported
O4 Communication	→ L2 Contribution	1.121	***	Supported
O5 Individual-contribution	→ L2 Contribution	1.000		
O6 Conclusion	→ L3 Feedback	1.000		
O7 Convinceness	→ L3 Feedback	.937	***	Supported
O9 Start smoothly	→ L4 Collaborative Learning	1.000		
O8 Member-contribution	→ L4 Collaborative Learning	1.005	***	Supported
O10 End smoothly	→ L4 Collaborative Learning	.859	***	Supported
O11 Easiness	→ L5 Learning Experience	1.000		
O12 Comfortability	→ L5 Learning Experience	1.054	***	Supported
O13 Participation	→ L6 Satisfaction	1.000		
O14 Satisfying	→ L6 Satisfaction	1.086	***	Supported

Notes. *** p < .001

Online. $df = 71$, p = .000, TLI = .926. Face-to-face. $df = 71$, p = .000, TLI = .942.

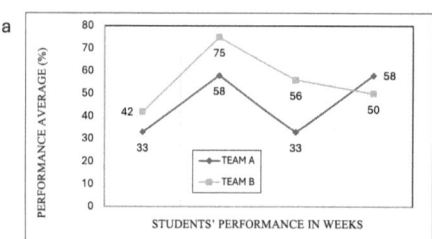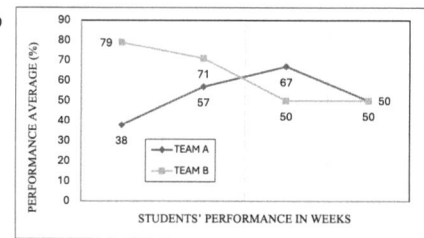

Fig. 4. (a) online; (b) face-to-face

6 Discussions

Our findings showcase that collaborative learning through group discussion can improve learning performance. We suggest that collaborative learning can create conditions, such as better student performance in group discussions, that initially increase their satisfaction.

First, we found significant differences in participation rates between online and face-to-face discussions depending on the level of preparation (pre-assignment). In online format, participation was lowest in week 1 (39%) and increased steadily, reaching 97% by week 4. Similarly, in FTF discussions, the participation rate started at 44% in week 1 and rose to 98% in week 4. Overall, these findings confirm our RQ1 by showing the role of preparation (pre-assignment) in facilitating effective participation in online and face-to-face discussions. Students who have experienced a satisfying group discussion feel more prepared and engaged, contributing actively, and communicating effectively.

Second, we found that preparation (pre-assignment) supports the relationship of satisfactory factors, and it also has a positive effect on satisfaction and learning performance. Notably, when discussions were conducted *without* prior preparation (pre-assignment), their performance scores remained high, with online reaching 75% in Week 2, while FTF discussion reached a higher productivity score of 79% in Week 1. This suggests that in-person discussions may foster a stronger collaborative environment even when no preparation (pre-assignment) is provided. Conversely, when preparation was provided, both online and FTF exhibited a significant decline. The highest performance (Fig. 4) was achieved in groups that did not receive prior preparation, challenging the assumption that providing preparation would consistently lead to a better outcome may not always lead to higher satisfaction and improve student performance as expected. This finding confirms our RQ2 that the difference in their performance outcomes and satisfaction factor measurement between online and FTF formats underscores the complex dynamics of collaborative learning, suggesting that structured preparation may produce varying results depending on contextual factors.

Furthermore, our results show that preparation (pre-assignment) availability significantly impacts satisfaction and learning performance in both online and FTF collaborative learning. This finding emphasizes the importance of fostering an environment conducive to spontaneous interaction, whether online or FTF, to maximize student satisfaction and learning outcomes. Overall, across the models we ran, our findings suggest that preparation and student contribution are important factors in enhancing satisfaction and learning performance during group discussions.

7 Conclusions

The present findings investigated whether the availability of preparation (pre-assignment) affects students' performance by measuring satisfaction factors and their impact on learning performance. To our knowledge, the current study is the first to systematically compare the effects of preparation (pre-assignment) availability by measuring satisfaction factors in online and FTF collaborative learning contexts.

The findings confirm the significance of key satisfaction factors, including preparation (pre-assignment), contribution, and feedback. However, feedback was not found to have a significant relationship with collaborative learning, suggesting that further research is needed to explore its role in greater depth. Identifying more different factors and characteristics that improve group discussions is essential for improving the design and implementation of collaborative learning strategies.

Importantly, our analysis also showed that preparation (pre-assignment) moderates the relationship between satisfaction and learning experience. In particular, providing preparation (pre-assignment) before the class was found to have a positive effect on satisfaction, as well as on students' performance. This aligns with previous research by Allen et al. [1], suggesting that information availability partially mediates the relationship between satisfying meetings and learning outcomes. During collaborative learning, instructors can support collaborative learning by offering explicit instructions to prepare students for teamwork, facilitating group activities, and encouraging active participation (Lee et al. [11]).

Moreover, the present study found the differences between preparation (pre-assignment) availability conditions and how they impact students' performance in group discussions. We suggest that students who had given preparation before the lecture would show a higher learning performance compared to students who had not given any preparation. However, our results do not fully explain why unprepared groups sometimes perform better. For future research, we suggest exploring different discussion strategies, cultural influences, and diverse student populations to better understand these dynamics.

Finally, we conducted a survey where the participants consisted mainly of Japanese undergraduate students in their second to fourth year of school. A cross-cultural comparison could provide insights into how discussion behaviors vary across learning environments. Overall, our study underscores the importance of incorporating structured preparation and effective communication strategies into collaborative learning frameworks to improve students' performance and satisfaction.

Acknowledgments. In regard to the data collection, we would like to thank the undergraduate students of Shizuoka University who participated in the case study for their willingness to respond to the questionnaires.

Disclosure of Interests. The authors declare that they have no competing interest.

References

1. Allen, J.A., Lehmann-Willenbrock, N., Sands, S. J.: Meetings as a positive boost? How and when meeting satisfaction impacts employee empowerment. J. Bus. Res. (2016). https://doi.org/10.1016/j.jbusres.2016.04.011
2. Awang-Hashim, R., Yusof, N., Benlahcene, A., Kaur, A., Suppiah Shanmugam, S.K.: Collaborative learning in tertiary education classrooms: What does it entail? Malaysian J. Learn. Instruction **20**(2), 205–232 (2023). https://doi.org/10.32890/mjli2023.20.2.1
3. Babb, K.A., Ross, C.: The timing of online lecture slide availability and its effect on attendance, participation, and exam performance. Computers & Education (2008). https://doi.org/10.1016/j.compedu.2008.12.009
4. Bach, A., Thiel, F.: Collaborative online learning in higher education—quality of digital interaction and associations with individual and group-related factors. Front. Educ. **9**, 1356271 (2024). https://doi.org/10.3389/feduc.2024.1356271
5. Baker, J.P., Goodboy, A.K., Bowman, N.D., Wright, A.A.: Does teaching with PowerPoint increase students' learning? A meta-analysis. Comput. Educ. **126**, 376–387 (2018). https://doi.org/10.1016/j.compedu.2018.08.003
6. Chen, J., Lin, T.F.: Does Downloading PowerPoint Slides Before the Lecture Lead to Better Student Achievement? Int. Rev. Econ. Educ. **7** (2008) https://doi.org/10.1016/S1477-3880(15)30092-X

7. Chen, J., Wang, M., Kirschner, P.A., Tsai, C.-C.: The role of collaboration, computer use, learning environments, and supporting strategies in CSCL: A meta-analysis. Rev. Education. Res. **88**, 799–843 (2018). https://doi.org/10.3102/0034654318791584
8. Hopkins, W.G.: Quantitative Research Design. sportsci.org/jour/0001/wghdesign.html (2000). Accessed 3. Jan. 2012
9. Judd, T., Kennedy, G., Cropper, S.: Using wikis for collaborative learning: Assessing collaboration through contribution. Aust. J. Educ. Technol. **26**(3) (2010). https://doi.org/10.14742/ajet.1079
10. Khalid, K., Hilman, H., Kumar, D.: Get Along With Quantitative Research Process. Int. J. Res. Manag. **2**(2) (March-2012) ISSN 2249-5908 (2012)
11. Ku, H.Y., Tseng, H.W., Akarasriworn, C.: Collaboration factors, teamwork satisfaction, and student attitudes toward online collaborative learning. Computer Human Behav. **29**, 922–929 (2013). https://doi.org/10.1016/j.chb.2012.12.019
12. Kulkarni, C., Cambre, J., Kotturi, Y., Bernstein, M.S., Klemmer, S.R.: Talkabout: Making Distance Matter with Small Groups in Massive Classes. In: Proceedings of the 18th ACM Conference on Computer Supported Cooperative Work & Social Computing, ACM, Vancouver BC Canada, 1116–1128 (2015). https://doi.org/10.1145/2675133.2675166
13. Le, H., Janssen, J., Wubbels, T.: Collaborative learning practices: Teacher and student perceived obstacles to effective student collaboration. Camb. J. Educ. **48**(1), 103–122 (2018)
14. Lee, H., Mori, C.: Reflective practices and self-directed learning competencies in second language university classes. Asia Pacific Journal of Education **41**(1), 130–151 (2021)
15. León, S.P., Martinez, I.G.: Impact of the provision of PowerPoint slides on learning. Comput. Educ. **173**(2021) 104283 (2021). https://doi.org/10.1016/j.compedu.2021.104283
16. Linton, D.L., Farmer, J.K., Peterson, E.: Is peer interaction necessary for optimal active learning. CBE Life Sci. Educ. **13**, 243–252 (2014)
17. McLoughlin, C., Luca, J.: A learner-centred approach to developing team skills through web-based learning and assessment. Br. J. Edu. Technol. **33**(5), 571–582 (2002)
18. Olson, J.S., Olson, G.M.: Working Together Apart: Collaboration over the Internet. Synthesis Lectures on Hum.-Cent. Inform. **6**(5) (November 2013), 1–151 (2013). https://doi.org/10.2200/S00542ED1V01Y201310HCI020
19. Scager, K., Boonstra, J., Peeters, T., Vulperhorst, J., Wiegant, F.: Collaborative Learning in Higher Education: Evoking Positive Interdependence. CBE—Life Sci. Educ. **15**(4) (2016). https://doi.org/10.1187/cbe.16-07-0219
20. Salter, S., Douglas, T., Kember, D.: Comparing face-to-face and asynchronous online communication as mechanisms for critical reflective dialogue. Educational Action Research **25**(5), 790–805 (2017). https://doi.org/10.1080/09650792.2016.1245626
21. Sotto, R.B., Jr.: Collaborative Learning in the 21st Century Teaching and Learning Landscape: Effects to Students' Cognitive, Affective and Psychomotor Dimensions. Intern. J. Educ. Manag. Innov. 2(2), 136–152 (2021). e-ISSN: 2716-2338. https://doi.org/10.12928/ijemi.v2i2.3325
22. Suthers, D.D., Hundhausen, C.D., Girardeau, L.E.: Comparing the roles of representations in face-to-face and online supported collaborative learning. Comput. Educ. **41**(2003), 335–351 (2003)
23. Zhang, S., Che, S.P., Nan, D., Kim, J.H.: How does online social interaction promote students' continuous learning intentions? Front. Psychol. **14** (2023). https://doi.org/10.3389/fpsyg.2023.1098110
24. Zhang, S., Che, S.P., Nan, D., Li, Y., Kim, J.H.: I know my teammates: The role of Group Member Familiarity in Computer-Supported and face-to-face collaborative learning. Educ. Inf. Technol. **2023**(28), 12615–12631 (2023). https://doi.org/10.1007/s10639-023-11704-w

Factors Influencing Continuance Intention of Watching in VTuber Live Streaming: A Co-creation Experience Perspective

Kai Wang, Yen-Chan Lee, and Chih-Hsuan Yeh[✉]

Department of Information Management, National University of Kaohsiung, Kaohsiung, Taiwan
kwang@nuk.edu.tw, {m1123309,m1133308}@mail.nuk.edu.tw

Abstract. This study explores key factors influencing viewers' continuance intention of watching in VTuber livestreams from a co-creation experience perspective. Given VTubers' unique visual personification, prior virtual influencer research may not fully apply. This study examines how platform affordance and viewers' fantasy proneness, mediated by parasocial relationships and social interaction quality as mediators, leads to continuance intention for watching VTubers. Drawing upon the value co-creation perspective, viewers are active participants in creating experiential value, not merely passive consumers. This research addresses theoretical gaps in VTuber interaction and experience and offers strategic insights to enhancing viewer engagement and commercial outcomes in the VTuber context.

Keywords: VTuber · co-creation experience · continued watching intention · fantasy proneness · platform affordances

1 Introduction

While the term "celebrity" refers to individuals known for accomplishments across fields, "influencers" are primarily recognized for social media engagement [6]. Virtual influencers are digitally created characters with first-person personalities who wield online influence [3]. The number of virtual influencers grew from 9 in 2015 to over 200 by 2023, with at least 150 in Asia [42, 85].

VTubers, featuring anime-style 3D characters operated by real individuals, are a form of virtual influencers. Despite low visual personification, they offer unique entertainment and commercial values. While previous studies have indicated that higher personification in virtual characters lead to increased trust and engagement among audiences [4, 6, 31]. VTubers, despite their relatively low visual personification, remain highly popular. This phenomenon has sparked interests in studying VTubers.

Continuance intention of watching is crucial for assessing influencer impacts [25, 45, 50] and is particularly important for VTubers because VTuber audiences are volatile and thus pose challenges in retention [92]. Identifying influencing factors of viewer continuance is therefore essential. Previous studies found that platform factors and participant characteristics are important determinants of continuance intention of watching in live streaming contexts [76]. This study adopts the affordance theory to explore how platform and individual factors affect continuance intention of watching in VTuber live streams. Livestream platforms encourage interaction, and platform affordances help explain how features foster engagement [103]. At the individual level, VTubers offer rich character stories and create interactive anime-like experiences [90]. Fantasy facilitates emotional immersion even when audiences know the content is fictional [13, 22], particularly for those with higher fantasy proneness [99]. Yet, the role of fantasy as a personal factor that fosters emotional bonds and engagement in VTuber contexts remains underexplored.

This study investigates fantasy proneness as a key viewer characteristic. However, platform affordances and fantasy proneness alone do not fully explain continuance intention of watching. VTuber livestreams involve mutual engagement and value co-creation [60] and constitute a service ecosystem where value emerges through interactions [33, 79].

To attract and retain audiences, platforms must support value co-creation and compelling experiences [79, 98]. Service-dominant logic emphasizes user experience as the center of value creation [77, 88]. Although prior research links experience with watching behavior [8], most focuses on individual perspectives [32, 62, 96] and thus overlooks the dynamics of value co-creation. Co-creation arises from viewer-streamer and viewer-viewer interactions [78, 94]. The former fosters parasocial relationships (PSRs) [45], which may further encourage viewer-viewer connections [52]. These ties enhance shared experiences, described as social interaction quality [46]. This study explores both as experiential components.

This study further investigates how co-creation experiences influence continuance intention of watching and thus proposes two questions: (1) How do co-creation experiences influence continuance intention of watching? (2) How do platform affordance and fantasy proneness affect continuous watching through co-creation experiences? This research addresses psychological as well as technical gaps in VTuber engagement and offers practical insights for improving viewer loyalty. Accordingly, the objectives of this study are as follows:

1. To clarify the formation of co-creation experiences in the context of VTuber livestreaming to address the gap in prior research on VTubers.
2. To identify factors that promote continuance intention of watching in VTuber livestreams to offer implications for VTubers in a highly competitive market.
3. To explore the roles of fantasy proneness and platform affordance in this context and provide guidance for service providers to target audiences more effectively and improve interactive and social experiences during livestreams.

2 Theoretical Foundation and Literature Review

2.1 Definition of VTuber

VTubers, or virtual streamer, are a type of virtual influencers. With the advancement and wider access of motion capture, 3D graphics, and artificial intelligence (AI) technologies, VTubers have gained increasing visibility [1]. VTubers are streamers who create audiovisual content using virtual avatars and offer life-like, vivid performances that distinguish them from traditional live streamers. While YouTube remains the major platform, many VTubers also stream on Twitch, Facebook, and Bilibili [15]. Starting from 2020, the VTuber industry has grown rapidly, featuring more diverse character styles and including animal-based avatars. The domestic market in Taiwan has also expanded, with various local VTuber streamers emerging.

2.2 Platform Affordance in Live Streaming

Affordances refer to perceived action possibilities rather than inherent object features [40, 41]. Hutchby [48] linked this concept to technological determinism, suggesting that technologies offer distinct affordances. Grange and Benbasat [35] extended the affordance perspective to e-commerce and social media, forming a basis for studying livestream platform affordances. Affordance also applies to AI, where robots perceive environmental affordances [72]. In social commerce, Dong and Wang [26] investigated the impacts of affordances on user reactions, and Sun et al. [87] identified visibility and metavoicing as two forms of IT affordances.

This study adopts Karahanna et al.'s [53] framework of four social media affordances, which includes metavoicing, visibility, relationship formation, and communication. Jia et al. [51] applied these affordances to livestreaming and investigated how they affect user engagement. Metavoicing includes actions such as sharing, commenting, or gifting [51, 53, 70]. Visibility refers to users' ability to view all content shared by participants within the live streaming environment through the platform [51, 53]. Relationship formation includes joining fan clubs or following others [51, 53, 93]. Communication involves real-time interactions through chat, comments, and messaging [51, 53].

2.3 Fantasy Proneness

Fantasy proneness, introduced by Wilson and Barber [102], refers to a personality trait involving frequent imaginative activity [71]. Fantasy proneness is a personality trait that reflects one's tendency to engage in imaginative activities. Driven primarily by mental imagery, fantasy plays a compensatory role in filling perceptual gaps, as the cognitive processes involved are neurologically similar to actual perception [59]. Research in anime, film, and virtual worlds shows that fantasy enhances emotional involvement and immersion in a world of fictional elements [13, 22, 29, 64]. Individuals with higher fantasy proneness engage more deeply in virtual settings [22]. In VTuber contexts, fantasy proneness stems from anime-style visuals, narratives, and real-time interactions—making it a crucial yet understudied factor in emotional connection and immersion.

2.4 Co-Creation Experience

Co-creation is a key concept emphasizing joint value creation between firms and customers [77, 81]. Vargo and Lusch [95] noted that firms offer value propositions, while customers co-create actual value. Füller et al. [30] highlighted customers' crucial role in product development. Co-creation reflects active user involvement in value generation [77], with firms enabling personalized experiences [77]. Co-creation experience refers to the psychological state that arises from users participating in value co-creation processes

[58]. Nambisan and Nambisan [73] identified four types of co-creation experiences in virtual environments, namely pragmatic, social, usability, and hedonic. Kohler et al. [58] emphasized pragmatic, social, and hedonic aspects in virtual contexts, and Zhang et al. [108] categorized co-creation experiences as learning, social integration, and hedonic values.

Horton and Wohl [49] defined parasocial relationships (PSRs) as one-sided emotional bonds from repeated mediated interactions, resembling imagined interpersonal ties. Initially linked to traditional media, PSRs enhance closeness in digital contexts [65] and are widely applied to both traditional and digital celebrities [7, 18, 80, 86]. VTubers interact with viewers in real time and strengthen emotional bonds and familiarity [45], making livestreams more conducive to PSR development. This study adopts PSRs to explain viewer-VTuber interactions.

Social interaction is a basic human need [69, 104]. Blau [12] proposed that relationships are formed through cost-benefit analysis [34]. Social interaction drives social media engagement [19, 101] and influences intention to watch in live commerce [43, 69]. Social interaction ties aid information sharing and content understanding [34], described by Jaegher et al. [28] as complex, involving communication, context, and shared activities. This study defines social interaction to viewer-viewer communication, integrated into the co-creation framework.

2.5 Continuance Intention of Watching

Continuance intention of watching reflects viewers' ongoing preferences for live streaming content and represents a form of loyalty [43, 45]. Continuance of watching may include consistent watching, engagement, and attachment to specific streams or streamers [106]. This study defines continuance intention of watching as a long-term behavioral intention, highlighting both temporal consistency and potential value generation—such as attracting traffic and enabling monetization [43]. Fostering continuance intention of watching is more crucial than one-time purchases or gifting behaviors. In the competitive VTuber market, enhancing continuance intention of watching is a key strategic goal.

3 Research Model and Hypothesis Development

VTubers represent a recent trend in live streaming, and this study investigates how platform affordances and one's fantasy proneness influence continuance intention of watching. Specifically, this research examines how co-creation experiences in VTuber livestreams are formed and how they affect viewers' intentions to continue watching. The research model is illustrated in Fig. 1.

Previous studies show that social media use enhances engagement, including immersion and participatory behaviors such as liking or sharing [55, 56]. Social media affordances increase visibility [10, 37], enable connections [23], and aid information retention [93], fostering interaction and participation [66]. Livestream platforms share these features. The affordances of the platforms (e.g., commenting, liking, sharing) help building social ties and enhance belonging and knowledge sharing [51].

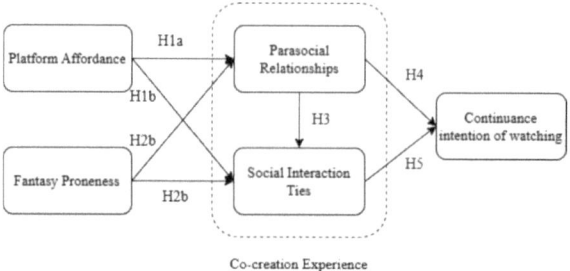

Fig. 1. Research Model

Metavoicing affordance allows users to perceive and respond during interactions through reposts, comments, votes, and likes. It also contributes to creating a more satisfying livestream experience for all participants [27, 75].

Communication affordance includes real-time tools in live chat sessions, comments, and messaging. Communication affordance also enables public interaction and streamer feedback [61, 92]. Watching others' reactions promotes engagement, emotional resonance, and stronger social ties [51, 93].

Relationship formation affordance allows users to connect around shared interests [53], encouraging interaction between users, content, and creators via features of fan clubs or chatrooms [93]. These social interaction elements in livestreams foster belonging and make watching more engaging [53]. Interactions with viewers and streamers help building deeper relationships and emotional support [51].

H1a: Platform affordances positively influence parasocial relationships.
H1b: Platform affordances positively influence social interaction ties.

Prior studies show that individuals with higher fantasy proneness vividly imagine interactions and deepen emotional engagement and parasocial bonds [64, 74]. VTubers are usually like anime characters with fictional backgrounds, and thus makes fantasy crucial for viewer engagement and relationship building.

H2a: Fantasy proneness positively influences parasocial relationships.
H2b: Fantasy proneness positively influences social interaction ties.

With the rapid growth of social media, users form parasocial relationships with media figures, thus leading to greater community participation [9]. Fan culture for virtual idols mirrors traditional fandoms, reshapes emotional bonds, and fosters fan interaction [107]. These emotional ties not only deepen engagement but also encourage discussion and build stronger social connections [105].

H3: Viewers' parasocial relationships with VTubers positively influence social interaction ties.

Parasocial relationships more strongly affect repeated watching than content itself [24, 39, 89]. Labrecque [65] found such relationships link to customers' brand loyalty. Other studies also support the role of repeated exposure in building behavioral loyalty

[16]. In livestreaming, parasocial ties influence emotional, instrumental, and financial support [100]. Based on previous findings, we propose the following hypothesis:

H4: Parasocial relationships positively influence viewers' continuance intention of watching for VTuber livestreams.

Social interaction ties are crucial to community building. Livestreams act as places where viewers engage in real-time chats with streamers and others [46] and thus foster friendship and inclusion [11, 82]. Prior studies show such ties influence platform attitudes and continued use [20, 21, 57]. Thus, this study proposes the following hypothesis:

H5: Social interaction ties positively influence viewers' continuance intention of watching for VTuber livestreams.

4 Research Methodology

4.1 Operationalization

This study defines platform affordances in livestreaming as four components: metavoicing, visibility, relationship formation, and communication [51]. These dimensions are configured as first-order formative constructs representing livestreaming platform affordances. All items are measured using a 7-point Likert scale (1 = strongly disagree and 7 = strongly agree). Metavoicing refers to the extent to which viewers participate in discussions and respond to others. [26]. Visibility refers to the extent to which viewers access and observe content provided by other users within the livestreaming environment. [53]. Relationship formation captures how viewers form social ties through platform features while watching VTubers, and the items are adapted from Karahanna et al. [53]. Communication measures real-time interactions viewers have with others during livestreams, with the measurement items adapted from Karahanna et al. [53]. Fantasy proneness is defined as viewers' tendency to imagine or fantasize during VTuber livestreams [36]. Parasocial relationship refers to the emotional connection between viewers and media figures in VTuber contexts [83]. Social interaction ties indicate the extent of interpersonal exchanges among viewers during livestreams [68]. Continuance intention of watching is viewers' willingness to keep watching VTuber livestreams [68].

4.2 Data Collection

This study investigates factors influencing viewers' continuance intention of watching in the context of VTuber livestreams. The target population includes users of VTuber content on livestream platforms, with no demographic restrictions.

An online questionnaire will be distributed via Bahamut (https://www.gamer.com.tw/), Facebook, and VTuber communities. Expert reviews and a pilot test serve to ensure clarity and readability will be conducted before official announcement of the survey.

5 Expected Contributions

This study examines how fantasy proneness and platform affordances affect viewers' continuance intention of watching in VTuber livestreams. Given livestreaming's interactive nature, co-creation experience perspective is adopted to explore underlying mechanisms. Academically, this research fills a gap in co-creation in VTuber livestream contexts. By integrating individual traits (fantasy proneness) and platform features (affordances), this study provides practical insights for service providers to better identify audiences and enhance interactive experiences. The proposed model offers VTubers guidance on how to sustain viewer engagement, which is crucial in today's competitive livestreaming environment.

References

1. Chen, C.-T.: A casual talk on NFTs and VTubers: New trends in digital development. Taiwan Economic Research Monthly **44**(8), 98–106 (2021)
2. Taiwan Virtual Influencer Association: The 3rd Virtual Influencer Design Competition – Local Taiwan. Lee, Hsiang-Tung. https://www.taiwanvtuber.org/vtubercontest (2020)
3. Audrezet, A., Koles, B.: Virtual influencer as a brand avatar in interactive marketing. In The Palgrave Handbook of Interactive Marketing, pp. 353–376 (2023)
4. Arsenyan, J., Mirowska, A.: Almost human? A comparative case study on the social media presence of virtual influencers. Int. J. Hum Comput Stud. **155**, 102694 (2021)
5. Audrezet, A., De Kerviler, G., Moulard, J.G.: Authenticity under threat: When social media influencers need to go beyond self-presentation. J. Bus. Res. **117**, 557–569 (2020)
6. Belanche, D., Casaló, L.V., Flavián, M.: Human versus virtual influences, a comparative study. J. Bus. Res. **173**, 114493 (2024)
7. Breves, P., Amrehn, J., Heidenreich, A., Liebers, N., Schramm, H.: Blind trust? The importance and interplay of parasocial relationships and advertising disclosures in explaining influencers' persuasive effects on their followers. Int. J. Advert. **40**(7), 1209–1229 (2021)
8. Bründl, S., Matt, C., Hess, T.: Consumer use of social live streaming services: The influence of co-experience and effectance on enjoyment. In: Proceedings of the 25th European Conference on Information Systems (ECIS), pp. 1775–1791, Guimarães, Portugal, 5–10 June 2017
9. Blight, M.G., Ruppel, E.K., Schoenbauer, K.V.: Sense of community on Twitter and Instagram: Exploring the roles of motives and parasocial relationships. Cyberpsychol. Behav. Soc. Netw. **20**(5), 314–319 (2017)
10. Boyd, D.: Social network sites as networked publics: Affordances, dynamics, and implications. In: A networked self, pp. 47–66. Routledge (2010).
11. Brandtzæg, P. B., Heim, J.: Why people use social networking sites. In: Online Communities and Social Computing: Third International Conference, OCSC 2009, Held as Part of HCI International 2009, pp. 143–152, San Diego, CA, USA, 19–24 July 2009. Proceedings 3. Springer Berlin Heidelberg. (2009).
12. Blau, P.: Power and Exchange in Social Life, John Wiley & Sons, New York, NY (1964)
13. Can, A.S., Ekinci, Y., Dilek-Fidler, S.: Fantasy or reality? Unveiling the power of realistic narratives in tourism social media advertising. Tour. Manage. **106**, 104998 (2025)
14. Chen, J., Liao, J.: Antecedents of viewers' live streaming watching: a perspective of social presence theory. Front. Psychol. **13**, 839629 (2022)
15. Chen, C.P.: Digital gifting in personal brand communities of live-streaming: Fostering viewer–streamer–viewer parasocial relationships. J. Mark. Commun. **27**(8), 865–880 (2021)

16. Cohen, J., Holbert, R.L.: Assessing the predictive value of parasocial relationship intensity in a political context. Commun. Res. **48**(4), 501–526 (2021)
17. Casaló, L.V., Flavián, C., Ibáñez-Sánchez, S.: Influencers on Instagram: Antecedents and consequences of opinion leadership. J. Bus. Res. **117**, 510–519 (2020)
18. Chung, S., Cho, H.: Fostering parasocial relationships with celebrities on social media: Implications for celebrity endorsement. Psychol. Mark. **34**(4), 481–495 (2017)
19. Cheung, C.M., Liu, I.L., Lee, M.K.: How online social interactions influence customer information contribution behavior in online social shopping communities: A social learning theory perspective. J. Am. Soc. Inf. Sci. **66**(12), 2511–2521 (2015)
20. Curras-Perez, R., Ruiz-Mafe, C., Sanz-Blas, S.: Determinants of user behaviour and recommendation in social networks: An integrative approach from the uses and gratifications perspective. Ind. Manag. Data Syst. **114**(9), 1477–1498 (2014)
21. Chiang, H.S.: Continuous usage of social networking sites: The effect of innovation and gratification attributes. Online Inf. Rev. **37**(6), 851–871 (2013)
22. Choi, B., Huang, J., Jeffrey, A., Baek, Y.: Development of a scale for fantasy state in digital games. Comput. Hum. Behav. **29**(5), 1980–1986 (2013)
23. Chen, J., Geyer, W., Dugan, C., Muller, M., Guy, I.: Make new friends, but keep the old: recommending people on social networking sites. In: Proceedings of the SIGCHI conference on human factors in computing systems, pp. 201–210, April 2009
24. Conway, J.C., Rubin, A.M.: Psychological predictors of television viewing motivation. Commun. Res. **18**(4), 443–463 (1991)
25. Dabiran, E., Farivar, S., Wang, F., Grant, G.: Virtually human: anthropomorphism in virtual influencer marketing. J. Retail. Consum. Serv. **79**, 103797 (2024)
26. Dong, X., Wang, T.: Social tie formation in Chinese online social commerce: The role of IT affordances. Int. J. Inf. Manage. **42**, 49–64 (2018)
27. Dong, X., Wang, T., Benbasat, I.: IT Affordances in Online Social Commerce: Conceptualization Validation and Scale Development. Amcis (August 2016)
28. De Jaegher, H., Di Paolo, E., Gallagher, S.: Can social interaction constitute social cognition? Trends Cogn. Sci. **14**(10), 441–447 (2010)
29. Erickson, S.E., Dal Cin, S.: Romantic parasocial attachments and the development of romantic scripts, schemas and beliefs among adolescents. Media Psychol. **21**(1), 111–136 (2018)
30. Füller, J., Mühlbacher, H., Matzler, K., Jawecki, G.: Consumer empowerment through internet-based co-creation. J. Manag. Inf. Syst. **26**(3), 71–102 (2009)
31. Gao, W., Jiang, N., Guo, Q.: How do virtual streamers affect purchase intention in the live streaming context? A presence perspective. J. Retail. Consum. Serv. **73**, 103356 (2023)
32. Gu, Y., Cheng, X., Shen, J.: Design shopping as an experience: Exploring the effect of the live-streaming shopping characteristics on consumers' participation intention and memorable experience. Information & Management **60**(5), 103810 (2023)
33. Giertz, J.N., Weiger, W.H., Törhönen, M., Hamari, J.: Content versus community focus in live streaming services: How to drive engagement in synchronous social media. J. Serv. Manag. **33**(1), 33–58 (2022)
34. Ghahtarani, A., Sheikhmohammady, M., Rostami, M.: The impact of social capital and social interaction on customers' purchase intention, considering knowledge sharing in social commerce context. J. Innov. Knowl. **5**(3), 191–199 (2020)
35. Grange, C., Benbasat, I.: The value of social shopping networks for product search and the moderating role of network scope (2013)
36. Garrido, S., Schubert, E.: Individual differences in the enjoyment of negative emotion in music: A literature review and experiment. Music. Percept. **28**(3), 279–296 (2011)

37. Grudin, J. Enterprise knowledge management and emerging technologies. In: Proceedings of the 39th annual Hawaii international conference on system sciences (HICSS'06), Vol. 3, pp. 57a–57a. IEEE, January 2006
38. Gefen, D., Straub, D., Boudreau, M.C.: Structural equation modeling and regression: Guidelines for research practice. Commun. Assoc. Inf. Syst. **4**(1), 7 (2000)
39. Grant, A.E., Guthrie, K.K., Ball-Rokeach, S.J.: Television shopping: A media system dependency perspective. Commun. Res. **18**(6), 773–798 (1991)
40. Gibson, J.J.: The Ecological Approach to Visual Perception. Houghton Mifflin, New York (1979)
41. Gibson, J.J.: Perceiving, acting, and knowing: Toward an ecological psychology. The Theory of Affordances, 67–82 (1977)
42. Hiort, A.: How Many Virtual Influencers Are There? virtualhumans.org. https://www.virtualhumans.org/article/how-many-virtual-influencers-are-there (24 June 2022)
43. Hou, F., Guan, Z., Li, B., Chong, A.Y.L.: Factors influencing people's continuous watching intention and consumption intention in live streaming: Evidence from China. Internet Res. **30**(1), 141–163 (2020)
44. Hou, M.: Social media celebrity and the institutionalization of YouTube. Convergence **25**(3), 534–553 (2019)
45. Hu, M., Zhang, M., Wang, Y.: Why do audiences choose to keep watching on live video streaming platforms? An explanation of dual identification framework. Comput. Hum. Behav. **75**, 594–606 (2017)
46. Hamilton, W. A., Garretson, O., Kerne, A.: Streaming on twitch: fostering participatory communities of play within live mixed media. In *Proceedings of the SIGCHI conference on human factors in computing systems*, pp. 1315–1324, April 2014
47. Hair, J.F., Ringle, C.M., Sarstedt, M.: PLS-SEM: Indeed a silver bullet. Journal of Marketing theory and Practice **19**(2), 139–152 (2011)
48. Hutchby, I.: Technologies, texts and affordances. Sociology **35**(2), 441–456 (2001)
49. Horton, D., Richard Wohl, R.: Mass communication and para-social interaction: Observations on intimacy at a distance. Psychiatry, **19**(3), 215–229 (1956)
50. Inan, D.I., et al.: Because follower experience matters: the continuance intention to follow recommendation of the influencer. Human Behavior and Emerging Technologies **2022**(1), 3684192 (2022)
51. Jia, M., Zhao, Y.C., Song, S., Zhang, X., Wu, D., Li, J.: How vicarious learning increases users' knowledge adoption in live streaming: The roles of parasocial interaction, social media affordances, and knowledge consensus. Inf. Process. Manage. **61**(2), 103599 (2024)
52. Kim, J., Liu, J.T., Chang, S.R.: Trans-Parasocial Relation Between Influencers and Viewers on Live Streaming Platforms: How Does it Affect Viewer Stickiness and Purchase Intention? Asia Marketing Journal **24**(2), 39–50 (2022)
53. Karahanna, E., Xu, S.X., Xu, Y., Zhang, N.: The needs–affordances–features perspective for the use of social media. MIS Q. **42**(3), 737-A23 (2018)
54. Koles, B., Nagy, P.: Avatars as transitional objects: The impact of avatars and digital objects on adolescent gamers. Journal of Gaming & Virtual Worlds **8**(3), 279–296 (2016)
55. Kim, Y., Wang, Y., Oh, J.: Digital media use and social engagement: How social media and smartphone use influence social activities of college students. Cyberpsychol. Behav. Soc. Netw. **19**(4), 264–269 (2016)
56. Kabadayi, S., Price, K.: Consumer–brand engagement on Facebook: liking and commenting behaviors. J. Res. Interact. Mark. **8**(3), 203–223 (2014)
57. Kim, Y.H., Kim, D.J., Wachter, K.: A study of mobile user engagement (MoEN): Engagement motivations, perceived value, satisfaction, and continued engagement intention. Decis. Support Syst. **56**, 361–370 (2013)

58. Kohler, T., Fueller, J., Matzler, K., Stieger, D., & Füller, J. (2011). Co-creation in virtual worlds: The design of the user experience. *MIS quarterly*, 773–788
59. Kosslyn, S.M., Ganis, G., Thompson, W.L.: Neural foundations of imagery. Nat. Rev. Neurosci. **2**(9), 635–642 (2001)
60. Liu, H., Chung, L., Tan, K.H., Peng, B.: I want to view it my way! How viewer engagement shapes the value co-creation on sports live streaming platform. J. Bus. Res. **170**, 114331 (2024)
61. Lu, Y., He, Y., Ke, Y.: The influence of e-commerce live streaming affordance on consumer's gift-giving and purchase intention. Data Science and Management **6**(1), 13–20 (2023)
62. Li, Y., Peng, Y.: What drives gift-giving intention in live streaming? The perspectives of emotional attachment and flow experience. International Journal of Human-Computer Interaction **37**(14), 1317–1329 (2021)
63. Liu, Y., Pan, Y.H.: An empirical study on the influencing factors of learning through knowledge sharing live streaming-Based on live streaming platform in China. Journal of the Korea Convergence Society **12**(12), 197–211 (2021)
64. Liebers, N., Straub, R.: Fantastic relationships and where to find them: Fantasy and its impact on romantic parasocial phenomena with media characters. Poetics **83**, 101481 (2020)
65. Labrecque, L.I.: Fostering consumer–brand relationships in social media environments: The role of parasocial interaction. J. Interact. Mark. **28**(2), 134–148 (2014)
66. Liu, Y.: Developing a scale to measure the interactivity of websites. J. Advert. Res. **43**(2), 207–216 (2003)
67. Merckelbach, H., Otgaar, H., Lynn, S.J.: Empirical research on fantasy proneness and its correlates 2000–2018: A meta-analysis. Psychol. Conscious. Theory Res. Pract. **9**(1), 2 (2022)
68. Ma, X., Zou, X., Lv, J.: Why do consumers hesitate to purchase in live streaming? A perspective of interaction between participants. Electron. Commer. Res. Appl. **55**, 101193 (2022)
69. Ma, Y.: To shop or not: Understanding Chinese consumers' live-stream shopping intentions from the perspectives of uses and gratifications, perceived network size, perceptions of digital celebrities, and shopping orientations. Telematics Inform. **59**, 101562 (2021)
70. Majchrzak, A., Faraj, S., Kane, G.C., Azad, B.: The contradictory influence of social media affordances on online communal knowledge sharing. J. Comput.-Mediat. Commun. **19**(1), 38–55 (2013)
71. Merckelbach, H., Horselenberg, R., Muris, P.: The Creative Experiences Questionnaire (CEQ): A brief self-report measure of fantasy proneness. Personality Individ. Differ. **31**(6), 987–995 (2001)
72. Murphy, R.R.: Case studies of applying Gibson's ecological approach to mobile robots. IEEE Transactions on Systems, Man, and Cybernetics-Part A: Systems and Humans **29**(1), 105–111 (1999)
73. Nambisan, S., Nambisan, P.: How to profit from a better virtual customer environment. MIT Sloan Manag. Rev. **49**(3), 53–61 (2008)
74. Newitz, A.: Anime otaku: Japanese animation fans outside Japan. Bad Subjects **13**(11), 1–14 (1994)
75. Park, H., Luo, Y., Yang, Y.: The social media entrepreneurship of inexperienced people: From the learning process perspective. In: Academy of Management Proceedings, Vol. 2022, No. 1, p. 11191. Briarcliff Manor, NY 10510: Academy of Management (2022)
76. Park, J.Y., Jang, S.S.: Revisit and satiation patterns: Are your restaurant customers satiated? Int. J. Hosp. Manag. **38**, 20–29 (2014)
77. Prahalad, C.K., Ramaswamy, V.: Co-creation experiences: The next practice in value creation. J. Interact. Mark. **18**(3), 5–14 (2004)

78. Qian, T.Y., Seifried, C.: Virtual interactions and sports viewing on social live streaming platforms: The role of co-creation experiences, platform involvement, and follow status. J. Bus. Res. **162**, 113884 (2023)
79. Qian, T.Y.: Watching sports on Twitch? A study of factors influencing continuance intentions to watch Thursday Night Football co-streaming. Sport Management Review **25**(1), 59–80 (2022)
80. Reinikainen, H., Munnukka, J., Maity, D., Luoma-Aho, V.: 'You really are a great big sister'–parasocial relationships, credibility, and the moderating role of audience comments in influencer marketing. J. Mark. Manag. **36**(3–4), 279–298 (2020)
81. Ramaswamy, V.: It's about human experiences… and beyond, to co-creation. Ind. Mark. Manage. **40**(2), 195–196 (2011)
82. Ridings, C. M., & Gefen, D. (2004). Virtual community attraction: Why people hang out online. *Journal of Computer-mediated communication*, *10*(1), JCMC10110
83. Rubin, R. B., & McHugh, M. P. (1987). Development of parasocial interaction relationships
84. Stein, J.P., Linda Breves, P., Anders, N.: Parasocial interactions with real and virtual influencers: The role of perceived similarity and human-likeness. New Media Soc. **26**(6), 3433–3453 (2024)
85. Sookkaew, J., Saephoo, P.: "Digital influencer": Development and coexistence with digital social groups. Int. J. Adv. Comput. Sci. Applications **12**(12) (2021)
86. Sokolova, K., Kefi, H.: Instagram and YouTube bloggers promote it, why should I buy? How credibility and parasocial interaction influence purchase intentions. J. Retail. Consum. Serv. **53**, 101742 (2020)
87. Sun, Y., Shao, X., Li, X., Guo, Y., Nie, K.: How live streaming influences purchase intentions in social commerce: An IT affordance perspective. Electron. Commer. Res. Appl. **37**, 100886 (2019)
88. Seifried, C.: Sport facilities as a broadcast studio for human extensibility? Geographic information system-based diagrams of a high-and low-identified sport fan. J. Sport Manag. **25**(6), 515–530 (2011)
89. Skumanich, S.A., Kintsfather, D.P.: Individual media dependency relations within television shopping programming: A causal model reviewed and revised. Commun. Res. **25**(2), 200–219 (1998)
90. Tambunan, O.N., Setiawan, A.B.: Factors in the Growth of Vtuber Livestreaming Entertainment Phenomenon in Japanese Society. Literacy: Int. Scientific J. Soc. Educ. Humanities **2**(2), 78–85 (2023)
91. Tukachinsky, R., Stever, G.: Theorizing development of parasocial engagement. Commun. Theory **29**(3), 297–318 (2019)
92. Trowbridge, C. Twitch's viewership now rivals CNN. The Esports Observer (2018)
93. Treem, J.W., Leonardi, P.M.: Social media use in organizations: Exploring the affordances of visibility, editability, persistence, and association. Ann. Int. Commun. Assoc. **36**(1), 143–189 (2013)
94. Verleye, K.: The co-creation experience from the customer perspective: its measurement and determinants. J. Serv. Manag. **26**(2), 321–342 (2015)
95. Vargo, S.L., Lusch, R.F.: Service-dominant logic: continuing the evolution. J. Acad. Mark. Sci. **36**, 1–10 (2008)
96. Wu, D., Wang, X., Ye, H.J.: Transparentizing the "Black Box" of live streaming: impacts of live interactivity on viewers' experience and purchase. IEEE Trans. Eng. Manage. **71**, 3820–3831 (2023)
97. Wang, D., Luo, X.R., Hua, Y., Benitez, J.: Big arena, small potatoes: A mixed-methods investigation of atmospheric cues in live-streaming e-commerce. Decis. Support Syst. **158**, 113801 (2022)

98. Wymer, S., Naraine, M.L., Thompson, A.J., Martin, A.J.: Transforming the fan experience through livestreaming: A conceptual model. J. Interact. Advert. **21**(2), 79–92 (2021)
99. Weibel, D., Martarelli, C.S., Häberli, D., Mast, F.W.: The fantasy questionnaire: A measure to assess creative and imaginative fantasy. J. Pers. Assess. **100**(4), 431–443 (2018)
100. Wohn, D. Y., Freeman, G., McLaughlin, C.: Explaining viewers' emotional, instrumental, and financial support provision for live streamers. In: Proceedings of the 2018 CHI conference on human factors in computing systems, pp. 1–13, April 2018
101. Wang, J.C., Chiang, M.J.: Social interaction and continuance intention in online auctions: A social capital perspective. Decis. Support Syst. **47**(4), 466–476 (2009)
102. Wilson, S. C., & Barber, T. X. (1982). The fantasy-prone personality: Implications for understanding imagery, hypnosis, and parapsychological phenomena. *Psi Research*
103. Yan, Y., Chen, H., Shao, B., Lei, Y.: How IT affordances influence customer engagement in live streaming commerce? A dual-stage analysis of PLS-SEM and fsQCA. J. Retail. Consum. Serv. **74**, 103390 (2023)
104. Yin, X., Wang, H., Xia, Q., Gu, Q.: How social interaction affects purchase intention in social commerce: A cultural perspective. Sustainability **11**(8), 2423 (2019)
105. Yin, Y., Xie, Z.: The bounded embodiment of fandom in China: Recovering shifting media experiences and fan participation through an oral history of Animation-Comics-Games lovers. Int. J. Commun. **12**, 18 (2018)
106. Zheng, S., Chen, J., Liao, J., Hu, H.L.: What motivates users' viewing and purchasing behavior motivations in live streaming: a stream-streamer-viewer perspective. J. Retail. Consum. Serv. **72**, 103240 (2023)
107. Zheng, X.: Borderless fandom and contemporary popular cultural scene in Chinese cyberspace. Doctoral dissertation (2016)
108. Zhang, H., Lu, Y., Wang, B., Wu, S.: The impacts of technological environments and co-creation experiences on customer participation. Information & Management **52**(4), 468–482 (2015)

Entropy-Based Anomaly Detection for Cybersecurity Threats in Network Traffic

Han-Wei Hsiao and Yun-Zhen Lee[✉]

Department of Information Management, National University of Kaohsiung, Kaohsiung, Taiwan
hanwei@nuk.edu.tw, m1133316@mail.nuk.edu.tw

Abstract. In modern networks, with the growing variety of cybersecurity threats, network anomaly detection has become increasingly important. Its primary goal is to promptly identify abnormal traffic to prevent catastrophic consequences such as data breaches and service disruptions. However, many traditional anomaly detection methods rely on traffic characteristics such as packet frequency, traffic volume, or known attack signatures. These approaches often face limitations when dealing with complex attack behaviors due to the variability of traffic patterns and evolving attack techniques. To address these challenges, entropy, a metric originally derived from information theory to quantify data unpredictability, has recently been applied in the field of cybersecurity, particularly in network anomaly detection, showing promising results. Compared to traditional methods, analyzing changes in entropy within packet content can reveal hidden suspicious activities more effectively. To explore the practical effectiveness of entropy-based indicators in anomaly detection, this study proposes a network anomaly detection method based on entropy variation. Specifically, it utilizes Shannon Entropy of packet-level statistical features to observe abnormal fluctuations in traffic predictability, thereby identifying potential attack activities and evaluating the impact on detection accuracy. To validate the performance of the proposed method, we use real network traffic data from an academic unit and conduct experiments simulating scanning behavior for evaluation.

Keywords: Network Anomaly Detection · Entropy-Based Anomaly · Cybersecurity

1 Introduction

In today's digital world, information infrastructure is vital for nations, enterprises, and individuals. Yet, systems face constant threats like data breaches and outages, causing financial and reputational harm. The 2025 Global Cybersecurity Report by Check Point noted a 44% rise in cyberattacks from the previous year [1]. With rising attack frequency and impact, cybersecurity's importance continues to grow, affecting not only organizational operations but also societal and national stability. ENISA's 2024 report identified eight persistent threat types from June 2023 to July 2024, including ransomware, malware, social engineering, data threats, DoS, and manipulation, with ransomware and DDoS being most prominent [2]. The increasing scale and complexity of attacks pose risks to critical infrastructure and personal data. A core challenge is detecting and responding to abnormal behavior quickly. Network Anomaly Detection is crucial in identifying unusual traffic to prevent severe consequences. It detects potential threats by analyzing network traffic patterns, even when firewalls or antivirus systems are in place. Unlike signature-based methods, it focuses on behavior anomalies, helping detect unknown or zero-day attacks.

Anomaly detection systems generally involve traffic collection, feature extraction, and behavior analysis. Techniques fall into three categories: Signature-based Detection, Anomaly-based Detection, and Stateful Protocol Analysis [3]. Signature-based methods match traffic against known attack patterns, effective for known threats but blind to unknown ones. Anomaly-based detection flags behavior deviating from a learned normal baseline, useful for unknown threats but prone to false positives and baseline manipulation. Stateful Protocol Analysis builds baselines from protocol specifications and logs deviations, detecting complex attacks but requiring more resources and failing when attacks don't violate protocol rules.

Encryption, diverse traffic patterns, and advanced attack techniques challenge traditional detection methods. More sensitive, generalizable detection techniques are needed. Based on this context, this study explores the potential application of entropy as a novel anomaly detection indicator. The concept of entropy was first introduced by Shannon in 1948 to quantify the uncertainty and diversity of information distribution [4]. Entropy reflects the degree of dispersion in the information of a system, where higher entropy values indicate a more uniform probability distribution of events, while lower values represent a more concentrated distribution. The following is the mathematical representation of entropy, where $p(x_i)$ denotes the probability of occurrence of event x_i.

$$H(X) = -\sum_{i=1}^{n} p(x_i) \log_2 p(x_i)$$

In a normally functioning and undisturbed network environment, the statistical characteristics of packet traffic, such as source IP, destination port, packet size, and so on, typically exhibit stable and predictable distribution patterns. This is because most user activities and application system operations are repetitive and regular, resulting in relatively fixed traffic distributions. However, when an attacker performs scanning, probing, or other atypical actions on the network, these behaviors disrupt the original statistical distributions, causing packet characteristics to become abnormally dispersed or concentrated, which in turn leads to anomalous changes in entropy values. Take port scanning as an example: an attacker usually attempts to rapidly connect to a large number of different destination ports in an effort to discover which services are active on the system. This behavior causes the previously stable distribution of destination ports to become suddenly more scattered, resulting in a significant increase in the entropy value. On the other hand, in the case of a DDoS attack, the attacker typically sends a large number of requests to a specific target host, making the distribution of destination IPs extremely concentrated, which leads to a drop in the entropy value. These abnormal changes are rarely seen during normal network operations; therefore, by monitoring entropy variations, potential anomalous behaviors can be detected.

Compared to content-based analysis or machine learning methods that require large amounts of training data, entropy offers a lightweight, real-time, and relatively sensitive approach to detecting unknown attacks. It is particularly suitable as a preliminary screening and alerting tool. This study aims to design and implement a network anomaly detection method based on entropy variation. By calculating the entropy of packet features, such as source IP, port, and time distribution, we analyze whether there are abnormal changes in traffic distribution diversity. To evaluate the effectiveness of this method, real network traffic data from an academic institution was used as the test dataset, and scanning scenarios were simulated for experimental validation. It is expected that this method can provide a network anomaly detection solution that does not rely on predefined signatures, and is generalizable and highly adaptive.

2 Literature Reviews

Anomaly detection methods can be categorized into several major types based on their technical principles and data formats. Statistical methods are among the earliest techniques proposed. These methods typically assume that data follows a known distribution, such as a normal distribution, and use statistical thresholds to identify outliers [5]. Yu et al. [6] proposed a network traffic anomaly detection method based on the Gaussian Mixture Model (GMM), assuming that normal network behavior can be modeled by a combination of Gaussian distributions. Using the Expectation-Maximization (EM) algorithm to learn model parameters, the method calculates the probability that an observation belongs to the "normal" distribution to determine whether it is anomalous. However, when facing high-dimensional or unstructured data, the effectiveness of such methods drops significantly [7]. To overcome these limitations, machine learning methods have gradually become mainstream. These methods include supervised learning, unsupervised learning, and semi-supervised learning, emphasizing model construction and feature learning, which are particularly suitable for handling large datasets and dynamically changing patterns. Zhou et al. [8] proposed a network traffic anomaly detection model based on feature grouping and multiple autoencoders to address the limitations of traditional autoencoders in capturing complex features from high-dimensional data. This approach not only enhances the model's ability to learn diverse traffic behaviors but also significantly improves detection accuracy.

In addition, for time-series data such as network traffic or system logs, researchers have proposed many anomaly detection techniques that incorporate temporal dependencies. For example, using models such as sliding windows, Recurrent Neural Networks (RNN), and Long Short-Term Memory (LSTM) networks to capture behavior trends or sudden changes. Psychogyios et al. [9] proposed a network intrusion detection model based on LSTM neural networks, specifically designed to handle complex intrusion patterns in large-scale network traffic. This model can capture long-term dependencies in time-series data, thereby improving the detection of unknown attacks. When anomalies are deeply hidden and evade traditional detection, researchers focus on data's structural features and information complexity. Although machine learning and statistical methods are established, the increasing volume and complexity of data challenge reliance on sample distributions or distance measures alone. Consequently, information-theoretic and complexity-based approaches to anomaly detection have gained attention in recent years.

Entropy analysis has recently been regarded as a promising technique in the field of information security, particularly demonstrating significant value in network anomaly detection. Berezinski et al. [10] proposed an entropy-based network anomaly detection method that identifies abnormal behaviors by calculating the entropy values of network traffic features. This method is especially suitable for detecting stealthy malicious activities such as botnets, and has been validated on multiple public datasets, showing good detection performance. Hassan et al. [11] proposed a network attack detection model combining entropy features and machine learning for detecting DDoS attacks in Software-Defined Networking (SDN). This approach comprehensively analyzes the entropy of packet features such as source IP, destination IP, source and destination ports to identify abnormal changes in network traffic. Experimental results show superior recognition ability of unknown attacks compared to traditional machine learning methods. Ibrahim and Gajin [12] proposed an anti-evasion entropy-based network anomaly detection method that combines multidimensional entropy features with the Kolmogorov-Smirnov test to effectively identify disguised traffic. Their approach achieved a detection accuracy of 96.2% on the CIC-IDS2017 dataset and improved anti-evasion performance by 17.8% over conventional methods. This study provides a practical solution for APT defense but still has limitations in handling encrypted traffic.

Besides the classical Shannon entropy, recent studies have adopted other generalized entropy measures to enhance the sensitivity and robustness of anomaly detection, such as Tsallis entropy [13] and Renyi entropy [14]. Compared to traditional Shannon entropy, the most notable characteristic of Tsallis entropy is its non-additivity: when two events are correlated, the total uncertainty is not necessarily equal to the sum of their individual uncertainties. It introduces a tunable parameter q that flexibly controls sensitivity to high- or low-probability events. When q is not equal to 1, Tsallis entropy exhibits behaviors distinct from Shannon entropy. The following is the mathematical representation of Tsallis entropy.

$$S_q = \frac{1}{q-1}\left(1 - \sum_{i=1}^{W} p_i^q\right)$$

When q equals 1, it converges to Shannon entropy; when q is greater than 1, it emphasizes high-probability events; conversely, when q is less than 1, it emphasizes low-probability events. This property makes Tsallis entropy particularly useful for data distributions that are uneven or skewed, such as long-tail distributions. Berezinski et al. [15] found that Tsallis entropy has higher sensitivity than Shannon entropy for long-tail distribution data. This characteristic implies that Tsallis entropy may have greater discriminative potential when facing anomalies with long latency or lacking obvious features, such as data leaks or intermittent abnormal connections.

Renyi entropy offers a more flexible measure of information quantity compared to Shannon entropy. It introduces an adjustable parameter α, which can be used to enhance sensitivity to specific probability ranges. When α approaches 0, Renyi entropy treats all events equally; however, as α increases, it assigns greater weight to high-probability events, making it particularly effective in identifying certain types of anomalies, especially those caused by high-frequency abnormal behaviors. Conversely, it can emphasize low-probability events when α is smaller. The following is the mathematical representation of Renyi entropy

$$H_\alpha = \frac{1}{1-\alpha} \log\left(\sum_{i=1}^{W} p_i^\alpha\right)$$

In the field of information security, Renyi entropy has been applied in various scenarios. For example, Khodjaeva et al. [16] observed statistical distribution changes in network flow features within encrypted DNS traffic, such as packet length and flow direction, finding that Renyi entropy can effectively distinguish between normal and malicious behaviors. Even when only a small number of packets show abnormalities in the early stages of an attack, it can detect subtle statistical shifts and provide early warnings before the anomaly escalates. Yu et al. [17] proposed an anomaly detection model combining Renyi entropy with dynamic thresholding. This method uses an Exponentially Weighted Moving Average (EWMA) model to track real-time network traffic changes and determines anomalies based on variations in Renyi entropy. Experimental results indicate that this model maintains high accuracy and low false alarm rates when facing sudden or gradual traffic changes, demonstrating the potential of Renyi entropy in dynamic network environments.

Shannon entropy focuses on the average information content, while Tsallis entropy emphasizes the non-additive properties of systems. Renyi entropy offers a regulatory mechanism in statistical weighting, allowing emphasis or suppression of specific probability distributions, which is particularly advantageous for recognizing diverse attack patterns. Compared to Tsallis entropy, Renyi entropy is better suited for emphasizing high-probability anomalies and stable detection of subtle changes. Tsallis entropy, on the other hand, is more adept at handling long-tail distributions and low-frequency, high-risk events. Therefore, although both belong to the extended applications of generalized entropy measures, they can be flexibly chosen in practice according to the nature of the data and detection requirements.

According to the research by Unit 42 [18], common types of network attacks currently include ransomware and disruption attacks, phishing, cloud and supply chain attacks, AI-assisted automated attacks, insider threats, and nation-state intrusions, among others. These attack types differ in techniques, targets, and impact levels. Phishing and insider threats rely heavily on social engineering skills and human vulnerabilities, making them difficult to detect directly from traditional network traffic features. Conversely, ransomware and destructive intrusions targeting cloud resources are often accompanied by significant spikes in packet volume, abnormal commands, or resource deletions, making them more suitable as experimental targets for anomaly detection systems. Moreover, with the widespread use of AI by attackers, automated generation of phishing emails and malicious payloads has made attacks more scalable and covert, posing new challenges to existing cybersecurity defenses. Among various attack methods, scanning and DDoS attacks are frequently used as core test subjects for anomaly detection system development due to their distinctive behavioral patterns and observable characteristics, and their common appearance in large public datasets such as CICIDS and NSL-KDD.

Among common attack types, active scanning and probing behaviors are the most frequent preliminary steps for intrusion, aiming to identify open hosts, ports, and network services. Staniford et al. [19] noted that scanning traffic usually originates from a single or a few source IPs and sends connection requests to a large number of different destination hosts or ports within a short time frame. The spatial and temporal distributions of these requests differ significantly from normal usage scenarios. DDoS attacks are another common destructive attack type, aiming to exhaust the resources of the target host or network by sending a large volume of high-frequency requests from distributed sources, thereby denying legitimate users access. Lakhina et al. [20] pointed out that DDoS traffic features include sudden spikes in packet rates, increased concentration of connections to a single or few destination IPs, resulting in abnormal overall traffic distribution shifts, and potentially leading to degraded service availability.

3 System Architecture of Anomaly Detection

In this study, we selected Shannon entropy as the primary feature indicator based on several key considerations. First, Shannon entropy is computationally simple and efficient, making it well-suited for real-time processing in high-traffic network environments. Second, scanning attacks tend to cause a rapid expansion in the distribution of packet attributes such as source IP addresses and destination port numbers. This results in significant and stable changes in Shannon entropy values during such attacks, effectively capturing anomalous behaviors. Furthermore, compared to Renyi entropy, which requires additional parameter tuning, and Tsallis entropy, which is more suitable for detecting long-tail distributions, Shannon entropy provides clearer feature variation in the presence of abrupt distribution changes. Therefore, considering computational efficiency, feature sensitivity, and stability, this study adopts Shannon entropy as the core feature indicator for detecting scanning attacks (Fig. 1).

The following section provides a layered explanation of the system architecture, with a detailed introduction to each component block.

- Network Traffic Collector

The raw packets transmitted and managed by routers in the network environment are captured in real-time using tools like Wireshark and saved for subsequent analysis. These tools can record complete packet data, including fields such as source and destination IP addresses, transport layer protocols, packet size, timestamps, and more.

- Preprocessing

The captured raw packets are preprocessed, and to enable real-time detection of abnormal activities, feature data related to anomaly detection are extracted every 30 s for subsequent analysis. The features include source IP, source port, and timestamp.

- Traffic Data Set

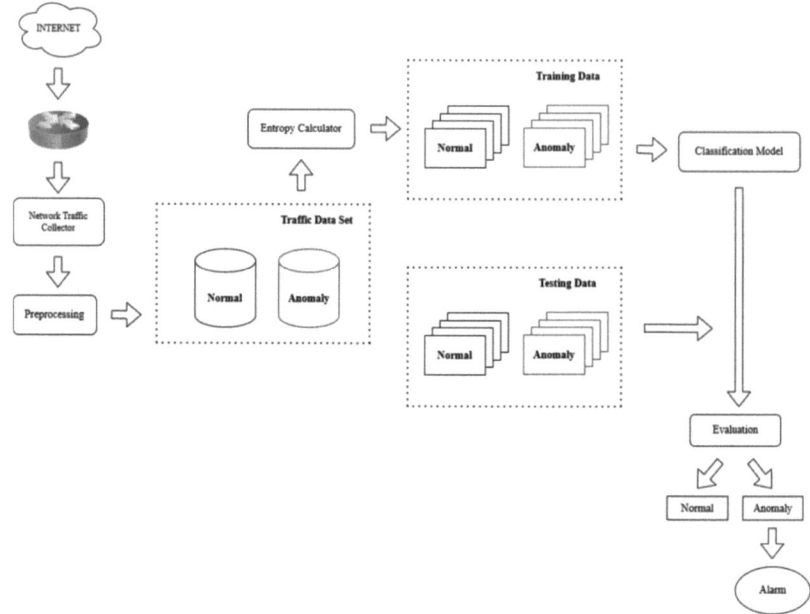

Fig. 1. An Entropy-Based Anomaly Detection Model

After preprocessing, the network traffic data is categorized and organized into a complete traffic dataset. In this study, the dataset is divided into two main categories based on behavioral characteristics: normal and anomalous. Normal traffic represents legitimate and expected network activities, such as browsing websites and sending or receiving emails. Anomalous traffic includes potentially malicious behaviors, such as DDoS attacks, scanning activities, or unauthorized access. These labeled data can be sourced from real network usage records of an academic institution. This classification provides labeled data for entropy calculation and model training, forming the foundation for supervised learning.

- Entropy Calculator

After dataset classification, entropy is calculated from statistical features, such as source IP, source port, packet count, and time distribution, to help identify abnormal behavior. Time distribution reflects how frequently packets occur over a period, indicating traffic concentration or disorder. This study uses Shannon Entropy to measure these patterns. Abnormal behaviors often cause sudden changes in feature values, leading to entropy fluctuations. These features are converted into entropy-based vectors and used as input for classification models. Different scanning behaviors affect feature distributions in distinct ways. In horizontal scanning, a single source IP sends packets to many destination IPs, causing high destination IP entropy but low source IP entropy. In vertical scanning, a single destination IP is targeted on multiple ports, increasing destination port entropy while source and destination IP entropy remain low. In hybrid scanning, the source IP stays constant while destination IPs span a subnet; this raises destination IP entropy and, if multiple ports are scanned, increases destination port entropy too. These scans alter the concentration or dispersion of IP and port distributions, producing clear entropy shifts that signal anomalies.

- Training Data

Based on the aforementioned traffic dataset, supervised learning training data is constructed using labeled normal and abnormal traffic flows, along with corresponding statistical features and entropy values. Prior to model training, the data undergoes feature extraction and selection, transforming it into representative vectors incorporating entropy and other features such as source/destination IPs and ports, packet size, protocol type, and packet frequency. Each sample is labeled to enable the model to learn the mapping between feature patterns and traffic categories. To ensure the model can accurately identify a range of abnormal behaviors, the training set includes diverse types of scanning attacks alongside a substantial volume of normal traffic. Given that dataset diversity and representativeness directly influence model performance and generalizability, particular care is taken to balance the distribution of normal and abnormal samples and avoid overrepresentation of specific attack types. This design minimizes model bias and supports more robust detection across varied scenarios.

- Classification Model

In the classification model, the system uses pre-labeled data for the "normal" and "anomalous" groups, enabling supervised learning. After converting the calculated entropy values into multidimensional feature vectors, along with packet quantity and other relevant features, a Random Forest classifier is employed to classify the entropy data. Instead of manually setting static thresholds for anomaly determination, the Random Forest model automatically learns the associations between entropy variations and anomalous behaviors from the training data. It then predicts the class of the feature vector based on the learned characteristics from normal and abnormal samples. Finally, the traffic flow is classified as either normal behavior or anomalous behavior according to the prediction results.

- Testing Data

The Testing Data is a subset of the overall Traffic Data Set that has been separated and not used for training the model. Its purpose is to objectively evaluate the classification model's generalization ability on unseen data. The testing data also includes two types of network traffic events: Normal and Anomalous. Testing Data shares the same feature format as the Training Data, allowing it to be input into the classification model for inference. This setup simulates real deployment scenarios, where the model's performance in judging real-time or historical network traffic is assessed. The evaluation verifies whether the model can effectively distinguish between normal and anomalous traffic, and quantifies its performance through metrics such as accuracy, precision, and recall. If the model performs well on the testing data, it is considered to have practical application value and can be further used in real-time anomaly detection and alert systems.

- Evaluation

After training, the system inputs testing data into the model to predict whether each traffic instance is normal or anomalous. Predictions are compared with true labels to assess the model's generalization ability. This step validates performance and supports anomaly detection. The system evaluates Accuracy, Precision, and Recall to measure performance from multiple perspectives. If anomalous traffic is detected, the system activates the alert module to notify administrators or initiate responses, providing real-time protection. This evaluation not only verifies performance but also integrates detection with practical application, forming a key part of the system architecture.

- Alarm

When the classification model identifies a network traffic instance as anomalous, the system immediately activates the alert mechanism as the final step of anomaly detection. Its main purpose is to promptly notify administrators to prevent or reduce potential damage from the abnormal behavior. Depending on deployment needs, alerts can be delivered through various channels, such as monitoring dashboards, email notifications, system logs, or visual prompts on interfaces. This module enables the system to move from passive detection to active defense, enhancing network security and real-time response efficiency.

4 Empirical Study

The experimental data come from actual network traffic of an academic institution provided, which uses packet mirroring between the campus network core and boundary firewall. This duplicates inbound and outbound traffic, recording it in PCAP format, preserving key packet details such as timestamps, source/destination addresses, protocols, and sizes for anomaly detection. Data covers both high-frequency daytime and low-frequency nighttime traffic to ensure diversity. To protect privacy, IP addresses are hashed to conceal real values while maintaining communication relationships, and application-layer content like user messages and credentials are excluded. The dataset also includes three types of simulated scanning attacks—horizontal, vertical, and subnet scans—automatically injected and logged within the real traffic, providing both normal and abnormal samples for accurate detection and classification evaluation.

This study captures key network variables every 10 s, divided into raw statistical features extracted from traffic and derived variables for anomaly detection. Basic variables include Source IP, Destination IP, Source Port, Destination Port, Packet Size, Packet Quantity, Timestamp, and Transport Layer Protocol Type, such as TCP, UDP, and ICMP. These discrete variables are used to calculate probability distributions and Shannon entropy values within each time window. Packet Quantity, which is a continuous variable, tracks traffic surges but is not included in the entropy calculations; its trends often correlate with entropy changes and can aid in judgment. The derived entropy variables consist of the entropy of Source IP, Destination IP, Source Port, Destination Port, and Packet Size, which serve as the primary indicators for anomaly detection in this study.

This study uses Shannon entropy as the main feature, calculating entropy values within 10-s windows for packet attributes such as Source IP, Destination IP, Source Port,

Destination Port, Protocol Type, and Packet Size. These values form a multidimensional feature vector. To detect anomalies, we apply the Random Forest classifier on labeled data containing normal traffic and simulated scanning behaviors, with anomalies clearly labeled by injection time and type. Model performance is evaluated using Accuracy, Precision, and Recall. Accuracy measures the overall correctness of the model, Precision indicates the correctness of detected anomalies, and Recall measures the proportion of actual anomalies identified.

Although the empirical results of this study are still in progress, the preliminary evaluation framework has been fully established. It is expected that using a Random Forest classification model to identify anomalies based on the Shannon Entropy values of packet attributes within each time window will successfully detect three types of simulated scanning behaviors and demonstrate strong performance across multiple classification metrics. This study also plans to further compare detection accuracy under two conditions: one using only packet statistical features for classification, and the other incorporating entropy-based features as well. It is anticipated that including entropy information will improve the detection of anomalous patterns, especially when attack behaviors have not yet caused significant changes in traffic volume. In such cases, changes in packet predictability can provide early warnings, leading to better detection accuracy than scenarios without entropy features.

5 Conclusion

The primary objective of this study is to explore the use of entropy in network anomaly detection and verify its effectiveness in identifying scanning attacks. Experiments show that entropy, which measures irregularity in network traffic, can effectively detect typical scanning behaviors. Its strength lies in quantifying traffic uncertainty, enabling better identification of abnormal patterns than traditional statistical or rule-based methods. We implemented an entropy-based detection system that accurately identifies scanning patterns in various traffic types by highlighting anomalies. Unlike fixed-threshold methods, entropy allows dynamic evaluation that adapts to traffic changes, reducing false positives and negatives.To validate entropy's impact, we conducted experiments comparing two models: one using entropy features and one without. The expected outcome is that the entropy-based model showed improved detection accuracy and better identification of subtle traffic changes. In contrast, the model without entropy features struggled with such variations. These findings support the use of entropy in network anomaly detection and highlight its potential as an effective tool in this field.

Although this study is expected to yield positive results, several limitations exist. First, the network traffic dataset mainly comes from an academic institution, which may not fully represent the diverse traffic of commercial or large enterprise networks. Thus, the generalizability of findings requires further validation with data from various domains. Second, a fixed 10-s time window was used for anomaly detection. This may lack sensitivity for slow or low-rate scanning attacks, as attack behaviors might not fully appear within such a short period, potentially missing anomalies during gradual attacks. Finally, while entropy effectively identifies scanning attacks, its effectiveness varies with attack type. Some application-layer or social engineering attacks may not cause noticeable changes in traffic statistics, making entropy less effective. Persistent or low-frequency attacks are also more covert and may not be well captured by entropy. Therefore, entropy's suitability depends on attack characteristics and is not universally applicable.

This study aims to validate the effectiveness of entropy in detecting certain network attacks and to reveal its application limitations. Future research can extend this work in several directions. First, it is suggested to examine the effect of multi-scale time windows on detection performance. A fixed time window may miss characteristics of persistent or low-frequency attacks. Adopting dynamic or sliding windows could improve sensitivity and stability. Second, while entropy captures randomness and distribution changes, it may struggle to detect subtle or concealed attacks like application-layer intrusions or APTs. Future work could combine entropy with other techniques, such as statistical models, time series analysis, deep learning, or semantic analysis, to improve adaptability to complex attacks. Finally, studies should expand data sources and application contexts, such as enterprise networks, cloud environments, and IoT devices, to evaluate entropy's generalizability. In large-scale, high-frequency settings, reducing entropy's computational cost while maintaining real-time detection will be key for practical deployment.

References

1. Check Point: The State of Cyber Security 2025. Check Point Software Technologies, Tel Aviv (2025)
2. European Union Agency for Cybersecurity: ENISA Threat Landscape. ENISA, Heraklion (2024)
3. Scarfone, K., Mell, P.: Guide to Intrusion Detection and Prevention Systems (IDPS). NIST Special Publication 800-94, NIST, Gaithersburg (2007)
4. Shannon, C.E.: A Mathematical Theory of Communication. Bell Syst. Technical J. **27**, 379–423, 623–656 (1948)
5. Barnett, V., Lewis, T.: Outliers in Statistical Data, 3rd edn. Wiley, Chichester (1994)
6. Yu, B., Zhang, Y., Xie, W., Zuo, W., Zhao, Y., Wei, Y.: A Network Traffic Anomaly Detection Method Based on Gaussian Mixture Model. Electronics **12**(6), 1397 (2023)
7. Thudumu, S., Branch, P., Jin, J., Singh, J.: A Comprehensive Survey of Anomaly Detection Techniques for High Dimensional Big Data. Journal of Big Data **7**(1), 42 (2020)
8. Zhou, Y., Zeng, H., Zheng, Z., Zhang, W.: Network Traffic Anomaly Detection Model Based on Feature Grouping and Multi-Autoencoders Integration. Electron. Lett. **60**(23), e70103 (2024)
9. Psychogyios, K., Papadakis, A., Bourou, S., Nikolaou, N., Maniatis, A., Zahariadis, T.: Deep Learning for Intrusion Detection Systems (IDSs) in Time Series Data. Future Internet **16**(3), 73 (2024)
10. Bereziński, P., Jasiul, B., Szpyrka, M.: An Entropy-Based Network Anomaly Detection Method. Entropy **17**(4), 2367–2408 (2015)
11. Hassan, A.I., Abd El Reheem, E., Guirguis, S.K.: An Entropy and Machine Learning Based Approach for DDoS Attacks Detection in Software Defined Networks. Scientific Rep. **14**(1), 18159 (2024)
12. Ibrahim, J., Gajin, S.: Entropy-Based Network Traffic Anomaly Classification Method Resilient to Deception. Comput. Sci. Inf. Syst. **19**(1), 87–116 (2021)

13. Tsallis, C.: Possible generalization of Boltzmann-Gibbs statistics. J. Stat. Phys. **52**(1–2), 479–487 (1988)
14. Renyi, A.: On Measures of Entropy and Information. In: Neyman, J. (ed.) Proceedings of the Fourth Berkeley Symposium on Mathematical Statistics and Probability, vol. 1, pp. 547–561. University of California Press, Berkeley (1961)
15. Berezinski, P., Jasiul, B., Szpyrka, M.: An Entropy-Based Network Anomaly Detection Method. Entropy **17**(4), 2367–2408 (2015)
16. Khodjaeva, Y., Zincir-Heywood, N.: Network Flow Entropy for Identifying Malicious Behaviours in DNS Tunnels. In: Reinhardt, D., Müller, T. (eds.) ARES 2021, p. 72:1–72:7. ACM, New York (2021)
17. Yu, H., Yang, W., Cui, B.: Renyi Entropy-Driven Network Traffic Anomaly Detection with Dynamic Threshold. Cybersecurity **7**, 64 (2024)
18. Unit 42: 2025 Incident Response Report, https://www.paloaltonetworks.com/resources/research/unit-42-incident-response-report. Accessed 12 May 2025
19. Staniford, S., Hoagland, J.A., McAlerney, J.: Practical Automated Detection of Stealthy Portscans. In: Proceedings of the 7th USENIX Security Symposium, pp. 251–270. USENIX Association (2002)
20. Lakhina, A., Crovella, M., Diot, C.: Mining Anomalies Using Traffic Feature Distributions. ACM SIGCOMM Computer Communication Review **35**(4), 217–228 (2005)

Designing an Agentic AI–Driven Framework for Decision-Making in Industrial Maintenance

Miguel Alonso-González[✉][iD], Vicente García-Díaz[iD], and Benjamín López Pérez[iD]

Department of Computer Science, University of Oviedo, 33007 Oviedo, Spain
alonsomiguel@uniovi.es

Abstract. Maintenance has become critically important for industrial sectors due to the increasing complexity of interactions among various production activities. The shift from predictive maintenance to prescriptive maintenance (PsM) marks a significant evolution, as PsM goes beyond forecasting to provide specific, actionable recommendations. It prescribes what actions should be taken, when they should be performed, and how they should be executed, leveraging advanced analytics and machine learning (ML) techniques. Prescriptive maintenance is considered the most advanced and intelligent form of maintenance, offering a holistic approach that integrates planning, decision-making, and adaptive learning within the context of smart manufacturing. This transition introduces a new perspective on flexibility, customization, and resilience in production planning. These capabilities are made possible through the application of advanced data analytics and ML models, which are capable of processing vast amounts of data from both IoT (Internet of Things) devices and historical records, thereby enhancing the precision and reliability of proposed maintenance interventions. To address these objectives, this work proposes a decision-making framework whose core foundation integrates agentic artificial intelligence (Agentic AI) with the digital twin paradigm. Compared to alternative approaches, the integration of Digital Twin technology and Agentic AI represents a notable advancement in the field of prescriptive maintenance systems. This integration enables the creation of virtual testing environments where maintenance strategies can be simulated, evaluated, and optimized prior to their implementation in real-world industrial contexts.

Keywords: Prescriptive maintenance · agentic AI · production planning · intelligent agents · conceptual framework · machine learning · reinforcement learning · IoT · predictive maintenance

1 Introduction

Intelligent maintenance has emerged as a consequence of Industry 4.0. Thanks to the widespread use of machine learning (ML) techniques in this research area, several intelligent maintenance paradigms have been developed, primarily Condition-based Maintenance (CBM) and Predictive Maintenance (PdM) [18]. However, Prescriptive Maintenance (PsM) has emerged as a challenge within the scope of intelligent maintenance. PsM is an emerging approach in Industry 4.0 that leverages cyber-physical production systems to transform maintenance management into automated decision-support systems [3].

The emergence of intelligent maintenance is closely linked to the advancement of Industry 4.0, characterized by the widespread adoption of digital technologies, advanced sensing, and machine learning (ML) techniques that enable more efficient and sophisticated use of data. This context has driven the evolution of maintenance paradigms, shifting from reactive approaches toward increasingly proactive and data-driven strategies: 1) **Condition-Based Maintenance (CBM):** Focuses on monitoring operational parameters (such as temperature, vibrations, acoustic emissions, etc.) to detect abnormal conditions that may indicate accelerated wear or the likelihood of an impending failure. Decision-making is based on thresholds or predefined rules according to the asset's condition, allowing interventions to be triggered when significant deterioration is observed [9]. 2) **Predictive Maintenance (PdM):** Employs machine learning or statistical models to predict the probable time of failure, estimate the remaining useful life (RUL), or assess the future evolution of specific health indicators of equipment. PdM combines historical data with prediction algorithms to provide early warnings regarding maintenance needs, thereby reducing unexpected downtimes and optimizing resource planning [23]. 3) **Prescriptive Maintenance (PsM):** Represents the next frontier in maintenance digitalization. It not only predicts when and how a failure might occur, but also recommends concrete actions to optimize both asset availability and overall plant operations. PsM integrates optimization techniques, simulation, and scenario analysis to suggest the most effective response to failure risks: for example, determining the most cost-effective time to intervene, identifying which spare parts to stock, or even adjusting production parameters to avoid greater damage [9].

Maintenance processes and Production Planning and Control (PPC) maintain a balance whose direct impact is reflected in operational efficiency and the responsiveness of industrial enterprises. The synergy between maintenance and PPC is a key enabler of smart manufacturing. While PPC aims to optimize resource allocation and ensure demand fulfillment, maintenance ensures the reliability and availability of equipment. Although maintenance activities may temporarily disrupt production and affect pre-established plans, it is precisely their proper integration with PPC that ensures these interruptions are minimal, planned, and effectively controlled.

In recent years, digital twins have become a prominent concept within Industry 4.0, particularly in the design and development of industrial assets, where multiple research fields converge, including the Internet of Things (IoT), simulation, and artificial intelligence. The digital twin of an industrial asset collects real-time data from the asset's sensors to build a dynamic digital model that mirrors its behavior and condition. This ability to synchronize the physical reality with its virtual counterpart enhances the management of manufacturing systems and processes [13].

Agentic AI is a novel approach to artificial intelligence systems designed to autonomously complete complex and long-duration tasks without direct human supervision, by assimilating context and making independent decisions—systems that exhibit "agency" [1]. Thanks to this capability, they can adapt to changing environments, manage unforeseen situations, and continuously optimize performance. What fundamentally distinguishes these systems is their autonomy and their capacity to adapt to goal-oriented processes. In industrial contexts, Agentic AI leverages the capabilities of predictive maintenance systems to continuously monitor machine conditions, detect and analyze patterns indicating potential bearing failures [2], and subsequently implement the necessary corrective actions on the production line autonomously, without requiring direct human intervention in the process.

This research proposes a conceptual model for an intelligent decision-making system for industrial maintenance, driven by the principles of Agentic AI and the Digital Twin. In this model, maintenance strategies are tested and refined within a virtual environment before being translated into actionable plans.

The remainder of this paper is organized as follows: Sect. 2 briefly presents a review of the scientific literature on intelligent industrial maintenance. Section 3 introduces the conceptual framework and details the architecture of the decision-making system. Section 4 describes a use case that applies the proposed framework. Finally, Sect. 5 outlines the conclusions and directions for future work.

2 Literature Review

Recent research has explored prescriptive maintenance (PsM) within Industry 4.0 contexts [20,22]. The PriMa model integrates data management, predictive analytics, and semantic reasoning to optimize maintenance planning in cyber-physical production systems (CPPS) [3]. Reinforcement learning (RL) approaches have been developed for multi-agent maintenance scheduling, outperforming traditional methods in failure prevention and downtime reduction [17].

Furthermore, deep learning techniques are being applied to adaptive fault prediction and maintenance, incorporating novel feature extraction and gating modules to enhance accuracy in complex manufacturing environments [16]. Additionally, a sensor-based learning approach has been proposed to estimate time-dependent parameters for prescriptive analytical models in dynamic environments. This method addresses challenges related to uncertainty, non-stationary data, and sensor noise in streaming environments, enabling more reliable decision-making [4].

PsM is an advanced approach to equipment maintenance that leverages machine learning and artificial intelligence to optimize maintenance schedules and reduce costs. Recent studies have explored various aspects of PsM, including the use of reinforcement learning (RL) for sequential decision-making in maintenance management [12]. Regarding the integration of predictive maintenance

(PdM) and production planning using advanced techniques, Hassan Dehghan Shoorkand et al. [8] propose a deep learning approach combining long short-term memory (LSTM) models. They introduce a real-time optimization system that jointly optimizes maintenance planning and production scheduling in response to factors such as incoming jobs, unexpected due date changes, machine degradation, random failures, minimal repairs, and condition-based maintenance.

Mageed Ghaleb et al. [11] present a real-time optimization system based on genetic algorithms and hybrid rescheduling policies. Ana Esteso et al. [10] review the applications of reinforcement learning in production planning and control, highlighting promising results compared to traditional methods, particularly for problems involving uncertainty or nonlinear characteristics. These studies demonstrate the potential of adaptive learning techniques to enhance integrated PdM and production planning, improving efficiency and reducing costs in manufacturing environments.

A holistic framework for PsM has also been proposed, integrating various components such as maintenance needs analysis, predictive technologies, and knowledge management [7]. This comprehensive approach aims to minimize unplanned equipment downtime while optimizing costs, representing a significant advancement over traditional reactive and proactive maintenance strategies.

Other recent studies explore the integration of Prescriptive Maintenance (PsM) and Production Planning and Control (PPC) using adaptive learning techniques in smart manufacturing environments. Padovano et al. [15] present a decision support system prototype that leverages digital twins to integrate PsM and PPC in a make-to-order manufacturing context. Chiurco et al. [6] introduce a workflow-based architecture for developing intelligent digital twins, employing Deep Q-Networks (DQN) for adaptive PPC.

The literature review reveals a set of characteristics that define a Digital Twin (DT). In general terms, all sources agree that a DT is a digital representation of the physical world that may encompass models of the structure, functionalities, and behavior of its real-world counterpart [13]. The success and widespread adoption of Digital Twins (DT) in the industrial sector [21] is largely attributed to the continuous interaction and persistent connection between the physical entity and its digital replica.

In summary, the Digital Twin offers a virtual reproduction that not only faithfully mirrors the physical asset but also enables continuous learning through simulation capabilities. In this way, monitoring, simulating, and predicting the behavior of machines and processes becomes a key approach to optimizing and maintaining physical assets within Industry 4.0.

Agentic AI refers to AI systems designed to operate autonomously, with minimal human supervision, and capable of planning, decision-making, and adapting to changing contexts [19]. Unlike traditional language models, where responses fall within a predefined scope, these systems have the ability to redefine their goals and actions as they receive new information or feedback [14]. To achieve this, they integrate the power of large language models with modular mechanisms, such as real-time search tools and connections to specialized databases [5].

Thanks to this architecture, Agentic AI systems can verify and update their responses based on current data, which is particularly valuable in applications requiring up-to-date and system-specific information. For instance, such systems can continuously monitor the operational availability of equipment or entire production lines, seamlessly integrating data from predictive maintenance systems [1].

The future evolution of Agentic AI in smart manufacturing will critically depend on addressing these limitations while leveraging the lessons learned, with particular emphasis on improving interoperability with existing systems and developing feedback mechanisms that incorporate human oversight to ensure optimal and ethical operation [1].

The aim of this article is to propose a decision-making system architecture that integrates Digital Twins and Agentic AI as feedback and continuous improvement components within a conceptual model of intelligent maintenance.

3 Conceptual Model for a Decision Support Agent

This paper proposes an innovative architecture for an advanced decision-making agent in the context of industrial maintenance, integrating Digital Twin technologies with Agentic AI.

The proposed approach aims to deliver a comprehensive framework for maintenance and production planning, with prescriptive capabilities and adaptive learning.

The architecture, illustrated in Fig. 1, is built upon two main pillars: 1) **Decision-Making System**, which includes a diagnostic module that collects failure predictions (PdM), and a recommendation module that acts on dynamic planning and integration with MES (Manufacturing Execution Systems), ERP (Enterprise Resource Planning), and CMMS (Computerized Maintenance Management System) platforms. 2) **Feedback and Continuous Improvement System.** This system is composed of a Digital Twin and Agentic AI, and is designed to simulate, evaluate, and refine in real time the maintenance and production actions recommended by the Agentic AI. It is tightly interconnected with the diagnostic module.

These two systems are coupled with the acquisition, preprocessing, and predictive modeling layers of the diagnostic module to create a system capable of evolving from classical PdM to a prescriptive maintenance (PsM) approach with direct integration into production planning.

3.1 Decision-Making System

The core modules are structured around the Diagnostic Module (PdM), which predicts failures with associated probabilities, estimates the remaining useful life (RUL) of components, and evaluates criticality to prioritize interventions.

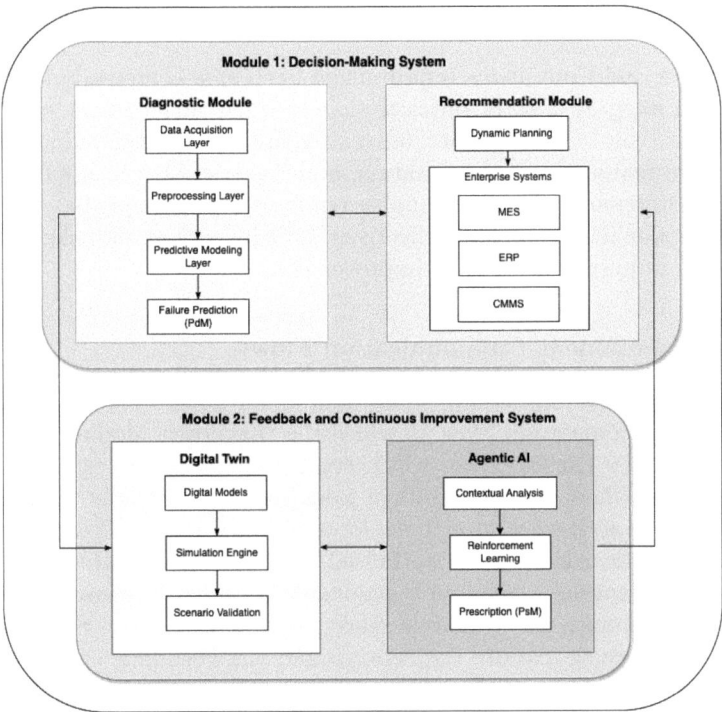

Fig. 1. Architecture of the Decision Support Agent.

This module is organized into several fundamental system layers, including: the Data Acquisition Layer, responsible for real-time information collection through IoT sensors, the use of SCADA (Supervisory Control and Data Acquisition) systems for monitoring and control, and the consolidation of historical data to analyze long-term trends. Next, the Preprocessing Layer handles data cleaning and normalization, feature engineering to extract relevant indicators, and information quality validation, ensuring data reliability. Finally, the Predictive Modeling Layer applies machine learning or deep learning (ML/DL) algorithms to identify patterns, perform trend analysis to detect progressive degradation, and conduct anomaly detection to recognize unusual behaviors.

On the other hand, the Recommendation Module focuses on optimizing maintenance resources, managing dynamic planning that adapts to changing conditions, and establishing priorities based on operational impact. Finally, Integration with Enterprise Systems is achieved through MES, which adjusts production scheduling; ERP, which manages resources and costs; and CMMS, which coordinates maintenance activities.

3.2 Feedback and Continuous Improvement System

The Feedback and Continuous Improvement System is composed of the Digital Twin, which integrates 3D or physical models of industrial assets, enabling the simulation of their behavior under different conditions and the validation of scenarios before real-world implementation; and Agentic AI, which autonomously generates maintenance strategies, applies reinforcement learning based on historical outcomes, and evaluates the effectiveness of its actions to make continuous adjustments in pursuit of ongoing improvement.

3.3 Intercomponent Communication Flows

The architecture establishes five main interaction flows between components: 1) Predictive-to-Simulation Flow, in which the Diagnostic Module sends failure predictions to the Digital Twin, which creates scenarios to test strategies; 2) Learning Flow, where the Digital Twin feeds simulation results to the Agentic AI, which refines strategies through reinforcement learning; 3) Prescriptive Flow, where the Agentic AI generates optimized strategies for the Recommendation Module, which translates them into actionable plans; 4) Implementation Flow, where the Recommendation Module sends plans to Enterprise Systems (MES, ERP, CMMS), which execute the actions; and the Feedback Loop, where the outcomes of the implemented actions generate new data that is fed back into the Diagnostic Module.

Operational flows between architectural components, the Digital Twin and Agentic AI represents a significant advancement in prescriptive maintenance systems, as it enables the creation of a virtual "sandbox" where strategies can be evaluated and refined prior to deployment in real-world environments. This synergy offers multiple advantages: 1) Risk-Free Optimization: The Digital Twin–Agentic AI tandem enables complete trial-and-error cycles to be executed without affecting actual production. The Digital Twin provides an accurate representation of equipment behavior, while Agentic AI can test various maintenance strategies under different conditions. This means that even highly innovative or potentially risky solutions can be evaluated without compromising operations, equipment, or safety. 2) Adaptive Learning Acceleration: The simulation environment acts as an accelerator for Agentic AI learning. In physical systems, collecting sufficient data to train advanced algorithms could take months or years, especially for infrequent events such as catastrophic failures. The Digital Twin can simulate thousands of scenarios in compressed time, allowing the Agentic AI to accumulate "experience" much more quickly and develop more robust models.

This architecture facilitates the evolution from predictive maintenance (PdM) to prescriptive maintenance (PsM) through a multi-layered integration framework. The system implements vertical data flow mechanisms connecting failure prediction algorithms with enterprise resource planning systems, enabling automated maintenance scheduling and resource allocation.

4 Case Study: Prescriptive Maintenance and Adaptive Planning in an Office Chair Assembly Line

This use case is set in a factory specialized in the production of ergonomic office chairs, which manufactures three product lines: Executive Model ($\mathcal{E}$), Operational Model ($\mathcal{O}$), and Gaming Model ($\mathcal{G}$). The plant consists of four main workstations, defined as the set $\mathcal{S} = S_1, S_2, S_3, S_4$: 1) S_1: **Frame Assembly Station** – Assembles the metal skeleton and mechanical components. 2) S_2: **Upholstery Station** – Applies foam, fabrics, and finishing materials. 3) S_3: **Mechanism Installation Station** – Installs lifting, reclining, and adjustment systems. 4) S_4: **Quality Control Station** – Verifies functionality and aesthetic aspects.

Production follows a mixed model that combines make-to-order (for corporate clients) and make-to-stock (for distributors). The processing times $T(m, s)$ for each model $m \in \mathcal{E}, \mathcal{O}, \mathcal{G}$ at each station $s \in \mathcal{S}$ are defined in Table 1 as:

Table 1. Processing Times and Daily Production per Model

Model	$T(m, S_1)$	$T(m, S_2)$	$T(m, S_3)$	$T(m, S_4)$	Daily Production $P(m)$
$\mathcal{E}$	20 min	45 min	25 min	10 min	40–50 units
$\mathcal{O}$	15 min	30 min	15 min	8 min	70–90 units
$\mathcal{G}$	25 min	50 min	30 min	15 min	30–40 units

4.1 Operational Challenges

The production line faces several specific challenges: 1) The upholstery station S_2 represents a critical bottleneck, with industrial sewing machines $M_1, M_2, ..., M_k \in \mathcal{M}_{S_2}$ requiring frequent maintenance. 2) The availability of special materials (fabrics, foam, etc.) is variable and follows a probability distribution $p(m_i, t)$, which indicates the likelihood of availability for material m_i at time t. 3) Model switching involves significant setup times, denoted as $C(m_i, m_j)$, to transition from model m_i to model m_j, particularly in the upholstery and mechanism installation stations. 4) Corporate orders have strict delivery deadlines D_i, with penalties for delays defined by the function $\text{Pen}(t) = \alpha \cdot \max(0, t - D_i)^2$. Where:

- t represents the actual delivery time.
- D_i is the committed delivery date for order i.
- α is a penalty coefficient that may vary depending on the client type or contract:

$$\alpha = \begin{cases} \alpha_{\text{premium}} = 150 \text{ u.m.}/\text{day}^2, & \text{if } i \in \mathcal{C}_{\text{premium}} \\ \alpha_{\text{standard}} = 80 \text{ u.m.}/\text{day}^2, & \text{if } i \in \mathcal{C}_{\text{standard}} \\ \alpha_{\text{retail}} = 40 \text{ u.m.}/\text{day}^2, & \text{if } i \in \mathcal{C}_{\text{retail}} \end{cases} \quad (1)$$

– The function $\max(0, t - D_i)$ ensures that penalties are only applied when $t > D_i$ (i.e., for late deliveries).

4.2 Implementation of the Decision Agent

Decision-Making System

- **Diagnostic Module: 1) Acquisition Layer:** Collects sensor data from sewing machines (temperature, vibrations, electrical consumption), stitch counters, and data from the SCADA system on line speeds. That is, it collects a state vector $\boldsymbol{X}(t) = [x_1(t), x_2(t), ..., x_n(t)]$ where each x_i represents an operational parameter of the machines. **2) Preprocessing Layer:** Normalizes data, detects anomalous stitch patterns, and monitors sewing quality through machine vision. Transforms $\boldsymbol{X}(t)$ into a feature vector $\boldsymbol{F}(t) = \Phi(\boldsymbol{X}(t))$ where Φ is a transformation function. **3) Predictive Modeling Layer:** Employs ML algorithms to predict sewing machine failures 24 h in advance and estimates the useful life of needles and feed mechanisms. Predicts the probability of failure $p(f_i|\boldsymbol{F}(t))$ for each component i and estimates the time to failure $\text{TTF}_i = g(\boldsymbol{F}(t))$.
- **Recommendation Module: 1) Dynamic Planning:** Adjusts production sequences and operator allocation according to predictions. Defines an optimal sequence $\sigma^* = \arg\min_{\sigma \in \Sigma} J(\sigma)$ where J is a cost function and Σ is the set of all possible sequences. **2) Enterprise Systems:** MES: Coordinates changes in production scheduling; ERP: Manages material inventories and prioritizes orders; CMMS: Schedules preventive maintenance interventions. Integration through a mapping function $\psi : \sigma^* \rightarrow (A_{\text{MES}}, A_{\text{ERP}}, A_{\text{CMMS}})$ that translates the optimal sequence into specific actions for each system.

Feedback and Continuous Improvement System

- **Digital Twin:** Models the behavior of the complete line, with special attention to upholstery machines. Simulates different line configurations and production/maintenance strategies. Models the line behavior through a set of differential equations $\frac{d\boldsymbol{S}}{dt} = f(\boldsymbol{S}, \boldsymbol{A}, t)$ where $\boldsymbol{S}$ is the system state vector and $\boldsymbol{A}$ is the action vector. Simulates different scenarios $\{E_1, E_2, ..., E_n\}$ and evaluates their performance using an evaluation function $V(E_i)$.
- **Agentic AI:** Applies reinforcement learning algorithms to optimize planning decisions, evaluates the impact of different strategies on key indicators (OEE, deadline compliance, costs) and recommends specific actions to balance production and maintenance. 1) Uses a policy $\pi : \mathcal{S} \rightarrow \mathcal{A}$ that maps states to actions. 2) Optimizes the policy through reinforcement learning according to $\pi^* = \arg\max_\pi \mathbb{E}[R|\pi]$ where R is the accumulated reward. Where:
  - $\pi : \mathcal{S} \rightarrow \mathcal{A}$ is a policy that maps states to actions
  - π^* represents the optimal policy
  - $\mathbb{E}[R|\pi]$ is the expected value of the accumulated reward R under policy π

  3) Generates recommendations based on the optimal policy π^*.

4.3 Interaction Flow in a Specific Scenario

1. **Initial Event:** At time t_0, the Diagnostic Module detects anomalous patterns in machines M_3 and M_7, with $p(f_{M_3} \mid \boldsymbol{F}(t_0)) = 0.92$ and $\text{TTF}_{M_3} = 18$ hours.
2. **Predictive Flow → Simulation:** The Diagnostic Module sends the triplet $(M_3, p(f_{M_3}), \text{TTF}_{M_3})$ to the Digital Twin. The Digital Twin simulates the impact of these failures on the objective function J. For example, it evaluates their effect on the scheduled production, which includes an urgent order of 120 Executive chairs for delivery.
3. **Learning Flow:** The Digital Twin generates multiple alternative scenarios, such as: Immediate intervention on both machines, Staggered intervention (one machine now, the other at the end of the shift), Night-shift intervention on both machines and Temporary adjustment of operational parameters to extend remaining useful life. Agentic AI evaluates each scenario by considering delivery deadlines, maintenance costs, and final product quality. The Digital Twin generates a scenario set $\mathcal{E} = \{E_1, E_2, ..., E_n\}$. Agentic AI evaluates each scenario by computing $V(E_i) = \mathbb{E}[J \mid E_i]$.
4. **Prescriptive Flow:** Agentic AI determines that the optimal strategy consists of: 1) Immediately rescheduling the production sequence to complete the urgent Executive chair order first. 2) Redistributing the workload to more reliable sewing machines, Scheduling preventive maintenance on both machines during the night shift and Temporarily reducing sewing speed by 15% to lower mechanical stress. Formally, Agentic AI selects the optimal strategy $E^* = \arg\max_{E_i \in \mathcal{E}} V(E_i)$. The optimal strategy is decomposed into a tuple of actions (a_1, a_2, a_3, a_4) where:
   - a_1: Rescheduling of the production sequence
   - a_2: Workload redistribution based on $W(M_i) = \frac{\text{TTF}_i}{\sum_j \text{TTF}_j} \cdot W_{\text{total}}$
   - a_3: Preventive maintenance scheduled within the optimal interval $[t_s, t_e]$
   - a_4: Adjustment of operational parameters according to the function $\phi(p) = p \cdot (1 - \delta)$ with $\delta = 0.15$
5. **Implementation Flow:** The Recommendation Module translates E^* into concrete actions through the function ψ:
   - $\psi(E^*)_{\text{MES}}$: Reorganization of the production schedule
   - $\psi(E^*)_{\text{CMMS}}$: Scheduling of the technical maintenance team
   - $\psi(E^*)_{\text{ERP}}$: Adjustment of material allocation
6. **Feedback Loop:** New data is collected, updating $\boldsymbol{X}(t)$ to $\boldsymbol{X}(t + \Delta t)$. The prediction error is calculated as $\epsilon = \|\text{TTF}_{\text{actual}} - \text{TTF}_{\text{predicted}}\|$. This error is used to update the predictive models via the operator $\Theta : (f, \epsilon) \to f'$, which adjusts model f into a refined version f'.

4.4 Demonstrated Advantages

This case illustrates the specific advantages of the system through quantifiable metrics: 1) **Reduction in unproductive times**: $\text{DT}_{\text{new}} = (1 - \alpha) \cdot \text{DT}_{\text{old}}$ where $\alpha = 0.37$, meaning a 37% reduction in downtime. 2) **OEE optimization**:

OEE = Availability × Performance × Quality improved by a factor of 1.23 (23% improvement). 3) **Production-quality balance**: The system finds the optimal point by solving $\max_p \text{Prod}(p) \cdot \text{Qual}(p)$ subject to operational constraints. 4) **Adaptation to variability:** The sensitivity function $S(p) = \frac{\partial J}{\partial p}$ allows identification of the most critical parameters for adaptation. 5) **Domain-specific learning:** The learning curve follows $L(t) = L_\infty \cdot (1 - e^{-\lambda t})$ where L_∞ is the asymptotic performance.

The implementation of this architecture allows the chair factory to evolve from a reactive approach (responding to failures) and basic predictive approach (anticipating failures) towards a truly prescriptive one, where the system intelligently orchestrates the entire operation to simultaneously optimize production, quality, and maintenance. This architecture will enable a quantifiable improvement defined by the triplet: (ΔAvailability, ΔCompliance, ΔCosts) = $(0.23, 0.18, -0.15)$, representing improvements of 23% in availability, 18% in delivery compliance, and 15% reduction in operating

5 Conclusions

The integration of Digital Twins with Agentic AI for prescriptive maintenance represents a truly innovative proposal that is currently unknown in both academic and industrial domains. While existing systems incorporate isolated elements of these technologies, they lack the full integration and level of autonomy that characterize our approach.

The absence of similar solutions is evident in the current scientific literature and available commercial products, where fragmented approaches prevail and fail to fully harness the transformative potential of this technological combination. Existing platforms tend to limit their scope to passive simulation or basic predictive analysis, without reaching the level of autonomous prescription envisioned by our proposal.

The conceptual model presented establishes a continuous feedback loop in which the Digital Twin provides accurate virtual representations of physical assets, while Agentic AI components autonomously analyze, learn, and optimize maintenance strategies. This architecture goes beyond traditional predictive maintenance approaches by incorporating dynamic learning and proactive decision-making capabilities. The system not only anticipates potential failures in industrial equipment but also continuously evolves by assimilating new operational data, refining its predictive models and intervention strategies.

In the near future, this approach is expected to be implemented in a real manufacturing environment, with the aim of developing innovative capabilities for intelligent agents that support decision-making in an increasingly effective and autonomous manner.

Disclosure of Interests. The authors have no competing interests to declare that are relevant to the content of this article.

References

1. Acharya, D.B., Kuppan, K., Divya, B.: Agentic AI: autonomous intelligence for complex goals—a comprehensive survey. IEEE Access **13**, 18912–18936 (2025). https://doi.org/10.1109/ACCESS.2025.3532853. https://ieeexplore.ieee.org/document/10849561/
2. Alonso-González, M., Díaz, V.G., Pérez, B.L., G-Bustelo, B.C.P., Anzola, J.P.: Bearing fault diagnosis with envelope analysis and machine learning approaches using CWRU dataset. IEEE Access **11**, 57796–57805 (2023). https://doi.org/10.1109/ACCESS.2023.3283466. https://ieeexplore.ieee.org/document/10145440/
3. Ansari, F., Glawar, R., Nemeth, T.: PriMa: a prescriptive maintenance model for cyber-physical production systems. Int. J. Comput. Integr. Manufact. **32**(4-5), 482–503 (2019). https://doi.org/10.1080/0951192X.2019.1571236. https://www.tandfonline.com/doi/abs/10.1080/0951192X.2019.1571236
4. Bousdekis, A., Papageorgiou, N., Magoutas, B., Apostolou, D., Mentzas, G.: Sensor-driven learning of time-dependent parameters for prescriptive analytics. IEEE Access **8**, 92383–92392 (2020). https://doi.org/10.1109/ACCESS.2020.2994933
5. Chan, A., et al.: Harms from increasingly agentic algorithmic systems. In: ACM International Conference Proceeding Series, vol. 16, no. 23, pp. 651–666 (2023). https://doi.org/10.1145/3593013.3594033. https://dl.acm.org/doi/10.1145/3593013.3594033
6. Chiurco, A., Elbasheer, M., Longo, F., Nicoletti, L., Solina, V.: Data modeling and ML practice for enabling intelligent digital twins in adaptive production planning and control. Procedia Comput. Sci. **217**, 1908–1917 (2023). https://doi.org/10.1016/J.PROCS.2022.12.391
7. Choubey, S., Benton, R., Johnsten, T.: Prescriptive equipment maintenance: a framework. In: Proceedings - 2019 IEEE International Conference on Big Data, Big Data 2019, pp. 4366–4374 (2019). https://doi.org/10.1109/BIGDATA47090.2019.9006213
8. Dehghan Shoorkand, H., Nourelfath, M., Hajji, A.: A deep learning approach for integrated production planning and predictive maintenance. Int. J. Prod. Res. **61**(23), 7972–7991 (2023). https://doi.org/10.1080/00207543.2022.2162618. https://www.tandfonline.com/doi/abs/10.1080/00207543.2022.2162618
9. Errandonea, I., Beltrán, S., Arrizabalaga, S.: Digital twin for maintenance: a literature review. Comput. Ind. **123**, 103316 (2020). https://doi.org/10.1016/J.COMPIND.2020.103316
10. Esteso, A., Peidro, D., Mula, J., Díaz-Madroñero, M.: Reinforcement learning applied to production planning and control. Int. J. Prod. Res. **61**(16), 5772–5789 (2023). https://doi.org/10.1080/00207543.2022.2104180. https://www.tandfonline.com/doi/abs/10.1080/00207543.2022.2104180
11. Ghaleb, M., Taghipour, S., Zolfagharinia, H.: Real-time integrated production-scheduling and maintenance-planning in a flexible job shop with machine deterioration and condition-based maintenance. J. Manufact. Syst. **61**, 423–449 (2021). https://doi.org/10.1016/J.JMSY.2021.09.018
12. Goby, N., Brandt, T., Neumann, D.: Deep reinforcement learning with combinatorial actions spaces: an application to prescriptive maintenance. Comput. Ind. Eng. **179**, 109165 (2023). https://doi.org/10.1016/J.CIE.2023.109165
13. Grieves, M., Vickers, J.: Digital twin: mitigating unpredictable, undesirable emergent behavior in complex systems. In: Kahlen, F.-J., Flumerfelt, S., Alves, A. (eds.)

Transdisciplinary Perspectives on Complex Systems, pp. 85–113. Springer, Cham (2017). https://doi.org/10.1007/978-3-319-38756-7_4
14. Maia, E.H.B., Assis, L.C., de Oliveira, T.A., da Silva, A.M., Taranto, A.G.: Structure-based virtual screening: from classical to artificial intelligence. Front. Chem. **8**, 481382 (2020). https://doi.org/10.3389/FCHEM.2020.00343/PDF. www.frontiersin.org
15. Padovano, A., Longo, F., Nicoletti, L., Gazzaneo, L., Chiurco, A., Talarico, S.: A prescriptive maintenance system for intelligent production planning and control in a smart cyber-physical production line. Procedia CIRP **104**, 1819–1824 (2021). https://doi.org/10.1016/J.PROCIR.2021.11.307
16. Pang, J.L.: Adaptative fault prediction and maintenance in production lines using deep learning. Int. J. Simul. Model. (2023). https://doi.org/10.2507/IJSIMM22-4-CO20. https://doi.org/10.2507/IJSIMM22-4-CO20
17. Ruiz Rodríguez, M.L., Kubler, S., de Giorgio, A., Cordy, M., Robert, J., Le Traon, Y.: Multi-agent deep reinforcement learning based predictive maintenance on parallel machines. Robot. Comput.-Integr. Manufact. **78**, 102406 (2022). https://doi.org/10.1016/J.RCIM.2022.102406
18. Shaheen, B.W., Németh, I.: Integration of maintenance management system functions with industry 4.0 technologies and features—a review. Processes **10**(11), 2173 (2022). https://doi.org/10.3390/PR10112173. https://www.mdpi.com/2227-9717/10/11/2173/htm. https://www.mdpi.com/2227-9717/10/11/2173
19. Shavit, Y., et al.: Practices for governing agentic AI systems. Research Paper, OpenAI (2023). https://cdn.openai.com/papers/practices-for-governing-agentic-ai-systems.pdf
20. Soller, S., Holzl, G., Kranz, M.: Predicting machine errors based on adaptive sensor data drifts in a real world industrial setup. In: Annual IEEE International Conference on Pervasive Computing and Communications (2020). https://doi.org/10.1109/PERCOM45495.2020.9127357
21. Tao, F., Zhang, H., Liu, A., Nee, A.Y.: Digital twin in industry: state-of-the-art. IEEE Trans. Ind. Inform. **15**(4), 2405–2415 (2019). https://doi.org/10.1109/TII.2018.2873186
22. Wang, Y., Wang, D., Li, M., Sun, K.: A EWC-based online-learning algorithm for ordinary differential equation network. In: Asian Control Conference (2024)
23. Zenisek, J., Wolfartsberger, J., Sievi, C., Affenzeller, M.: Modeling sensor networks for predictive maintenance. In: Debruyne, C., Panetto, H., Guédria, W., Bollen, P., Ciuciu, I., Meersman, R. (eds.) OTM 2018. LNCS, vol. 11231, pp. 184–188. Springer, Cham (2019). https://doi.org/10.1007/978-3-030-11683-5_20

Comparing Human and Machine-Labeled Sentiment Analysis for Disaster Response

Hafiz Budi Firmansyah[(✉)] and Aidil Afriansyah

Institut Teknologi Sumatera, Lampung, Indonesia
`hafiz.budi@if.itera.ac.id`

Abstract. Social media plays a crucial role in providing timely information during disasters. On social media platforms, people share information about victims conditions, impacted infrastructure, and the severity of damage. Given the vast amount of information available on social media, machine learning techniques can accelerate the analysis process. Despite the usefulness of social media data, deploying it in operational settings remains challenging. One of the key challenges is the availability of labeled data for analysing the sentiment. This paper compares human-labeled and machine-labeled data. The data are evaluated using various machine learning algorithms, including Support Vector Machine, Logistic Regression, and Naive Bayes. The experimental results demonstrate that human understanding captures nuances more effectively when assigning labels, leading to better classifier performance. This research is expected to assist policymakers in considering the proposed approach for disaster response.

Keywords: Sentiment Analysis · Machine Learning · Crowdsourcing · Disaster Response

1 Introduction

Social media can play an important role in improving disaster response. During disasters, people often post real-time updates about the conditions of victims, damaged infrastructure, and the severity of the situation. These posts can provide valuable information, particularly during the early hours of a crisis. However, using this information effectively remains a challenge.

One of the main problems is the lack of labeled data needed for sentiment analysis. Since the volume of social media posts is so large, analyzing them manually takes significant time and effort. To accelerate the process, machine learning can be used to automatically analyze social media content. However, most machine learning methods rely on labeled data to work well, and manually labeling large-scale datasets is both costly and labor-intensive.

Moreover, existing automated sentiment labeling tools—such as lexicon-based and rule-based systems—tend to exhibit a positivity bias. These tools often overestimate neutral or positive sentiment while under-representing negative emotions, which are particularly crucial in disaster-related contexts. This bias can lead to an inaccurate understanding of public distress and hinder timely and appropriate responses. While this problem is often discussed anecdotally, it has not been systematically quantified or compared with human annotation in the context of disaster response.

In previous studies, we explored how human judgment and automated methods could be integrated to make social media data more useful in crisis scenarios. For example, we developed tools to automatically classify and geolocate social media content during emergencies [2–5].

Building on this work, the current study compares human-labeled and machine-labeled data in sentiment analysis tasks. Specifically, we examine how the known positivity bias of automated labeling tools affects sentiment distribution and classification performance in disaster-related datasets. This study aims to understand how human interpretation compares to machine-generated labels and how these differences influence the effectiveness of machine learning models in identifying critical emotional signals during emergencies.

The remainder of the paper is structured as follows: Sect. 2 discusses related work. Section 3 presents the dataset and proposed approach. Section 4 shows the experimental results using real-world social media data from four disaster events. Section 5 discusses the findings. Finally, Sect. 6 reports the conclusion and future works.

2 Related Work

Sentiment analysis on social media has emerged as a critical component in understanding public reactions and emotional responses during disaster events. This task involves classifying user-generated content such as tweets and posts, into categories like positive, negative, or neutral. The tasks provides valuable insights for emergency responders and decision-makers. With increasing reliance on both human annotators and machine learning models to label sentiments, recent research has begun exploring the reliability and accuracy of these two approaches. In the context of disaster response, comparing human-labeled and machine-labeled sentiment analysis offers a deeper understanding of how emotional signals are interpreted and utilized in real-time crisis management. To contextualize this study, the following section reviews key contributions and methodologies in the existing literature on sentiment analysis for disaster-related social media data.

Social media platforms have become important channels for communication during disasters, offering real-time insights into public sentiment and emotional states. Sentiment analysis enables emergency responders to better understand evolving public concerns, emotional distress, and support needs. Verma et al. (2011) conducted one of the earliest studies, demonstrating how sentiment and subjectivity features can enhance situational awareness through the classification of disaster-related tweets [11]. Nagy et al. (2012) proposed a Bayesian network-based approach that combined sentiment ontologies and emoticons to assess tweet sentiment during the California gas explosion, highlighting the role of hybrid sentiment signals [9]. Caragea et al. (2014) extended this approach by integrating geolocation data and sentiment polarity, offering visual insights into how regional sentiment fluctuates with proximity to disaster zones [1]. While Vo et al. (2013) tracked emotional trends such as fear and anxiety over time during the Japan earthquake, illustrating the dynamics of public mood using emotion-aware classifiers [12]. These works collectively emphasize the value of sentiment analysis in enriching disaster situational awareness.

In most situations, sentiment analysis is conducted by training machine learning models using human-labeled data, for instance, CrisisNLP and CrisisLex. Several studies have leveraged these datasets to explore sentiment dynamics during crisis events. Paul et al. (2023) employed the CrisisLex dataset in an active learning framework to reduce annotation efforts while fine-tuning transformer-based representations for sentiment classification, demonstrating the feasibility of semi-automated labeling in crisis scenarios [10]. Another study by Jaiswal and Kaushik (2016) utilized the CrisisLexT26 dataset and introduced a segmentation approach to improve sentiment classification accuracy by decomposing tweets into more meaningful segments [6]. More recently, a study published in 2024 by Myint et al. proposed a multi-task learning model with an attention mechanism and subject-based intent prediction, using both CrisisNLP and CrisisLex datasets to uncover the emotional and sentiment dynamics during crisis events [8]. Despite the usefulness of human-labeled data, the scalability problem exist since it demands huge human resource.

Building on this foundation, this study explores how human-labeled and machine-labeled sentiment analyses differ in effectiveness and reliability.

3 Dataset and Proposed Approach

In this section, we describe the dataset used to validate our proposed approach, which was carefully selected to ensure a comprehensive and representative evaluation. The dataset consists of multilingual social media content, allowing for a diverse analysis. This section also introduces our approach, which aims to compare the labeling process between human experts and machine learning models, and to analyze sentiment in social media posts through an innovative and structured workflow. This comparison helps to evaluate the reliability of automated labeling and supports a deeper understanding of sentiment analysis performance across different methods.

3.1 Dataset

The dataset consists of 5,434 text curated by Joint Research Center European Commission through the platform Social Media For Disaster Risk Management (SMDRM) [7]. The dataset encompasses four disaster events: the Catania floods of 2021 (207 post), the European Union floods of 2021 (1120 post), the Croatia earthquake of 2020 (869 post), and the Haiti earthquake of 2010 (3238 post).

The dataset contains texts written in 44 different languages, including Italian, English, Swahili, Spanish, Portuguese, Tagalog, Arabic, Catalan, Turkish, Dutch, Polish, Slovenian, Greek, French, Hungarian, Traditional Chinese, German, Albanian, Indonesian, Finnish, Latvian, Croatian, Afrikaans, Swedish, Norwegian, Danish, Romanian, Somali, Urdu, Russian, Persian, Japanese, Estonian, Hebrew, Vietnamese, Hindi, Thai, Ukrainian, Kannada, Tamil, Czech, Korean, and Welsh. Figure 1 shows the data distribution used in this experiment.

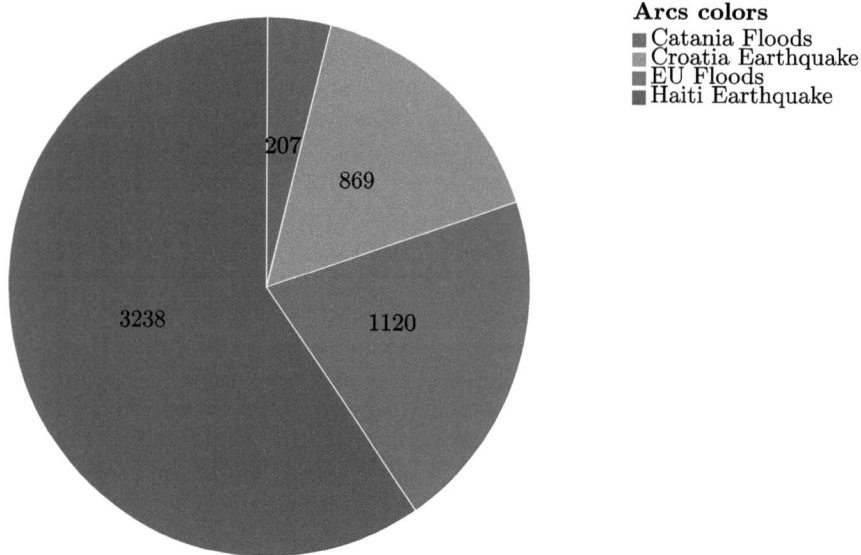

Fig. 1. Data Distribution.

3.2 Proposed Approach

This subsection outlines the proposed approach employed in this study to analyze the collected data. The approach is designed to ensure a systematic and replicable analysis aligned with the study's objectives. It consists of five key steps: translation, pre-processing, labeling, classification, and evaluation. Figure 2 illustrates the approach proposed in this study.

- **Translation.** Since the dataset contains text in various languages (in total 44 languages), this step involves translating all content into English. The aim is to use a common language to ensure consistent annotation, enabling consistent labeling in the next step.
- **Pre-processing.** This step removes stopwords and irrelevant characters to ensure the data is clean and ready for labeling. Proper pre-processing enhances the accuracy and effectiveness of subsequent analysis.

- **Labeling.** In this step, the data is annotated using two approaches. The first approach involves manual annotation by linguistic and English language experts to ensure high-quality labeling.
- **Classification.** This step involves training a machine learning model to classify the text into three sentiment categories: positive, neutral, and negative.
- **Evaluation.** The goal of this step is to evaluate the model's performance using appropriate metrics. It also includes assessing the quality of the labeling process by measuring inter-annotator agreement.

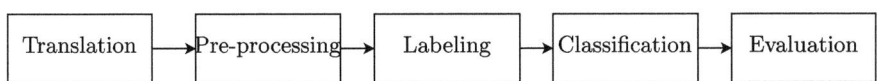

Fig. 2. Proposed Approach.

4 Experimental Results

This section presents the experimental results based on two configurations: manual annotation and automatic annotation.

The experiment begins with a translation step using the deep-translator library, which offers several advantages, including unlimited translations, free access, and typo detection[1]. The translation target language is English.

The pre-processing phase involves several steps to clean and prepare the text data. It starts with data cleansing, where irrelevant characters such as hashtags, symbols, and URLs are removed. This is followed by case folding, which converts all letters to lowercase to ensure consistency. Next, tokenization is performed using the Natural Language Toolkit (NLTK),[2] breaking down the text into smaller units or tokens for easier processing. Subsequently, stop word removal is applied to eliminate common, non-informative words, allowing the analysis to focus on more meaningful content. Finally, stemming is used to reduce words to their root form, helping to unify different word variants under a common base.

Table 1. Before and After Pre-processing

Step	Haiti Earthquake	Croatia Earthquake	EU Floods	Catania Floods
Before pre-processing	3238	869	1120	207
After pre-processing	281	18	431	172

---

[1] https://pypi.org/project/deep-translator/.
[2] https://www.nltk.org/.

After pre-processing, the dataset was reduced to a total of 902 data points. The decrease from 5,434 to 902 was primarily due to the removal of duplicates, as well as irrelevant or empty posts. A common duplication pattern was found in the Croatia earthquake dataset, where many posts consisted of retweets (RT) followed by a URL. This significant reduction was necessary to ensure a more balanced and non-redundant dataset for training and evaluation.

Table 1 shows the number of data before and after pre-processing. After pre-processing, we continue the process to provide label.

The labeling phase incorporates manual and automatic labeling. The manual method incorporates human to label the data. We ask three different annotators for providing the label. The annotators are available on a crowdsourcing platform called projects.co.id. The qualification of annotators are understanding english and having linguistic background. To ensure annotation quality, we provided clear guidelines to the annotators prior to labeling and manually reviewed the submitted results before accepting their work. After obtaining the labels, we use majority vote to determine the ground truth. A small portion of data faces no agreement where every annotator gives different label (positive, negative, and neutral). At that case, we manually give the label for having ground truth.

The automatic method leverages TextBlob, Vader, and Afinn to label the data automatically. These three libraries are mostly used by other researchers and demonstrate affirmative result. After obtaining the label, the majority vote is conducted to have the ground truth. A small portion has no agreement. In this case, we also manually label the data.

Table 2 presents a comparative analysis of sentiment label distributions and agreement levels between manual and automatic labeling methods. In the manual labeling setting, the majority of instances were labeled as neutral (489 instances), followed by negative (271) and positive (142). In contrast, the automatic labeling setting, generates from the ensemble of TextBlob, Vader, and Affin, produces a higher number of positive labels (227) and a noticeably lower number of negative labels (172), suggesting a potential positivity bias among automated tools. While using automatic label produces 489 neutral.

In terms of inter-rater agreement, the manual annotations yields a high partial agreement rate (78%), suggesting that while annotators often shared similar perspectives, they occasionally diverged on exact sentiment classifications. However, full agreement among annotators was relatively low (17.1%), reflecting the inherent subjectivity and ambiguity in human sentiment interpretation.

For fleiss kappa, the experimental results shows that manual labeling remains the preferred approach for generating high-quality ground truth data, our analysis of inter-annotator agreement reveals some variability among human annotators. The Fleiss' Kappa score for the manual labeling was 0.061, indicating slight agreement. This outcome may reflect the inherent subjectivity of sentiment interpretation, particularly in nuanced or context-dependent text. In contrast, the automatic labeling methods achieved a Fleiss' Kappa of 0.455, suggesting moderate consistency across the tools used. Although automatic methods offer more stable agreement, they may lack the depth and contextual understanding of human judgment. These findings highlight the trade-off between consistency and interpretative nuance, reinforcing the importance of careful annotation design and aggregation methods when constructing sentiment-labeled datasets.

Conversely, the automatic labeling process achieved a much higher full agreement rate (51.55%), due to the consistency among the sentiment analysis tools. While this high agreement may initially appear favorable, it does not necessarily reflect higher label quality. In fact, the lower partial agreement rate (44.35%) and skewed class distribution suggest that the tools may consistently exhibit biases, particularly overestimating positive sentiment.

Table 2. Manual and Automatic Labeling

Labeling method	Positive	Neutral	Negative	Full Agreement	Partial Agreement	No Agreement	Fleiss Kappa
Manual	142	489	271	17.1%	78%	4.9%	0.061
Automatic	227	503	172	51.55%	44.35%	4.1%	0.455

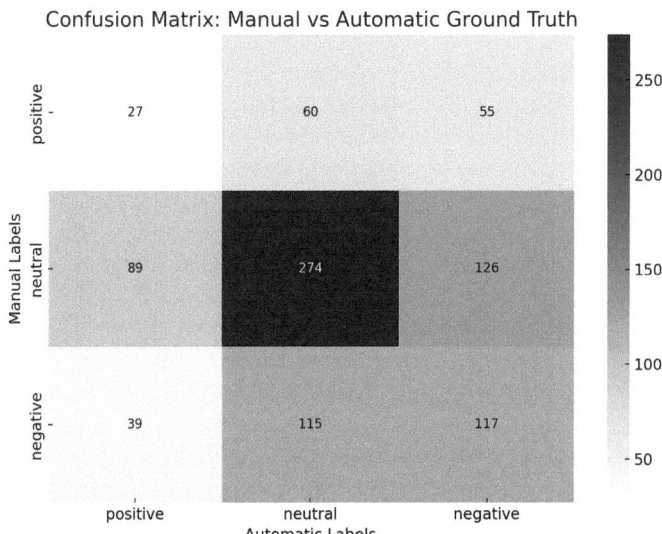

Fig. 3. Confusion matrix manual vs automatic.

The confusion matrix in Fig. 3 offers deeper insight into the nature of agreement and disagreement between manual and automatic sentiment labels. A clear pattern emerges: while the majority of predictions align along the diagonal—indicating agreement between manual and automatic labeling—there is a disproportionate concentration of automatic labels in the positive class. This aligns with the earlier finding of a positivity bias in automated tools. For instance, a

Table 3. Performance Comparison - Manual

Labeling method	SVM	LR	NB
Accuracy	71.95%	69.73%	65.74%
Balanced Accuracy	65.23%	58.10%	50.04%
Macro Precision	0.750	0.772	0.772
Macro Recall	0.652	0.581	0.500
Macro F1 Score	0.685	0.615	0.497

Table 4. Performance Comparison - Automatic

Labeling method	SVM	LR	NB
Accuracy	54.88%	59.88%	59.88%
Balanced Accuracy	43.01%	45.51%	45.88%
Macro Precision	0.416	0.403	0.397
Macro Recall	0.430	0.455	0.459
Macro F1 Score	0.411	0.417	0.419

notable portion of instances that were manually labeled as neutral or negative were automatically classified as positive, potentially distorting the emotional reality reflected in social media posts during disasters.

Such misclassification has practical consequences. In crisis scenarios, underestimating negative or neutral sentiment could result in delayed recognition of distress, dissatisfaction, or unmet needs. The confusion between neutral and positive further suggests that automated tools may overestimate optimism or satisfaction, possibly due to lexical cues like polite language that are misinterpreted as positive sentiment. These findings reinforce the importance of fine-tuning sentiment models in domain-specific contexts and highlight the limitations of out-of-the-box tools when used for high-stakes decision-making in disaster response.

The next step is classification and evaluation. To classify the sentiment class, we train a model using several machine learning algorithm. To distribute data training, we conduct 10-folds cross validation.

Table 3 and 4 present the classification accuracy of three supervised machine learning models—Support Vector Machine (SVM), Logistic Regression (LR), and Naive Bayes (NB)—evaluated using 10-fold cross-validation with shuffle and determined random state to make the experiment reproducible. The algorithm trains data under two labeling conditions: manual and automatic. When trained on manually labeled data, SVM achieved the highest accuracy (71.95%), followed by Logistic Regression (69.73%) and Naive Bayes (65.74%). In the automatic labeling scenario, Naive Bayes and Logistic Regression produced the highest accuracy (59.88%), followed closely by Support Vector Machine (54.88%).

In terms of balanced accuracy, which accounts for class imbalance by averaging recall across all classes, the models trained on manually labeled data outperformed those trained on automatically labeled data. The highest balanced accuracy was observed for SVM (65.23%), indicating its ability to better generalize across underrepresented classes. In contrast, the balanced accuracy scores in the automatic labeling scenario dropped significantly for all models, with SVM achieving only 43.01.

For macro-averaged precision, which gives equal weight to all classes regardless of their frequency, all models exhibited stronger performance with manual labels. Both Logistic Regression and Naive Bayes achieved the highest macro precision (0.772), closely followed by SVM (0.750). In the automatic setting, macro precision values declined substantially across all models—SVM scored 0.416, LR 0.403, and NB 0.397—indicating less precise predictions, especially for minority classes.

When comparing macro-averaged recall, the results further emphasize the advantage of manual labeling. SVM attained the highest macro recall (0.652), reflecting its strong ability to retrieve relevant instances across all sentiment classes. With automatic labeling, macro recall values were uniformly lower across models, ranging between 0.430 and 0.459, which implies that many instances, particularly in the positive class, may have been misclassified or overlooked.

Lastly, macro-averaged F1 scores, which balance precision and recall across all classes, echoed the trends observed in the previous metrics. The F1 score for SVM reached 0.685 under manual labeling—substantially higher than the scores for Logistic Regression (0.615) and Naive Bayes (0.497). Under automatic labeling, F1 scores dropped for all models, with the highest score being 0.419 (NB). These results suggest that manual labels offer a more reliable foundation for training models to capture sentiment distinctions effectively, despite some annotation disagreement observed earlier.

5 Discussion

The experimental results demonstrate distinct differences between human-labeled and machine-labeled sentiment data, each with its own strengths and limitations. Manual labeling, although associated with lower full agreement (17.1).

In contrast, the automatic labeling method produced a higher full agreement rate (51.55).

Model performance under different labeling conditions further emphasizes the importance of annotation quality. When trained on manually labeled data, all three models—SVM, Logistic Regression, and Naive Bayes—achieved significantly higher accuracy, with SVM performing best (71.95).

Although human-labeled data is often considered the gold standard, it is not without limitations. Disagreement among annotators can stem from subjectivity in sentiment interpretation, differences in cultural or linguistic backgrounds, or varying levels of annotator expertise. For instance, sarcasm, irony, or context-dependent cues may be interpreted differently, leading to label divergence. Emotional fatigue or cognitive overload may also influence annotator judgment, especially when processing large volumes of distressing content. Rather than viewing this disagreement as annotation error, it may reveal the complexity and richness of emotional expression in real-world scenarios. Recognizing and accounting for such subjectivity is essential when designing annotation protocols, interpreting agreement metrics, and assessing model reliability.

Given the trade-off between annotation quality and scalability, future work should consider hybrid or semi-supervised labeling strategies. These approaches aim to combine the consistency of machine-generated labels with the contextual depth of human insight. For example, active learning methods can be employed to identify and manually review the most ambiguous or impactful samples, while weak supervision frameworks can leverage multiple noisy sources—such as lexicons, heuristics, or pretrained classifiers—to generate probabilistic labels. These hybrid strategies are especially relevant in time-sensitive domains like disaster response, where rapid yet reliable analysis is crucial for operational decisions.

Several limitations of this study should be acknowledged. First, the final dataset size (902 samples) was constrained by preprocessing steps, which may limit the statistical generalizability of the findings. Second, the multilingual nature of the original posts required translation into English for sentiment analysis, which may have introduced distortion in sentiment nuances, particularly for idiomatic expressions. Third, the study focused on four disaster events, which may not represent the full diversity of emotional expression across crisis types, cultural settings, or communication platforms. As such, further validation in real-time and multilingual contexts is warranted.

Finally, the implications of this study extend to the design of future disaster informatics systems. While automated tools offer efficiency and scalability, they must be carefully evaluated for emotional fidelity—particularly in high-stakes environments. A hybrid annotation model that incorporates human oversight, feedback loops, and bias correction may offer a more balanced approach. Additionally, future sentiment classification systems should consider integrating language-specific emotion lexicons and contextual models to better capture the emotional landscape of affected populations.

This study contributes to a growing body of research in crisis informatics by providing a detailed comparison between human and machine-generated sentiment labels. Beyond accuracy, it highlights the ethical and practical consequences of relying on automated tools with known biases. As these systems increasingly inform early warning platforms, dashboards, and aid distribution strategies, their design must prioritize not only computational performance but also the responsible representation of public emotion and need.

6 Conclusions and Future Work

This study compared sentiment labeling using human annotators and automated tools for disaster-related social media posts. The results show that human-labeled data leads to more accurate and reliable classification, especially when using models like Support Vector Machines. While automatic tools offer faster processing and higher consistency, they often miss subtle emotional cues, particularly in urgent or negative contexts.

These findings highlight the critical role of human judgment in analyzing crisis communication. Relying solely on automated sentiment tools could lead to misinterpretations, especially during emergencies when detecting public distress is critical.

Future work will focus on leveraging more advanced language models, such as BERT, and improving translation and labeling techniques to better handle multilingual data. Combining automated methods with human oversight could offer a more balanced and effective approach for disaster response systems.

References

1. Caragea, C., et al.: Mapping moods: geolocation-based sentiment analysis during hurricane sandy. In: ISCRAM (2014)
2. Firmansyah, H.B., Bono, C.A., Lorini, V., Cerquides, J., Fernandez-Marquez, J.L.: Improving disaster response by combining automated text information extraction from images and text on social media. In: FAIA - Frontiers in Artificial Intelligence and Applications, vol. 369, pp. 320–329. IOS Press (2023)
3. Firmansyah, H.B., Fernandez-Marquez, J.L., Cerquides, J., Serugendo, G.D.M.: Single or ensemble model? A study on social media images classification in disaster response. In: Proceedings of the 10th Multidisciplinary International Social Networks Conference (MISNC 2023), Phuket, Thailand, pp. 48–54. ACM (2023)
4. Firmansyah, H.B., Fernandez-Marquez, J.L., Mulayim, M.O., Gomes, J., Lorini, V.: Accelerating crisis response: automated image classification for geolocating social media content. In: 2023 IEEE/ACM International Conference on Advances in Social Networks Analysis and Mining (ASONAM), pp. 77–84. IEEE/ACM (2023)
5. Firmansyah, H.B., Lorini, V., Mulayim, M.O., Gomes, J., Fernandez-Marquez, J.L.: Improving social media geolocation for disaster response by using text from images and chatgpt. In: Proceedings of the 11th Multidisciplinary International Social Networks Conference (MISNC 2024), Bali, Indonesia, p. 6. ACM (2024)
6. Jaiswal, A., Kaushik, S.: Sentiment classification of crisis related tweets using segmentation. In: Proceedings of the 9th ACM India Compute Conference, pp. 1–5. ACM (2016)
7. Lorini, V., Panizio, E., Castillo, C.: Smdrm: a platform to analyze social media for disaster risk management in near real time. In: Workshop Proceedings of the 16th International AAAI Conference on Web and Social Media ICWSM (2022)
8. Myint, P.Y.W., Lo, S.L., Zhang, Y.: Unveiling the dynamics of crisis events: sentiment and emotion analysis via multi-task learning with attention mechanism and subject-based intent prediction. Inf. Process. Manag. **61**(4), 103695 (2024)
9. Nagy, C., Grasso, V., Hartley, K.: Using Bayesian networks for sentiment analysis of tweets during the California gas explosion. In: ISCRAM (2012)
10. Paul, N.R., Balabantaray, R.C., Sahoo, D.: Fine-tuning transformer-based representations in active learning for labelling crisis dataset of tweets. ResearchGate (2023)
11. Verma, S., et al.: Natural language processing to the rescue? Extracting "situational awareness" tweets during mass emergency. In: ICWSM (2011)
12. Vo, B., Nhuhuan, N., Nguyen, H.: Tracking real-time crowd emotions during earthquakes. In: International Conference on Information Systems for Crisis Response and Management (ISCRAM) (2013)

Reconfiguring Knowledge Management Systems for the Web 3.0 Ecosystem

Yu Cui[1(✉)], Hiroki Idota[2], and Masaharu Ota[3]

[1] Faculty of Sociology, Kansai University, Osaka, Japan
yucui@kansai-u.ac.jp
[2] Creative Management and Innovation Research Institute, Kindai University, Osaka, Japan
[3] Falcuty of Business Administration, Osaka Gakuin University, Osaka, Japan

Abstract. The convergence of Web 3.0 technologies with knowledge management (KM) is reshaping how knowledge is created, validated, and shared across organizational and sectoral boundaries. Drawing on an extensive review of KM theory and the affordances of blockchain, decentralized storage, and smart contracts, this study develops a rigorous decentralized knowledge management framework that reconceptualizes technical trust and token-mediated governance as dual pillars of next-generation KM. Two emblematic cases are analyzed through a qualitative multiple-case design to elucidate how distributed ledgers, cryptoeconomic incentives, and community-driven decision processes operate in practice. Findings show that (i) token staking and slashing can partially substitute for editorial hierarchies in mitigating low quality content, (ii) Decentralized Autonomous Organization (DAO) voting structures redistribute but do not eradicate power asymmetries, and (iii) interoperability protocols unlock cross-platform knowledge recombination while introducing new technical bottlenecks. A comparative evaluation against traditional, centrally managed KMS highlights fundamental trade-offs in governance, incentive design, quality assurance, and scalability.

Keywords: Knowledge Management · Web 3.0 · Blockchain · Decentralized Autonomous Organization · Token Incentives

1 Introduction

In recent years, knowledge has come to be widely recognized as one of the most critical resources in both business and society, and its effective management and utilization are viewed as key sources of competitive advantage [1]. Research on the creation and sharing of organizational knowledge has expanded rapidly since the 1990s, extending traditional resource-based views of the firm to include knowledge. In particular, knowledge management (KM) is defined as "the process of systematically acquiring and organizing an individual's knowledge so that it can be communicated for use by others," with its importance stressed in information systems and business research. Early works, notably Nonaka [2], point out that effective sharing of knowledge is the key to continuous innovation; meanwhile, other studies suggest that social capital, such as reputation and reciprocity, strongly influences people's willingness to contribute knowledge in electronic communities [3]. Historically, KM research focused primarily on how organizations or discrete communities create and share knowledge internally, but its scope has lately expanded beyond firm boundaries to business ecosystems. For example, in an ecosystem where multiple independent organizations loosely collaborate for shared value creation, modular design can allow coordination without strict hierarchical dominance [4]. In such environments, knowledge flows autonomously and in a distributed manner across firms. This trend necessitates rethinking KM frameworks that were originally premised on hierarchical organizational structures.

Moreover, recent advances in Web3.0 technologies are bringing more profound transformations in the generation, sharing, and management of knowledge. Web3.0 signifies an emerging internet paradigm characterized by decentralization, autonomy, and intelligence, exemplified by blockchain technology, which replaces the trust function historically provided by centralized platforms with automated protocols. Blockchain's distributed ledger enables data consistency and reliability to be guaranteed even in the absence of a single managing authority, while users themselves hold ownership and control over their data and content. Such a decentralized digital infrastructure presents new possibilities and challenges for open knowledge sharing communities and business ecosystems.

The purpose of this study is to provide a comprehensive analysis of how Web3.0 reshapes business ecosystems and KM, and to propose a new "decentralized knowledge management framework" aligned with the autonomous, distributed, and intelligent nature of these technologies. After reviewing relevant literature to establish the theoretical underpinnings of KM, this paper defines the self-governing and decentralized knowledge environment of the Web3.0 era and analyzes its structure. Subsequently, we examine two case studies, Encyclopedia Network (i.e., Everipedia) and Decentralized Science (DeSci), both pioneering examples of distributed knowledge management in distinct contexts: collaborative knowledge bases and scientific knowledge production. We integrate insights from these cases into a newly constructed decentralized knowledge management framework and then discuss issues of knowledge quality assurance, governance, and incentive design under this framework. We conclude with our theoretical contributions and future directions for research.

In undertaking this analysis, we also draw explicit comparisons between Web3.0-based knowledge management approaches and traditional KM frameworks, highlighting how decentralization alters key processes and outcomes.

2 Conventional Knowledge Management Theories

Research on knowledge management has developed across various fields, including information systems and organizational science. Knowledge is broadly viewed as "justified true belief," a concept that differs from data or information in its capacity to enhance an agent's effectiveness [1]. Alavi and Leidner (2001) classify organizational knowledge in multiple ways: as state, object, process, access, or capability, each definition converging on the idea that knowledge raises one's effective capacity for action. Meanwhile, KM involves the systematic processes of creating, sharing, and applying that knowledge. In practice, organizations have long employed KM initiatives such as best-practice sharing, training, or building internal knowledge bases.

A well-known theory of knowledge creation within organizations is the SECI model, introduced by Nonaka, which distinguishes tacit knowledge (unarticulated know-how) and explicit knowledge (articulated, codified) [2]. In the SECI model, new knowledge is generated through four modes of knowledge conversion: (1) socialization (tacit-to-tacit), (2) externalization (tacit-to-explicit), (3) combination (explicit-to-explicit), and (4) internalization (explicit-to-tacit). This cyclical process forms a "spiral" in which individual-level knowledge expands and is eventually institutionalized into an organization's collective knowledge base [2]. Nonaka's framework, grounded in analyses of Japanese corporate examples, has become widely accepted as a principal approach to knowledge creation. He and his collaborators also emphasize the concept of "Ba," or shared context, as the setting where knowledge conversion takes place. Taken together, these ideas offer guidance on how an organization can manage knowledge to foster continuous innovation.

Many scholars categorize KM processes into four phases: knowledge creation, knowledge storage/retention, knowledge sharing/transfer, and knowledge application/utilization [1]. For example, Alavi and Leidner propose organizational knowledge processes as: acquisition (creation), storage (retention), dissemination (sharing), and use (application), each supported by relevant technologies and management practices. Knowledge creation entails the emergence of new ideas or skills through interactions among individuals or teams. Knowledge retention involves capturing knowledge assets in databases or organizational memory. Knowledge sharing refers to processes that make an individual or team's knowledge accessible for use by others, for instance, through collaborative platforms, training, or document circulation [1]. Knowledge application or utilization is the final stage, where stored or shared knowledge is actually leveraged for problem solving, product development, and decision making, generating real value. Systems known as knowledge management systems (KMS) are frequently introduced to facilitate these processes, ranging from repository type KMS (central databases or document management systems) to network-type KMS (social media or community tools).

3 Knowledge Sharing and Inter-Organizational Knowledge Networks

One of the core themes in KM is knowledge sharing. The act of transmitting one's knowledge to others is by no means automatic; it has been the subject of extensive study in organizational behavior and the IS field to identify motivational factors. Prior works highlight both extrinsic motivators (monetary rewards, promotion prospects) and intrinsic motivators (joy of helping others, job satisfaction) as influencing people's intent to share knowledge, alongside social/psychological elements like perceived fairness and management support [3]. In online knowledge communities, internal rewards such as altruism, enjoyment, or reputation gains often play major roles in sustaining knowledge contribution. At the same time, concerns about losing one's unique value by sharing knowledge, or about insufficient incentives ("why share if there is no return?") can deter knowledge sharing. Many firms have thus implemented strategies, such as financial incentives, recognition programs, or fostering a collaborative culture, to encourage knowledge exchange.

Although early studies on knowledge sharing focused on problems of information transfer within teams or departments, the scope has now widened to inter-organizational networks and online communities. In communities of practice (CoPs) or electronic networks of practice, specialists across organizational boundaries exchange expertise [3]. For instance, one study on a legal advice online forum explained how participants voluntarily offered knowledge out of altruism and community identification, with reputation and the anticipation of reciprocal assistance acting as strong motivations [3]. Similarly, other research posits that social capital, comprising structural, relational, and cognitive dimensions, facilitates knowledge creation and sharing by forming trust and a shared vision. These insights guide firms seeking to harness communities inside and outside the firm for collaborative knowledge creation. Indeed, many global companies have launched internal social networks and knowledge communities to promote knowledge exchange among employees and with external partners. Meanwhile, large scale collaborative editing projects like open-source software communities or Wikipedia exemplify autonomous knowledge networks unconfined by formal organizational structure. Wikipedia demonstrates that peer review on discussion pages can help maintain a certain level of content quality, despite decentralized editing.

Thus, KM research initially concentrated on knowledge flows within organizations, then progressively extended toward collaborative knowledge activities across organizational boundaries. Yet many of these works assume some central coordinating role (e.g., corporate managers or community leaders) [5]. By contrast, we still know relatively little about how knowledge management functions in a fully autonomous and decentralized environment with minimal hierarchical oversight. Addressing this gap, our study constructs a theoretical framework of decentralized knowledge management by incorporating Web3.0 technologies such as blockchain. We begin by clarifying the distinctive features of knowledge environments in the Web3.0 era.

4 Definition and Characteristics of a Decentralized Knowledge Environment

A decentralized knowledge environment refers to a setting in which knowledge is generated, shared, and retained over a distributed network without relying on a centralized managing authority. In this study, we focus on knowledge platforms and communities built on Web3.0 technologies such as blockchain, decentralized storage, and smart contracts. Traditionally, knowledge management systems, whether inside a firm or Wikipedia style nonprofits, relied on a central administrator to oversee quality control and access permissions. By contrast, in a self-governing decentralized environment, a blockchain or other distributed ledger technology provides a guarantee of transactional validity and data integrity, shifting the trust anchor from human administrators to algorithmic mechanisms. For example, blockchains like Ethereum use consensus protocols to confirm the legitimacy of each transaction among a wide set of participants, making the ledger robust against tampering. This design prevents arbitrary censorship or manipulation of data by any single entity. Some scholars even liken blockchain networks themselves to "knowledge commons," referencing Elinor Ostrom's concept of communal resource governance [6]. In such networks, all participants collectively manage the ledger, and rules are determined by open, transparent consensus.

An organizational structure known as a Decentralized Autonomous Organization (DAO) is often adopted within such networks. A DAO encodes organizational rules into smart contracts on a blockchain, enabling automated governance. Decision making is carried out collectively through token-based voting, so no single manager monopolizes control. If applied to knowledge communities, a DAO could theoretically replace the editorial authority and dispute resolution once held by central administrators with code-based self-governance.

Another key feature of Web3.0-based knowledge environments is interoperability and network effects. Emerging blockchain protocols are increasingly enabling cross-chain communication, letting distinct platforms share messages and data. For example, systems like Polkadot, designed for blockchain interoperability, can interlink heterogeneous blockchains, thus mitigating "silo" issues [7]. This integration suggests the possibility of unifying knowledge resources previously locked in isolated repositories, forming expansive knowledge networks. Such synergy can facilitate recombination of knowledge, spurring innovation. Further, as Web3.0 evolves, it promises to incorporate AI and semantic technologies in distributed contexts, enabling advanced knowledge discovery and reasoning.

In summary, decentralized knowledge environments in the Web3.0 era have four distinguishing traits: (1) trust and transparency assured by distributed ledgers, (2) DAO mechanisms for self-governance, (3) expanded knowledge ecosystems via interoperability, and (4) potential for AI-driven intelligence. These traits diverge significantly from traditional centralized KM models and underscore the need for new theoretical frameworks.

5 Case Study Analysis

To illustrate these characteristics of self-governing, decentralized knowledge environments, we analyzed two prominent cases: (1) Encyclopedia Network (e.g., Everipedia/IQ.wiki), a decentralized encyclopedia project, and (2) Decentralized Science (DeSci), a movement to transform scientific research through blockchain-based collaboration. Each operates in distinct contexts, collaborative knowledge bases and knowledge creation in science, yet both exemplify how decentralized, autonomous KM is explored. We collected secondary data from white papers, academic articles, and public web materials, focusing on governance structures, incentive mechanisms, and their implications for knowledge management.

This case analysis was qualitative in nature, relying on publicly available documentation rather than primary fieldwork. The two cases were selected because each represents an early, prominent application of Web3.0 principles to knowledge management in a different domain (general knowledge and scientific research). Using multiple sources (white papers, technical documentation, community blogs, and prior studies), we triangulated facts about each platform's design and operations. We acknowledge that this methodology has limitations; deeper field studies or interviews with participants would likely yield additional insights beyond what secondary data can provide. Nonetheless, the chosen cases and data enable an initial exploration of Web3.0-based KM models, which this study synthesizes and evaluates.

5.1 Encyclopedia Network and Decentralized Knowledge Sharing

Wikipedia is a celebrated example of an open knowledge platform, yet it operates under a centralized organizational structure, the Wikimedia Foundation. In recent years, "blockchain-based encyclopedias" have emerged as decentralized equivalents. A leading project is Everipedia, now often referred to as IQ.wiki, a blockchain-powered online encyclopedia [8]. Everipedia gained attention in 2018 when Larry Sanger (co-founder of Wikipedia) joined the team, announcing the world's first on-chain encyclopedia. Its guiding principle is to give everyone the right to edit and participate in governance, enforced by blockchain technology to ensure censorship resistance and to reward participants with incentives. Specifically, Everipedia issues a cryptocurrency called IQ tokens to contributors who write or edit encyclopedia articles. The article approval process itself uses a token-based voting system, so the entire community evaluates submissions for accuracy and appropriateness. This approach attempts to maintain content quality through economic incentives and collective decision making, rather than relying on an editorial board. For instance, Everipedia implemented a rule requiring users to stake a certain amount of IQ tokens when submitting a new article or edit; if the community accepts the submission, the user's stake is returned with additional token rewards, but if rejected, the staked tokens are forfeited. This token-staking mechanism is intended to deter spam and low quality contributions by attaching a monetary cost to malicious or frivolous edits.

Everipedia initially used the EOS.IO blockchain, with IPFS (InterPlanetary File System) providing decentralized storage. Hence, article metadata is recorded on-chain in hashed form, while the full text is stored on IPFS, eliminating dependence on a single centralized server. Xu et al. discuss a similar decentralized wiki platform, arguing that traditional wiki systems struggle with opacity and trust issues (risk of hidden edits, arbitrary deletions) [8]. They demonstrate that incorporating blockchain can ensure transparency and immutability of editing records, boosting user trust and motivation to contribute. Their "DecWiki" framework logs edit histories on-chain, uses IPFS for article hosting, and secures wireless network environments via a Trusted Execution Environment (TEE), all forming a robust distributed architecture. Xu et al. report that a prototype performed well and was well-received by users, offering strong evidence that a decentralized encyclopedia is technically feasible [8].

In practice, Everipedia's governance model is somewhat hybrid. Although a foundation and development team maintain the core infrastructure, article editing in daily operations is delegated to a token-based economy. For example, any user can propose an edit, which is accepted or rejected by majority vote of IQ token holders. However, large token holders could disproportionately influence outcomes, risking a concentration of decision power. Another open question is how thoroughly users adhere to editorial guidelines resembling Wikipedia's, such as neutral point of view, verifiability, etc. Larry Sanger himself has acknowledged the difficulties of maintaining a high quality encyclopedia in a purely decentralized network and has proposed an "Encyclosphere" vision, in which multiple encyclopedias (e.g., Wikipedia and Everipedia) interconnect via a distributed network [9]. Users would be free to choose the interface they prefer, while aggregated data from different encyclopedia sites would be integrated. This approach seeks to build a knowledge commons without dependence on any single platform, representing a further step toward decentralized knowledge management.

Overall, Everipedia illustrates the intersection of collaborative editing and decentralized consensus, revealing how the principle of co-creating knowledge (central in conventional KM) merges with the Web3.0 notion of autonomous governance. It also highlights practical difficulties: for instance, balancing incentives (token rewards) with editorial quality, and ensuring equitable distribution of voting power. Nonetheless, it demonstrates that a distributed approach to knowledge curation is technologically achievable and can motivate contributor participation. One advantage of this fully distributed architecture is that the encyclopedia becomes effectively uncensorable, as users in countries where Wikipedia is blocked can still access and contribute Everipedia content, since no single server can be shut down.

5.2 Decentralized Science (DeSci) and the Transformation of Knowledge Production

Decentralized Science (DeSci) is a newly emerging movement that aims to reshape how academic research is conducted by leveraging blockchain and decentralized organization. Traditional science has faced criticisms regarding centralized funding structures (government grants, large foundations), monopolized scholarly publishing, and restricted access to research outcomes (expensive journal paywalls or patent systems). By contrast, DeSci seeks to democratize everything from research funding to results sharing through community-driven mechanisms. One vision is to form blockchain-based research communities in which scientists and backers can connect directly, bypassing intermediaries. Research outcomes, such as papers, datasets, can be managed as NFTs (non-fungible tokens) or governed by smart contracts on an open knowledge platform accessible to all. Additionally, DeSci advocates for peer review or patent evaluation by community-run DAOs that compensate reviewers with tokens, aiming for faster and more equitable processes.

A practical example is VitaDAO, a decentralized autonomous organization funding longevity research. Its token holders (researchers, investors, general community members) vote on project proposals and allocate funds to those selected. Intellectual property resulting from these studies is owned by the DAO and may later be released via open access or licensed to pharmaceutical companies, with proceeds returning to the DAO [10]. Thus, the entire community shares in both the dissemination and commercialization of research findings. Some efforts also involve timestamping research data on a blockchain, guaranteeing an immutable record of contributions and allowing researchers to prove precisely when they made a discovery. DeSci's philosophy continues the principles of open science and citizen science but leverages blockchain's economic and transparency features as a critical enabler.

Recent discussions by Unfried [10] contend that DeSci could serve as a remedy for sluggish progress in conventional science, where research investments climb but discovery rates stagnate, akin to "Eroom's Law," which describes how pharmaceutical R&D productivity halves every nine years. By shifting funding from centralized agencies to community-led initiatives, DeSci may enable riskier yet potentially transformative ideas to receive support. Opening research data and results as open source resources may accelerate knowledge flow, cut duplicate investments, and foster collaboration. Placing the entire research workflow on a blockchain ensures transparency and reproducibility, potentially reducing fraud and bias. Furthermore, Weidener and Spreckelsen propose that DeSci should center on transparency, open access, decentralized governance, and inclusivity [11]. However, they note that most scholarly references to DeSci remain sparse, with the conversation driven largely by practitioners rather than academia, reflecting DeSci's nascent status.

From these observations, we can identify some shared dimensions of decentralized knowledge environments: (1) the use of blockchain to prevent data tampering and build trust among participants (seen in both Everipedia and DeSci), (2) token-based incentives that reward knowledge contributions or fund governance activities (IQ tokens, VitaDAO tokens), (3) community-led governance based on proposals and votes (Everipedia's editorial voting, VitaDAO's research funding), and (4) open access to knowledge assets, facilitated by decentralized storage and NFT-based ownership. These insights guide our development of a new decentralized knowledge management framework in the following section.

6 Proposed Decentralized Knowledge Management Framework

Figure 1 presents the "decentralized knowledge management framework" proposed in this study. At its core, a blockchain ledger (or DAO platform) acts as the foundation for knowledge registration, sharing, and governance, surrounded by multiple participants (individual or organizational nodes). Within this structure, participants autonomously generate knowledge, record it on the blockchain, and retrieve or utilize existing knowledge from the ledger [7]. Since the ledger is sustained by consensus among all nodes, recorded knowledge (e.g., documents, transactions, evaluations) is highly resistant to alteration. The arrows in Fig. 1 illustrate knowledge flow: from participants to the ledger (submitting new content) and from the ledger to participants (retrieving previously recorded assets). Smart contracts on the ledger automate several functions involved in this flow, such as awarding token rewards for knowledge contributions, verifying the content, and enforcing access controls. As a result, a high-trust KM process can be implemented with minimal human mediation. Participants who hold tokens can also engage in ledger governance, forming consensus on broader policies like content moderation rules or reward distribution algorithms. In this manner, our framework conceptualizes decentralized knowledge management around technical trust (blockchain) and collective governance (token economy) as dual pillars.

The main components in this framework are:

1. Decentralized Knowledge Ledger (Blockchain / DAO):

    A blockchain or distributed ledger storing knowledge content and metadata (revision history, ratings, ownership). All nodes update it via consensus, ensuring immutability and availability. Smart contracts automate business rules, such as content approval workflows and access permissions, while DAO functionality manages governance via community voting.

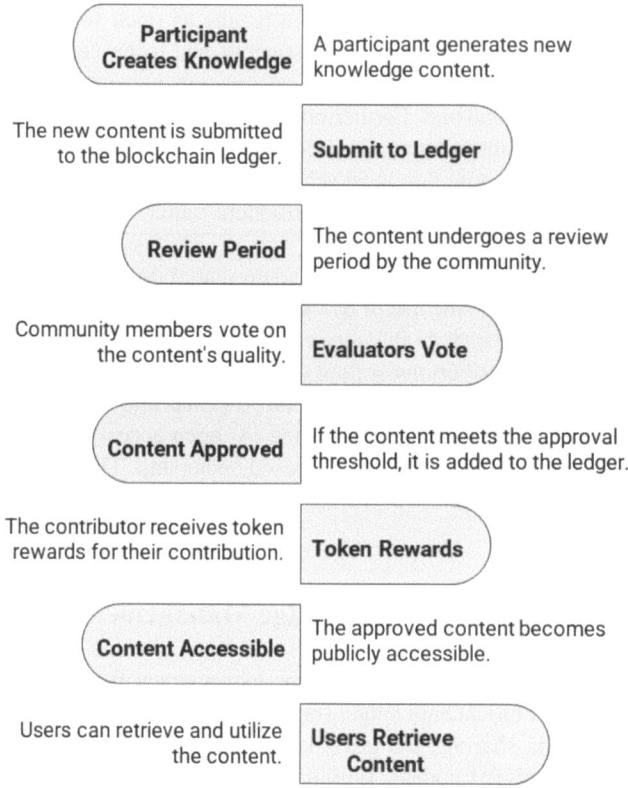

Fig.1. Decentralized Knowledge Management Framework

2. Participant Nodes:

Users, experts, and organizations that provide or consume knowledge. Each participant has read/write privileges to the ledger and acts autonomously. They connect as peers without a central manager. Participants have linked token wallets enabling them to earn and spend tokens based on their contributions or voting activity.

3. Token / Incentive Mechanism:

A native token used for value exchange within the network, such as awarding knowledge contributions, weighing governance votes, or controlling access to certain knowledge. While tokens can motivate participation, careless incentive design risks creating low-quality "spam" or gaming of the system, so careful calibration is essential.

4. 4.Knowledge Evaluation and Quality Control:

Evaluation data, such as user trust scores, review comments, and view counts, are recorded for each knowledge item. The community manages the review process, and only content surpassing a consensus threshold can be officially posted on the ledger. A token-slashing penalty can deter malicious contributors. This distributed review aims to maintain content quality in lieu of a central editor.

5. Decentralized Storage / Knowledge Repository:

The blockchain holds only transaction data, while large files (articles, images, videos) are stored in IPFS or other distributed systems. Hashes and reference links reside on-chain, ensuring consistency and tamper-detection. This architecture keeps the ledger from becoming unwieldy while guaranteeing broad availability of content.

Collectively, these elements produce a KM lifecycle suited to decentralized settings. The typical flow is:

1. A participant creates knowledge (e.g., an article or dataset) and submits a transaction to the ledger.
2. The submission enters a review period, governed by a smart contract that signals community members to evaluate the content.
3. Evaluators cast up/down votes, which the contract logs on-chain. If the submission meets a threshold of support within a set time, it is added to the ledger. The contributor automatically receives token rewards. Unapproved submissions either expire or await revision.
4. Approved content is publicly accessible via the ledger and distributed storage; other users can retrieve it, build upon it, and optionally contribute new knowledge or improvements, thus continuing the cycle [7].

We emphasize a hybrid approach of human judgment and automated control, acknowledging that fully autonomous decentralization can lead to rampant low-quality information or the inability to remove harmful content. Key decisions remain vested in the community's collective intelligence (voting, feedback) but are codified into the system to reduce reliance on any single gatekeeper. Likewise, we highlight the two-sided nature of incentives: while token rewards can fuel engagement, they may crowd out intrinsic motivators such as curiosity or altruism if overemphasized [12]. A balanced design that respects participants' reputational and intrinsic motivations is crucial.

7 Comparative Analysis of Decentralized vs. Traditional Knowledge Management

The proposed Web3.0-based knowledge management model differs fundamentally from traditional, centrally managed KMS along several dimensions. First, in terms of governance, conventional KMS platforms (e.g., corporate intranets or Wikipedia) rely on hierarchical administration or a central authority (such as the Wikimedia Foundation and its moderator team) to make decisions and enforce rules. By contrast, our decentralized approach replaces hierarchical oversight with community-driven governance via smart contracts and token-holder voting. This means that whereas Wikipedia's content policies are ultimately overseen by an elite group of editors and foundation staff, which has led one co-founder to argue the site has become an "oligarchy" of insiders, a Web3.0 community encodes editorial decisions into transparent rules that no single entity can unilaterally control.

Second, the incentive structures diverge. Traditional knowledge sharing systems motivate contributors through intrinsic rewards, such as personal satisfaction, community reputation, or altruism, and, in some cases, extrinsic recognition, such as employee bonuses or academic credit. Mainstream open platforms like Wikipedia explicitly eschew financial rewards; its millions of editors are volunteers driven by a mix of passion and reputation. In the decentralized model, however, participation is also fueled by cryptoeconomic incentives: users earn transferable tokens for their contributions. For example, Wikipedia's most active editors work for free, whereas Everipedia allows prolific contributors to gain a financial stake in the platform via IQ tokens. This tokenization can broaden participation by attracting those who might not contribute without monetary reward and align individual incentives with the community's success. At the same time, there is a risk that extrinsic incentives could encourage quantity of contributions over quality, a concern absent in purely volunteer-based systems.

Third, content curation and quality control follow different mechanisms. Traditional KMS, including institutional knowledge repositories, typically entrust quality assurance to designated experts or moderators who review submissions against established standards. Wikipedia, for instance, relies on community norms and a hierarchy of volunteer admins to patrol edits, an approach that has proven surprisingly effective. A well-known study found Wikipedia's science articles only marginally less accurate than those of Encyclopædia Britannica [13]. In decentralized knowledge networks, by contrast, quality control is collectivized and encoded: contributions are vetted by the community through consensus protocols (e.g., token-weighted voting or staking mechanisms) rather than by editor authority. This approach promotes openness and resistance to censorship, but it also poses new challenges in ensuring that voting outcomes truly reflect content reliability rather than the influence of large token holders. Our model attempts to mitigate this by coupling community review with token stakes (slashing bad actors' deposits), but it remains an open question whether such decentralized vetting can match the rigor of traditional expert review in practice.

Finally, the two paradigms differ in their underlying infrastructure and scalability. A conventional KMS often operates on centralized databases or servers, yielding fast transaction processing and straightforward data management within a single organization's control. In a Web3.0-based system, operations are distributed across a blockchain network, trading off some performance and simplicity for greater trust and transparency. For example, recording knowledge transactions on Ethereum or similar blockchains introduces latency and throughput limits that would not exist in a local corporate database. Thus, the decentralized system may be preferable in environments where data integrity, user sovereignty, or censorship-resistance are top priorities, such as cross-organizational or public knowledge communities, whereas a traditional KMS might be more practical for closed, small-scale contexts that demand efficiency and centralized oversight. In summary, the Web3.0 approach offers unique strengths, such as trustless collaboration, global participation, tamper-proof records, but also comes with limitations, including technical complexity, uncertain content quality dynamics that must be weighed against those of conventional knowledge systems.

8 Discussion

From our analysis and the proposed framework, three major challenges emerge for decentralized KM.

1. Ensuring Knowledge Quality

Without a central authority, controlling low quality or misinformation becomes critical. Encouragingly, prior experiences with Wikipedia suggest that collective peer review may suffice to maintain acceptable quality levels [14]. A famed study in Nature found that the accuracy of Wikipedia articles on scientific topics nearly matched Encyclopaedia Britannica, implying that a sufficiently large volunteer community can produce quality content rivaling centralized editorial teams. Still, in a fully decentralized context, sustaining the community's editing workforce and motivation is itself a challenge. Blockchain-based systems often provide strong user anonymity, necessitating robust mechanisms to filter bad actors and assess reliability of content. Our framework proposes a token-based evaluation and penalty system, but how well it works requires empirical validation. Additionally, a decentralized knowledge community must clarify lines between public vs. private information, ensuring legal or ethical compliance (e.g., copyright, privacy).

2. Governance

A second challenge is clarifying who makes decisions and how in a self-governing system. Without careful design, a token-weighted vote could permit "vote buying" or large holders to dominate. Scholars of blockchain governance often advocate for polycentric structures, involving multiple stakeholder groups to broaden representation [6]. This might mean weighting votes by reputation or expertise as well as token holdings. There is also the distinction between technical governance, such as protocol changes and content governance, such as rules on appropriate content. Each domain may need separate processes or specialized DAOs. Meanwhile, legal status can be ambiguous when a community operates purely on-chain. Since many DAOs lack formal corporate personhood, accountability for disputes or liabilities is murky. With ongoing regulatory changes surrounding crypto assets and DAOs, knowledge communities must ensure compliance while preserving decentralization.

3. Incentive Design

A third challenge is designing incentives that spur valuable contributions rather than superficial or exploitative behaviors. Token rewards can stimulate knowledge sharing yet risk overshadowing intrinsic drivers (e.g., curiosity or altruism) [12]. In Everipedia, for instance, some worry that IQ tokens could prompt quantity-over-quality edits. Token volatility could also disrupt stable engagement if rewards suddenly drop in value. Thus, best practices might include mixing financial and non-financial incentives, offering not just tokens but also badges, reputation scores, or recognition. Communities can reward intangible factors like user pride or self-efficacy. A multi-layer approach is likewise needed to curb spam or manipulative behavior; for example, imposing deposit requirements for new article submissions that can be slashed if the content is deemed malicious, or employing AI-based anomaly detection.

4. Technical Constraints

ting a global, high-volume KMS on-chain. It is important to recognize these practical limitations; our proposed model, while conceptually sound, wouldA final challenge lies in the technical limitations of current blockchain infrastructure. Public blockchains typically suffer from limited transaction throughput, latency issues, storage bloat, and, in some cases, high energy consumption [15]. Writing every knowledge transaction to a distributed ledger could be significantly slower and costlier than using a centralized database, which may hinder user experience if the system scales up. For instance, without further innovation, a blockchain-based encyclopedia might process edits more slowly than a conventional wiki. Encouragingly, improvements are underway. Ethereum's recent transition from proof-of-work to proof-of-stake has slashed its energy usage by over 99%, and ongoing developments such as layer-2 networks and data sharding promise to boost blockchain throughput in the near future. Nevertheless, until such scalability solutions fully mature, the performance and cost constraints of blockchain technology remain a barrier to implemen require careful engineering choices or future blockchain advancements to be feasible on a large scale.

Despite these hurdles, decentralized KM can deliver significant benefits. Open and distributed frameworks enable knowledge to flow across organizational and national boundaries, fostering truly global co-creation of knowledge. If blockchain provides trustworthy means for storing and verifying data, the fragmented knowledge scattered in diverse repositories can converge into a single giant knowledge base. This has the potential to dramatically enhance collective intellectual productivity. Further, the explicit representation of knowledge ownership or contribution shares via tokens may clarify and reward contributions that used to go unnoticed. An engineer who records an innovative idea on a blockchain can prove authorship, potentially boosting professional credibility. Moreover, these platforms can be adapted within firms or research institutes to enrich organizational learning and open innovation.

9 Conclusions

This research examined how Web3.0 technologies, including blockchain and DAOs, transform knowledge management, revisiting conventional KM theories and proposing a decentralized knowledge management framework. Traditional KM scholarship frequently focuses on knowledge processes within individual firms; our framework, however, integrates cutting-edge developments in blockchain-based autonomy to present a governance model for cross-boundary knowledge commons. In this design, knowledge creation, sharing, and utilization occur on a distributed ledger with token-based incentives and community-driven governance. Our approach extends an earlier blockchain-KM model by Akhavan et al. [16], specifically adapting it for business ecosystems and open science in the Web3.0 age. While innovative, our model does not fully resolve all problems facing next-generation KM. Instead, it acknowledges persistent issues, such as content quality assurance and scalability, that will require continued effort to address.

This paper also highlighted several key challenges, notably knowledge quality, governance, and incentive design, for which socio-technical approaches that marry technical solutions with institutional or cultural strategies are necessary. For instance, leveraging AI to evaluate content, drafting community charters, or providing stakeholder education. Fortunately, the flexibility of emerging Web3.0 frameworks leaves ample room for experimentation. Future research should apply this framework to real communities, assessing how decentralized KM affects knowledge-creation speed or quality. Additional impetus is provided by DeSci's rise, suggesting decentralized models can serve as alternatives to established, centralized paradigms of scientific knowledge dissemination. Our proposal is part of that broader movement and can be advanced further through empirical tests.

Ultimately, knowledge is a shared asset of humanity, evolving in tandem with societal changes. The decentralized knowledge management model proposed here, aiming at the democratization and maximization of knowledge use, is one approach that we hope will mature through further theoretical refinement and practical experimentation. Continuous exploration by the research community promises to clarify the value principles and success factors of decentralized KM, possibly paving the way for a new knowledge ecosystem appropriate to the digital era.

References

1. Alavi, M., Leidner, D.E.: Review: Knowledge management and knowledge management systems: Conceptual foundations and research issues. MIS Q. **25**(1), 107–136 (2001)
2. Nonaka, I.: A dynamic theory of organizational knowledge creation. Organ. Sci. **5**(1), 14–37 (1994)
3. Wasko, M.M., Faraj, S.: Why should I share? Examining social capital and knowledge contribution in electronic networks of practice. MIS Q. **29**(1), 35–57 (2005)
4. Jacobides, M.G., Cennamo, C., Gawer, A.: Towards a theory of ecosystems. Strateg. Manag. J. **39**(8), 2255–2276 (2018)
5. Beck, R., Pahlke, I., Seebach, C.: Knowledge exchange and symbolic action in social media-enabled electronic networks of practice. MIS Q. **38**(4), 1245–1270 (2014)
6. Murtazashvili, I., Murtazashvili, J.B., Weiss, M.B.H., Madison, M.J.: Blockchain networks as knowledge commons. Int. J. Commons **16**(1), 108–119 (2022)
7. Frozza, T., de Lima, E.P., da Costa, S.E.G.: Knowledge Management and Blockchain Technology for Organizational Sustainability: Conceptual Model. Brazilian Journal of Operations & Production Management **20**(1), e20231354 (2023)
8. Xu, Z., Liu, C., Zhang, P., Lu, T., Gu, N.: A blockchain-enabled wiki framework based on participatory design. Digital Communications and Networks **8**(6), 995–1004 (2022)
9. Abbas, H., Caprolu, M., Di Pietro, R.: Analysis of polkadot: Architecture, internals, and contradictions. arXiv:2207.14128 (2022)
10. Unfried, M.: Advancing longevity research through decentralized science. Frontiers in Aging **3**, 11317406 (2024)
11. Weidener, L., Spreckelsen, C.: Decentralized science (DeSci): definition, shared values, and guiding principles. Frontiers in Blockchain, 7, Article 1375763 (2024)
12. Ballandies, M.C.: To incentivize or not: Impact of blockchain-based cryptoeconomic tokens on human information sharing behavior. IEEE Access **10**, 74112–74127 (2022)
13. Arthur, C.: Wikipedia no worse for science info than Britannica, study finds. The Guardian. https://www.theguardian.com/technology/blog/2005/dec/14/wikipedianowo Accessed 1 July 2025
14. Stvilia, B., Twidale, M.B., Smith, L.C., Gasser, L.: Information quality work organization in Wikipedia. J. Am. Soc. Inform. Sci. Technol. **59**(6), 983–1001 (2008)
15. Wijesekara, P.A.D.S.N., Gunawardena, S.: A review of blockchain technology in knowledge-defined networking, its application, benefits, and challenges. Network **3**(3), 343–421 (2023)
16. Akhavan, P., Philsoophian, M., Rajabion, L., Namvar, M.: Developing a block-chained knowledge management model (BCKMM): beyond traditional knowledge management. In Proceedings of the 19th European Conference on Knowledge Management (ECKM) (pp. 18–27). Academic Conferences International. (2018)

A LLM-Driven Agent System for Automotive Fault Diagnosis with Integrated Reasoning and Expertise Sharing

Tzu-Ting Weng, Jun-Teng Sun, Zhen-Xin Fu, and Chung-Hong Lee(✉)

National Kaohsiung University of Science and Technology, Kaohsiung City 80778, Taiwan
leechung@mail.ee.nkust.edu.tw

Abstract. This study investigates the potential of large language models (LLMs) and intelligent agent technology in the field of automobile maintenance. With the development of artificial intelligence, the automotive maintenance process is gradually shifting from traditional manual operation to the use of AI agents to complete tasks. Especially in vehicle fault diagnosis, troubleshooting, and customer service, the use of LLMs and intelligent agent technology can significantly improve work efficiency. In this work, we developed an automotive fault-diagnosis agent system that incorporates large language models (LLMs) for analyzing user-supplied multimodal information queries and performs expert-like reasoning and questioning in the troubleshooting process, in order to make precise diagnostic and maintenance recommendations for vehicle repair and maintenance. In addition, in this study, we also evaluated the performance of various LLMs such as GPT, Gemini, Llama, and DeepSeek based models on system implementation. Based on several criteria, we also compare the applicability and performance of the LLM models for this application. The experimental results show that the developed system is capable of performing automotive fault diagnosis and providing appropriate maintenance recommendations.

Keywords: Large Language Model · Agent System · Fault Diagnosis

1 Introduction

With the rapid advancement of artificial intelligence (AI), many industries have begun adopting AI-powered systems to enhance efficiency and accuracy. Among these fields, automotive maintenance is undergoing a significant transformation, evolving from traditional manual troubleshooting toward automated fault diagnosis assisted by AI systems. Automotive fault diagnosis has long been a critical domain in intelligent system studies due to the increasing complexity of modern vehicles and the growing demand for efficient, accurate, and adaptive maintenance support. Accurate automotive fault diagnosis traditionally relies heavily on experienced mechanics who manually inspect, diagnose, and repair vehicle faults based on accumulated knowledge and personal expertise. However, this process is often time-consuming, costly, and prone to human error. Therefore, integrating AI technology into automotive fault diagnosis has become increasingly crucial. While commercial onboard diagnostics (OBD) systems [1] provide standardized fault code reporting, they offer limited interpretability and lack contextual reasoning capabilities. In addition, existing systems often have difficulty handling natural language input or integrating unstructured data, such as driver complaints or technician records, which are critical for diagnosing complex problems. As automobiles are increasingly software-defined and sensor-driven, there is a greater need for intelligent diagnostic systems capable of multimodal, adaptive reasoning tasks.

Recent advances in large language models (LLMs), such as OpenAI's ChatGPT [2], Gemini, Llama, and DeepSeek, have shown exceptional capabilities in natural language understanding, reasoning, and generating responses that closely resemble human-like expert reasoning. Such models possess the ability to interpret complex multimodal queries, provide contextual understanding, and make expert-level suggestions and diagnostic recommendations, significantly enhancing diagnostic accuracy and efficiency. At the same time, the growing volume of user-generated content, from repair records to forum discussions and social media threads, provides a rich corpus of crowd-sourced diagnostic insights that remains largely untapped by traditional automotive systems. By integrating these elements, a new class of automotive diagnostic agents can be constructed that not only makes inferences using LLMs, but also extracts viable knowledge from real-world discourse, with great potential for utility in the automotive industry.

The view is taken, therefore, in this study we explore the capacity of applying LLMs in developing an intelligent automotive fault-diagnosis agent system. Specifically, we propose an agent framework that integrates various state-of-the-art LLMs for analyzing multimodal user-supplied information, performing expert-like reasoning, dynamically engaging with users for further questioning, and ultimately delivering accurate diagnostic and maintenance recommendations. In addition, the study systematically evaluates and compares the performance of multiple LLM-based models to determine their respective applicability and effectiveness within the automotive domain. The proposed system is designed to (1) interpret user inputs and sensor data using LLM-powered natural language understanding; (2) generate diagnostic hypotheses through autonomous multi-step reasoning; and (3) enhance inference quality by retrieving and synthesizing relevant information from community-driven sources such as repair forums and technical discussion threads.

2 Related Work

Recently, AI approaches such as machine learning and LLM models are emerging as powerful tools for automotive fault diagnosis, offering new capabilities for interpreting complex vehicle data, reasoning about interconnected component failures, and providing transparent, actionable insights for both technicians and car owners [3]. Hossain have assessed the impact of AI-driven vehicle fault diagnosis systems, emphasizing their transformative role in modern automotive maintenance [4]. Beyond general diagnostics, specialized studies have been conducted on various automotive subsystems. For instance, Viswanathan et al. [5] proposed an innovative dry clutch fault diagnosis method that integrates image transformation techniques with Vision Transformers (ViT), significantly improving diagnostic accuracy. Tao et al. [6] developed an LLM-based framework using fine-tuning strategies like LoRA and QLoRA to enhance generalization in bearing fault diagnosis. Zhai and Wang [7] proposed a neural network-based model aimed at diagnosing external deformations in modern vehicles. In the context of new energy vehicles, Zhang et al. [8] introduced a RAG frame-work that integrates LLMs with knowledge graphs, effectively improving the accuracy and reasoning capabilities of fault classification and information retrieval. Regarding predictive failure analysis, Mandala [9] combined Kalman filtering with AI models to optimize remaining useful life (RUL) estimation and preventive maintenance strategies. Furthermore, the applications of LLMs in specialized domains have received increasing attention for applications in specialized domains. In particular, models such as GPT, Gemini, Llama, and DeepSeek have demonstrated excellent capabilities in areas such as semantic understanding [10], logical reasoning [11, 12], and multimodal information processing [13]. These advances have led to significant breakthroughs in areas such as healthcare [14, 15], education, and technical support, making LLM one of the core technologies for intelligent agent systems.

In comparison, traditional vehicle fault diagnosis primarily relied on rule-based systems and machine learning models [16, 17]. In several LLM applications, the developed systems are increasingly being used to process multimodal data, such as sensor readings, repair manuals, and technician notes, transforming these sources into human-interpretable insights. For example, Qaid et al. [18] proposed FD-LLM, which uses LLMs to interpret sensor data for fault diagnosis. Similarly, Ojima et al. [19] developed a retrieval-augmented generation (RAG) system that combines LLMs with knowledge graphs (KGs) to enhance knowledge management in automotive fault analysis. Pavlopoulos et al. [20] explored the application of multilingual pre-trained Transformer models in automotive fault diagnosis, demonstrating improvements in multilingual text classification efficiency within this domain.

While existing studies have demonstrated the potential of LLM-based fault diagnosis systems [21, 22], few systems have effectively integrated real-time inference with social knowledge in the domain of automotive maintenance. This study aims to address that gap by proposing an LLM-driven system that incorporates social media information to enhance real-world automotive fault diagnosis.

3 System Framework and Implementation

Figure 1 illustrates the architecture of the proposed vehicle fault diagnosis agent system centered around a Large Language Model (LLM). When a user inputs a vehicle-related issue, the system first converts the text into vectors and uses a retrieval module to search for relevant information within a database consisting of repair manuals, troubleshooting guides, and content extracted from social media. The retrieved results are then combined with predefined instructions (e.g., providing expert responses exclusively related to automotive repair), forming an augmented prompt for the LLM to generate a response. The output from the language model is processed and optimized through a data analysis module and can be stored in a memory module to retain valuable information, enhancing the accuracy and consistency of future interactions. Finally, the system delivers specific

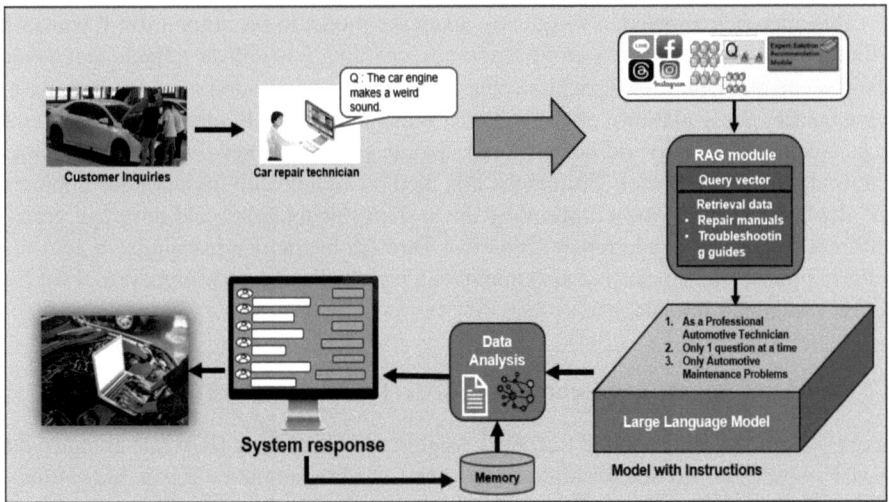

Fig. 1. System Framework

diagnostic insights and repair recommendations to the user. The functional module for gathering troubleshooting expertise from social networks to enhance automotive repair knowledge will be described later.

3.1 Adaptive Instruction Tuning for Automotive Diagnostics

In this study, we adopted the instruction learning approach to train large language models (LLMs) to better adapt to the specific task of automotive diagnostics and repair support. Among the various methods for training LLMs, instruction learning emerged as our preferred choice due to its advantages, such as not relying on large-scale datasets and its strong generalizability across different tasks.

When training models, one option is to fine-tune them using extensive datasets, which can make the model more specialized and accurate for a single task. However, this approach has significant drawbacks, including the need for substantial training time and computational resources. Moreover, if the diagnostic target or repair process changes, the dataset must also be updated, which can be costly and impractical in fast-paced service environments.

In contrast, the advantage of instruction learning lies in its independence from large amounts of training data. Instead, it guides the model to generate results by providing clear and specific instructions. This method requires much less data than large-scale fine-tuning and offers greater flexibility in responding to different vehicle problems.

In our system, we first needed to ensure that the model clearly understood its role as an automotive repair assistant. To achieve this, we designed a set of instructions to help the model focus on diagnostics, repair recommendations, and safety guidelines. These instructions not only include fundamental rules, such as prohibiting responses to unrelated topics, but also impose constraints on the response length to ensure they are clear, actionable, and efficient for mechanics or vehicle owners.

This approach allowed us to quickly adapt the model to the automotive repair task without requiring extensive training resources, enabling it to deliver reliable diagnostic insights and maintenance suggestions. Furthermore, the application of instruction learning enhances the scalability of our system. When expanding to other vehicle systems (e.g., electric motors, hybrid powertrains), we can simply add new instructions without retraining the entire model. Ultimately, this method significantly reduces development and deployment costs while improving the system's adaptability and coverage across different vehicle types and repair scenarios. Through instruction learning, we not only achieve precise diagnostics but also maintain a practical and robust support tool for the automotive repair domain.

3.2 Retrieval-Augmented Generation (RAG) for Automotive Fault Diagnosis

In our system, we incorporated Retrieval-Augmented Generation (RAG) to enhance the model's ability to provide accurate and context-aware automotive repair suggestions. RAG is a hybrid framework that combines the power of retrieval-based methods and generative models. Instead of relying solely on pre-trained knowledge, RAG allows the system to retrieve relevant documents or repair manuals in real time and then use that information to generate responses tailored to specific issues.

In the RAG framework, both unstructured and structured data can serve as supporting evidence, depending on the use case. Unstructured data lacks a predefined organizational format and includes sources such as PDF files, Google or Office documents, Wiki pages, images, and videos. These sources are rich in narrative or visual content and are suitable for tasks involving summarization or contextual understanding. In contrast, structured data is organized in a tabular format with defined rows and columns and typically exists in a database. Examples include customer records from business intelligence (BI) or data warehouse systems, transactional data from SQL databases, and output from enterprise application APIs such as SAP or Salesforce. The choice of data type in an RAG application should match the nature of the task: unstructured data is ideal for extracting insights from descriptive content, while structured data provides precise and reliable support when dealing with factual or transactional information. In this study, we use CSV files as the data format for RAG.

This is particularly beneficial in automotive repair, where up-to-date technical references, vehicle-specific schematics, and manufacturer guidelines are essential. For example, when diagnosing a warning light or engine fault code, the system can retrieve the latest service bulletins or parts diagrams and incorporate that information into its explanation. By grounding the model's responses in real-world documents, RAG significantly improves the reliability and relevance of its outputs.

Additionally, the integration of RAG offers scalability and adaptability. As new vehicles or technologies emerge, the system does not need to be re-trained; instead, it simply accesses newly added repair documents. This approach ensures that the system remains efficient, accurate, and aligned with the evolving automotive landscape.

3.3 Gathering Troubleshooting Expertise from Social Networks to Enhance Automotive Repair Knowledge

In particular, we use a hybrid approach that applies social media to improve automotive repair expertise. By building a knowledge base from online platforms such as automotive specialty forums on social media, we can better understand the practical experiences shared by professional technicians and automotive enthusiasts. We analyze the dataset using natural language processing techniques to identify common repair problems, emerging repair techniques, and user-reported failure cases for different car models. The knowledge base developed is combined with traditional training modules to provide technicians with instant access to solutions relevant to real-world situations. This approach not only improves the accuracy of system diagnostics, but also promotes the ability for continuous learning and an effective interface between formal education and practical experience. The system provides a highly adaptable knowledge support platform for automotive maintenance by introducing social media and data-driven knowledge integration. As shown in Fig. 2, the platform adopts the Two-Pass system structure. Given that social platforms in general contain groups of friends with both strong and weak ties, in this study, we utilize various social media platforms, such as Facebook and Line, to build a platform that answers users' questions through "social connections". By combining the "Expert-Solution Recommendation" function, we can increase automotive repair knowledge in the RAG module.

In the RAG Module, we utilize the hot topics of professional questions collected in recent years to build the "Q&A Knowledge Base"; through the external network, the system will continue to collect the answers to professional questions and hot topics online throughout the year, and then store the information related to these topics in the Q&A Knowledge Base, so the ability to answer questions will be enriched with the expansion of professional knowledge; the ability to answer questions will be enriched through the expansion of professional knowledge. Therefore, the ability to answer questions is enriched with the expansion of professional knowledge; through the real-time and rapid diffusion of community information, the LLM model is strengthened to generate more real-time and correct answers.

3.4 Adaptive Learning for Customized Automotive Maintenance

In this study, we developed an adaptive automotive maintenance system designed to provide real-time, vehicle-specific diagnostic recommendations. Given the highly variable nature of vehicle usage patterns and individual driving behaviors, a one-size-fits-all maintenance approach often falls short. While large language models (LLMs) have shown impressive capabilities in natural language understanding and general automotive knowledge, their generic nature limits their effectiveness in handling vehicle-specific conditions and histories. To address this gap, our system integrates historical maintenance records with real-time sensor data and diagnostic outputs to deliver customized recommendations. At the core of our design is a Memory-Augmented Neural Network (MANN), which enables the system to retain and utilize long-term contextual information specific to each vehicle.

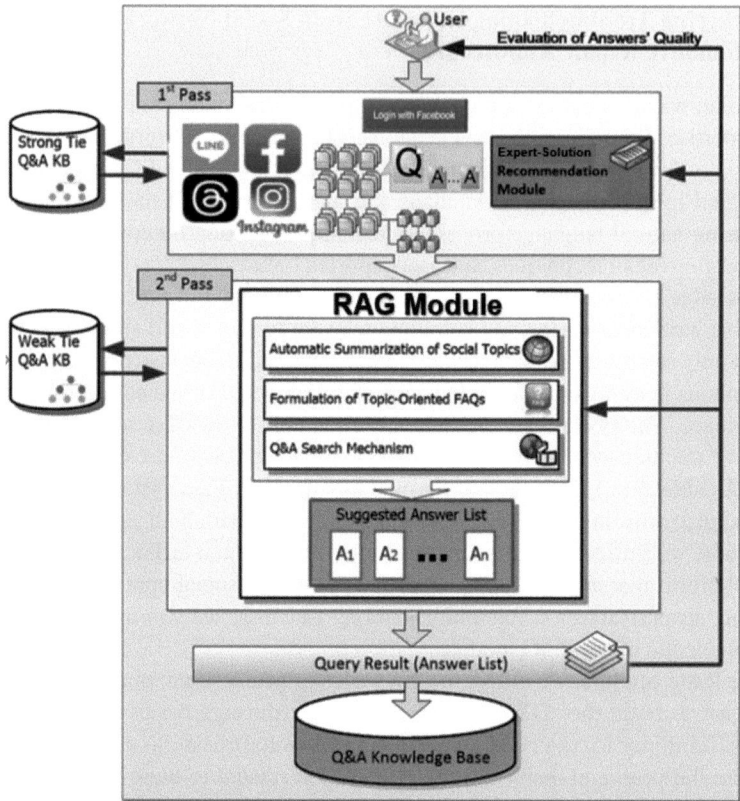

Fig.2. Gathering expertise from social networks to enhance professional knowledge of automotive troubleshooting

The MANN architecture includes an external memory module interfaced with a controller, allowing the system to store and retrieve key information such as past maintenance activities, driving patterns, and common failure signatures. As user interaction continues, a memory controller incrementally updates the stored representations, enabling more accurate and personalized responses over time.

4 Experimental Result and Evaluation

To evaluate the system's performance, we designed five key evaluation indicators: accuracy, F1-Score, multimodal capability, cost, and average response time. The selection of these indicators is based on the basic needs of practical applications. Accuracy and F1-Score evaluate the prediction correctness and stability of the model, ensuring high reliability when diagnosing faults or providing repair suggestions. Multimodal capability evaluates the model's ability to identify damaged vehicle parts and part conditions, crucial for enhancing multimodal functions. The choice of cost takes into account the model's expenditure on token input and output. The average response time reflects the system's operating efficiency, directly affects the user experience, and is especially important in time-sensitive or emergencies (Table 1).

Table 1. The version of LLM we tested in our experiment

LLMs	Model Type	Version
ChatGPT-4o	Closed Source	2025-03-26
ChatGPT o4 mini	Closed Source	2025-04-16
Gemini 2.0	Closed Source	Flash thinking-exp-01-21
Deepseek	Open Source	R1
Llama4	Open Source	Scout-17B-16E-Instruct

As shown in Fig. 3 and Table 2, GPT o4 mini achieved high scores in accuracy and F1-Score, and demonstrated excellent multimodal processing capabilities. In comparison, although GPT-4o's scores in multimodal processing capabilities, latency, and F1-Score are slightly higher than those of GPT o4 mini, when considering the cost price, GPT-4o is not considered because of its high price. Therefore, we finally chose GPT-o4 mini as the core model of the smart maintenance system.

In addition to natural language processing capabilities, we also conducted a preliminary comparison of the performance of each model in multimodal capabilities. Although GPT o4 mini's performance is slightly inferior to Gemini 2.0 Flash in some image tasks, GPT o4 mini has shown obvious advantages in semantic understanding and decision support. With its high accuracy and stable performance, GPT o4 mini can provide instant, reliable and practical suggestions in car maintenance applications to help users make wise decisions.

In summary, GPT o4 mini has demonstrated stability, instant response capabilities and adaptability in actual car service scenarios, fully verifying its feasibility and effectiveness as a core component of the intelligent maintenance support system. It is worth mentioning that, the model tested in the experiment is only for preliminary use in the system architecture, and other latest LLM models can still be selected in the future.

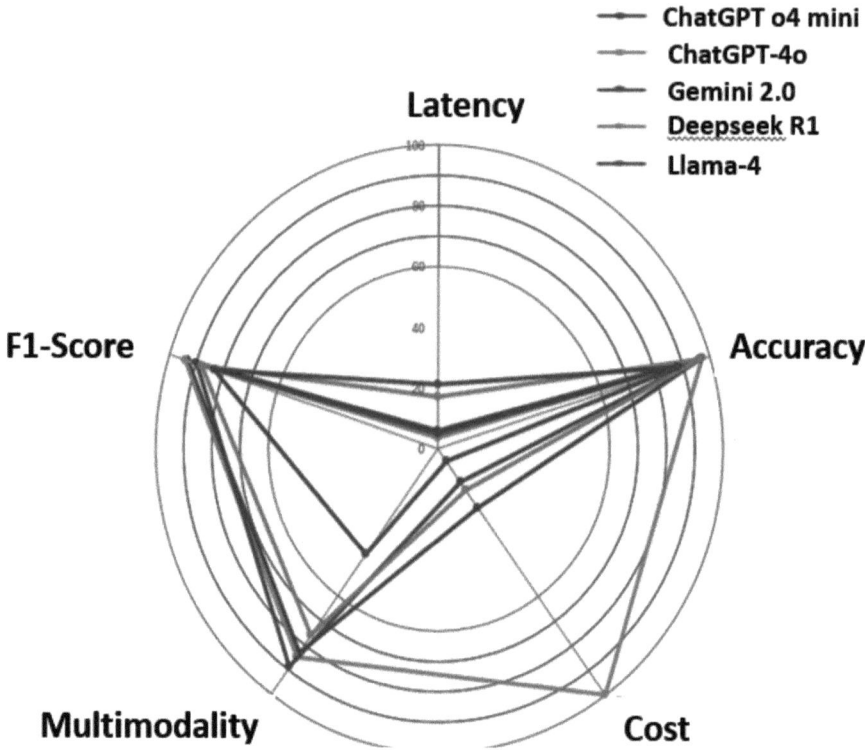

Fig. 3. Comparative Evaluation of Multimodal Language Models Across Core Performance

Table 2. Experimental Result of all tested LLMs

LLMs	Accuracy	Cost	Latency(Avg)
GPT-4o	97.3	100	3.5s
GPT o4 mini	97.2%	24	6s
Gemini 2.0	95.4%	13.5	5s
Deepseek	94.8%	16.9	17s
Llama4	88.3%	5	21s

5 Conclusion

This study demonstrates the feasibility and effectiveness of integrating large language models (LLMs) with intelligent agent technologies to create an AI-powered expert system for automotive diagnostics. By leveraging the advanced natural language understanding, reasoning capabilities, and instruction-following flexibility of LLMs, the proposed system can interact with users in a natural and context-aware manner, providing accurate and actionable maintenance suggestions.

Through the adoption of instruction learning, the system can efficiently adapt to various diagnostic scenarios without the need for extensive retraining, significantly reducing development costs and improving scalability. The integration of Retrieval-Augmented Generation (RAG) further enhances the model's relevance and precision by incorporating up-to-date repair manuals and technical documents during response generation. Additionally, the incorporation of social intelligence—by mining insights from automotive communities—enables the system to capture real-world repair experiences and emerging issues, effectively bridging the gap between formal training and practical knowledge.

Moreover, the implementation of adaptive learning and memory-augmented neural networks allows the system to provide personalized maintenance support based on vehicle-specific data and usage history. This dynamic, user-centric design ensures high diagnostic accuracy and contextual adaptability, even in complex and evolving vehicle environments.

Experimental results validate the system's strong performance, with GPT o4 mini achieving high diagnostic accuracy in the tested LLMs, while having low latency and cost. These findings affirm the system's potential as a robust, efficient, and smart solution for modern automotive diagnostics, paving the way for wider application of LLM-based agent systems in real-world service environments.

References

1. Saibannavar, D., Math, M.M., Kulkarni, U.: A survey on on-board diagnostic in vehicles. In: Raj, J.S. (eds) International Conference on Mobile Computing and Sustainable Informatics. ICMCSI 2020. EAI/Springer Innovations in Communication and Computing. Springer, (2021). https://doi.org/10.1007/978-3-030-49795-8_5
2. Roumeliotis, K.I., Tselikas, N.D.: ChatGPT and Open-AI models: A preliminary review. Future Internet **15**, 192 (2023). https://doi.org/10.3390/fi15060192
3. Hulbert, S., Mollan, C., Pandey, V.: Fault diagnosis and prediction in automotive systems with real-time data using machine learning. SAE Technical Paper 2022-01-0217 (2022). https://doi.org/10.4271/2022-01-0217
4. Hossain, M.N., Rahman, M.M., Ramasamy, D.: Artificial intelligence-driven vehicle fault diagnosis to revolutionize automotive maintenance: A review. Computers, Materials & Continua (2024), https://doi.org/10.32604/cmes.2024.056022
5. Viswanathan, P.C., Banerjee, A., Sridharan, N.V., Chakrapani, G., Vaithiyanathan, S.: Advancing automobile dry clutch fault diagnosis through innovative imaging techniques and vision transformer integration. Measurement **220**, 115975 (2024). https://doi.org/10.1016/j.measurement.2024.115975
6. Tao, L., Liu, H., Ning, G., Cao, W., Huang, B., Lu, C.: LLM-based Framework for Bearing Fault Diagnosis. arXiv preprint arXiv:2411.02718 (2024). https://arxiv.org/abs/2411.02718
7. Zhai, S., Wang, Z.: Research on automobile fault diagnosis model based on neural network. In: 2024 IEEE 4th International Conference on Power, Electronics and Computer Applications (ICPECA), pp. 411–418. IEEE, Shenyang, China (2024). https://doi.org/10.1109/ICPECA60615.2024.10471060
8. Zhang, H., Zhao, Y., Sun, B., Wu, Y., Fu, Z., Xiao, X.: Large language model based intelligent fault information retrieval system for new energy vehicles. Appl. Sci. **15**(7), 4034 (2025). https://doi.org/10.3390/app15074034

9. Mandala, V.: Predictive failure analytics in critical automotive applications: Enhancing reliability and safety through advanced AI techniques. Journal of Artificial Intelligence and Big Data **3**(1), 4–16 (2023)
10. Zhang, Z., Wu, Y., Zhao, H., Li, Z., Zhang, S., Zhou, X. and Zhou, X. 2020. Semantics-Aware BERT for Language Understanding. Proceedings of the AAAI Conference on Artificial Intelligence. 34, 05 (Apr. 2020), 9628–9635. https://doi.org/10.1609/aaai.v34i05.6510
11. Creswell, A., Shanahan, M., Higgins, I.: Selection-inference: Exploiting large language models for interpretable logical reasoning. arXiv preprint arXiv:2205.09712 (2022), https://doi.org/10.48550/arXiv.2205.09712
12. Pan, L., Albalak, A., Wang, X., Wang, W.Y.: Logic-LM: Empowering large language models with symbolic solvers for faithful logical reasoning. Findings of EMNLP 2023, arXiv preprint arXiv:2305.12295 (2023), https://doi.org/10.48550/arXiv.2305.12295
13. Zhu, Y., Yuan, H., Wang, S., Liu, J., Liu, W., Deng, C., Chen, H., Liu, Z., Dou, Z., Wen, J.-R.: Large language models for information retrieval: A survey. arXiv preprint arXiv:2308.07107 (2024). https://arxiv.org/abs/2308.07107
14. Mehandru, N., Miao, B.Y., Almaraz, E.R., et al.: Evaluating Large Language Models as Agents in the Clinic. npj Digital Medicine 7, 84 (2024). https://doi.org/10.1038/s41746-024-01083-y
15. Qiu, J., Lam, K., Li, G., et al.: LLM-based agentic systems in medicine and healthcare. Nature Machine Intelligence **6**, 1418–1420 (2024). https://doi.org/10.1038/s42256-024-00944-1
16. Denton, T.: Advanced automotive fault diagnosis: Automotive technology: Vehicle maintenance and repair. 5th edn. Routledge, London (2020). https://doi.org/10.1201/9780429317781
17. Su, H., Xiang, L., Hu, A.: Application of deep learning to fault diagnosis of rotating machineries. Meas. Sci. Technol. **35**(4), 042003 (2024). https://doi.org/10.1088/1361-6501/ad1e20
18. Qaid, H.A.A.M., Zhang, B., Li, D., Ng, S.-K., Li, W.: FD-LLM: Large language model for fault diagnosis of machines. arXiv preprint arXiv:2412.01218 (2024), https://arxiv.org/abs/2412.01218
19. Ojima, Y., Sakaji, H., Nakamura, T., Sakata, H., Seki, K., Teshigawara, Y., Yamashita, M., Aoyama, K.: Knowledge management for automobile failure analysis using graph RAG. arXiv preprint arXiv:2411.19539 (2024), https://arxiv.org/abs/2411.19539
20. Pavlopoulos, J., Romell, A., Curman, J., Steinert, O., Lindgren, T., Borg, M.: Automotive multilingual fault diagnosis. arXiv preprint arXiv:2210.06918 (2022), https://arxiv.org/abs/2210.06918
21. Alsaif, K.M., Albeshri, A.A., Khemakhem, M.A., Eassa, F.E.: Multimodal large language model-based fault detection and diagnosis in context of industry 4.0. Electronics **13**(24), 4912 (2024), https://doi.org/10.3390/electronics13244912
22. Zheng, S., Pan, K., Liu, J., Chen, Y.: Empirical study on fine-tuning pre-trained large language models for fault diagnosis of complex systems. Reliab. Eng. Syst. Saf. **251**, 110382 (2024). https://doi.org/10.1016/j.ress.2024.110382

Sales Predictive Model with Customer Segmentation Integration

Alex Mejía and Priscila Valdiviezo-Diaz[✉]

Department of Computer Science, Universidad Técnica Particular de Loja,
San Cayetano Alto, Loja 1101608, Ecuador
{armejia1,pmvaldiviezo}@utpl.edu.ec

Abstract. Predictive analytics is a fundamental tool for optimizing business decisions in contexts such as e-commerce and marketing. This paper proposes a sales prediction model that integrates customer segmentation information generated through clustering techniques. Three customer segments are identified from an e-commerce transaction dataset: the first characterized by customers who make recent, low-volume purchases; the second group related to customers with high historical investment but no recent activity; and the third group, customers who purchase frequently at medium levels. This segmentation was an additional predictor variable in the regression model to predict future sales. Two clustering algorithms, DBSCAN and K-means, were tested and evaluated with the Silhouette metric, while the prediction model using Random Forest was evaluated with conventional metrics such as Root Mean Square Error and R-Squared. The results indicate that the integration of clusters into supervised models can enhance prediction models, demonstrating good performance.

Keywords: Machine learning · Sales prediction · Clustering

1 Introduction

In today's business environment, a vast amount of data is generated daily, and understanding and anticipating customer behavior has become critical to ensuring organizational success. Companies must analyze information about their products and services to optimize the allocation of marketing and sales resources over time. In this context, predictive analytics emerges as a useful tool that, leveraging historical data and machine learning algorithms, allows the identification of customer patterns, trends, and preferences and the prediction of their future actions [1]. A key strategy to improve the accuracy and relevance of predictive analytics is user segmentation [2]. This allows a better understanding of customer needs and the adaptation of marketing strategies and product offerings to specific segments [3]. In addition, companies can perform more accurate predictive analyses and gain more detailed insight into each segment's behavior.

While most prediction systems estimate preferences based on historical data, they can also predict these preferences by considering user similarities. Companies can identify common traits and behaviors of their users by analyzing historical data and applying predictive models. This knowledge can be used to develop targeted marketing strategies and personalized offers to retain and attract similar customers.

Some works in the literature have focused on the application of predictive analytics by applying machine learning algorithms to predict customer behavior [1] and to classify new customers into existing segments [2]. While other studies use clustering techniques to segment customers based on similar characteristics to improve marketing and product offerings [4].

Although predictive analytics provides tools for forecasting future actions, efficiently allocating resources and making informed decisions based on generalized predictions can be problematic. The need for relevant predictions considering customer segments becomes crucial for more effective strategies. Therefore, it is a challenge to effectively integrate segmentation with predictive analytics to improve the accuracy and applicability of customer behavior predictions.

The objective of this paper is to apply predictive analytics, integrating customer segmentation as a feature addition to the customer's individual features, to gain valuable insights into user behavior and preferences. Using machine learning techniques, knowledge is inferred based on similar characteristics between customers, and sales are predicted considering the identified cluster segments.

The paper is organized as follows. Section 2 presents related work. Section 3 includes the materials and methods used in this study. Section 4 presents the experimental results with the clustering and prediction algorithms. Finally, Sect. 5 presents the conclusions and future work.

2 Related Work

Predictive analytics applied to user segmentation has been widely studied to support different disciplines, including digital marketing [5], business [6], e-government [7], and others.

Segmentation approaches that use machine learning models enable more precise and dynamic classification of users based on their behaviors and preferences. For example, numerous studies have investigated the application of clustering algorithms, such as K-Means [8] and DBSCAN [9], to identify patterns in user data. Moreover, supervised models, such as decision trees, have proven effective in predicting the assignment of new users to predefined segments.

Table 1 presents a series of previous works, highlighting a description of the paper, the machine learning method, and the dataset used.

Related work demonstrates the growing interest in applying machine learning techniques for sales prediction in e-commerce and retail settings. Most of the research focuses on employing supervised algorithms such as linear regression, random forests, and neural networks to estimate sales volumes based on historical variables.

Table 1. Previous work analyzed in this paper

Author	Description	Algorithms	Data used
[10]	Authors present customer segmentation in retail through predictive analytics for marketing budgeting, customer targeting, and customized offers	K-Means Markov Model	Historic retail transactional data is used
[3]	The paper focuses on the clustering of customers using recency, frequency, and monetary (RFM) analysis and K-means clustering	K-means, Boosting tree	Marketing campaign dataset, that encompasses several key demographic attributes
[11]	Authors explore the application of clustering techniques to group customers with similar behavioral patterns and include explainability using LIME (Locally Interpretable Model-Agnostic Explanations) to understand the factors influencing the groups	K-means	Mall customer dataset, including demographic attributes, annual income, and spending scores
[12]	Authors use clustering algorithms to group users with similar behaviors and characteristics by combining related association rules with multivalued discrete features	K-mode, Association Rules	two specific datasets are used, which contain user characteristics data
[13]	This paper analyzes customer purchasing behavior in the retail sector and utilizes association rules to identify frequent itemsets and purchasing patterns in order to predict customer behavior	Apriori algorithm, Large Language Model (LLM)	Authors do not specify the dataset used, they mention the use of AWS S3 for data storage
[14]	Authors use machine learning models such as linear regression, Random Forest regressor, XGBoost, and the LSTM algorithm to forecast future sales based on historical data	Random Forest, XGBoost, LSTM	"Store Item Demand Forecasting" dataset includes sales data for items corresponding to ten stores
[15]	The authors employ clustering techniques to segment data on plants and crops and then analyze the correlation between individual products for Plant Crop Sales Prediction	Correlation Clustering model	dataset on fresh plants and crops
[16]	The authors investigate machine learning methods for predicting revenue for a retail company Moreover, they find crucial elements that have a big effect on purchases	Neural networks, Decision trees, Random Forest, and linear regression	past sales dataset from a retail company
[17]	Authors use data mining to analyze and predict user consumption behavior, moreover, the entropy weight method is used to segment e-commerce consumers based on RFM	Bayesian method	consumer purchase data and social media data

Likewise, other studies have addressed customer segmentation using unsupervised algorithms, such as K-means, to improve marketing personalization or customer retention. However, in many of these approaches, segmentation is presented as an isolated step, without being formally integrated into the predictive modeling process.

Therefore, this study is based on integrating the segmentation variable as an additional input to the sales prediction model. This incorporation enhances the representation of customer behavior within the supervised model, capturing non-linear relationships between segments and sales that traditional approaches often overlook. Thus, the proposed approach combines supervised and unsupervised machine learning techniques, offering good predictive accuracy and contributing to generating hybrid models for business decision making.

On the other hand, previous studies also show that variables such as the level of interaction on social networks and sentiment analysis to analyze consumer behavior can complement segmentation and prediction models [17,18]. Thus, in the future, we plan to incorporate additional data, such as social media information, into the segmentation and sales forecasting process to enrich the customer context and enhance the model's predictive capabilities.

3 Materials and Methods

This section presents the algorithms tested on the selected dataset and the conventional metrics selected to measure the performance of the supervised and unsupervised models. The components of the proposed prediction approach are also described.

3.1 Dataset

The dataset used in this study is available in the Kaggle repository[1] and refers to information on transactions made in an online retail company in the UK over eight months. The dataset contains 541,909 records and 8 variables refer to transactional data on sales of unique gifts for all occasions. Table 2 shows the variables included in the dataset.

The dataset was processed as follows.

- Records with empty values in the CustomerID column were removed, which in this case were less than 24%.
- Quite a lot 4% of the transactions were found to have unusual quantities, while around 6% of the unit prices were outliers. For this reason, the outlier values of the Quantity and UnitPrice variables were eliminated, as they can affect the algorithm results.
- Negative values were removed from the Quantity and UnitPrice variables, as there are no negative purchase quantities.

---

[1] https://www.kaggle.com/code/amirmotefaker/predicting-sales-e-commerce/notebook.

Table 2. Variables description

Variable	Description
InvoiceNo	Unique invoice number
StockCode	Unique product code in stock
Description	Product description
Quantity	Number of products purchased
InvoiceDate	date and time of the transaction
UnitPrice	Unit price of each product
CustomerID	Customer identifier
Country	Country where the purchase was made

– Two new variables were created: totalValue, which results from multiplying the attributes Quantity and UnitPrice; and lastPurchaseGap, which results from calculating the number of days since the last purchase.

Consequently, the final dataset is prepared by maintaining unique customer records using the CustomerID variable to eliminate duplication of multiple purchases reflected in different invoice codes. To avoid losing valuable information one different purchases of a customer, the lastPurchaseGap attribute was created. This process ensures that each customer is represented only once in the clustering model, thus ensuring a segmentation representative of individual behavior.

As a result, a final dataset with 168,649 records was obtained. Four variables were selected for experiments with supervised and unsupervised machine learning algorithms: Quantity, UnitPrice, totalValue, and lastPurchase. The Country and Description variables are not considered for segmentation for several reasons: The country variable contains many different countries, which can generate an excessive number of categories, complicating the analysis. The description variable, since it contains free text, cannot be easily quantified, creates noise, and has excessive variability that is not relevant for segmentation.

Considering that the selected variables are on different scales, the sales data was normalized using a natural logarithmic transformation [19] to reduce the dispersion or bias of the data. This type of transformation minimizes the scale of large values to prevent them from affecting the linear regression model used in this study. Furthermore, logarithmic transformation enables linearization of the relationship between the dependent and independent variables, contributing to improving model performance in terms of sales prediction.

3.2 Algorithms and Metrics

This section details the clustering algorithms and prediction algorithms used to predict sales once user segmentation has been performed. For segmentation, the K-means [20] and DBSCAN [9] algorithms were tested using the Silhouette metric.

K-means and DBSCAN were selected because they allow exploration of different structures in the data, are easy to implement, and can handle large datasets. Another reason is that DBSCAN is more effective at detecting well-separated clusters and handling noise, while K-means produces more compact clusters, but can be sensitive to outliers. Experimentation with these two algorithms allowed the results to be compared to select the most appropriate segmentation algorithm.

To predict sales, the Random Forest (RF) algorithm is used based on the obtained purchasing behavior segments. According to [21], the RF algorithm is used for both classification and regression problems. RF performance is measured using the R-squared to determine how well the regression model fits the data, and Root Mean Squared Error metrics to measure the average difference between the model's predicted values and the actual values [22].

3.3 Proposed Prediction Approach

The proposed approach to sales prediction integrates user segmentation techniques using clustering algorithms before predictive modeling. In this approach, each user is assigned to a cluster that represents a segment with homogeneous behavioral characteristics. The assigned cluster label is explicitly incorporated as a characteristic (predictive variable) into the sales prediction model, along with the other variables used in the segmentation.

Figure 1 shows the schematic of the sales prediction model based on user segmentation. The process highlights the explicit incorporation of clustering into the prediction.

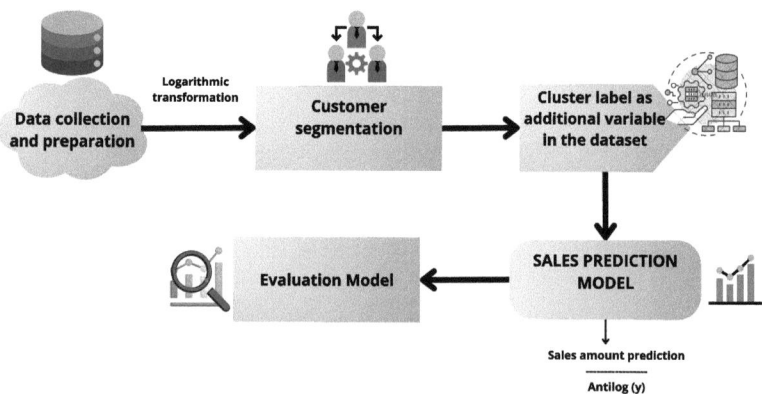

Fig. 1. Proposed approach to sales prediction

Since the predictions are calculated with the transformed model, that is, generated with the data set that was normalized using logarithmic transformation (natural logarithm), as shown in Eq. 1. To obtain the prediction on the original scale, we apply the antilogarithm. (see Eq. 2).

$$ln(y) = \beta_0 + \beta_1 ln(X_1) + \beta_2 ln(X_2) + \beta_3 ln(X_3) + \beta_4 ln(X_4) + e \qquad (1)$$

$$\hat{y} = exp(y) \qquad (2)$$

4 Results

This section consists of two parts. The first presents a comparison of the performance of the clustering algorithms using the Silhouette metric. The second part analyzes the results of the Random Forest algorithm using metrics to measure prediction quality.

4.1 Segmentation Results

The Fig. 2 shows the silhouette values for different values of k (number of clusters). As can be seen, the maximum silhouette value is 0.69 at k = 3, which indicates a good grouping, with a good separation and cohesion, and this means that the cluster labels could be used as a predictor variable.

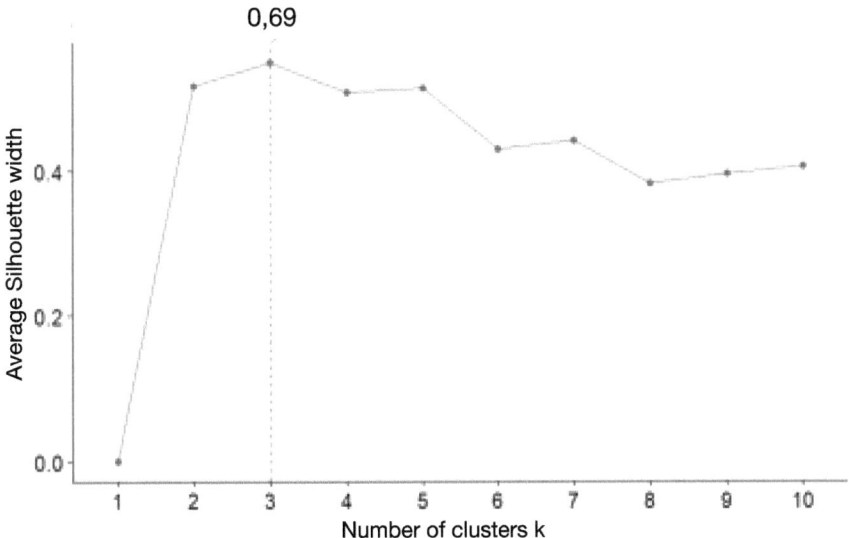

Fig. 2. Silhouette values for different numbers of clusters

On the other hand, in the DBSCAN algorithm, two parameters, epsilon (eps) and the minimum number of points (minPts) must be set. In this case, the best

Table 3. Clustering results

Algorithm	Silhouette	#Cluster
K-means	0.69	3
DBSCAN	0.4	4

clustering results were obtained with eps = 0.5 and minPts = 5, resulting in a silhouette value of 0.4.

Table 3 summarizes the results obtained with both clustering algorithms.

From the Table 3, it can be seen that the best clustering result is with K-means. This algorithm obtained three groups: Cluster 1 contains customers who have made purchases recently, but in relatively low quantities. This group of customers might be thought to be in an early stage of interaction with the brand and may still be testing the products. Cluster 2, is made up of customers who make large, but not recent, purchases; they represent a high potential value for the company. Finally, Cluster 3 is made up of customers who buy frequently and in moderate quantities. These are stable customers, committed to the brand.

Figure 3 shows the distribution of customers across the three clusters, revealing the distinct purchasing behaviors:

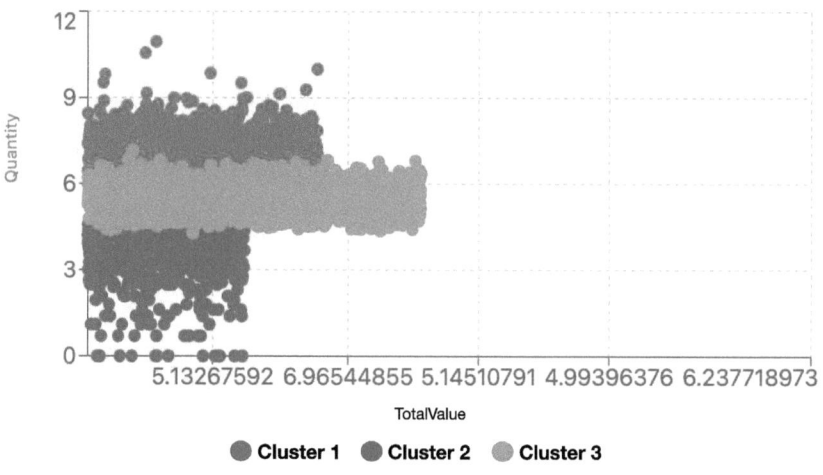

Fig. 3. Distribution of data in the three groups for TotalValue vs. Quantity

From Fig. 3, we can observe a clear distinction in purchase volume (Quantity) driving TotalValue. The segmentation results indicate that while clusters capture significant segments, some customers may exhibit characteristics across different clusters. These patterns suggest that companies could target high-volume clusters with loyalty programs and low-volume clusters with promotions.

4.2 Prediction Results

For the prediction process, first the cluster label is added as a feature to the cleaned dataset, which allows incorporating information about the client's behavior that is not derived from the individual variables. Then, the dataset is divided into 70% training set and 30% testing set. The model is generated by applying the Random Forest algorithm to the training set.

Table 4 shows the results in the prediction quality metrics using the test set.

Table 4. Metric values for Random Forest algorithm

Metrics	Value
R-Squared	0.885
Adjusted R-Squared	0.884
RMSE	0.052

From Table 4, we can conclude that the model explains approximately 88% of the variability in the target variable, reflecting an excellent level of fit. Furthermore, the RMSE value obtained indicates that the predictions are highly accurate. These results demonstrate the predictive power of the model, which can be used to support strategic decisions of the company.

Taking into account the calculated predictions, each group's performance relative to projected sales can be identified by grouping sales results by segment. Therefore, each segment contributes distinct patterns to the model, demonstrating a segmentation with differentiated consumption dynamics that enriches the modeling. These results would help customize marketing strategies based on anticipated purchasing potential; moreover, would allow e-commerce companies to tailor offers to the characteristics and behaviors of each customer group, likewise, adjust their inventory levels more precisely, and develop products and services tailored to specific needs and preferences.

Importance of Characteristics: To estimate the importance of the predictor variables, two approaches are considered: 1) the calculation of the importance of each variable permutation and 2) the importance of the Gini variable. The first approach refers to the increase in the mean square error (MSE) in the model when the particular variable is permuted, and the second is based on the changes in the impurities of the nodes at each split in each tree in the random forest [23].

As a result of the application of these two approaches, it was determined that TotalValue and UnitPrice were the most relevant variables in the predictive model, aligned with sales, since these variables are directly related to the economic value of transactions. Their high significance suggests that the customer's historical spending behavior is a strong predictor of their future purchasing behavior.

In contrast, the variables lastPurchase and cluster reflect less importance in the model. However, the cluster variable provides additional information about the types of customers by combining behavioral patterns from other variables. Therefore, its inclusion enhances the model by providing a categorical view of customer heterogeneity. Furthermore, the cluster value enables one to associate the model's results with specific segments, facilitating the making of strategic decisions such as personalizing offers or targeting campaigns.

5 Conclusion and Future Work

This study presents the integration of segmentation techniques using clustering algorithms within supervised predictive models. Incorporating the cluster variable as a predictor enriches the model by providing a categorical view of customer heterogeneity and allows model results to be associated with specific segments. Segmentation based on derived attributes, such as purchase recency and total value, adds value to the interpretation of the clustering results.

The results show the superiority of the K-means algorithm in the segmentation of customers with respect to the DBSCAN in the silhouette metric. On the other hand, the Random Forest algorithm showed a strong predictive capacity of the sales prediction model in terms of R-squared and RMSE metrics, suggesting that segmentation adds value to the model and can support marketing or inventory planning decisions.

This study focused on transactional and purchasing behavior data; therefore, one limitation is that it does not incorporate social media metrics for modeling. However, in future work, we propose: 1) exploring more complex models, such as deep neural networks, to compare predictive performance; 2) incorporating temporal variables and social media data to enrich customer context and evaluate their contribution to model performance.

References

1. GhorbanTanhaei, H., Boozary, P., Sheykhan, S., Rabiee, M., Rahmani, F., Hosseini, I.: Predictive analytics in customer behavior: anticipating trends and preferences. Results Control Optim. **17**, 100462 (2024)
2. Levin, N., Zahavi, J.: Predictive modeling using segmentation. J. Interact. Mark. **15**(2), 2–22 (2001)
3. Kasem, M.S., Hamada, M., Taj-Eddin, I.: Customer profiling, segmentation, and sales prediction using AI in direct marketing. Neural Comput. Appl. **36**(9), 4995–5005 (2024)
4. Salazar-Santander, C., Cawley, A.F.M., Martinez-Troncoso, C.: An optimal effectiveness-driven target segment selection modeling approach for marketing campaign management. Comput. Ind. Eng. **202**, 110945 (2025)
5. Soni, H.K., Sharma, S.: Big data analytics for market prediction via consumer insight. In: Sharma, S., Rahaman, V., Sinha, G.R. (eds.) Big Data Analytics in Cognitive Social Media and Literary Texts, pp. 23–46. Springer, Singapore (2021). https://doi.org/10.1007/978-981-16-4729-1_2

6. Abdullahi, A.M., Htet, S.A., Naing, N.N.N., Hossain, M.S., Ismail, S., Zaaba, M.A.M.: A prediction of customer segmentation using random forest in rapidminer, pp. 33–38 (2023). https://ieeexplore.ieee.org/document/10563451
7. Cestnik, B., Kern, A.: Using predictive analytics for user segmentation in e-government, vol. Part F130282, pp. 125–129 (2017). https://doi.org/10.1145/3129757.3129779
8. Nandapala, E., Jayasena, K.: The practical approach in customers segmentation by using the k-means algorithm. In: 2020 IEEE 15th International Conference on Industrial and Information Systems (ICIIS), pp. 344–349 (2020)
9. Yan, X., Li, Y., Nie, F., Li, R.: Bank customer segmentation and marketing strategies based on improved DBSCAN algorithm. Appl. Sci. **15**(6) (2025)
10. Harish, A., Malathy, C.: Customer segment prediction on retail transactional data using k-means and Markov model. Intell. Autom. Soft Comput. **36**(1), 589–600 (2023)
11. Alijoyo, F.A., et al.: Personalized marketing: leveraging AI for culturally aware segmentation and targeting. Alexandria Eng. J. **119**, 8–21 (2025). https://www.sciencedirect.com/science/article/pii/S1110016825000997
12. Zhang, B., Wang, L., Li, Y.: Precision marketing method of e-commerce platform based on clustering algorithm 2021 (2021). https://dl.acm.org/doi/10.1155/2021/5538677
13. Farheen, Z., Dharani, A.: Prediction of customer purchasing patterns for retail optimization using market basket techniques. In: 2024 8th International Conference on Computational System and Information Technology for Sustainable Solutions (CSITSS), pp. 1–5 (2024)
14. Loukili, M., Messaoudi, F., Ghazi, M.E., Azirar, H.: Predicting future sales: a machine learning algorithm showdown. In: Farhaoui, Y., Hussain, A., Saba, T., Taherdoost, H., Verma, A. (eds.) ICAISE 2023. LNNS, vol. 837, pp. 26–31. Springer, Cham (2024). https://doi.org/10.1007/978-3-031-48465-0_4
15. Zhao, H., Bai, L., Yang, Y., Li, Y.: Research on plant crop sales prediction based on correlation clustering and product attributes, pp. 730 – 735 (2023)
16. Yadav, P.K., Kumar, V., Bhushan, R., Singh, P.K.: Analysis of machine learning model for predicting sales forecasting (2023)
17. Zhou, M.: Social media-based e-commerce consumer behavior prediction model in marketing strategy **9**(1) (2024). https://sciendo.com/article/10.2478/amns-2024-2655
18. Iezzi, D.F., Monte, R.: Social media effects on sales: consumer sentiment in a state-space model. In: Giordano, G., Misuraca, M. (eds.) JADT 2022, pp. 77–90. Springer, Cham (2024). https://doi.org/10.1007/978-3-031-55917-4_7
19. Fornieles, A.: Transformaciones de datos en la elaboración de estudios salariales. Revista de Psicología del Trabajo y de las Organizaciones **29**(2), 75–82 (2013). https://www.sciencedirect.com/science/article/pii/S1576596213700112
20. Zhang, J., Zhang, Y., Tang, F., Song, Y., Deng, Y., He, S.: E-commerce retail merchandise based on optimized k-means algorithm and multi-model fusion demand forecasting research, pp. 512–516 (2024)
21. Sabeena, K., Anitha, G.: Sales prediction using machine learning algorithms (2025)
22. Prathiba, L., Raja, S., Umadevi, A., Sucharitha, M.M., Basha, M.S.A.: Leveraging ensemble methods for accurate prediction of customer spending scores in retail (2024)
23. Aldrich, C.: Process variable importance analysis by use of random forests in a shapley regression framework. Minerals **10**(5) (2020). https://www.mdpi.com/2075-163X/10/5/420

Study on the Use of Generative AI by Japanese IT Freelance Engineers: Effectuation as a Theory of Entrepreneurial Behavior

Kentaro Goto

Department of Informatics, Graduate School of Integrated Science and Technology, Shizuoka University, 3-5-1, Johoku, Chuo-ku, Hamamatsu-shi, Shizuoka 432-8011, Japan
goto-k@inf.shizuoka.ac.jp

Abstract. This study investigates the integration of generative artificial intelligence (AI) into the professional practices of Japanese freelance information technology (IT) engineers. To support the analysis, an entrepreneurial orientation index was employed in the survey. Moreover, this study examines the use of generative AI between two groups: those who consider freelancing a primary occupation and those who consider it a secondary occupation. An ordered logistic regression analysis investigates the factors associated with a decision-making style rooted in effectuation, a concept commonly tied to entrepreneurial behavior. The responses of the questionnaire survey indicate that freelance IT engineers who consider freelancing as their primary occupation are likelier to adopt generative AI in their professional activities than those who view it as a secondary occupation. An ordered logistic regression analysis revealed that in the primary occupation group, higher engagement with generative AI was significantly associated with a tendency toward causation orientation. However, the secondary occupation group showed no statistically significant relation with effectuation orientation.

Keywords: Effectuation · Generative artificial intelligence · Freelance engineer

1 Introduction

Freelancing refers to individuals performing tasks by outsourcing rather than under conventional employment contracts, which has become increasingly prevalent. This trend is particularly pronounced in the rapidly evolving information technology (IT) field, where improving work efficiency has become a pressing concern for freelance professionals. Generative artificial intelligence (AI) has attracted significant attention as a means of enhancing productivity across various business domains. A recent McKinsey survey found that 78% of organizations have adopted AI in at least one business function, reflecting an increase from 72% in early 2024 and 55% in the previous year [1].

Efficiency is crucial for freelancers who work independently, and recent data show that 73% of freelancers currently use generative AI tools in their professional activities [2]. Therefore, freelance engineers are adopting generative AI to streamline their workflows and improve performance. However, despite the growing interest in AI, research has not sufficiently explored the specific practices and applications of generative AI among IT freelance engineers.

The "effectuation" concept proposed by Sarasvathy has recently attracted considerable attention as a theoretical framework for entrepreneurial decision-making [3]. Effectuation refers to a set of decision-making principles derived from the behaviors of expert entrepreneurs. These entrepreneurs are individuals or teams who have founded at least one startup and worked full-time as founders or entrepreneurs for over 10 years. They are also distinguished by having successfully led at least one startup to an initial public offering.

Effectuation's key feature lies in its flexible approach, wherein decisions are made based on the entrepreneur's available means rather than predetermined goals. Conversely, the competing model of "causation" follows a more goal-oriented and systematic process. Clear objectives are established first, followed by organizing necessary resources. These two concepts represent distinct approaches to entrepreneurial decision-making and are applied following different contextual factors.

Studies on effectuation in the IT sector have explored its role in developing wireless integrated circuit (IC) tags [4]. One such study underscores a case of a robotics researcher at MIT and a product manager from a cosmetics company collaborated, ultimately leading to the emergence of a low-cost wireless IC tag market. Inspired by the researcher's presentation, the product manager identified a novel application for an inexpensive microchip and contributed to market creation. This adaptability is consistent with the effectuation principles and demonstrates the importance of entrepreneurial flexibility in the IT domain.

Chandler et al. [5] and Brettel et al. developed measurement indices to assess entrepreneurial orientation more systematically [6]. Using Chandler et al.'s measure, prior research has shown that the application of effectuation and causation differs across phases of the business process and by demographic variables such as gender, parenting experience, and career stage [7, 8].

Additionally, to thrive amid high uncertainty, IT companies have been advised to adopt a balanced approach that integrates both effectuation and causation along with design thinking [9]. Therefore, it is essential to have a comprehensive understanding of the effectuation and causation decision-making models.

As the use of generative AI continues to grow rapidly, it is vital to examine its application in real-world professional contexts. This study focuses on freelance IT engineers in Japan, highlighting the differences between primary and secondary occupation groups and explores how generative AI is being incorporated into their work. A questionnaire-based survey was conducted to examine the current status of generative AI utilization using an established index for evaluating entrepreneurial orientation.

2 Effectuation Theory

Sarasvathy identifies effectuation as focusing on both (1) the micro-mechanisms that generate new ventures and markets by utilizing the entrepreneur's characteristics, knowledge, and contacts and (2) the micro-processes that encourage voluntary participation of those involved and that create new networks [3]. Expert entrepreneurs use their resources and practices to form new markets by interacting with stakeholders.

2.1 Effectuation and Causation

Sarasvathy explains the causation model compared to the effectuation model [3]. The causation model is shown in Fig. 1 [10].

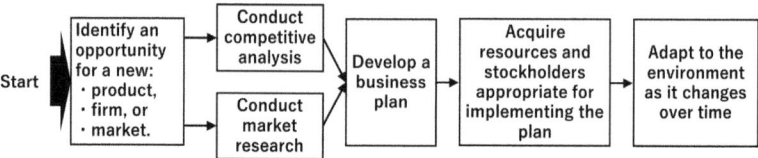

Fig. 1. Schematic diagram of the causation process [10].

In the causation model, the market is first defined and segmented based on factors such as customer age, income, and interests. Targeting is then conducted based on evaluation criteria, and positioning clarifies the company's position, informing strategy development. This process is based on standard STP marketing techniques.

The causation model assumes a process that begins with identifying new products and market opportunities, followed by competitive analysis and market research, and formulating a business plan. This model emphasizes the importance of acquiring resources and engaging stakeholders while remaining responsive to environmental changes. It is characterized by having the goals and means predetermined before an action is taken.

The effectuation model's ends and means are acquired later (Fig. 2) [3, 10]. The skilled entrepreneur starts with "who I am," "what I know," and "who I know" to define what can be done. Subsequently, "new means" and "new goals" are created through effectual commitment and stakeholder interactions. New means lead to a cycle of resource expansion, and new goals to new markets and other effectual artifacts. This way, adept entrepreneurs evolve their business while expanding their practice. The effectuation model shows the process of evolving a business through this interaction.

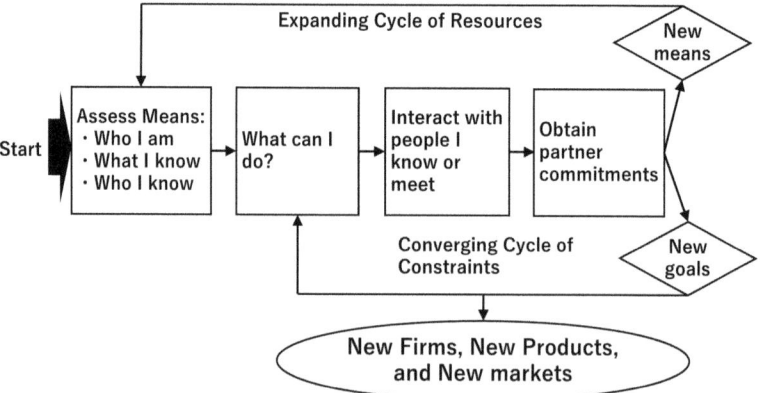

Fig. 2. Schematic diagram of the effectuation process [3, 10].

2.2 Five Principles of Effectuation

Sarasvathy also identified the five principles of entrepreneurial expertise: "bird in hand," "affordable loss," "crazy quilt," "lemonade," and "pilot in the plane" [3].

The bird-in-hand principle refers to leveraging available resources rather than seeking new means to achieve goals.

The affordable loss principle emphasizes decision-making based on the amount of loss one can tolerate rather than expected profits.

The crazy quilt principle focuses on building relationships with collaborators rather than relying solely on market analysis, allowing project members to lead in determining project goals.

The lemonade principle encourages embracing uncertainty, adapting flexibly to unexpected situations, and turning challenges into opportunities.

The pilot-in-the-plane principle emphasizes creating business opportunities through human connections rather than relying on external factors. This principle comprises the other four principles and was thus not specifically included in this survey.

3 Research Method

This study aims to investigate the practical use of generative AI by IT freelance engineers through the perspective of entrepreneurial orientation. Two widely recognized measures for assessing entrepreneurial orientation are those developed by Chandler et al. [5] and Brettel et al. [6].

Chandler et al.'s scale is one-sided, which may lead to notable differences in responses regarding effectuation and causation [5]. There is an ongoing academic debate as to whether effectuation and causation represent opposing concepts. Some researchers have argued that the two are not mutually exclusive and may coexist or even overlap. This overlapping nature suggests that effectuation and causation should not necessarily be treated as conflicting models [5].

However, Brettel et al. view effectuation and causation as fundamentally contrasting approaches [6]. Their scale aligns closely with the original theoretical framework proposed by Sarasvathy [3] and has been applied to various business, project, and corporate settings [11]. Our study aims to clarify the distinction between effectuation and causation in the context of IT freelance engineers; therefore, the measure developed by Brettel et al. [6] was deemed more appropriate.

The survey targeted IT freelance engineers and employed a questionnaire based on the effectuation measure developed by Brettel et al. [6]. Specifically, 16 items were drawn from the five core principles of effectuation. These included four questions related to the bird-in-hand principle, four for affordable loss, three for crazy quilt, and five for lemonade. The questionnaire content was further informed by Ikeda [12], who adapted these principles for the study on entrepreneurial orientation in Japan.

The survey was administered through CrowdWorks [13], an online platform facilitating access to freelance workers for research. Respondents indicated the extent to which they agreed with various statements using a 7-point Likert scale, ranging from "strongly disagree" to "strongly agree," for both effectuation- and causation-oriented items. Table 1 summarizes the contents of the questionnaire.

Table 1. Contents of the questionnaire.

Overall Cronbach's alpha = 0.82	
Causation	**Effectuation**
Bird in Hand (Cronbach's alpha = 0.62)	
Purpose starting point	Mean starting point
1) Our project was specified on the basis of given project targets. 2) The target of our project was clearly defined in the beginning. 3) Given the project, targets have been the starting point. 4) The project specification was predominantly based on given targets.	Our project was specified on the basis of given means/resources. The target of our project was vaguely defined in the beginning. Given means/resources have been the starting point for the project. The project specification was predominantly based on given resources.
Affordable Loss (Cronbach's alpha = 0.69)	
Potential returns	Affordable loss
5) Considerations about potential returns were decisive for the selection of the project option. 6) Project budgets were approved based on calculations of expected returns (e.g., ROI). 7) The selection of the project option was mostly based on analyses of future returns. 8) We mainly considered the potential odds of the project.	Considerations about potential losses were decisive for the selection of the project option. Project budgets were approved on the basis of considerations about acceptable losses. The selection of the project-option was mostly based on a minimization of risks and costs. We mainly considered the potential risks of the project.
The Crazy Quilt (Cronbach's alpha = 0.57)	
Market and competitor analysis	Partnership
9) We tried to identify risks of the project through thorough market and competitor analyses. 10) We have analyzed the market and external trends to better assess future developments. 11) We have taken our decisions on the basis of systematic market analyses.	We tried to reduce risks of the R&D project through internal or external partnerships and agreements. Our focus was rather on the reduction of risks by approaching potential partners and customers. In order to reduce risks, we started partnerships and received pre-commitments.
The Lemonade (Cronbach's alpha = 0.65)	
Surviving unexpected events	Accepting unexpected events

(*continued*)

Table 1. (*continued*)

Overall Cronbach's alpha = 0.82	
Causation	**Effectuation**
12) We only integrated surprising results and findings when the original project target was at risk. 13) Our project process focused on reaching the project target without any delay. 14) The project planning was basically carried out at the beginning of the project. 15) We first of all took care of reaching our initially defined project targets without delays. 16) By the use of upfront market analyses we tried to avoid setbacks or external threats.	We always tried to integrate surprising results and findings during the project process-even though this was not necessarily in line with the original project target. Our project process was flexible enough to be adjusted to new findings The project planning was carried out in small steps during the project implementation. Despite of potential delays in project execution we were flexible and took advantage of opportunities as they arose. Potential setbacks or external threats were used as advantageous as possible.

A Cronbach's alpha value of 0.70 or higher is generally considered the benchmark for acceptable reliability; however, the values for each effectuation principle, as shown in Table 1, ranged from 0.57 to 0.69. Specifically, the crazy quilt principle exhibited a Cronbach's alpha below 0.60, which may be attributed to the limited number of items associated with this dimension. Similarly, the lower reliability coefficients for the remaining principles are also presumed to result from the small number of measurement items; however, the overall Cronbach's alpha for the entire effectuation scale was 0.82, indicating satisfactory internal consistency. Accordingly, this study utilizes the effectuation score as the dependent variable in the analysis.

This research examines the practical use of generative AI among IT freelance engineers by formulating and testing hypotheses. An ordered logistic regression analysis examined these hypotheses by categorizing the respondents into two groups based on whether freelancing was their primary or secondary occupation. The questionnaire instrument included a question specifically designed to distinguish between these two types of freelance engagement. The effectuation score is used as the dependent variable in this study. It represents the degree to which decision-making is aligned with the principles of effectuation.

The dummy variable for generative AI usage is the focus of this study; thus, it is designated as the primary independent variable, allowing us to identify the key determinants of the effectuation tendencies. Studies have shown that environmental factors, such as marital status and gender, can influence entrepreneurial orientation [7, 8]. Building on this insight, we explore whether similar factors influence effectuation tendencies among IT freelance engineers, focusing on age, marital status, and gender. Generational differences may influence individuals' experiences with and attitudes toward AI adoption; thus, the age dummy is an independent variable. However, marital status and gender dummies are incorporated as control variables to mitigate potential confounding effects. This modeling approach incorporates age as an independent variable, whereas marital status and gender are treated as control variables. The design enhances the precision in assessing the impact of the primary independent variable. Accordingly, four dummy variables are employed following the research hypotheses: generative AI usage, age, marital status, and gender. To empirically test the proposed framework, this study tested the following hypotheses:

(1) Hypothesis regarding the use of generative AI: Freelance engineers who actively integrate generative AI into their work are presumed to have a higher tendency toward effectuation, given their openness to adopting new technologies. The generative AI usage dummy is treated as the independent variable.

(2) Hypothesis regarding age (over 40 years): Freelance engineers aged 40 and above, likely to possess diverse professional experiences, are assumed to be more adaptable to environmental changes and thus more oriented toward effectuation. Age (over 40 years) dummy is treated as the independent variable.

(3) Hypothesis regarding marital status: Married freelance engineers are expected to have a higher effectuation orientation, as they must accommodate dynamic changes in work and family life. The marital status dummy is the control variable related to the effectuation score.

(4) Hypothesis regarding gender: Chandler et al. have indicated a positive association between being female and a higher orientation toward causation [8]. Accordingly, this study hypothesizes being female freelancers have a higher orientation toward causation. Gender dummy is included as the control variable related to the effectuation score.

Additionally, a questionnaire survey was conducted to achieve the research objective of clarifying the usage patterns of generative AI among freelance IT engineers in Japan. Participants were asked to indicate how frequently they use generative AI by responding to a 5-point Likert scale: 1 = not used at all, 2 = rarely used (several times per year), 3 = occasionally used (several times per month), 4 = frequently used (several times per week), and 5 = very frequently used (daily). The survey also included items concerning the specific purposes for using generative AI. The findings aim to provide a clearer understanding of the practical application of generative AI in the Japanese IT freelance engineering sector.

4 Analysis Method

This study employed ordered logistic regression analysis as the analytical method. This approach is appropriate because the effectuation score is measured on a 7-point Likert scale and is treated as ordinal categorical data. The analysis was conducted separately for groups based on whether freelancing was the respondent's primary or secondary occupation.

Four dummy variables were included as explanatory variables in the regression model to test the hypotheses concerning the practical work of IT freelance engineers.

(1) The generative AI usage dummy indicates whether the respondent uses generative AI in practice.
(2) The age dummy indicates whether the respondent is over 40.
(3) The marital status dummy indicates whether the respondent is married.
(4) The gender dummy indicates whether the respondent is female.

The analysis incorporated generative AI and environmental factors to comprehensively examine the practical work environment of IT freelance engineers. Table 2 presents a summary of the variables used in the regression analysis.

Table 2. Variables in this study.

Variable name	Explanation
effectuation_mpg	Effectuation orientation was measured using the average score of 16 items, each rated on a 7-point scale. The internal consistency (Cronbach's alpha) was 0.82. Effectuation_mpg is treated as the dependent variable.
generativeAI_dummy	A dummy variable is set to 1 if the respondent reported using generative AI in practice. GenerativeAI_dummy is treated as the independent variable.
over40years_dummy	A dummy variable is set to 1 if the respondent was over 40. Over40years_dummy is treated as an independent variable.
married_dummy	A dummy variable is set to 1 if the respondent was married. married_dummy is included as the control variable.
gender_dummy	A dummy variable is set to 1 if the respondent was female. Gender_dummy is included as the control variable.

5 Results

Table 3 presents the descriptive statistics for the main variables used in the analysis; these include the number of observations (n), and means with standard errors (SE). All dummy variables are binary. A total of 291 valid responses were collected through CrowdWorks [13], comprising 198 primary occupation freelancers and 93 secondary occupation freelancers. Overall, more than 80% of respondents reported using generative AI in their professional work, with approximately 90% among primary occupation freelancers and about 70% among secondary occupation freelancers.

Regarding areas of expertise, the most commonly reported specializations among IT freelance engineer respondents were system engineer (81 respondents), programmer (58 respondents), and web engineer (50 respondents).

Table 3. Descriptive statistics table of the survey results.

Variable name	All		The primary occupation		The secondary occupation	
	n	Mean (SE)	n	Mean (SE)	n	Mean (SE)
effectuation_mpg	291	3.35 (0.05)	198	3.20 (0.05)	93	3.69 (0.07)

(*continued*)

Table 3. (*continued*)

Variable name	All		The primary occupation		The secondary occupation	
	n	Mean (SE)	n	Mean (SE)	n	Mean (SE)
generativeAI_dummy	247	0.84 (0.02)	180	0.91 (0.02)	67	0.72 (0.05)
over40years_dummy	142	0.48 (0.03)	96	0.48 (0.04)	46	0.49 (0.05)
married_dummy	196	0.67 (0.03)	140	0.71 (0.03)	56	0.60 (0.05)
gender_dummy	100	0.34 (0.03)	75	0.38 (0.03)	25	0.27 (0.05)

Table 4 presents the survey results related to the effectuation score (effectuation_mpg). The effectuation score (all) among the IT freelance engineers surveyed was below the mid-value of 4.0. Similarly, the effectuation score for those whose primary occupation was freelancing and those who engaged in freelancing as a secondary occupation were both below 4.0. These results suggest that the IT freelance engineers targeted in this study exhibit a relatively low degree of effectuation orientation.

Furthermore, a comparison between respondents engaged in freelance work as their primary occupation and those involved in it as a secondary occupation revealed a notable difference. The effectuation scores of the secondary occupation group were significantly higher than those of the primary occupation group, exceeding the standard error margin. This finding indicates that individuals pursuing freelance work as their primary occupation tend to demonstrate a lower effectuation orientation.

Table 4. Survey results related to the effectuation score.

Group	n	effectuation_mpg (SE)	The bird in hand (SE)	The affordable loss (SE)	The crazy quilt (SE)	The lemonade (SE)
All	291	3.35 (0.05)	3.14 (0.06)	3.46 (0.06)	3.30 (0.06)	3.49 (0.06)
Primary occupation	198	3.20 (0.05)	2.92 (0.06)	3.27 (0.07)	3.16 (0.07)	3.41 (0.07)
Secondary occupation	93	3.69 (0.07)	3.6 (0.1)	3.9 (0.1)	3.6 (0.1)	3.6 (0.1)

The results regarding the frequency of generative AI usage are presented below. The average frequency of generative AI use for the primary occupation group, measured on a 5-point scale, was 3.79 (SE = 0.06). The score is substantially above the score's midpoint of 3.0, suggesting that the respondents use generative AI relatively frequently. However, the average frequency of generative AI use for the secondary occupation group was 3.5 (SE = 0.1). These results suggest that freelance engineers engaged in their work as a primary occupation tend to use generative AI more frequently than those for whom freelancing is a secondary occupation.

The three most frequently reported applications for generative AI were (1) writing and editing text (e.g., emails, reports, and proposals), (2) creating images and designs, and (3) programming and code generation. These findings suggest that generative AI is widely utilized across various professional tasks.

6 Analysis and Discussion

Table 5 reports the results of the ordered logistic regression analysis examining the relation between individual characteristics and effectuation orientation. The results are presented separately for the primary and secondary occupation groups. Value of estimates, SE, odds ratios (OR), 95% confidence intervals, and p-values are also presented in Table 5. All explanatory variables' variance inflation factor values were confirmed to be below 3.0, indicating no evidence of multicollinearity.

Table 5. Results of the ordered logistic regression analysis.

Group	Variable name	Estimate	SE	OR	95% confidence intervals	p-value
Primary occupation (n = 198)	generativeAI_dummy	− 0.9986	0.23	0.14	0.05–0.34	< 0.0001
	over40years_dummy	0.0343	0.12	1.07	0.65–1.76	0.7861
	married_dummy	− 0.3305	0.13	0.51	0.30–0.89	0.0171
	gender_dummy	0.1694	0.12	1.40	0.85–2.31	0.1838
Secondary occupation (n = 93)	generativeAI_dummy	− 0.3234	0.21	0.52	0.23–1.20	0.1264
	over40years_dummy	0.4839	0.19	2.63	0.18–0.79	0.0097
	married_dummy	− 0.5619	0.20	0.33	0.15–0.72	0.0059
	gender_dummy	− 0.4980	0.21	0.37	0.16–0.86	0.0204

For primary occupation freelancers, the regression model yielded a coefficient of determination (R^2) of 0.0199 and a p-value of less than 0.0001. For secondary occupation freelancers, the model yielded an R^2 of 0.0339 and a p-value of 0.0002. These results evince that the regression models employed in this study are statistically significant.

As shown in Table 5, for Hypothesis 1 concerning generative AI usage, the corresponding dummy variable showed a statistically significant negative effect at the 1 percent level in the primary occupation group (OR = 0.14, p < 0.0001). Conversely, in the secondary occupation group, the effect was not statistically significant at the 5 percent level (OR = 0.52, p = 0.12). These results suggest that the hypothesis, which proposes that freelancers who incorporate generative AI into their work have a higher orientation toward effectuation, is not supported in either group.

For Hypothesis 2, the age dummy variable in the primary occupation group did not reach the 5 percent significance level (OR = 1.07, p = 0.7861), indicating no significant relation. However, the age dummy variable showed a statistically significant positive effect at the 5 percent level in the secondary occupation group (OR = 2.63, p = 0.0097). Therefore, the hypothesis is supported in the secondary occupation group but not in the primary occupation group.

Regarding Hypothesis 3, the dummy variable for marital status had a statistically significant negative effect at the 5 percent level in both groups. Specifically, in the primary occupation group, the OR was 0.51 (p = 0.0171), and in the secondary occupation group, the OR was 0.33 (p = 0.0059). These results indicate that the hypothesis that married freelancers have a higher orientation toward effectuation is not supported in either group.

Finally, regarding Hypothesis 4 on gender, no statistically significant results were observed in the primary occupation group (OR = 1.40, p = 0.1838). However, the results in the secondary occupation group showed a statistically significant negative effect at the 5 percent level (OR = 0.37, p = 0.0204). These findings support the hypothesis that female freelancers exhibit a higher orientation toward causation in the secondary occupation group.

The relation between generative AI usage and effectuation orientation was also analyzed. Individuals in the primary occupation group used generative AI more frequently on a daily basis than those in the secondary occupation group. In the primary occupation group with higher generative AI usage, greater engagement was associated with a tendency toward causation orientation, whereas no significant relationship was found in the secondary occupation group. These findings align with previous studies that have shown the growing adoption of emerging technologies, such as generative AI, across various business contexts [1].

Understanding effectuation orientation is crucial in rapidly evolving technological environments, as it emphasizes adaptability and responsiveness to uncertainty; however, this study suggests that high generative AI usage may instead reinforce causation-oriented behavior, particularly among primary occupation freelance engineers. Practical work demands may shape this tendency. In response to specific client requirements, freelance engineers are typically required to complete tasks with accuracy and efficiency. They are also expected to apply generative AI in a goal-oriented manner, which is considered to be closely associated with causation-based decision-making. This professional setting may inherently foster a causation-oriented mindset. Primary occupation freelancers tend to favor causation-oriented decision-making. To better understand this tendency, future studies should adopt qualitative methods, such as interview-based investigations.

The survey for this study was conducted via CrowdWorks, with participation limited to individuals residing in Japan. Future research may benefit from using multiple survey platforms to enhance data reliability. Furthermore, similar surveys should be conducted in countries outside Japan to improve the generalizability and international relevance of the findings.

Studies on entrepreneurship in the IT sector have emphasized that achieving a balance between effectuation and causation approaches is essential for promoting business growth and innovation [9]. In particular, research has demonstrated that design thinking is effectively integrated with both effectuation and causation theories, thereby supporting the development of entrepreneurial ventures [9]. In practical applications, introducing design thinking through structured training programs can provide meaningful support for fostering this balance. For instance, design thinking workshops can be implemented as part of capacity-building efforts to enhance problem-solving capabilities and strengthen strategic adaptability among freelance engineers. To contribute to the ongoing advancement of the IT sector, freelance engineers must acquire a comprehensive understanding of effectuation principles and cultivate a mindset that aligns with these principles. Future research should explore specific support systems and educational interventions, such as experiential learning involving design thinking, to enable freelance engineers to develop and apply an approach grounded in effectuation. These efforts represent promising avenues for further investigation and practical implementation.

7 Conclusion

This study aims to investigate how IT freelancers in Japan utilize generative AI in their professional work. The questionnaire survey was conducted from the perspective of effectuation orientation to understand the practical application of generative AI. The survey was administered using the crowdsourcing platform CrowdWorks [13], and the questionnaire items were developed based on the effectuation scale proposed by Brettel et al. [6]. The collected data were analyzed using an ordered logistic regression analysis, with a particular focus on comparing freelancers whose primary occupation is freelancing to those whose freelancing was a secondary occupation.

Based on the responses obtained from the questionnaire survey regarding the frequency of generative AI usage, the results suggest that freelance IT engineers who consider freelancing their primary occupation are more likely to adopt generative AI in their professional activities than those for whom freelancing is a secondary occupation. This study investigated the effects of several factors, including generative AI usage in professional tasks, age, marital status, and gender, on effectuation orientation among IT freelance engineers, comparing the results between primary and secondary occupation groups using an ordered logistic regression analysis. In the primary occupation group, greater engagement with generative AI was associated with a tendency toward causation orientation. In contrast, in the secondary occupation group, no statistically significant relation with effectuation orientation was observed.

These results reflect the working conditions faced by freelance engineers. In rapidly evolving technological environments, adaptability is essential; however, these results suggest that frequent use of generative AI may reinforce causation-oriented behavior, particularly among those in the primary occupation group. This situation may be attributed to the goal-driven demands of professional tasks, in which freelance engineers are expected to deliver accurate and efficient results. Such work settings often require the application of generative AI in a manner aligned with predefined objectives, thereby promoting causation-based decision-making. Future research can further explore this trend by incorporating qualitative approaches, such as in-depth interviews, to uncover the contributing factors.

In the rapidly evolving IT sector, IT freelance engineers' understanding and application of effectuation principles may contribute to developing new markets. In particular, aspects of the effectuation framework, such as building partnerships with diverse stakeholders, are expected to play a critical role in market creation. In this context, exploring potential support measures and training programs that can create an effectuation-oriented mindset among IT freelance engineers would be beneficial. Integrating design thinking into such programs may offer practical value in reinforcing effectuation-based decision making.

Acknowledgments. This study was funded by Shizuoka University. I am deeply appreciative of the funding provided by Shizuoka University. I am grateful to all IT freelance engineers for their cooperation in conducting this survey through CrowdWorks. I would like to thank the anonymous reviewers for their valuable comments and suggestions, which greatly improved the quality of this paper.

Disclosure of Interests. The authors have no competing interests to declare that are relevant to the content of this article.

References

1. Quantum Black AI by McKinsey: The state of AI: How organizations are rewiring to capture value, https://www.mckinsey.com/capabilities/quantumblack/our-insights/the-state-of-ai, last accessed 2025/4/9
2. Freelancer: Generative AI is boosting freelancers pay, finds global survey, https://s3.amazonaws.com/press.freelancer.com/Generative%20AI%20Is%20Boosting%20Freelancers%20Pay%2C%20Finds%20Global%20Survey.docx.pdf#:~:text=earnings,to%20never%20using%20AI%20tools, last accessed 2025/4/9
3. Sarasvathy, S.D.: Effectuation: Elements of entrepreneurial expertise. Edward Elgar, Cheltenham, Gloucestershire, UK (2008)
4. Sarasvathy, S.D.: Nicholas Dew on effectuation and new markets. In: Effectuation: Elements of entrepreneurial expertise, Chapter 12. 1, 215–227, Edward Elgar, Cheltenham, Gloucestershire, UK (2008)
5. Chandler, G.N., DeTienne, D.R., McKelvie, A., Mumford, T.V.: Causation and effectuation processes: A validation study. J. Bus. Ventur. **26**(3), 375–390 (2011)
6. Brettel, M., Mauer, R., Engelen, A., Küpper, D.: Corporate effectuation: Entrepreneurial action and its impact on R&D project performance. J. Bus. Ventur. **27**(2), 167–184 (2012)
7. Melo, F.L.N.B.D., Silva, R.R.D., Almeida, T.N.V.D.: Gender and entrepreneurship: A comparative study between the causation and effectuation approaches. BBR. Brazilian Business Review **16**(3), 273–296 (2019)
8. de Villiers Scheepers, M.J., Boshoff, C., Oostenbrink, M.: Entrepreneurial women's cognitive ambidexterity: Career and cultural influences. South African Journal of Business Management **48**(4), 21–33 (2018)
9. Kamble, S., Rana, N.P., Gupta, S., Belhadi, A., Sharma, R., Kulkarni, P.: An effectuation and causation perspective on the role of design thinking practice and digital capabilities in platform-based ventures. Technol. Forecast. Soc. Chang. **193**, 122646 (2023)
10. Read, S., Dew, N., Sarasvathy, S.D., Song, M., Wiltbank, R.: Marketing under uncertainty: The logic of an effectual. Approach Journal of Marketing **73**, 1–18 (2009)

11. McKelvie, A., Chandler, G.N., DeTienne, D.R., Johansson, A.: The measurement of Effectuation: Highlighting research tensions and opportunities for the future. Small Bus. Econ. **54**(3), 689–720 (2020)
12. Ikeda, M.: Effectuaito-shikou wo koujyousaseru eikyou-innsi ni kannsuru kennkyu: Effectuation-shikou to Causation-shikou ha keiken ya kankyou niyotte kotonarunoka (A Study on factors influencing the improvement of effectuation orientation: Whether effectuation orientation and causation orientation differ by experience and environment). Waseda Univ, Master of Business Administration, Waseda (2024). (In Japanese)
13. CloudWorks Inc Homepage.: https://crowdworks.co.jp/, last accessed 2025/4/09

Relational Taxonomy of Cyberattacks: A Model for Threat Classification and Connection in Digital Environments

Mikel Ferrer Oliva(✉), José Amelio Medina Merodio,
José Javier Martínez Herraiz, and Carlos Cilleruelo Rodriguez

Computer Science Department, University of Alcalá, 28801 Madrid, Spain
`{mikel.ferrer,josea.medina,josej.martinez,`
`carlos.cilleruelo}@uah.es`

Abstract. Cyberattacks have evolved into coordinated strategies where multiple techniques are combined to amplify their impact on digital infrastructures. However, current approaches have limitations in representing their interdependencies, making early detection and risk management difficult. This work highlights the need to move towards relational taxonomies and ontologies that model the progression of attacks and their interconnections. Therefore, a relational taxonomy of cyberattacks with eight main categories and twenty key relationships is proposed, providing an adaptable framework to analyze the dynamics of threats. Its open structure allows integration of new tactics, improving risk anticipation and strengthening cybersecurity in critical and corporate environments.

Keywords: Cybersecurity · taxonomy · threats · risk modeling · cyber defense

1 Introduction

The advancement of digital transformation has increased the attack surface in technological environments, exposing organizations and governments to increasingly sophisticated threats. Cyberattacks have evolved beyond isolated incidents to become highly coordinated strategies, where multiple techniques are combined to maximize the impact on compromised systems. Despite advances in cybersecurity, current attack classification methodologies have limitations in modelling their interdependencies, making it difficult to anticipate complex risk scenarios.

Current taxonomies such as MITRE ATT&CK [1], as well as widely used schemes for sharing threat intelligence such as STIX2 [2], TAXII [3], and operational platforms such as MISP [4], provide a detailed description of individual offensive techniques and facilitate the operational categorization of incidents; however, they lack a structured approach that clearly represents the interaction dynamic and strategic be-tween the different types of attacks. Numerous documented incidents show how an initial attack acts as a facilitator for the subsequent execution of more advanced offensive techniques, thus allowing privilege escalation, persistence in com-promised networks, and extension of reach over critical infra-structures [5]. The absence of an explicitly relational model in these frameworks significantly limits the ability to identify strategic patterns, reducing effectiveness in early detection and proactive response to emerging threats.

The use of social engineering has been a key factor in the spread of malware and credential theft, facilitating unauthorized access to strategic digital infrastructures [6, 7]. Likewise, the exploitation of vulnerabilities in software has been a recurring technique to establish persistence in networks, allowing malicious actors to maintain long-term access in high-value government and business environments [8, 9]. However, current methodologies do not adequately capture the dynamics of these attacks in real scenarios, which prevents predictive analysis of their evolution.

Therefore, this work aims to propose a taxonomy of cyberattacks, which classifies threats according to the attack group and the connections between the different groups. This taxonomy will enable structured analysis of attacks and will provide a foundation for improving risk management and incident response through an approach that models the connections between threats in the digital ecosystem.

The article is organized as follows: Sect. 2 reviews the state of the art in cyberattack taxonomies and the current limitations in their modeling. Section 3 introduces the proposed taxonomy, describing the eight attack groups and the relationships between them. Section 4 examines how the proposed taxonomy allows for modeling the progression and strategic interactions between attacks through documented cases and comparisons with previous approaches. Finally, Sect. 5 presents the conclusions and future lines of re-search.

2 Theoretical Framework

The rapid evolution and sophistication of cyberattacks have driven the development of various taxonomies and frameworks aimed at their classification and analysis. Initiatives such as CASE [10] have proposed standardized ontologies for security events, but their inactivity has greatly diminished their practical application. Widely adopted platforms such as MISP [4] offer excellent capabilities for operational categorization and structured sharing of threat information, although they do not fully cover the needs related to strategic and relational understanding between attacks.

This section presents a review of the current literature that supports the need to move towards a relational taxonomy, which overcomes these existing limitations and improves applicability in practical cybersecurity scenarios.

2.1 Limitations of Static Taxonomies

Traditional taxonomies in cybersecurity, such as the MITRE ATT&CK [1] framework, are based on static classifications that organize cyberattack techniques and tactics into clearly defined and separate categories. While such classifications provide an accurate description of individual techniques, they have significant limitations in that they do not capture the dynamic relationships between different threats or represent the strategic transitions that often occur in com-plex, multifaceted attacks. This structural rigidity negatively affects the ability to anticipate sophisticated attacks and makes it difficult to proactively identify alternative vectors [8].

Specific studies have delved into these limitations through contextual threat analysis. Such is the case with watering hole attacks, in which malicious actors compromise legitimate sites frequented by specific groups to distribute malware selectively [11]. Another context studied is the maritime sector, in which exposure to critical threats such as distributed denial-of-service (DDoS) attacks and ransomware demands comprehensive and dynamic taxonomies due to the specificity of their operational scenarios [9, 12]. Additionally, recent research documents the accelerated evolution of ransomware, influenced by global conditions such as the massive adoption of teleworking and the proliferation of compromised credentials, reaffirming the urgent need for relational taxonomies that allow visualizing how an initial incident can be a precursor to subsequent threats [13]. In summary, the lack of conceptual and dynamic interconnection between categories significantly limits the effectiveness of proactive cyber risk management [14, 15].

2.2 Limitations in the Representation of Intrusion Chains

Various studies have found that cyber incidents usually develop as articulated sequences, where an initial event constitutes the gateway to successive, increasingly critical phases [16]. Techniques such as phishing and vishing often represent the initial phase of complex attacks that culminate in advanced spear phishing intrusions, subsequently facilitating the distribution of sophisticated modular malware, such as Trick- Bot [6, 17, 18].

Likewise, the hybrid malware Lucifer clearly exemplifies this problem by combining various offensive techniques in the same attack, including cryptojacking, DDoS and exploitation of vulnerabilities, allowing scaling within compromised infrastructures [19]. In IoT environments, botnets have demonstrated how initial infections on devices with low security can quickly escalate into massive attacks, exposing critical relationships between consecutive offensive phases [20, 21]. Although the evidence supports the existence of these intrusion chains, current models in literature fail to capture their complexity with sufficient flexibility.

The absence of integrated relational taxonomies makes it difficult to accurately anticipate the strategic progression of adversaries, restricting the organizational ability to design proactive, adaptive, and efficient defenses against evolving threats [22].

2.3 Implications for Risk Assessment and Incident Response

Organizations and industries have widely recognized the need to share threat intelligence using standard formats such as STIX2 [2] and TAXII [3]. However, these formats lack explicit schemes for modeling strategic and progressive relationships between attacks, being mainly limited to structured and isolated descriptions of offensive techniques [14, 23]. Other platforms such as MISP [4] facilitate the classification and sharing of incident intelligence information using specific taxonomies, but they also do not explicitly contemplate these dynamic relationships. While initiatives such as CASE [10] have attempted to provide a common ontology for security research, their low adoption and recent inactivity currently limit their operational effectiveness.

Recent research has shown that incidents that initially seem isolated are often part of complex articulated attacks [5]. Social engineering techniques, for example, often facilitate more severe later stages such as lateral movements or advanced exploitation of vulnerabilities [9, 16, 24]. This strategic sequence, not clearly represented in standard formats such as STIX2, is key to anticipating multifaceted threats and proactively managing risk.

Along these lines, specific studies on emerging threats, particularly attacks against the supply chain, underscore the need for modular approaches that clearly represent how different techniques are strategically integrated without relying on rigid categories [5]. Established analytical models, such as the Diamond Model [25] and VERIS [26], agree that the explicit representation of these relationships significantly improves comprehensive risk assessment and optimizes strategic anticipation, strengthening the defensive capacity against sophisticated and emerging attacks [23].

While there have been attempts to create unified ontologies in this field, initiatives such as CASE [10] highlight the importance of relational schemas and the need for collaboration for current frameworks to evolve in that direction.

3 Proposed Cyberattack Taxonomy

The constant evolution of cyberspace shows that attacks are not isolated events, but interconnected actions that enhance each other in the digital environment. Therefore, conceptual models are required to represent the complexity and dependencies between the various offensive techniques.

The proposed taxonomy addresses this complexity through an intrinsically relational design, which links eight main categories of attacks, forming an analytical framework that allows elucidating how a specific attack facilitates or amplifies another. Fundamentally, the taxonomy is, by de-sign, extensible, allowing for the incorporation of new attack patterns and modalities as they emerge, thus ensuring their continued relevance and offering a deep understanding of the tactics employed by attackers.

3.1 Classification of Attack Groups

Accurate identification of the initial vector and specific target allows cyber incidents to be effectively grouped into clearly defined categories. The analysis of attacks documented in scientific literature provides solid criteria for constructing such groups, establishing fundamental differences and similarities between them.

This detailed classification improves understanding of at-tack patterns, thus favoring early detection and effective application of specific defensive strategies according to the nature of each identified category. This model facilitates the identification of recurring patterns in the evolution of threats, allowing us to anticipate possible escalation strategies used by attackers in complex cyber incidents. Table 1 presents the eight proposed attack groups indicated below. The report [7] indicates that social engineering, especially phishing, is the second most successful attack case, second only to malware-based attacks, where ransomware is the most successful.

Table 1. Attack Groups in the Proposed Taxonomy.

Strike Group	Description
Social Engineering	Use of psychological manipulation to gain access to credentials or resources. Phishing attacks have been identified as one of the leading causes of security breaches in companies [6].
Malware based attacks	Use of malware to compromise systems, facilitating attacks such as ransomware, Trojans, and botnets. The analysis of new malware variants has proven its adaptability to evade detections [12].
Network Infrastructure Attacks	Attacks targeting network devices, servers, and interconnected systems to disrupt services or facilitate unauthorized access. Recent research has looked at the impact of IoT botnets on massive DDoS attacks [20].
Exploiting Software Vulnerabilities	Exploiting flaws in software to escalate privileges, install malware, or compromise critical systems. The persistence of unpatched vulnerabilities has been a key factor in multiple attack campaigns [21].
Attacks on Protocols and Communications	Compromise of communication protocols through techniques such as Man-in-the-Middle attacks and DNS hijacking, facilitating the interception of sensitive data. Studies have shown the vulnerability of industrial protocols to targeted attacks [16].
Identity and Authentication Attacks	Compromise of credentials and authentication systems through attacks such as credential stuffing, brute force, and dictionary attacks. Exploiting credentials in corporate environments remains one of the main initial access tactics [7].
Attacks on critical IT/OT infrastructure	Attacks targeting industrial networks and SCADA systems for sabotage and espionage. Device tampering in critical environments has been documented as a significant risk to operational safety [9].
APTs and Cyberespionage	Persistent infiltration and espionage operations that combine multiple attack vectors to maintain long-term access to strategic government and enterprise networks. Cyberespionage continues to evolve with more sophisticated persistence tactics [8].

The exploitation of vulnerabilities in software focuses on exploiting flaws in operating systems, applications or firmware to facilitate unauthorized access [19]. Targeted attacks against network infrastructure affect devices such as routers, servers, and other interconnected equipment with the goal of degrading services or making lateral movements within an organization [20]. Attacks on protocols and communications seek to intercept, manipulate, or disrupt data transmission using techniques such as Man-in-the-Middle attacks [16]. Threats related to identity and authentication compromise credentials and access systems, facilitating intrusions without the need for technical vulnerabilities [7].

Critical infrastructures, such as industrial networks, have been targeted by advanced attacks due to their strategic impact. Thus, attacks on critical IT/OT infrastructure seek sabotage operations in sectors such as energy, finance or transport, where the interruption of services can have cata-strophic consequences [9]. Finally, advanced persistent threats (APTs) and cyberespionage are characterized by sophisticated and prolonged operations, designed to infiltrate government or business networks for espionage, sabotage or extortion [14].

3.2 Relationships Between Attack Groups

Each group represents techniques used in different phases of an attack. In most cases, attackers combine multiple approaches to maximize their impact and ensure persistence on a compromised network. Social engineering is one of the main gateways, allowing the distribution of malware and the theft of credentials through phishing and vishing [6, 24, 27]. In targeted campaigns, spear phishing has been documented to deploy modular malware that facilitates privilege escalation within the victim's infrastructure [17].

Malware is a key enabler for compromising systems and expanding attacks. The use of botnets to degrade services through DDoS attacks has been documented, while some variants have integrated cryptojacking and credential exfiltration to prolong their impact [19, 20, 28]. In this con-text, ransomware has evolved to extract sensitive data and exploit credentials, allowing lateral movement in the affected infrastructure [13].

The exploitation of vulnerabilities in software has been a recurring mechanism in chain attacks. Lack of system up-dates and software failures allow network devices to be compromised and critical systems to infiltrate. Zero-day exploits have been employed in cyberespionage campaigns, facilitating prolonged access to government networks with-out being detected [14]. Likewise, the manipulation of communication protocols has allowed the interception of data through attacks based on social engineering [16], evidencing the importance of security in communications.

Credential and authentication compromise remains one of the most effective strategies for unauthorized access. Attacks have been documented where techniques such as credential stuffing have allowed malware to be deployed without triggering security alerts [12]. In addition, the lack of multi-factor authentication has facilitated access to critical infrastructures, allowing interaction with industrial systems without exploiting technical vulnerabilities [7]. In turn, the compromise of network infrastructures has allowed the infiltration of advanced threats, facilitating espionage and sabotage in business and industrial networks [5].

DNS hijacking attacks have been documented to capture credentials in transit, while BGP hijacking techniques have been employed to redirect legitimate traffic to malicious servers [21]. APTs have been documented in campaigns targeting critical infrastructure, combining exploiting vulnerabilities with cyberespionage. The manipulation of industrial devices to compromise strategic control systems [7] has been recorded, facilitating persistence in government and business networks [14]. Attacks against critical IT/OT infrastructures have shown a direct connection to APTs, as they seek pro-longed infiltration into strategic systems. The use of stolen credentials allows access to networks without raising suspicion, making it possible to manipulate industrial processes without sophisticated techniques [29]. The exfiltration of sensitive data from industrial and government networks poses a significant threat to national security [8]. To better visualize these interconnections in real scenarios, the following Table 2 shows the 20 relationships between the groups.

Table 2. Relationships between Attack Groups in Taxonomy.

Initial Attack	Facilitated Attack	Description
Identity and Authentication Attacks	Social Engineering	Access to credentials facilitates manipulation techniques that increase the effectiveness of deception in corporate and government environments, allowing for more successful attacks [6, 7].
Identity and Authentication Attacks	Malware based attacks	Systems compromised by stolen credentials can allow unauthorized software to run, bypassing security restrictions and facilitating the spread of malicious code [12, 13].
Identity and Authentication Attacks	Exploiting Software Vulnerabilities	Improper authentication or privileged access allows attackers to identify security flaws in applications and services, increasing the risk of compromise in connected infrastructures [14, 21].
Identity and Authentication Attacks	Attacks on Protocols and Communications	Credential exposure allows access to communication channels where data can be intercepted or the integrity of messages can be altered without the need to exploit technical vulnerabilities [16, 21].

(continued)

Table 2. (*continued*)

Initial Attack	Facilitated Attack	Description
Identity and Authentication Attacks	Network Infrastructure Attacks	Improper access to administrative configurations facilitates the manipulation of networks, being able to establish persistence or modify the flow of information without being detected [15, 20].
Social Engineering	Malware based attacks	Persuasive deceptions remain one of the most effective methods of inducing users to run unverified programs, allowing for the execution of harmful actions within a trusted environment [6, 18].
Social Engineering	Exploiting Software Vulnerabilities	Persuasion through personalized information can lead victims to interact with insecure platforms, exposing them to unauthorized code execution [11, 14].
Malware based attacks	Attacks on Protocols and Communications	Some malware is designed to take advantage of communication between systems, allowing the manipulation or exfiltration of data without the need to exploit credentials [16, 21].
Malware based attacks	Network Infrastructure Attacks	Compromised devices can be used as a middle ground to degrade networks and facilitate unauthorized access to internal systems, without the need for direct human interaction [9, 20].
Malware based attacks	Attacks on critical IT/OT infrastructure	It has been documented how harmful software has been exploited to affect systems that control essential processes, interrupting their normal operation with strategic consequences [9, 22].

(*continued*)

Table 2. (*continued*)

Initial Attack	Facilitated Attack	Description
Malware based attacks	Exploiting Software Vulnerabilities	Automated access to compromised environments enables the search for faults in applications and services, increasing the ability to maintain persistence within a digital infrastructure [14, 21].
Exploiting Software Vulnerabilities	Attacks on Protocols and Communications	CommunicationsThe alteration of applications and services enables the manipulation of communications within affected networks, allowing access to data without the need for user intervention [16, 21].
Exploiting Software Vulnerabilities	Network Infrastructure Attacks	Deficiencies in software protection allow the compromise of key network devices, affecting the stability of services and facilitating unauthorized access [15, 21].
Exploiting Software Vulnerabilities	APTs and Cyberespionage	Persistent attacks often take advantage of flaws in software to infiltrate sensitive environments, allowing prolonged access undetected by standard security measures [8, 14].
Attacks on Protocols and Communications	APTs and Cyberespionage	It has been recorded how the manipulation of communication between devices facilitates the exfiltration of information without the need to directly compromise credentials, generating high security risks [14, 21].
Attacks on Protocols and Communications	Attacks on critical IT/OT infrastructure	The operation of communication systems in industrial environments allows the remote manipulation of processes, with impacts that can compromise the operational continuity of critical infrastructures [9, 21].

(*continued*)

Table 2. (*continued*)

Initial Attack	Facilitated Attack	Description
Attacks on Protocols and Communications	Network Infrastructure Attacks	It has been documented how unauthorized access to digital infrastructures allows persistent attackers to establish entry points for the extraction of sensitive information without being detected [8, 14].
Network Infrastructure Attacks	APTs and Cyberespionage	Compromised networks have been used to facilitate access to environments with industrial systems, where the impact can be significant at a strategic and operational level [9, 22].
Network Infrastructure Attacks	Attacks on critical IT/OT infrastructure	Prolonged infiltrations have been used for espionage purposes and process alteration in strategic sectors, representing a threat to operational stability and the confidentiality of information [8, 22].
APTs and Cyberespionage	Attacks on critical IT/OT infrastructure	It has been shown how prolonged access to high-impact environments has allowed the manipulation of processes and the theft of key information in infrastructures of global importance [9, 14].

3.3 Taxonomy Framework

To illustrate the relationships between the different attack groups, the taxonomy is represented by a diagram in which each category is connected to those techniques that can be derived from it. Fig. 1 shows the most frequent transitions between attack types, providing a clear view of how incidents unfold in real-world scenarios.

The model follows a hierarchical structure, where initial attacks, such as identity and authentication compromise or social engineering, act as primary vectors that facilitate more advanced attacks. As you go further in the diagram, you can see how these initial attacks can lead to more complex threats, such as the exploitation of vulnerabilities in software or the compromise of network infrastructures, progressively escalating to critical levels in IT/OT infrastructures and cyberespionage opera-ions and APTs.

The connections in the diagram reflect the most recurrent relationships in the evolution of cyberattacks, allowing us to visualize how an initial technique can facilitate the execution of another, generating chained attack scenarios. This model offers mechanisms to understand the dynamics of threats and anticipate possible escalation trajectories within com-promised infrastructures, facilitating the design of more effective strategies in the face of cyber threats.

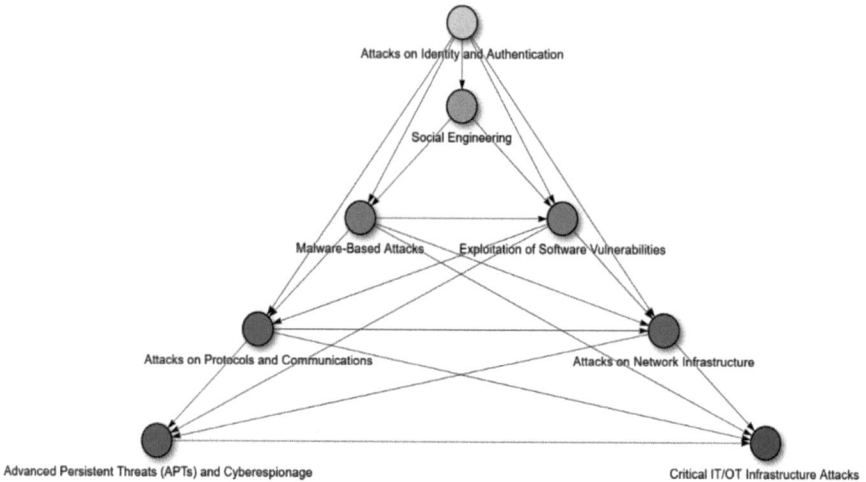

Fig. 1. Framework of the Proposed Relational Taxonomy.

4 Discussion

The growing sophistication and inherent dynamism of cyberattacks has motivated the development of conceptual frameworks. Traditionally, cybersecurity taxonomies have been predominantly static, describing individual techniques without adequately considering strategic and sequential interaction [1, 23]. These models, while offering a basis for cataloging threats, do not effectively represent transitions and dependencies between offensive stages, thus limiting predictive capacity [4, 5].

Numerous studies show that cyber incidents are complex chains where an initial attack facilitates larger offensives [8, 11]. For example, social engineering techniques like phishing facilitate malware propagation and credential compromise, acting as initial vectors in attacks targeting critical infrastructure [6, 24]. Additionally, vulnerability exploitation is recurrent for persistent infiltration into strategic networks, enabling industrial espionage or sabotage in APTs [9, 21]. Essentially, a model is needed that dynamically represents the relationships between techniques. This proposal addresses these limitations with an intrinsically relational taxonomy (eight main groups, twenty strategic relationships) that represents how one type of attack facilitates or amplifies another, improving anticipation and response.

A fundamental aspect is its expandability: unlike static models, it facilitates the integration of emerging threats (e.g., supply chain attacks) without redefining categories, by identifying their interactions. This sequencing allows for the identification of transitions between offensive phases without creating additional groups. Its inherent flexibility maintains validity and applicability in the face of dynamic threats, facilitating defensive strategies [5, 23].

Recent analyses support the relational approach. Hybrid threats like the Lucifer malware (which combines cryptojacking, DDoS, and vulnerability exploitation) [15, 19], and studies on IoT botnets [20] confirm the relevance of dynamically integrating offensive techniques. Attacks on communication protocols through social engineering facilitate stages like cyberespionage or industrial sabotage, validating the emphasis on these relationships [9, 16]. The importance of the human factor is supported by studies on awareness to reduce vulnerabilities from social engineering and credential management [7, 27]. Previous initiatives like CASE [10] or platforms like MISP [4] had limitations in adoption and flexibility. The proposed taxonomy is an advancement, offering a flexible, scalable, and adaptable conceptual scheme to dynamically represent real scenarios without frequent restructuring.

In conclusion, the proposed relational taxonomy is a significant contribution to cybersecurity, an effective tool to understand, anticipate, and manage the strategic complexity of modern cyberattacks. Its flexibility and scalability ensure its continued relevance, facilitating proactive defensive strategies.

5 Conclusions

This work shows cyberattacks are interrelated sequences enabling escalation, not isolated actions. Addressing prior taxonomy limits, we proposed a relational model (8 groups, 20 relationships) capturing the dynamic complexity of real incidents. The model contributes to understanding attack organization and progression by explaining technique interdependencies, helping identify patterns across various environments, including critical infrastructures.

Organizationally, the model aids early detection and response. Analyzing technique interaction helps anticipate attack evolution and mitigate impact, enabling efficient, coordinated responses and resource allocation, fostering preventative cultures.

Socially, the model highlights the human factor in social engineering and credential mismanagement. It underscores investigating human/technical vulnerability interactions and promotes awareness/training programs to foster good practices and reduce susceptibility to manipulation like phishing. Limitations exist: the theoretical basis needs practical validation for effectiveness and applicability. Constant threat evolution necessitates regular updates for ongoing adaptability. Future research includes modelling the identified groups to analyze characteristic attack behaviors practically, and establishing explicit STIX 2 (TC, 2025a) representations by defining entity relationships and rules.

Acknowledgments. This work has been developed within the "Recovery, Transformation and Resilience Plan", project C084/23 Ada Byron INCIBE-UAH, funded by the European Union (Next Generation).

References

1. MITRE Corporation.: ATT&CK: Adversarial tactics, techniques, and common knowledge. https://attack.mitre.org/. last accessed 2025/03/19

2. OASIS Cyber Threat Intelligence (CTI) TC.: STIX Version 2.1, Committee Specification 02. https://docs.oasis-open.org/cti/stix/v2.1/cs02/stix-v2.1-cs02.html, last accessed 2025/03/19
3. OASIS Cyber Threat Intelligence (CTI) TC.: TAXII Version 2.1, Committee Specification 01. https://docs.oasis-open.org/cti/taxii/v2.1/cs01/taxii-v2.1-cs01.html, last accessed 2025/03/19
4. CIRCL.: MISP taxonomies and classification as machine tags. https://www.misp-project.org/, last accessed 2025/03/19
5. European Union Agency for Cybersecurity (ENISA).: ENISA threat landscape for supply chain attacks
6. Bhardwaj, A., Sapra, V.: Why is phishing still successful? Comput. Fraud Secur. **2020**(9), 15–19 (2020)
7. Hellemann, D.N.: Human Risk Review 2023. SoSafe Awareness GmbH, Cologne (2023)
8. Connolly, L.Y., Wall, D.S.: The rise of crypto-ransomware in a changing cybercrime landscape: Taxonomising countermeasures. Comput. Secur. **87**, 101568 (2019)
9. Clavijo Mesa, M.V., Patiño-Rodríguez, C.E., Guevara Carazas, F.J.: Cybersecurity at sea: A literature review of cyber-attack impacts and defenses in maritime supply chains. Information 15(11) (2024)
10. CASE Community.: CASE: Cyber-investigation Analysis Standard Expression. https://caseontology.org/, last accessed 2025/03/19
11. Alrwais, S., et al.: Catching predators at watering holes: Finding and understanding strategically compromised websites. In: Proceedings of the ACM Conference on Computer and Communications Security (CCS 2016), pp. 153–166. ACM, New York (2016)
12. Ivanov, M.A., Kliuchnikova, B.V., Chugunkov, I.V., Plaksina, A.M.: Phishing attacks and protection against them. In: Proceedings of IEEE ElConRus 2021, pp. 1–5. IEEE, Moscow (2021)
13. Beaman, C., et al.: Ransomware: Recent advances, analysis, challenges and future research directions. Comput. Secur. **111**, 102490 (2021)
14. Rauf, U., Mohsen, F., Wei, Z.: A Taxonomic classification of insider threats: Existing techniques, future directions & recommendations. J. Cyber Secur. Mobility (2023)
15. Salim, M.M., Rathore, S., Park, J.H.: Distributed denial of service attacks and its defenses in IoT: A survey. J. Supercomput. **76**(7), 5320–5363 (2019)
16. Javeed, D., Mohammed Badamasi, U., Ndubuisi, C.O., Soomro, F., Asif, M.: Man in the middle attacks: Analysis, motivation and prevention. Int. J. Comput. Netw. Commun. Secur. **8**(7), 52–58 (2020)
17. Cybersecurity and Infrastructure Security Agency (CISA) and Federal Bureau of Investigation (FBI).: TrickBot Malware. https://www.cisa.gov/, last accessed 2025/03/05
18. Álvarez, A.L., et al.: Phishing as a threat in corporate cybersecurity of large companies. Research. Latin America. Eng. Arquit. **1**, 26–33 (2024)
19. Palo Alto Networks Unit 42.: Lucifer: New Cryptojacking and DDoS Hybrid Malware Exploiting High and Critical Vulnerabilities to Infect Windows Devices. https://unit42.paloaltonetworks.com/, last accessed 2025/03/05
20. Gelgi, M., et al.: Systematic literature review of IoT Botnet DDoS attacks and evaluation of detection techniques. Sensors 24(11) (2024)
21. Wu, Q., et al.: Intelligent reflecting surface-aided wireless communications: A tutorial. IEEE Trans. Commun. **69**(5), 3313–3351 (2021)
22. Niño, F.Y. Á.: Ransomware, a latent threat in Latin America. InterSedes 24(49) (2023)
23. Hutchins, E.M., Cloppert, M.J., Amin, R.M.: Intelligence-driven computer network defense informed by analysis of adversary campaigns and intrusion kill chains. Leading Issues Inf. Warfare Secur. Beef. **1**(1), 80–106 (2011)
24. Sedano Pinzón, J.J.: The current and historical context of social engineering. LATAM Rev. Latinoam. Cienc. Soc. Humanid. 5(5) (2024)

25. Caltagirone, S., Pendergast, A., Betz, C.: The diamond model of intrusion analysis. Tech. Rep. ADA586960 (2013)
26. VERIS Community.: VERIS: Vocabulary for Event Recording and Incident Sharing. https://verisframework.org/, last accessed 2025/03/19
27. Nassir, N.F.M., et al.: Revealing the multi-perspective factors behind insider threats in cybersecurity. J. Media Inf. Warfare **17**(2), 65–82 (2025)
28. Raghi, K.R., Paramarthalingam, A.: Proactive Detection of Mirai Botnet Threats: Leveraging XGBoost for Enhanced Cybersecurity. In: IET Conference Proceedings 2024, vol. 23, pp. 34–39. IET, London (2025)
29. Priya, P.M., Ranganathan, A.: Cyber Awareness Learning Imitation Environment (CALIE): A card game to provide cyber security awareness for various groups of practitioners. Adv. Netw. Appl. **14**(2), 5334–5341 (2022)

Taxonomy of Human Errors in Smart Energy Systems Cybersecurity

Mikel Ferrer Oliva(✉) ⓘ, José Amelio Medina Merodio ⓘ, Alberto Larena Luengo, and José Javier Martínez Herraiz ⓘ

Computer Science Department, University of Alcalá, 28801 Madrid, Spain
{mikel.ferrer,josea.medina,alberto.larena,josej.martinez}@uah.es

Abstract. Cybersecurity in smart energy systems, characterized by the integration of CPS and IoT technologies in critical infrastructures, faces growing risks derived from the human factor. This article proposes a theoretical taxonomy to classify the most relevant human errors in industrial energy environments, considering cognitive, organizational and technical aspects. The proposal combines classic models of human reliability with formal error analysis methodologies and international regulatory guidelines. Its multidimensional structure allows a practical application in risk assessment, the design of operational simulations and the development of training programs aimed at reducing the probability and impact of incidents linked to human failure.

Keywords: Cybersecurity · taxonomy · risk modeling · human error

1 Introduction

The digital transformation of the energy sector has favored the integration of cyber-physical technologies (CPS) and IoT devices into electricity grids, giving rise to the so-called smart energy systems. This evolution has enabled significant improvements in operational efficiency, demand management and the integration of renewable energies. However, it has also led to a substantial increase in the attack surface, due to the connection of heterogeneous components, the adoption of distributed architecture, and increased exposure to cyberthreats [1].

A more recent example that demonstrates how malicious actors continue to exploit vulnerabilities in energy systems to disrupt power supply through coordinated attacks is the study by Zhao et al. [2]. This article discusses how cyberattacks, combined with physical offensives, have significantly affected energy infrastructure, highlighting the evolution of cyber threats towards physical impacts on energy systems. The convergence between OT (Operational Technology) and IT technologies has introduced new vulnerabilities that affect all levels of energy infrastructure: from generation plants to transmission and distribution networks, as well as distributed storage systems [3].

Faced with this panorama, the human factor has emerged as a critical vector of risk. Verizon's annual report [4] estimates that approximately 82% of cybersecurity breaches involve human error, and in high-risk industrial sectors such as nuclear, chemical, or energy, human error has been documented to contribute a high percentage of major incidents [5]. In the specific context of power CPS systems, these failures can manifest as misconfigurations, misuse of credentials, or incorrect responses to security events [1].

Despite the importance of this problem, most cybersecurity approaches in CPS have focused on technical vulnerabilities, leaving the human dimension in the background [6]. Existing taxonomies on insider threats or operational errors have limitations when applied to smart energy environments, where advanced automation, critical human decisions, and strict availability requirements are combined [7]. Many taxonomies do not distinguish between unintentional errors (such as lapses or mistakes) and deliberate violations, nor do they consider the role that operational pressure, fatigue, or interface design play in the occurrence of human failures [5].

Faced with this theoretical and practical gap, this article proposes a theoretical taxonomy of human error in cybersecurity, specifically designed for application in smart energy systems. This proposal is based on the Reason [8] and Norman [9] models, on human reliability analysis methodologies such as THERP [10], HEART [11] and SPAR-H [12], as well as on recognized standards such as ISO/IEC 27019:2024[13], IEC 62443 [14] and NIST SP 800–82 Rev.3 [15].

Therefore, the main objective is to develop a theoretical taxonomy that systematically classifies human errors in the cybersecurity of smart energy systems, considering organizational and technological cognitive dimensions, to facilitate their analysis, mitigation and alignment with international industrial safety standards.

The article is organized as follows: Sect. 2 reviews the state of the art in human error and taxonomies applied to industrial cybersecurity. Section 3 presents the proposed taxonomy and its main dimensions. Section 4 discusses its applicability and advantages over previous approaches. Finally, Sect. 5 presents the conclusions and future lines of research.

2 Theoretical Framework

Human errors in complex industrial systems have been extensively studied by disciplines such as human factors, engineering and operational reliability. These disciplines have generated models that allow the systematic identification and classification of the types of errors that can arise during human-machine interaction. Its usefulness has been demonstrated in sectors such as aviation and nuclear energy, and more recently it has been extended to the field of industrial cybersecurity [5]. In the context of energy SPC, these theories are critical to understanding how cognitive failures, organizational mismatches, or poorly designed interfaces can translate into vulnerabilities exploitable by malicious actors [16].

The digital transformation of energy infrastructures has been driven by the development of cyber-physical systems (CPS), which integrate physical and digital processes in real time. This evolution has optimized the control of smart grids, while at the same time increasing the complexity and risks associated with human intervention.

2.1 Classical Models of Human Error Applicable to Industrial Environments and CPS

Reason [8] established a foundational model that differentiates human errors based on cognitive mechanisms and intentionality. He proposed a taxonomy that classifies unintentional errors into slips (errors in execution) and lapses (failures in memory or attention), and intentional errors (as mistakes) defined as planning or decision-making failures due to inadequate knowledge or faulty rule application. This classification has been fundamental in high-risk industrial settings due to its explanatory capacity across different phases of task performance [8].

Norman [9] extended Reason's taxonomy by focusing on human-system interaction, introducing a cognitive model that distinguishes between formulation and execution failures. His contribution allows for the mapping of errors to specific stages of interaction, identifying whether the failure originated in forming the intention or in the execution of the intended action. This distinction is especially relevant in control room environments where operator interfaces can be unintuitive or overly complex [9].

Pollini et al. [5] broadened the scope of error classification by incorporating violations deliberate deviations from established procedures. They identified two main types: malicious violations, involving intent to harm or subvert the system, and non-malicious violations, often motivated by contextual factors such as time constraints, stress or convenience. This perspective is essential for analysing human behaviour in critical infrastructures where operational demands may drive protocol breaches without hostile intent [5].

In CPS environments, this combined taxonomy facilitates the classification of incidents such as the use of unauthorised USB devices in SCADA terminals, firewall misconfigurations, or inadvertent clicks on phishing emails. These incidents can be interpreted through cognitive error categories (slips, lapses, mistakes) or motivational classifications (non-malicious or malicious violations), offering a richer understanding of the underlying causes. Almehmadi et al. [16] applied this framework to design detection systems for accidental operator lapses, illustrating its utility for proactive cybersecurity in CPS [16].

2.2 Human Reliability Analysis (HRA) Techniques in Industrial Cybersecurity

Human Reliability Analysis (HRA) comprises a set of methodologies designed to quantify the probability of human error and assess its consequences in complex systems. THERP (Technique for Human Error Rate Prediction) and HEART (Human Error Assessment and Reduction Technique) are two of the most established HRA techniques, originally applied in high-stakes domains such as nuclear energy and aerospace, where human error poses catastrophic risks [5, 7].

THERP [10] provides a probabilistic model that maps human tasks into sequences of actions, assigning error probabilities at each step. It enables analysts to compute cumulative error likelihoods and identify critical paths prone to failure. This modelling approach is particularly useful in industrial automation, where multistep tasks are common and system resilience depends on precise human responses [10].

HEART [11] differs in its methodology by using predefined task categories with associated baseline error probabilities, adjusted through error-producing conditions (EPCs). These EPCs include factors such as time pressure, cognitive overload, and interface complexity, which are particularly relevant in industrial cybersecurity scenarios involving control rooms and real-time responses. Its flexibility allows for rapid risk estimation while incorporating situational variables [11].

SPAR-H [12] introduces more advanced modelling by considering cognitive demand and uncertainty in decision-making processes. Its relevance in cybersecurity lies in its ability to evaluate human actions under high-pressure or ambiguous conditions, such as interpreting alerts or responding to real-time threats in smart grids. These capabilities have made SPAR-H increasingly attractive for analysing cognitive reliability in cyber-physical systems [12].

2.3 Limitations of Existing Taxonomies and Theoretical Gaps in Energy 4.0

Traditional taxonomies of human error and insider threats were developed in the context of centralised IT infrastructures, lacking adaptability to the distributed and dynamic nature of Operational Technology (OT) networks embedded in Energy 4.0 systems [7]. This mismatch creates a structural limitation in how human error is currently conceptualised and addressed in critical energy environments.

In these infrastructures, there is often a failure to distinguish between technical errors resulting from inadequate training or interface design and protocol violations driven by situational factors such as urgency or conflicting goals. For instance, mislabeling a relay during maintenance and deliberately disabling a firewall to speed up operations may both result in security vulnerabilities, yet their motivations and mitigation strategies differ significantly [5].

The IEC 62443 series recognises the centrality of human factors in OT security and calls for structured training and procedural controls to minimise human-related vulnerabilities. It emphasises the role of operator behaviour, awareness, and accountability in securing cyber-physical systems [14]. These standards highlight the limitations of one-size-fits-all classifications and stress the need for dynamic, context-aware approaches to human error.

Similarly, NIST SP 800–82 Rev.3 provides guidelines for reducing operator errors by promoting a strong organisational security culture. It recommends training tailored to common mistakes, such as weak password management, incorrect privilege assignments, and delayed incident reporting—all frequent in energy control environments [15].

Despite these institutional efforts, a theoretical and practical gap persists in the accurate categorisation of human error within CPS energy ecosystems. The absence of a domain-specific taxonomy complicates incident analysis, impedes comparative risk assessment, and undermines the effectiveness of countermeasures. This was emphasised by Almehmadi et al. [16], who argued that without precise error typologies adapted to CPS contexts, proactive mitigation remains elusive.

In conclusion, the development of error taxonomies tailored to CPS in Energy 4.0 is both a scientific necessity and an operational imperative. Such taxonomies would allow cybersecurity practitioners to understand the motivational and cognitive origins of operator actions, allocate resources more effectively, and implement training aligned with real-world error patterns. Without this refinement, the gap between technological sophistication and human-centred risk analysis will continue to grow, undermining the resilience of smart energy systems [4].

Although adapted to industrial cybersecurity contexts (including the analysis of maintenance practices, alarm responses and credential handling) these methods face inherent limitations. As noted by Pollini et al. [5], traditional HRA tools were not developed to address the intentionality behind operator behaviour nor the presence of adversaries who exploit human error. This restricts their utility in threat modelling for CPS environments, where human errors may be opportunistically targeted by attackers [5].

Nevertheless, their underlying analytical constructs (task decomposition, error-inducing factors, and contextual weighting) remain relevant. They serve as a basis for the development of more tailored frameworks that reflect the dual challenge in CPS: safeguarding against accidental failures and mitigating exploitation by external threats [7].

3 Proposal of the Taxonomy of Human Errors in Energy Cybersecurity

The classification of human error in smart energy systems required a rigorous methodological approach that combined consolidated theoretical foundations and adaptation to current regulatory frameworks. An iterative design methodology based on Nickerson et al. [19] was followed for its development, complemented by the recommendations of Kundisch et al. [17] regarding empirical validation and systematic documentation. Conceptually, the proposal was based on the models of Reason [8] and Norman [9], duly adapted to the current requirements of industrial cybersecurity set out in ISO/IEC 27019:2024 [13], IEC 62443–2-1 [14] and NIST SP 800–82 Rev.3 [15].

3.1 Human Error Classification Dimensions

The classification of human error was structured around four key dimensions, integrating well-established theoretical models with normative criteria applicable to industrial and cyber-physical system (CPS) environments. The first dimension addresses the intentionality of the error, drawing from the typologies proposed by Reason [8] and Norman [9]. These authors distinguished among execution errors (slips), memory or attention failures (lapses), planning or knowledge-based mistakes (mistakes), and non-malicious violations, understood as deliberate deviations from procedures without hostile intent. This last category aligns with the IEC 62443–2-1 standard [14], which recognises the critical role of human factors in the context of industrial cybersecurity.

The second dimension focuses on the phase of the operational cycle in which the error occurs. It adopts the structure proposed by NIST SP 800–82 Rev.3 [15], which differentiates between design, configuration, routine operation, maintenance and incident response. This classification allows for contextual analysis of the error based on when it emerges, which is crucial for designing targeted corrective measures. As highlighted by Reason [8], errors must be analysed in relation to their operational context rather than in isolation.

The third dimension concerns the impact of the error on the core principles of information security: Confidentiality, Integrity and Availability. This approach is consistent with ISO/IEC 27019:2024 [13], which outlines information security requirements specific to the energy sector. Assessing errors based on their effect on these principles supports the prioritisation of response actions and the alignment of countermeasures with actual risk exposure [13].

The fourth and final dimension involves an analysis of the causal context of the error, extending Reason's [8] concept of latent conditions. This perspective identifies three categories of contributing factors: human (such as distraction or fatigue), organisational (including procedural gaps or permissive culture) and technological (such as poor interface design or excessive automation). These categories are aligned with IEC 62443–2-1 [14] and the operational recommendations of NIST SP 800–82 Rev.3 [15], allowing for a comprehensive understanding of error causes that integrates both immediate failures and structural vulnerabilities.

3.2 Construction Methodology

The taxonomy was developed through an iterative process grounded in empirical evidence and theoretical frameworks. Initially, a comprehensive review of documented incidents was conducted using authoritative sources such as ICS-CERT advisories and Verizon's Data Breach Investigations Report (DBIR) [4]. These records provided a robust foundation for identifying recurring human error patterns in cyber-physical energy infrastructures, particularly in SCADA environments and distributed control systems.

In parallel, classification criteria were defined based on established human reliability principles. Reason's [8] Swiss cheese model served as the structural basis for identifying relationships between active failures and latent conditions across multiple organisational layers. Norman's framework [9], particularly his distinction between goal formulation and execution, was essential in analysing human-system interactions in industrial control environments.

Each dimension of the taxonomy was systematically validated against international standards to ensure technical consistency and practical relevance. ISO/IEC 27019:2024 [13] provided guidance on information security within energy systems, while IEC 62443–2-1 [14] offered insight into the integration of human and organisational factors into risk mitigation. NIST SP 800–82 Rev.3 [15] contributed an operational perspective by detailing the full cybersecurity lifecycle, allowing a functional mapping of error phases and associated risks.

This triangulated approach (based on incident analysis, theoretical structuring and regulatory alignment) ensured that the resulting taxonomy was conceptually coherent, empirically grounded and readily applicable to modern CPS environments in the energy domain.

3.3 Validation of the Taxonomy

The taxonomy's practical utility was verified through its application to real-world incidents in critical energy infrastructures. These examples demonstrated its capacity to classify different types of human error, identify underlying causes and suggest corrective actions aligned with established cybersecurity frameworks. By analysing events across different operational phases and attributing them to cognitive, organisational or technical factors, the model proved to be a valuable instrument for post-incident analysis and prevention planning. Its structure facilitates not only the classification of the error itself but also the tracing of its origin and its consequences, which is essential for designing systemic corrective strategies in complex cyber-physical environments.

One case involved an unplanned outage at a thermal power facility, caused by the accidental activation of a switch during routine operations. This event was categorised as a slip in the operational phase, with a direct impact on system availability. The root cause was linked to an ambiguous interface that lacked adequate visual feedback (a technological factor). Events of this nature are not uncommon and have been documented in numerous gap assessments across industrial domains [4]. Their recurrence has highlighted the need for better interface ergonomics and clearer signalling mechanisms. As noted in NIST SP 800–82 [15], interface redesign and usability improvements represent effective mitigation measures that contribute to the reduction of execution-based errors in high-reliability settings.

A second incident occurred during scheduled maintenance activities, where a technician inadvertently deactivated a pressure sensor, unintentionally compromising both the integrity and availability of the system. The error was classified as a lapse, having occurred in the maintenance phase, and was mainly attributed to the absence of a checklist or formalised verification process (an organisational factor). Failures of this kind have been consistently identified in operational technology (OT) environments, where procedural discipline is often undermined by informal practices [5]. Preventive strategies in this context require not only procedural documentation but also cultural reinforcement, ensuring that safety protocols are systematically respected. IEC 62443-2-1 [14] specifically advocates the use of formal checklists and structured work orders to reduce the occurrence of omissions in safety-critical operations.

A third example involved a misconfiguration in a firewall deployed within a SCADA environment. The error exposed internal network services to external traffic, resulting in a severe breach of confidentiality. This mistake occurred during the configuration phase and was attributed to insufficient technical training and a lack of review procedures (human and organisational factors). Similar misconfigurations have been repeatedly cited in cybersecurity incident reports and industrial literature [5], especially in sectors where rapid deployment pressures coexist with limited personnel specialisation. Effective mitigation, as indicated in ISO/IEC 27019:2024 [13], requires the implementation of change management controls, the validation of security rules prior to deployment and the reinforcement of training programmes tailored to technical roles.

The final case involved a situation in which technicians, under intense operational pressure, chose to disable the alarm system to prevent a process interruption. Although the action was intentional, it was not driven by malicious intent, and was therefore classified as a non-malicious violation in the incident response phase. The resulting compromise of system integrity was closely associated with a permissive organisational culture that tolerated circumvention of procedures under exceptional circumstances.

This behavioural pattern is recurrent in high-stress environments where production continuity is prioritised over procedural compliance [5]. IEC 62443-2-1 [14] addresses this challenge by recommending the enforcement of procedural controls, continuous operator training and the cultivation of a risk-aware organisational mindset that discourages informal workarounds.

To support operational understanding and ensure its applicability, Fig. 1 provides a visual representation of the human error taxonomy developed in this study. It illustrates how errors are categorised across the four proposed dimensions and highlights their connection to cognitive mechanisms, operational phases, affected security principles and contextual root causes. This visual structure serves as a practical tool for guiding risk assessments, informing training design and supporting incident prevention efforts in the domain of cyber-physical energy systems.

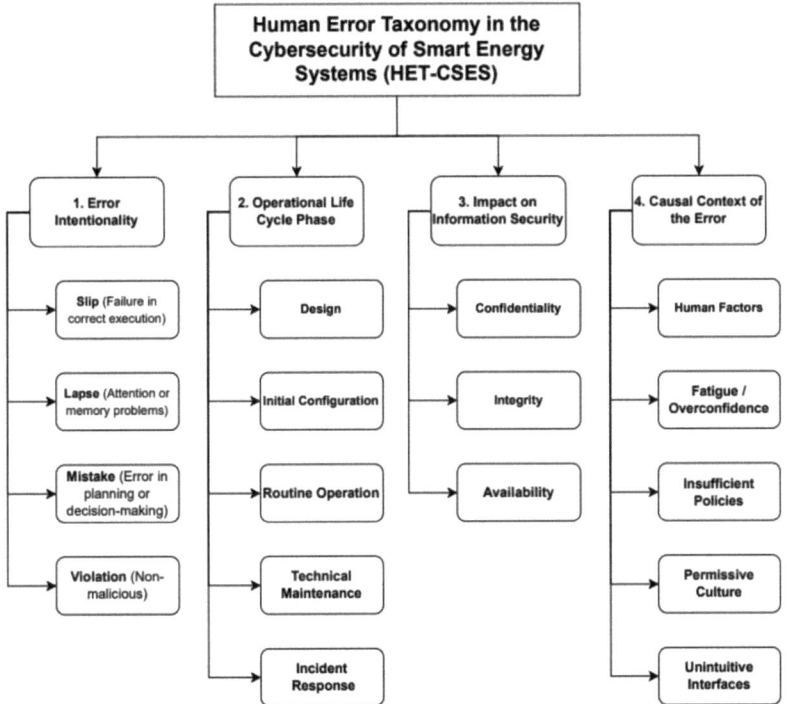

Fig. 1. Framework of the Proposed Taxonomy.

4 Discussion

The development of the proposed taxonomy has enabled a structured and systematic approach to analysing human error within the context of energy cybersecurity. It offers a multidimensional analytical framework that integrates the intentionality behind the action, the operational phase in which the error manifests, the specific impact on core security principles and the organisational and technological conditions in which the event takes place. This structured perspective facilitates a more accurate and holistic understanding of human error, supporting its integration into safety management systems within industrial contexts where human behaviour represents a critical vulnerability as well as a strategic point of intervention for improving cyber-physical resilience [5] [7].

A particularly relevant contribution of this work lies in the taxonomy's capacity to establish meaningful links between distinct types of human error and the environmental factors that enable or amplify their occurrence. Rather than addressing human failure as an isolated decision or accidental oversight, the framework helps to identify patterns of behaviour shaped by inadequate system design, organisational shortcomings or contextual constraints. This reinforces the perspective that errors are not purely the result of individual action but are often the consequence of multiple interacting elements within the socio-technical system [18]. This insight is especially valuable for security teams and risk managers tasked with developing interventions that move beyond blame-oriented approaches.

From an implementation perspective, the taxonomy is closely aligned with internationally recognised standards and guidance documents in the cybersecurity field. It is consistent with the requirements outlined in ISO/IEC 27019:2024 [13] and IEC 62443-2-1 [14], particularly in relation to procedures for root cause analysis, planning of technical and procedural training and the evaluation of internal organisational controls. Its flexible design allows for its application in a range of energy sector contexts, from conventional power generation facilities to distributed and increasingly automated smart grid environments. This adaptability is essential for responding to the heterogeneous nature of cyber-physical architectures in modern industrial ecosystems.

Furthermore, the proposed taxonomy is not intended to replace existing classification models or regulatory mechanisms. Instead, it is designed to complement them by introducing additional dimensions that enrich the understanding of human error within operational security. Its value lies in extending current perspectives through a more detailed and contextualised view of error dynamics, especially when dealing with hybrid infrastructures where human, organisational and digital elements interact continuously. This orientation is strongly aligned with international trends that promote a proactive, systemic and non-punitive approach to security management, where learning and resilience take precedence over individual sanction [7] [15].

In summary, this proposal marks a step forward in the understanding and categorisation of human error in complex cyber-physical environments. It contributes to building a more precise, context-sensitive and integrated framework for managing human risk in cybersecurity. Beyond its conceptual contribution, the taxonomy offers a robust and practical structure for identifying, classifying and analysing failure events in energy infrastructures. Its potential applications extend to operational risk assessment, the design of targeted training programmes and the development of safety policies explicitly oriented towards the human factor [8] [9] [19].

5 Conclusions

This study presents a taxonomy proposal that seeks to provide a clear, applicable and structured tool to classify human errors in the specific context of cybersecurity in smart energy systems. The taxonomy allows technical, organizational and cognitive aspects to be integrated into a common framework, facilitating the analysis of human failures from a contextual perspective and not merely behavioral.

From a social perspective, its application can contribute to a fairer and more understanding safety culture, in which mistakes are interpreted as opportunities for improvement and not only as individual failures. This vision can be key in critical sectors, where the human component coexists with high automation and operational pressure.

In the organizational field, the proposed structure allows aligning human error management with training processes, control design and performance evaluation. By classifying errors according to their origin and context, it is possible to establish more adjusted and efficient responses, optimizing the allocation of resources in industrial cybersecurity programs.

At a technological level, this taxonomy can be the basis for developing automated tools that integrate human analysis into SIEM systems, even correlation engines or simulation platforms. The proposed classification facilitates the connection between technical records and behavioral patterns, opening the door to detection and prevention systems that are more sensitive to the human factor.

Regarding its limitations, it should be noted that the taxonomy has not yet been empirically validated in real productive environments of the energy sector, which represents a necessary phase to strengthen its practical applicability. Likewise, the current structure does not contemplate complex temporal dynamics among the factors that influence error, an issue that can be addressed in future developments through approaches based on probabilistic inference or dynamic systems modelling.

As future lines of research, it is proposed to explore the development of human performance indicators based on the types of error defined in the taxonomy, with the aim of monitoring critical areas of exposure to human error in OT environments and providing quantitative support in audits and technical talent management. In addition, taxonomy could be used as an input in the design of adaptive training environments, allowing content and simulations to be adjusted to the most recurrent or high-impact error profiles, and thus favoring more precise and contextualized training in highly demanding operating environments.

This proposal is presented as an initial contribution with a vocation for continuous improvement and shared utility. Its purpose is to open new possibilities to address the role of the human factor in cybersecurity in the energy sector in a more precise and complete way, promoting comprehensive resilience that articulates the technical, the organizational and the human.

Acknowledgments. This work has been developed within the "Recovery, Transformation and Resilience Plan", project C084/23 Ada Byron INCIBE-UAH, funded by the European Union (Next Generation).

References

1. Diaba, S., Shafie-khah, M., Elmusrati, M.: "Cyber-physical attack and the future energy systems: A review," Energy Rep. **12**, 2914–2932, (2024), https://doi.org/10.1016/j.egyr.2024.08.060

2. Zhao, Y., Chen, J., Zhu, Q.: Integrated Cyber-Physical Resiliency for Power Grids Under IoT-Enabled Dynamic Botnet Attacks. IEEE Trans. Control Syst. Technol. **32**(5), 1755–1769 (2024). https://doi.org/10.1109/TCST.2024.3378993
3. T. Krause, R. Ernst, B. Klaer, I. Hacker, and M. Henze, "Cybersecurity in Power Grids: Challenges and Opportunities," Sensors (Basel), vol. 21, no. 18, Sep 16 2021, https://doi.org/10.3390/s21186225
4. Verizon, "Data Breach Investigations Report," 2022. [Online]. Available: https://www.verizon.com/business/en-gb/resources/2022-data-breach-investigations-report-dbir.pdf
5. A. Pollini et al., "Leveraging human factors in cybersecurity: an integrated methodological approach," Cognition, Technology & Work, Vol. 24, 05/01 2022, https://doi.org/10.1007/S10111-021-00683-Y
6. M. Nafees, N. Saxena, A. Cardenas, S. Grijalva, and P. Burnap, "Smart Grid Cyber-Physical Situational Awareness of Complex Operational Technology Attacks: A Review," ACM Computing Surveys, Vol. 55, 10/08 2022, https://doi.org/10.1145/3565570
7. W. B. Wan Ismail, A Classification Of Human Error Factors In Unintentional Insider Threats. 2022, pp. 667–676
8. Reason, J.: Human error: models and management. BMJ **320**(7237), 768 (2000). https://doi.org/10.1136/BMJ.320.7237.768
9. D. A. Norman, The Design of Every Things. 2013
10. Swain, G.H.: AD, "Handbook of Human Reliability Analysis with Emphasis on Nuclear Power Plant Applications," in "NUREG/CR-1278," Sandia National Laboratories / U.S. Nuclear Regulatory Commission, Washington, DC, SAND80–0200 / NUREG/CR-1278, August 1983 1983
11. Alexander, T.M.: "Human Error Assessment and Reduction Technique (HEART) and Human Factor Analysis and Classification System (HFACS)," presented at the 9th IAASS Conference, 2017
12. B. H. Gertman D, Marble J, Byers J, Smith C, "The SPAR-H Human Reliability Analysis Method," in "NUREG/CR-6883," Idaho National Laboratory, U.S. Nuclear Regulatory Commission, Washington, DC, INL/EXT-05–00509, August 2005 2005
13. Information security, cybersecurity and privacy protection — Information security controls for the energy utility industry, ISO/IEC 27019:2024, ISO/IEC, Geneva, 2024. [Online]. Available: https://www.iso.org/standard/85056.html
14. Security for industrial automation and control systems – Part 2–1: Security program requirements for IACS asset owners, IEC 62443–2–1:2024, IEC, Geneva, 2024. [Online]. Available: https://standards.iteh.ai/catalog/standards/clc/33949f56-abab-4203-9673-c80efb2bf6d0/en-iec-62443-2-1-2024
15. Guide to Operational Technology (OT) Security, NIST, 2023
16. A. Almehmadi, "Micro-Behavioral Accidental Click Detection System for Preventing Slip-Based Human Error," Sensors, **21**, 8209, (2021), https://doi.org/10.3390/s21248209
17. D. Kundisch et al., "An Update for Taxonomy Designers - Methodological Guidance from Information Systems Research," Business & Information Systems Engineering, **64**, 10/22 (2021), https://doi.org/10.1007/s12599-021-00723-x
18. P. Bhosale, W. Kastner, and T. Sauter, "Modeling Human Error Factors with Security Incidents in Industrial Control Systems: A Bayesian Belief Network Approach," presented at the Proceedings of the 19th International Conference on Availability, Reliability and Security, (2024)
19. R. Nickerson, U. Varshney, and J. Muntermann, "A Method for Taxonomy Development and its Application in Information Systems," European Journal of Information Systems, **22**, (2013), https://doi.org/10.1057/ejis.2012.26

Exploring New Business in Japanese SMEs from the Perspective of Employees' Engagement

Kazunori Minetaki[✉], Hiroki Idota, and Teruyuki Bunno

Creative Management and Innovation Research Institute, Kindai University, 3-4-1 Kowakae, Higashi-Osaka City, Osaka Prefecture, Japan
`kminetaki@bus.kindai.ac.jp`

Abstract. This study examines how Japanese small and medium-sized enterprises (SMEs) are leveraging digital transformation and employee empowerment to explore new business opportunities amid a severe labor shortage. As sales recover following the COVID-19 pandemic, enhancing productivity and maintaining global competitiveness have become critical. Beyond quantitative labor issues, qualitative approaches such as motivating younger employees and fostering inclusive workplaces are gaining importance. Through a case study of Iwasaki Co., Ltd., a manufacturing SME in the metal processing industry, this paper highlights initiatives that combine internal capabilities with digital tools, such as integrated supply chain systems and smartphone-based product promotion platforms. The company's efforts to empower foreign and female employees with advanced IT skills and to build a supportive work environment demonstrate how internal human capital can be aligned with innovation strategies. Drawing on these insights, this paper also conducts an empirical analysis using survey data from SMEs to explore the factors that contribute to successful business expansion. The findings emphasize that employee engagement and external collaboration, through joint product development, study sessions, and direct customer interactions, play crucial roles in driving entry into new business domains, especially in AI and digital fields. The study concludes that developing digitally competent personnel, including foreign professionals, will be essential for the sustained growth of SMEs in Japan.

Keywords: engagement · empowerment · small and medium-sized enterprises (SMEs)

1 Introduction

Japanese small and medium-sized enterprises (SMEs) are facing a severe labor shortage as sales recover from the COVID-19 pandemic's impact. Improving productivity is a crucial theme, necessitating the maintenance of international competitiveness through labor-saving investments and higher unit costs. Additionally, many companies are currently leveraging digital transformation to enhance operational efficiency and lower costs. Still, it is a means to create new businesses, add value, and drive the company's growth.

As many SMEs aim to expand into new markets, this paper will primarily provide an empirical analysis of exploring new business opportunities. In addition to the quantitative problem of labor shortages, it is also essential to consider how to utilize human resources qualitatively to tackle challenges in new fields. The interviews conducted by the authors also showed that a workplace where young employees are motivated to work leads to corporate growth. In this year's analysis, we have incorporated a human resource perspective.

This paper presents a case study of innovative SMEs that are challenged to explore new growth markets by empowering their employees, such as in the environmental and medical equipment sectors. Iwasaki Co., Ltd. Manufactures industrial products with expertise in product development, utilizing metal processing technology. It offers a comprehensive range of services, from 3D CAD-based planning and design to processing, welding, painting, and assembly, all of which are carried out in-house. The company places great importance on fostering an organizational culture that encourages employees to contribute ideas and suggestions proactively. This inclusive environment extends to five foreign professionals with advanced IT competencies, multilingual proficiency, and strong negotiation skills. In pursuit of digital innovation, the company has implemented a smartphone-based application to promote its products, enabling customers to submit inquiries and place orders directly through the app. Moreover, it has established an integrated supply chain management system encompassing order processing, production, and distribution, thereby achieving both increased sales and cost reduction. Additionally, the company is actively developing environmentally friendly heating and cooling systems, as well as medical product management systems that utilize AI-driven image processing technologies. Notably, foreign and female employees with advanced IT competencies constitute a pivotal force in advancing the company's digital transformation initiatives. The company has also established stylish cafeterias and private booths for post-lunch relaxation, thereby contributing to the creation of an attractive and employee-friendly work environment.

Building on the insights gained from this case study, the following section presents a review of relevant prior literature, which clarifies the research question of this paper. Subsequently, an empirical analysis is conducted using survey data.

2 Literature Review

In recent years, employee empowerment and empowering leadership have emerged as pivotal factors in fostering creativity and innovation in organizational contexts. Drawing on both theoretical frameworks and empirical studies, a growing body of literature highlights the importance of empowering practices in enhancing employee performance, satisfaction, and, most notably, creativity.

A comprehensive model linking empowering leadership to employee creativity through psychological empowerment, intrinsic motivation, and engagement in the creative process was introduced [1]. A proposed comprehensive model that links empowering leadership with employee creativity through psychological empowerment, intrinsic motivation, and engagement in the creative process. Their study highlighted the psychological mechanisms underlying the way leadership behaviors that delegate authority and convey confidence stimulate creative thinking. Importantly, their findings revealed that empowering leadership influences creativity directly and through enhanced cognitive and motivational states of employees.

Organization-based self-esteem and access to resources mediate the relationship between empowering leadership and creativity [2]. Empowering leadership has a positive impact on leader evaluations, employee motivation and resources, attitudes, and even performance outcomes [3]. Both structural and psychological empowerment positively influence employee-driven innovation (EDI), with the link between structural empowerment and EDI being partially mediated by psychological empowerment [4].

From a broader theoretical standpoint, Lee and Song [5] developed a causal model to illustrate both the direct and indirect effects of empowerment on performance, mediated by job satisfaction and innovativeness. A causal model to explain how empowerment affects employee attitudes and performance in the public sector. Their structural equation modeling demonstrated that empowerment directly enhances job performance and indirectly influences it through job satisfaction and innovativeness. This aligns with managerial perspectives that view empowerment as a relational practice involving the distribution of information, resources, and decision-making authority.

Building on empirical findings, Joo and Shim [6] examined psychological empowerment and its impact on organizational commitment, highlighting the moderating role of organizational learning culture. Their study emphasized the importance of aligning structural autonomy with employees' psychological readiness to take responsibility.

Foundational work identified key aspects of empowerment, including goal clarity and participation in decision-making, establishing a conceptual framework for its measurement [7]. This framework elaborates on four key dimensions: meaning, self-determination, competence, and impact. These elements form the core of psychological empowerment and underscore its significance in fostering innovative and high-performing workplaces. Spreitzer's research illustrates how these dimensions enhance individual autonomy and enable greater influence within organizational structures.

In recent developments in the empowerment–innovation literature, more focus has been given to employees' motivation as a key factor driving innovative behavior. He et al. [8] carried out a cross-level study showing that challenging stressors within an organizational innovation climate significantly boost sustained innovation behavior through the mediating effect of creative self-efficacy. Their results highlight the importance of motivational factors in connecting environmental influences with creative outcomes.

Similarly, Nurhaeda et al. [9] conducted an empirical investigation into how both intrinsic and extrinsic motivation contribute to employee creativity and innovation-related performance. Their findings highlight the critical role of leadership strategies, resource accessibility, and motivational support in fostering an innovation-friendly work environment.

Moreover, Anagha and Magesh [10] highlighted the mediating role of organizational commitment in the relationship between employee motivation and innovation behavior. Their research demonstrated that high levels of motivation are more likely to result in innovative engagement when aligned with strong organizational structures that foster autonomy and commitment.

Taken together, these findings reinforce the idea that employee motivation functions not only as a psychological enabler but also as a key mechanism that translates empowering leadership into tangible innovation, particularly in dynamic and resource-constrained environments, such as small and medium-sized enterprises (SMEs).

3 Theoretical Framework

This study draws on empirical evidence to examine the organizational factors that contribute to the successful development of new businesses and product innovations in Japanese small and medium-sized enterprises (SMEs). Based on the literature on empowering leadership, employee motivation, and innovation, as well as the case study of Iwasaki Co., Ltd., the following theoretical framework is proposed.

Accordingly, the following hypotheses are formulated:

Hypothesis 1: Enhancing employee motivation contributes to the exploration of new business fields.
Hypothesis 2: Employee involvement in the new business affects the current progress of the latest business initiative.
Hypothesis 3: Employee involvement has a positive impact on innovation in future-oriented sectors, including energy, healthcare, recycling, and AI/DT.

This framework bridges prior theoretical insights with empirical findings, contextualizing innovation as a multi-level process shaped by organizational assets, motivational systems, and employee agency. By integrating structural, motivational, and collaborative dimensions, it offers a comprehensive understanding of how SMEs can mobilize internal and external resources to navigate new growth frontiers.

4 Methodology

This paper examined the effects on new business by focusing on the company's strengths, employee motivation, and employee commitment to new business in growing fields as internal factors, and external collaboration as external factors.

The empirical analysis in this study is based on data derived from the "Survey on Business Environment and New Business Development," jointly administered by Creative Management and Innovation Research Institute of Kindai University and the Osaka Prefectural Government between July and August 2024. The survey targeted small and medium-sized manufacturing enterprises located in Osaka Prefecture, yielding a total of 494 valid responses (Out of 2,741 valid questionnaires, resulting in a valid response rate of 18.0%). The descriptive statistics of the main variables were summarized in Table 1.

The method was an extended ordinal probit model, and two-stage estimation was conducted using ordinal data. The main explanatory variables for each estimation equation were estimated in the first stage, using industry, company size, and average employee age. The main explanatory variables assessed in the second stage were then used to calculate the explained variables, such as the outcome and progress of the new business.

Table 1. Descriptive statistics

Variable	Number of obs.	Mean	Std. Dev.	Min	Max
exploration of new business fields	480	0.0520833	0.2224269	0	1
empower employees	477	0.1446541	0.352121	0	1
promoting active collaboration with external partners and customers	477	0.0419287	0.2006367	0	1
current progress of the latest business initiative	177	1.694915	0.4617495	1	2
collection of market trends and business-related information	170	2.435294	1.129979	1	4
planning of businesses, products, and services	170	2.358824	1.085438	1	4
development of businesses, products, technologies, and methods	170	2.082353	0.9876368	1	4
development of raw materials, parts, and outsourcing suppliers	169	2.284024	1.075614	1	4
initiatives in future growth (new energy sector)	176	3.386364	0.8061452	1	4
initiatives in future growth (medical, health, and nursing care sectors)	176	3.170455	0.9881769	1	4
initiatives in future growth (resource and recycling sectors)	176	3.113636	0.9249781	1	4
initiatives in future (growth AI and DT-related sectors)	176	3.198864	0.8881434	1	4

Focusing on outcomes and progress rather than whether new projects were undertaken, we examined the essential factors in achieving results in new business initiatives. The current progress of the latest business is a four-level variable: progress is exceeding objectives, progress is on target, progress is slower than objectives, and efforts are not progressing. We also focused on the results of efforts in new energy-related, medical/healthcare/nursing care-related, resource recycling, and AI/DT-related areas, which are among the areas where future growth is expected in new businesses.

The effect of enhancing employee motivation, which was empowering employees and promoting active collaboration with external partners and customers, on exploring new business fields is summarized in Table 2. The case of promoting active collaboration with external partners and customers was statistically significant($p < .01$). Hypothesis 1 was partially verified.

Table 2. Effect of Enhancing employees' motivation on an exploration of a new business field

Explained variable: Exploration of new business fields	[1] empower employees		[2] Promoting active collaboration with external partners and customers	
Enhancement of motivation	0.8878		5.0673	***
	(4.0470)		(0.4934)	
Number of observations	465		465	
Wald chi2	0.05		105.48	
Prob > chi2	0.83		0.00	

***:p<.01,**:p<.05,*:p<.1 same as below

In the following analyes, the effect of employee involvement in the new business included: (1) collecting market trends and business-related information, (2) planning businesses, products, and services, (3) developing businesses, products, technologies, and methods, and (4) developing raw materials, parts, and outsourcing suppliers examined (Tables 3, 4, 5, 6 and 7).

The effects of employees' involvement in the new business on its current progress were shown in Table 3. The case of the collection of market trends and business-related information was only significant($p < .05$). Hypothesis 2 was evaluated exclusively within a narrowly constrained set of cases.

Table 3. Effect of employee involvement in the new business on the current progress of the latest business initative

Explained variable: current progress of the latest business initiative	[3] Collection of market trends and business-related information		[4] Planning of businesses, products, and services	[5] Development of businesses, products, technologies, and methods	[6] Development of raw materials, parts, and outsourcing suppliers
employee involvement	0.5140	**	0.4225	0.2210	0.4502
	(0.2300)		(0.8605)	(1.3133)	(0.4205)
Number of obsevations	166		166	166	165
Wald chi2	4.99		0.24	0.03	1.15
Prob > chi2	0.03		0.62	0.87	0.28

The cases of employee involvement in initiatives related to future growth fields, such as new energy-related sectors, medical, health, and nursing care sectors, resource and recycling sectors, and AI and digital transformation (DT) sectors, were summarized in Table 4, 5, 6 and 7.

Employee involvement in the development of raw materials, parts, and outsourcing suppliers was significant($p < .05$) in Table 4. All cases were significant ($p < .05$ or $p < .01$) in the resource and recycling sectors in Table 6. The remarkable results were particularly evident in the AI and DT-related sectors, as shown in Table 7 ($p < .01$). Hypothesis 3 was verified, especially in both the resource and recycling sectors, as well as in the AI and DT-related sectors.

Table 4. Effect of employee involvement in new businesses on their initiatives for growth in the energy-related sectors

	[7] Collection of market trends and business-related information	[8] Planning of businesses, products, and services	[9] Development of businesses, products, technologies, and methods	[10] Development of raw materials, parts, and outsourcing suppliers
Explained variable: Initiatives in future growth (new energy sector)				
employee involvement	0.3861	0.4842	0.6691	0.5060 **
	(0.2435)	(0.4090)	(0.6083)	(0.2574)
Number of obsevations	165	165	165	164
Wald chi2	2.51	1.40	1.21	3.87

Table 5. Effects of employee involvement in new businesses on their initiatives for growth in the medical, health, and nursing care-related sectors

	[11] Collection of market trends and business-related information	[12] Planning of businesses, products, and services	[13] Development of businesses, products, technologies, and methods	[14] Development of raw materials, parts, and outsourcing suppliers
Initiatives in future growth (medical, health, and nursing care sectors)				
employee involvement	0.3123	0.4501	0.6609	0.3355
	(0.2912)	(0.4100)	(0.4092)	(0.4169)
Number of obsevations	165	165	165	164
Wald chi2	1.15	1.20	2.61	0.65
Prob > chi2	0.28	0.27	0.11	0.42

Table 6. Effect of employee involvement in new businesses on their initiatives for growth in the resources and recycling sectors

	[15] Collection of market trends and business-related information	[16] Planning of businesses, products, and services	[17] Development of businesses, products, technologies, and methods	[18] Development of raw materials, parts, and outsourcing suppliers
Explained variable: Initiatives in future growth (resource and recycling sectors)				
employee involvement	0.5559 ***	0.7225 ***	0.8240 ***	0.6026 **
	(0.1454)	(0.1532)	(0.1790)	(0.2933)
Number of obsevations	165	165	165	164
Wald chi2	14.62	22.24	21.19	4.22
Prob > chi2	0.00	0.00	0.00	0.04

Table 7. Effects of employee involvement in new businesses on their initiatives for growth in AI and DT-related sectors

Explained variable: Initiatives in future (growth AI and DT-related sectors)	[19] Collection of market trends and business-related information		[20] Planning of businesses, products, and services		[21] Development of businesses, products, technologies, and methods		[22] Development of raw materials, parts, and outsourcing suppliers	
employee involvement	0.7438	***	0.8814	***	0.9765	***	0.9506	***
	(0.1640)		(0.0902)		(0.1043)		(0.0876)	
Number of obsevations	165		165		165		164	
Wald chi2	20.58		95.51		87.68		117.71	
Prob > chi2	0.00		0.00		0.00		0.00	

5 Conclusion

New business initiatives are a major driver of future growth for SMEs. It is necessary to leverage the company's existing strengths while applying its technology and expertise to new fields. It has been demonstrated that promoting active collaboration with external partners and customers is effective for exploring a new business field. Practical examples include encouraging employees to actively participate in study sessions with external companies, engaging in joint product planning initiatives, and providing opportunities for direct customer interaction at exhibitions and similar events. Such initiatives have been shown to facilitate entry into new business domains.

One of the key findings of this study is that employee engagement plays a particularly significant role in advancing new business fields, especially in areas such as artificial intelligence (AI) and digital transformation. In the context of Japanese small and medium-sized enterprises (SMEs), the recruitment and development of personnel with expertise in AI is becoming increasingly necessary. As illustrated by the cases presented in this paper, the proactive employment of foreign professionals is expected to become an increasingly important strategy in the near future.

Disclosure of Interests.. This paper has no interest.

References

1. Zhang, X., Bartol, K.M.: Linking empowering leadership and employee creativity: The influence of psychological empowerment, intrinsic motivation, and creative process engagement. Acad. Manag. J. **53**(1), 107–128 (2010)
2. Zhang, S., Ke, X., Frank Wang, X.-H., Liu, J.: Empowering leadership and employee creativity: A dual-mechanism perspective. J. Occup. Organ. Psychol. **91**, 896–917 (2018)
3. Kim, M., Beehr, T.A., Prewett, M.S.: Employee responses to empowering leadership: A meta-analysis. J. Leadersh. Organ. Stud. **25**(3), 257–276 (2018)

4. Echebiri C., Amundsen S., Engen M.: Linking structural empowerment to employee-driven innovation: the mediating role of psychological empowerment. Administrative sciences. Jul 13. **10**(3), 42. (2020)
5. Lee, H.W., Song, J.H.: Causal relationships among empowerment, job satisfaction, and performance in the public sector. Public Manag. Rev. **19**(4), 479–500 (2017)
6. Joo, B.K., Shim, J.H.: Psychological empowerment and organizational commitment: The moderating effect of organizational learning culture. Hum. Resour. Dev. Int. **13**(4), 425–441 (2010)
7. Spreitzer, G.M.: Psychological empowerment in the workplace: Dimensions, measurement, and validation. Acad. Manag. J. **38**(5), 1442–1465 (1995)
8. He, P.X., Wu, T.J., Zhao, H.D., Yang, Y.: How to motivate employees for sustained innovation behavior in job stressors? A cross-level analysis of organizational innovation climate. Int. J. Environ. Res. Public Health **16**(23), 4608 (2019)
9. Nurhaeda, Z., Maryadi, M., Salim, M.S., Kitta, S.: The relationship between employee motivation, creativity and performance. Paradoks: J. Ilmu Ekon. **7**(4), 466–480 (2024)
10. Anagha, K., Magesh, R.: Employee motivation to innovate and resources management: The mediating role of organisational commitment. Int. J. Manag. Concepts Philos. **9**(3), 185–199 (2016)

Facial Expression-Based Monitoring of Fatigue and Stress in Care Facility Workers

Kazunori Minetaki[1](✉), I-Hsien Ting[2], Teruyuki Bunno[1], and Hiroki Idota[1]

[1] Creative Management and Innovation Research Institute, Kindai University, 3-4-1 Kowakae, Higashi-Osaka, Osaka, Japan
kminetaki@bus.kindai.ac.jp

[2] Social Networks Innovation Center, National University of Kaohsiung, 700, Kaohsiung University Rd., Nanzih District, 811 Kaohsiung, Taiwan

Abstract. This study explores the relationship between objective facial expression–based indicators and subjective perceptions of fatigue and stress among care workers in Japan. Caregiving involves significant physical and emotional demands, yet traditional survey-based assessments are often constrained by limited time and resources in care settings. To address this challenge, the study employed video recordings and automated facial analysis to obtain objective indicators such as the Percentage of Eyelid Closure over Time (PERCLOS) and selected Facial Action Units (AUs). These metrics were compared with structured questionnaire responses regarding perceived fatigue and stress. The results revealed consistent links between objective facial metrics and self-reported experiences. Elevated PERCLOS values were observed in participants who reported heavy workloads and task-related difficulties, suggesting that increased eye closure reflects physical fatigue. Higher scores for AU04 (brow lowerer) and AU07 (lid tightener)—markers of muscular tension—aligned with perceptions of sustained effort. A composite AU index comprising AU01, AU02, AU20, and AU23 corresponded with emotional exhaustion and anxiety. One notable case exhibited both the highest AU composite scores and elevated self-reported stress, supporting the potential of facial metrics as indicators of psychological strain. These findings demonstrate that AU intensity values, rather than binary presence indicators, enable the detection of subtle expressions linked to fatigue and stress. While no single metric proved superior, integrating facial metrics with subjective assessments offers a more holistic view of caregiver well-being. Overall, this study highlights facial expression analysis as a promising non-invasive tool for early detection of fatigue and stress, with practical implications for workplace interventions and future longitudinal research.

Keywords: PERCLOS · Action Unit · Stress

1 Introduction

Care workers face physically and emotionally demanding conditions that frequently result in mental strain, burnout, and reduced quality of care in Japan. Unlike other high-demand professions, such as nursing or teaching, the fatigue experienced by caregivers has not been consistently assessed or addressed, especially in light of recent policy changes that may have increased workplace pressures. Early detection of fatigue can play a crucial role in preventing serious health consequences for workers and maintaining care quality. Non-invasive and continuous monitoring tools are particularly needed in care facilities, where time and resource constraints limit the use of traditional survey-based methods.

Facial expression analysis has emerged as a promising method for monitoring both psychological and physiological states in real-time. Percentage of Eyelid Closure over the Pupil over Time (PERCLOS) and Action Units (AUs) provide measurable indicators of fatigue and stress. This study examines the application of facial metrics to evaluate fatigue levels in care workers using video data, thereby contributing to efforts to promote healthy work environments and personnel sustainability in care settings. Additionally, a questionnaire survey was conducted to assess participants' subjective perceptions of fatigue and stress. The main aim of this paper is to examine whether there is consistency between the objective fatigue indicators derived from facial expression analysis and the subjective evaluations reported in the questionnaire.

The remainder of this paper is structured as follows. Section 2 presents a review of the relevant literature. Section 3 describes the methodology, and Sect. 4 reports the results. Finally, Sect. 5 provides the conclusions.

2 Literature Review

Facial expressions are powerful indicators of internal states such as fatigue and stress. The Facial Action Coding System (FACS), developed by Ekman and Friesen, defines Action Units (AUs) that represent facial muscle movements associated with emotional and cognitive states [1]. Among these, AU04 (brow lower) and AU07 (lid tightener) are widely associated with stress and cognitive load. While these action units may contribute to fatigue detection, their relationship to fatigue is context-dependent and often supported by additional indicators such as PERCLOS or AU17.

Recent research highlights the potential of facial expression recognition in occupational health. Giannakakis et al. [2] demonstrated that AU-based models can effectively detect stress and emotional distress in workplace environments. Gavrilescu and Vizireanu [3] proposed a model for predicting depression, anxiety, and stress using AUs extracted from video recordings, reinforcing the importance of facial indicators in mental health monitoring. In parallel, PERCLOS has become a standard index for detecting drowsiness in the transportation and healthcare sectors. Chang et al. [4] developed a system combining PERCLOS and facial physiological signals, achieving high accuracy in detecting fatigue. PERCLOS has also been validated in studies examining the effects of sleep deprivation and prolonged attention tasks [5, 6].

In educational contexts, Li et al. [7] proposed a temporal deep learning model that monitors student fatigue by analyzing facial features, particularly around the eyes and mouth. Their findings suggest that specific AU-based indicators may hold potential for cross-domain applicability beyond education.

Beyond educational settings, the healthcare sector has also embraced facial expression analysis. Ofei-Dodoo et al. [8] examined how facial expressions and nonverbal cues function as indicators of burnout among nurses, particularly in high-stress environments such as intensive care units. Building on this, Chatterjee et al. [9] developed a CNN-based system to monitor nurse fatigue via facial emotion detection. Their results revealed that prolonged shifts are associated with heightened facial tension and increased activation of stress-related action units (AUs). These studies underscore the urgent need for reliable, non-invasive methods to assess fatigue in clinical environments.

Returning to the educational domain, D'Mello and Graesser [10] demonstrated that brow furrowing and eyelid tightening serve as reliable predictors of cognitive load during learning tasks, further reinforcing the significance of AU-based indicators in cognitive and emotional monitoring.

3 Methodology

Twenty care workers employed at an elderly care facility in Osaka City participated in this study. The facility was selected based on its cooperative attitude and willingness to support academic research. Data were collected through individual video interviews conducted via Zoom, a widely used video conferencing platform. The duration of the interviews varied between approximately 5 and 60 min, depending on each participant's availability and communication pace.

Prior to data collection, participants were thoroughly informed about the purpose of the study and the intended use of their facial expression data. Verbal consent was obtained from all participants. Ethical considerations—including privacy protection and data anonymization—were strictly observed. All video recordings were deleted immediately upon completion of the analysis.

This study focused on detecting fatigue and psychological stress using facial expression–based indicators. Two primary methods were employed. The first was the Percentage of Eye Closure (PERCLOS), a well-established metric representing the proportion of time a subject's eyes remain closed during a given observation period. The second involved the analysis of Facial Action Units (AUs) as defined by the Facial Action Coding System (FACS), particularly AUs 04 (brow lower) and 07 (lid tightener), both of which are commonly associated with stress-induced muscle activity.

Rather than analyzing each AU in isolation, composite intensity scores were calculated to capture more nuanced patterns of stress-related facial muscle tension. Specifically, two aggregated indices were computed:

(1) AU04 + AU07, representing muscular tension in the glabellar and periorbital regions, are indicative of physical strain or sustained concentration, such as brow lowering and tight eyelid closure commonly seen in fatigue or effortful tasks;

(2) AU01 + AU02 + AU20 + AU23, representing a constellation of upper facial expressions commonly linked to psychological distress or negative affect, such as inner brow raising, lip tension, and mouth corner depression.

In detail, AU01 (inner brow raiser) and AU02 (outer brow raiser) are typically associated with expressions of concern or fear. AU20 (lip stretcher) is often related to nervousness or emotional suppression, while AU23 (lip tightener) indicates tension around the mouth. By integrating these AUs, the study aimed to detect subtle facial cues of psychological stress that may not be captured through individual unit analysis, thereby enhancing the sensitivity of stress detection.

Facial expression data were extracted from the recorded video footage using an automated facial recognition and analysis system. For each participant, time-series plots and frequency histograms were generated to visualize patterns in both PERCLOS and AU activations.

Additionally, a questionnaire survey was administered during the video recording sessions to assess participants' self-reported perceptions of their work environment, levels of fatigue, and psychological stress. The results from the facial expression analysis were then compared with the survey responses to examine the relationship between observable facial indicators and subjective experiences.

4 Results

Figure 1, 2, 3 illustrate the temporal patterns of PERCLOS and composite Action Units (AUs), which served as objective indicators of ocular and facial signs of fatigue and psychological stress. As described in the methodology, AU intensity values (e.g., AU01_r, AU02_r, AU04_r, AU07_r, AU20_r, AU23_r) were used to capture continuous variations in facial muscle activity, enabling the detection of subtle tensions in the glabellar, periorbital, and perioral regions—areas known to reflect both physical strain and emotional discomfort.

This study examined two composite AU indices. The first combined AU04 and AU07, representing muscle activation associated with brow lowering and eyelid tightening, which are typical in physical fatigue or intense concentration. The second combined AU01, AU02, AU20, and AU23, reflecting emotional expressions such as inner brow raising and lip tension, which are often linked to psychological stress.

Participants with elevated PERCLOS values—such as ID14 (36.19%) and ID20 (31.80%)—also exhibited high AU04 and AU07 intensities (1.77 and 1.32, respectively), suggesting that increased eyelid closure and brow tension are reliable indicators of ocular fatigue. In contrast, ID16 showed the lowest PERCLOS (8.94%) and a low AU04 and AU07 score (0.34), indicating minimal physical fatigue.

Regarding psychological stress, ID14 also had the highest composite score for AU01, AU02, AU20, and AU23 (1.33), consistent with heightened emotional strain. Interestingly, ID16, despite its low physical fatigue indicators, recorded the second-highest score (0.73) for this psychological AU composite. This suggests that emotional fatigue or internal stress may be facially expressed even in the absence of overt physical symptoms.

These objective indicators were further compared with participants' self-reported questionnaire responses. ID14, who demonstrated the highest PERCLOS and AU activation levels, also reported elevated subjective symptoms, including "extremely tired" (3), "sluggish" (3), "tense" (3), and "eye fatigue" (5). This convergence of subjective and objective indicators reinforces the validity of facial expression metrics in assessing fatigue and stress.

In contrast, ID16 reported high subjective fatigue levels—"extremely tired" (5) and "eye fatigue" (4)—despite showing the lowest PERCLOS and low AU04 and AU07 values. The relatively high score for AU01, AU02, AU20, and AU23 (0.73), however, indicates that psychological stress was still observable through facial expressions. This discrepancy highlights the variability in how individuals perceive and manifest fatigue, emphasizing the importance of integrating both subjective reports and objective facial indicators.

Overall, these findings suggest that facial expression analysis, particularly using composite AU scores, offers a sensitive and non-invasive means of detecting both physical and psychological fatigue. When combined with self-report data, this multimodal approach provides a more comprehensive assessment framework. It holds considerable potential for early detection of fatigue-related stress among care workers and for informing occupational health interventions in high-demand environments (Table 1 and 2).

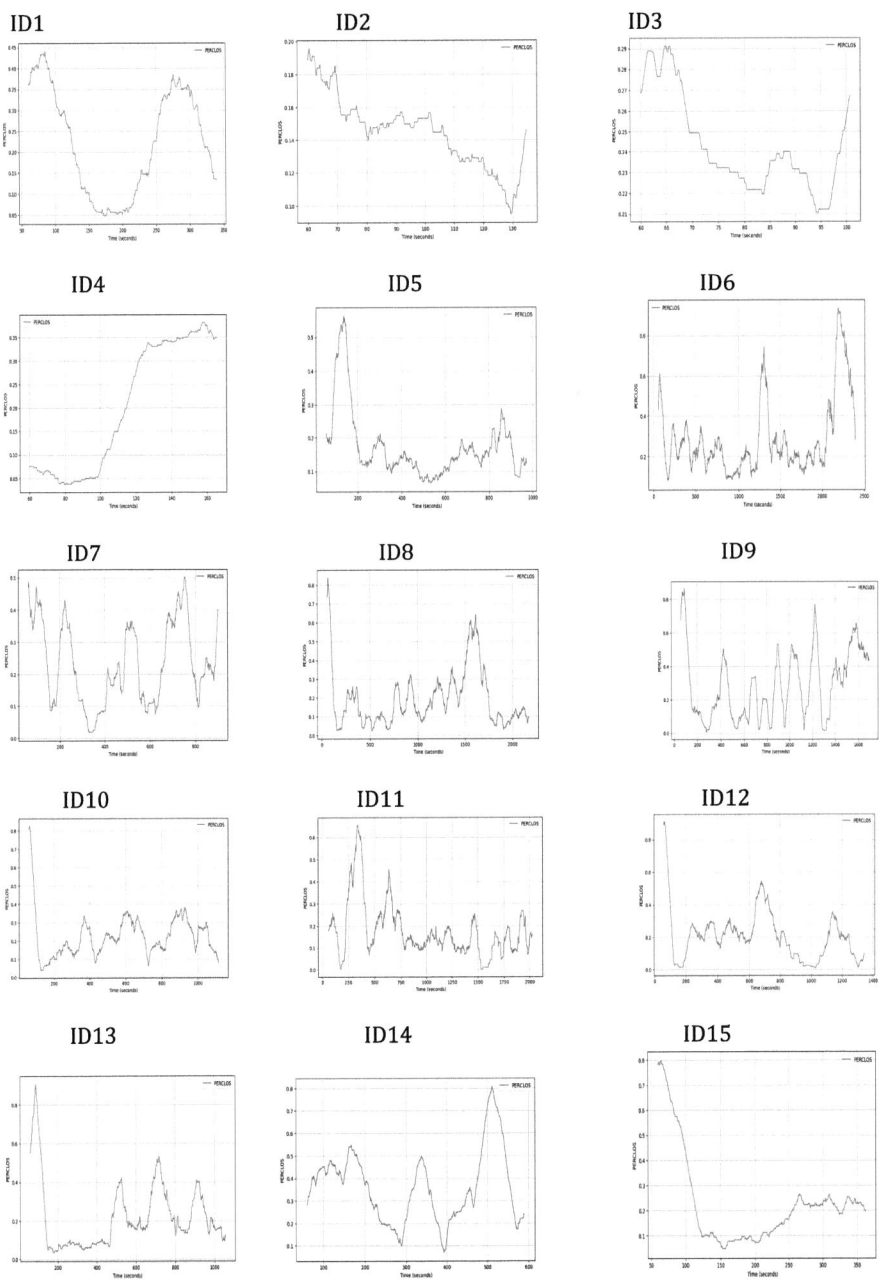

Fig. 1. PERCLOS

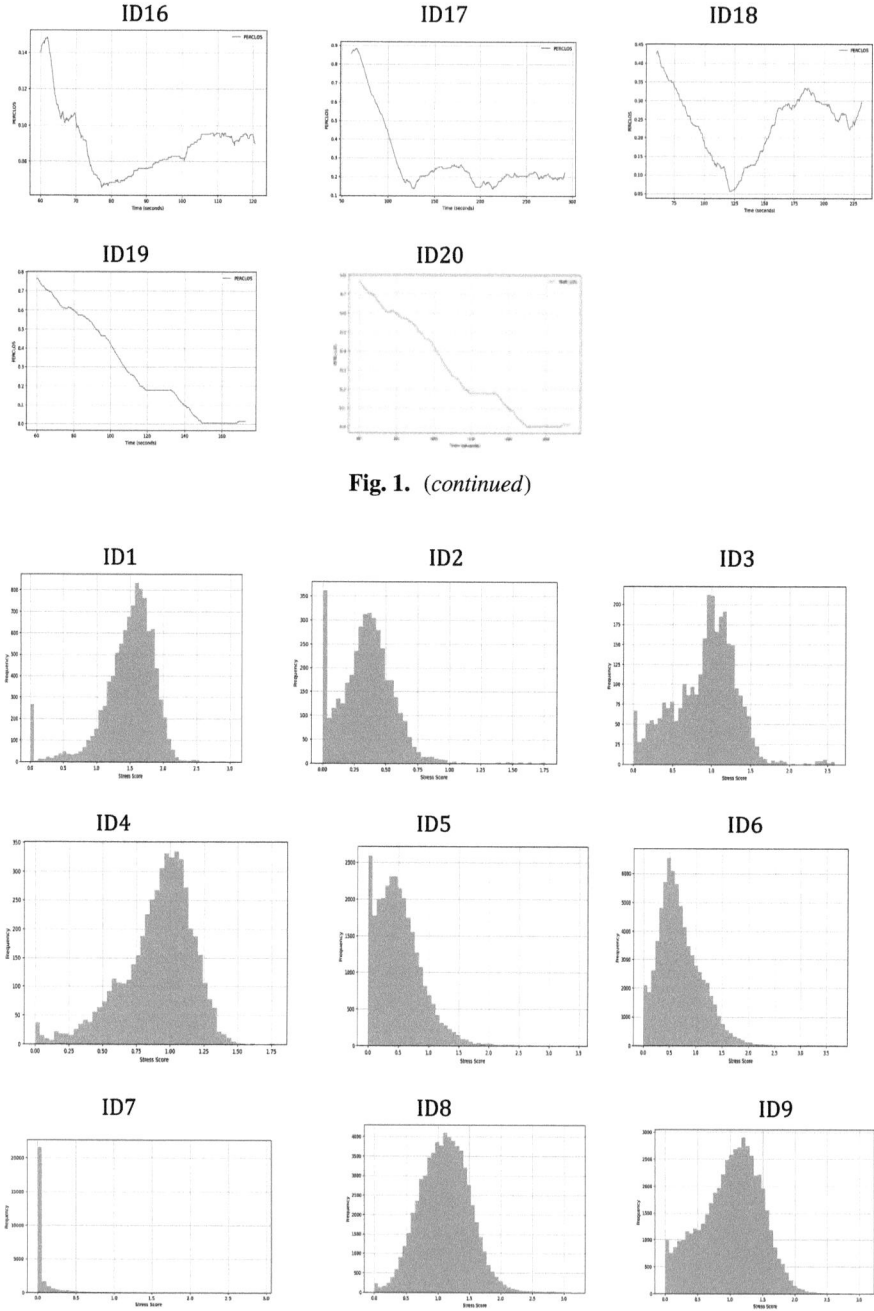

Fig. 1. (*continued*)

Fig. 2. AU04 + AU07

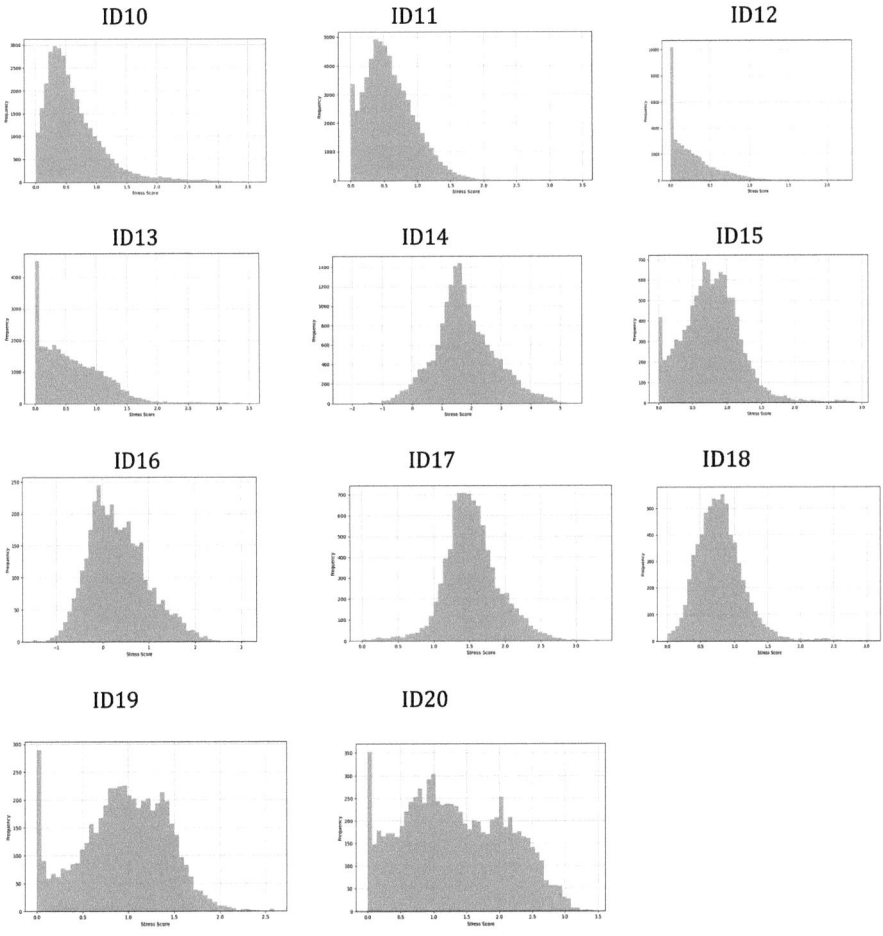

Fig. 2. (*continued*)

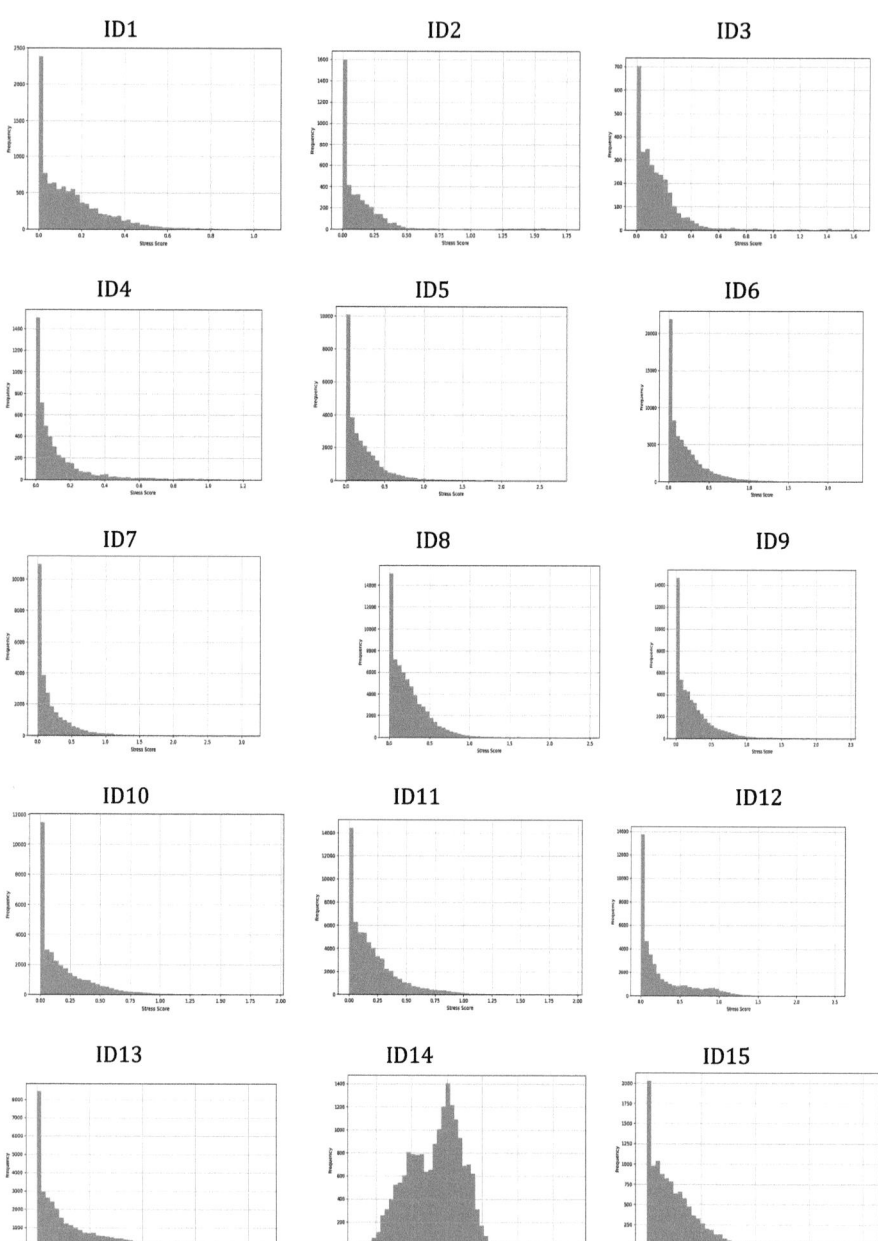

Fig. 3. AU01 + AU02 + AU20 + AU23

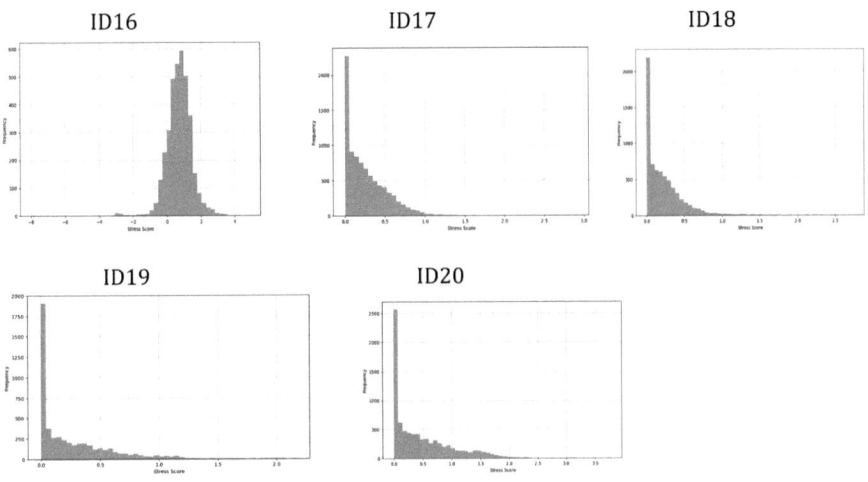

Fig. 3. (*continued*)

Table 1. Summary of questionnaire.

I Questionnaire about your job. (1: Applies, 5: Does not apply, 5-point scale)	ID1	ID2	ID3	ID4	ID5	ID6	ID7	ID8	ID9	ID10	ID11	ID12	ID13	ID14	ID15	ID16	ID17	ID18	ID19	ID20
I have to handle a very large amount of work.	1	2	2	2	2	2	4	1	2	2	2	2	2	1	1	2	2	2	3	
I cannot complete my work within the allotted time.	2	1	2	3	2	3	4	2	2	3	2	4	2	2	1	2	2	2	5	
I need to concentrate intensely on my work.	2	2	3	1	2	1	2	2	2	2	2	3	2	1	1	1	1	2	2	3
My job is difficult and requires advanced knowledge or skills.							2	2	2	1	3	1	3	2	1	2	3	2	1	4
	1	2	3	2	1	2														
I can express my opinions in the workplace's policy decisions.							3	1	2	3	2	3	2	1	4	1	2	2	3	
	2	2	2	2	2	2														
I rarely use my skills or knowledge at work.	5	4	3	4	4	4	5	5	4	5	4	4	3	4	4	4	5	4	5	3
There are differences of opinion within my department.	3	4	3	3	4	3	2	3	5	3	5	3	2	4	4	2	2	3	4	5
The atmosphere in my workplace is friendly.	2	2	1	2	1	2	2	3	1	2	3	2	1	1	1	2	1	1	2	1
The content of my job suits me.	1	2	3	2	2	2	1	1	2	2	3	2	1	2	2	2	1	1	2	1
My work is fulfilling.	1	2	2	2	1	2	1	1	1	2	3	1	2	1	1	2	1	1	2	1

II. Questionnaire about your condition over the past month. (1: Never, 5: Always, 5-point scale)	ID1	ID2	ID3	ID4	ID5	ID6	ID7	ID8	ID9	ID10	ID11	ID12	ID13	ID14	ID15	ID16	ID17	ID18	ID19	ID20
I feel irritated.	3	2	2	2	4	2	3	3	3	2	4	3	1	1	3	5	1	2	2	2
I feel extremely tired.	5	4	4	4	4	4	3	3	3	2	4	3	3	3	4	5	1	2	3	2
I feel sluggish.	1	4	1	2	4	2	3	3	3	1	3	2	3	1	4	5	1	3	2	3
I feel tense.	3	4	4	4	5	4	4	3	4	2	4	3	4	3	4	5	3	3	2	2
I feel anxious.	4	2	4	4	4	4	3	3	2	3	3	5	2	2	4	1	2	3	2	
I feel stiffness in my neck and shoulders.	3	2	5	4	5	4	3	2	2	4	4	5	2	5	4	4	3	3	4	
I have lower back pain.	1	2	1	4	4	4	5	3	1	4	2	1	5	4	5	4	1	3	2	4
I have eye fatigue.	5	4	5	5	5	5	2	2	1	1	4	2	4	5	5	4	3	3	2	2

Table 2. Summary of Facial expression analysis.

	ID1	ID2	ID3	ID4	ID5	ID6	ID7	ID8	ID9	ID10	ID11	ID12	ID13	ID14	ID15	ID16	ID17	ID18	ID19	ID20
PERCLOS	22.29%	14.46%	24.27%	19.96%	17.31%	29.08%	23.10%	19.21%	28.55%	22.38%	17.60%	20.67%	21.91%	36.19%	22.99%	8.94%	29.30%	23.68%	29.43%	31.80%
AU04&AU07	1.48	0.35	0.92	0.90	0.53	0.71	0.06	1.11	1.01	0.65	0.60	0.28	0.63	1.77	0.79	0.34	1.54	0.80	0.94	1.32
AU01&AU02&AU20&AU23	0.15	0.14	0.16	0.13	0.21	0.23	0.21	0.24	0.24	0.19	0.21	0.27	0.27	1.33	0.24	0.73	0.27	0.23	0.26	0.49

5 Conclusion

This study explored the relationship between facial expression–based indicators and self-reported perceptions of fatigue and stress among care workers in Japan. Using video recordings and automated facial analysis, objective measures such as PERCLOS and selected Facial Action Units (AUs) were examined in conjunction with subjective responses collected through structured questionnaires.

The findings revealed meaningful consistencies between subjective fatigue reports and physiological markers. Elevated PERCLOS values were associated with participants who reported high workloads and difficulty completing tasks, suggesting that ocular fatigue captured through eye closure metrics can reflect internal experiences of physical strain. Similarly, higher composite scores for AU04 and AU07—indicators of muscular tension in the brow and eye regions—aligned with self-reported perceptions of sustained effort and task-related fatigue.

Furthermore, a separate composite index comprising AU01, AU02, AU20, and AU23 successfully reflected psychological stress, particularly in participants who reported emotional exhaustion, anxiety, and general tension. One notable case (ID16) demonstrated high subjective stress scores alongside the highest AU composite values, underscoring the potential of facial expression metrics to detect psychological strain. These results support the use of AU intensity values, rather than binary occurrence indicators, to detect subtle yet meaningful facial expressions linked to internal states.

While each indicator offers unique strengths in assessing different dimensions of fatigue—physiological, muscular, and emotional—the study does not provide sufficient evidence to determine which metric is superior. Instead, the findings highlight the complementary value of combining subjective and objective assessments to form a more holistic understanding of fatigue and stress in care settings.

In conclusion, facial expression analysis offers a promising non-invasive method for monitoring caregiver well-being in environments where traditional survey methods are limited by time and resource constraints. By enabling early detection of fatigue and stress, such tools can inform workplace interventions aimed at sustaining the health and productivity of care personnel. Future research with larger samples and longitudinal designs is recommended to further validate and optimize the integration of facial metrics into occupational health monitoring systems.

References

1. Ekman, P., Friesen, W.V.: Facial Action Coding System. Consulting Psychologists Press, Palo Alto (1978)
2. Giannakakis, G., et al.: Automatic stress analysis from facial videos based on deep facial action units recognition. Pattern Anal. Appl. **25**, 521–535 (2022)
3. Gavrilescu, M., Vizireanu, N.: Predicting depression, anxiety, and stress levels from videos using FACS. IEEE Trans. Affect. Comput. **12**(2), 370–383 (2019)
4. Chang, R.C.-H., et al.: Drowsiness detection system using PERCLOS and facial physiological signal. Sensors **22**(14), 5380 (2022)
5. Dinges, D.F., et al.: Evaluation of ocular techniques for fatigue detection. DOT HS 808 762, U.S. Dept. of Transportation (1998)

6. Abe, T.: PERCLOS-based fatigue detection technologies. Sleep Adv. 4(1), zpad006 (2023)
7. Li, X., et al.: Explainable student fatigue monitoring with joint facial representation. Sensors **23**(7), 3602 (2023)
8. Ofei-Dodoo, S., et al.: Burnout, stress, and nonverbal communication among medical staff. J. Occup. Health **61**(3), 245–253 (2019)
9. Chatterjee, A., et al.: CNN-based facial emotion detection for monitoring nurse fatigue in ICU. IEEE Access **9**, 109876–109887 (2021)
10. D'Mello, S., Graesser, A.: AutoTutor and affect detection: Integrating cognitive and affective states. Int. J. Artif. Intell. Educ. **21**(1), 1–28 (2012)

Facial and Vocal Cues in Daily Life: A Multimodal Framework for Behavioral Insight

Kazunori Minetaki[1(✉)] and I-Hsien Ting[2]

[1] Creative Management and Innovation Research Institute, Kindai University, 3-4-1 Kowakae, Higashi-Osaka, Osaka, Japan
kminetaki@bus.kindai.ac.jp
[2] Social Networks Innovation Center, National University of Kaohsiung, 700, Kaohsiung University Rd., Nanzih District, 811 Kaohsiung, Taiwan

Abstract. This study proposes a multimodal framework for detecting cheerful affective states by integrating facial expression analysis and vocal signal processing. Facial features were extracted using the Facial Action Coding System (FACS). Cheerfulness was defined from AU06 (cheek raiser), AU12 (lip corner puller), and AU25 (lips part), while confidence was derived from AU05, AU06, AU07, AU12, and AU23. Fatigue was operationalized through PERCLOS, using AU45 (eye closure) as an indicator of vigilance loss. To validate these AU-based indices, Mel-Frequency Cepstral Coefficients (MFCCs) were computed from speech and tested for convergence with the facial composites. Results showed that cheerfulness aligned with MFCC7, MFCC12, and MFCC6, confidence with MFCC7 and MFCC12, and fatigue with MFCC3, MFCC10, MFCC11, and MFCC6. These findings suggest that cheerfulness and confidence exhibit distinct multimodal signatures, whereas fatigue primarily manifests as a facial expression. The integration of AU and MFCC features thus strengthens the reliability and ecological validity of affective assessment.

Keywords: Facial features were extracted using the Facial Action Coding System (FACS) · Mel-Frequency Cepstral Coefficients (MFCCs) · PERcentage of eye CLOSure(PERCLOS)

1 Introduction

Understanding emotional expression in everyday life requires a comprehensive approach that reflects the multimodal nature of human communication. Emotions are not conveyed through a single channel, but rather emerge from the dynamic interplay of facial expressions, vocal prosody, and contextual behavior. While traditional affective research has often emphasized facial cues or relied on self-reported emotions, recent advances in behavioral computing emphasize the integration of multiple sensory modalities to enhance the accuracy, ecological validity, and robustness of emotional assessment.

A multimodal framework is particularly crucial for capturing subtle and complex affective states such as cheerfulness, which may be manifested inconsistently or modestly through facial cues alone. Facial expressions are sometimes subject to voluntary control or social masking, whereas vocal patterns often reflect more involuntary and nuanced emotional signals. Therefore, combining facial and vocal information enables a more holistic understanding of emotional expressions, especially in naturalistic, daily-life interactions.

This study adopts a multimodal approach by integrating facial action coding with key acoustic indicators of vocal expression: Mel-Frequency Cepstral Coefficients (MFCCs), Fundamental Frequency (F0), and Root Mean Square (RMS) energy. Each of these features provides a distinct yet complementary perspective on the speaker's emotional state.

Mel-Frequency Cepstral Coefficients (MFCCs) are among the most widely used features in speech and emotion recognition systems due to their ability to model the perceptual characteristics of the human auditory system. The computation of MFCCs involves a sequence of signal transformations. First, the raw audio waveform is divided into short overlapping frames (typically 20–40 ms) to assume short-term stationarity. Each frame is subjected to a Fast Fourier Transform (FFT) to convert the signal into the frequency domain. This frequency spectrum is then filtered using a Mel-scaled filter bank, which applies triangular filters distributed non-linearly to reflect the ear's higher sensitivity to low-frequency sounds.

The logarithm of the energy in each Mel-filtered band is then computed to simulate the logarithmic perception of sound intensity by human listeners. Finally, a Discrete Cosine Transform (DCT) is applied to the log Mel spectrum to obtain the MFCCs. This transformation reduces dimensionality and ensures decorrelation among features, with typically the first 12 or 13 coefficients retained for analysis. These coefficients effectively summarize the spectral envelope of speech and are sensitive to articulatory configurations that change with emotion. In this study, we compute the standard deviation of MFCCs over time windows to capture vocal variability, which is then weighted (with greater emphasis on MFCC2–MFCC5) to derive an acoustic-based cheerfulness score. These mid-range coefficients are known to encode formant structure and have been empirically associated with affective expression.

Fundamental Frequency (F0), commonly perceived as pitch, represents the rate of vocal fold vibrations. It is a key prosodic feature and is strongly associated with emotional arousal and valence. Cheerful or excited speech tends to exhibit higher F0 values and greater pitch variability. F0 is extracted using the probabilistic YIN (pYIN) algorithm, which provides accurate tracking even in noisy or conversational speech. In this framework, both the mean and standard deviation of F0 are computed to reflect the speaker's dynamic vocal modulation.

MFCCs offer a rich, multidimensional representation of vocal expression. When integrated with facial Action Units (AUs)—specifically AU06 (cheek raiser), AU12 (lip corner puller), and AU25 (lips part), which are known to index genuine smiling—the result is a robust and interpretable multimodal framework for detecting cheerfulness.

This framework enables the identification and quantification of affective behavior in natural interactions, thereby contributing to broader applications in behavioral monitoring, occupational well-being, and human-computer interaction.

2 Literature Review

Recent advances in affective computing emphasize the importance of multimodal emotion recognition systems that combine both facial and vocal signals. While early approaches often focused on a single modality, such as facial expressions or speech prosody, more recent studies have demonstrated that integrating multiple sensory modalities significantly enhances emotional inference in terms of robustness and ecological validity [1].

Multimodal frameworks have shown particular effectiveness in capturing subtle affective states such as cheerfulness, which may not be reliably conveyed through facial cues alone. Visual expressions can be voluntarily controlled or socially masked, whereas vocal characteristics often reveal more spontaneous emotional information [2]. Consequently, combining facial Action Units (AUs) and vocal features such as pitch, energy, and spectral descriptors has become a common practice in recent literature [3].

Among vocal features, Mel-Frequency Cepstral Coefficients (MFCCs) are widely used to characterize the spectral envelope of speech and have been shown to correlate with emotional tone. MFCCs are derived through a sequence of transformations, including the Fast Fourier Transform (FFT), Mel-scale filter banks, and Discrete Cosine Transform (DCT), and are particularly sensitive to articulatory configurations associated with emotion [4]. Empirical studies have found that mid-range coefficients (MFCC2–MFCC5) are especially relevant for detecting affective variability [5].

In addition to spectral features, prosodic indicators such as Fundamental Frequency (F0) and Root Mean Square (RMS) energy provide complementary insights into vocal expressiveness. Higher mean F0 and greater pitch variability are typically observed in cheerful or excited speech, reflecting emotional arousal [6]. Similarly, RMS energy, which captures vocal intensity, has been associated with emotional engagement and communicative enthusiasm [7].

On the visual side, the Facial Action Coding System (FACS) remains a foundational tool for quantifying facial expressions. Specific AUs—particularly AU06 (cheek raiser), AU12 (lip corner puller), and AU25 (lips part)—are well-documented indicators of genuine smiling, often referred to as "Duchenne smiles" [8]. These AUs have been integrated into real-time analysis pipelines in multimodal emotion recognition systems.

By leveraging these prior findings, the present study proposes a multimodal framework that integrates MFCCs, F0, RMS energy, and selected AUs to detect and quantify cheerful behavior in natural conversational settings. This approach builds upon established methodologies in behavioral signal processing while addressing the need for interpretability and ecological validity in real-world applications.

3 Methodology

To explore behavioral manifestations of cheerfulness, confidence, and fatigue in everyday communication, this study adopts a multimodal analytical framework that integrates facial expression analysis and acoustic signal processing. The combined use of visual and auditory indicators enables a more comprehensive assessment of affective states, particularly in natural conversational settings. Participants are eleven Taiwanese.

3.1 Facial Expression

3.1.1 Cheerfulness

Facial expressions were quantified using the Facial Action Coding System (FACS), which is a widely recognized and extensively validated framework for the systematic description of visible facial movements. FACS decomposes complex expressions into discrete Action Units (AUs), each corresponding to the contraction of a specific set of facial muscles. This method has been employed across psychology, affective science, and human–computer interaction studies for decades, and its robustness makes it an ideal choice for quantifying subtle nonverbal cues in both experimental and naturalistic contexts.

In the present analysis, three AUs were selected based on their theoretically and empirically established associations with positive affect and cheerful emotional states: AU06_r (cheek raiser), AU12_r (lip corner puller), and AU25_r (lips part). AU06 and AU12, when co-activated, are particularly important as they signal so-called Duchenne smiles, which are widely regarded as reliable indicators of genuine enjoyment. AU25, while not uniquely diagnostic of happiness on its own, complements the configuration by reflecting vocalization or laughter-related expressivity, thereby enriching the interpretability of observed smiles. The decision to focus on this triad of AUs thus reflects both theoretical consensus in the literature and practical considerations of data availability in automated expression analysis. Extraction of these AUs was conducted from frame-level facial video data using automated computer vision algorithms trained on large-scale annotated databases.

Prior to downstream analysis, each dataset underwent a quality control step in which the presence, naming consistency, and scaling of the relevant AU columns were verified. This preprocessing ensured that no missing or mislabeled features could bias subsequent computations of the composite score. Additionally, temporal smoothing procedures were applied to minimize noise from transient misdetections, thereby producing a stable representation of the participant's expressive behavior across time. To compute a composite cheerfulness index, a weighted sum of the three AUs was constructed.

Following prior affective computing conventions, AU06_r and AU12_r were each assigned a weight of 0.4, reflecting their primary role in distinguishing genuine positive expressions. In contrast, AU25_r received a smaller weight of 0.2, acknowledging its supplementary role in enhancing expressiveness without being as central to the definition of happiness-related facial displays. This weighting scheme is grounded not only in empirical findings from experimental psychology but also in computational studies showing that such a configuration yields a reliable prediction of subjective reports of happiness. The resulting cheerfulness score provides a continuous, interpretable metric that captures moment-to-moment fluctuations in visible positive affect throughout the recorded interaction.

By condensing complex facial dynamics into a single numerical trajectory, the measure allows researchers to investigate how cheerfulness evolves in response to conversational events, contextual factors, or experimental manipulations. Moreover, because the score is built on theoretically meaningful components, it remains interpretable to both domain experts and applied practitioners, thereby bridging the gap between low-level computer vision outputs and high-level psychological constructs.

3.1.2 Confidence

The confidence score was developed as a composite index derived from specific Facial Action Units (AUs) that have been repeatedly linked in the literature to nonverbal displays of confidence, assertiveness, and self-assured affective states. Unlike cheerfulness, which is predominantly tied to positive valence, confidence is conceptualized as a socially evaluative emotion, often conveyed through posture, gaze, and distinct facial muscular patterns. For the present analysis, five AUs were selected to operationalize confidence: AU05_r (upper lid raiser), AU06_r (cheek raiser), AU07_r (lid tightener), AU12_r (lip corner puller), and AU23_r (lip tightener). Each of these contributes in complementary ways. AU05 and AU07, for example, are associated with eye openness and intensity, signaling alertness and determination. AU06 and AU12, while also central to smiles, can signal confidence-laden positivity when combined with other markers. AU23, in contrast, reflects a tightening of the lips, often interpreted as a display of controlled assertiveness rather than unrestrained joy.

A weighted sum of these AUs was computed to produce the confidence score. In the current implementation, equal weights (0.2 each) were assigned to all five AUs to reflect their joint and balanced contribution to the construct. This decision was motivated by prior affective computing research suggesting that no single AU alone is sufficient to capture confidence, but that their synergistic co-activation forms a reliable behavioral signature. The formula thus treats confidence as an emergent property of multiple subtle muscular cues, rather than being reducible to a single "confidence muscle."

From a methodological perspective, the use of multiple AUs reduces the risk of over-relying on expressions that could be ambiguous in isolation. For instance, AU12 alone might indicate friendliness rather than confidence, while AU23 alone might be misinterpreted as tension. However, when these cues are considered together in weighted combination, they yield a score that is both stable across individuals and sensitive to within-interaction changes in self-assured expression.

The resulting confidence score serves as a continuous, interpretable measure of nonverbal confidence, enabling researchers to track fluctuations in confidence levels during key moments of an interaction, such as speaking turns, responses to questions, or negotiation phases. By grounding the measure in both theoretical models of social signaling and empirical validation from computer vision tools, the score provides a robust bridge between automated facial analysis and psychological interpretations of interpersonal confidence.

3.1.3 PERCLOS (PERcentage of Eye CLOSure)

PERCLOS (PERcentage of eye CLOSure) is a well-established indicator of fatigue and reduced vigilance, initially introduced in the context of human factors and transportation safety (Wierwille et al., 1994). It quantifies the proportion of time during which the eyes are at least partially closed. It has since been widely adopted in occupational health and affective computing as a sensitive, non-invasive measure of drowsiness. Compared to simple blink rate, PERCLOS captures sustained eyelid closures and micro-sleeps, making it particularly effective for detecting early signs of fatigue.

In the present study, Action Unit 45 (AU45: blink) from the Facial Action Coding System (FACS) was employed to operationalize eye closure. Frame-level AU45_r intensity values were extracted using automated computer vision tools, and a threshold of AU45_r > 0.5 was applied to classify each frame as "eyes closed." All frames below this threshold were treated as "eyes open."

The analytical procedure proceeded as follows. First, frame-level classifications were aggregated within fixed observation windows to compute the proportion of closed-eye frames relative to the total number of frames, yielding the PERCLOS score: $PERCLOS = \frac{Number of frames with AU45\_r > 0.5}{total number of frames in the inerval}$

Second, for each participant, mean PERCLOS scores were computed across the entire session, providing a stable individual-level measure of fatigue. Finally, to identify a global cutoff for distinguishing fatigued from alert states, distribution-based thresholding methods (e.g., Otsu's method, Gaussian Mixture Models) were applied to the pooled PERCLOS values across participants.

The interpretation of the resulting metric follows established conventions: higher PERCLOS values indicate greater fatigue and diminished alertness, while lower values reflect sustained vigilance. Due to its solid foundation in the physiological mechanisms of eyelid closure and its extensive validation in real-world contexts, PERCLOS provides a transparent, interpretable, and theoretically meaningful measure suitable for both experimental research and practical monitoring systems.

3.2 Mel-Frequency Cepstral Coefficients (MFCCs)

Vocal expressions of emotion were systematically analyzed using Mel-Frequency Cepstral Coefficients (MFCCs), which represent a widely adopted feature set for speech analysis due to their ability to approximate the spectral envelope of human speech in a perceptually meaningful manner. The raw audio recordings were first preprocessed by normalizing the sampling rate to 16 kHz. They then segmented the data into fixed-length overlapping windows (1.0 s with a 0.5 s hop size), ensuring both temporal resolution and robustness to local fluctuations in the signal. Within each segment, 13 coefficients (MFCC0–MFCC12) were extracted using the LibROSA Python library, which applies a series of steps including short-time Fourier transformation, Mel-scale filtering, and discrete cosine transformation.

To capture not only the static spectral characteristics but also the dynamic variability of speech, the standard deviation of each coefficient across all frames within a segment was computed. This statistical measure reflects the degree of fluctuation in vocal tract configurations over time and provides sensitivity to emotional states, which are often conveyed through prosodic variation and articulation dynamics. For instance, heightened arousal or emotional intensity may manifest as greater variability in higher-order MFCCs, whereas lower variability tends to correspond to more neutral or monotonic speech delivery.

The resulting feature set thus encoded both spectral shape (through mean MFCC patterns) and temporal variability (through standard deviations), offering a comprehensive representation of vocal expressions. These features served as the basis for subsequent machine learning analyses, which aimed to map acoustic patterns to emotional states, thereby allowing for the quantitative assessment of vocal affect.

3.3 Multimodal Integration

We previously operationalized three affective constructs—cheerfulness, confidence, and fatigue—from facial Action Units (AUs). While these AU-based indices are theoretically grounded, they remain proxies rather than direct evidence of internal states. Our starting point was therefore to test whether they exhibit convergent patterns in the acoustic channel. Specifically, we examined whether Mel-Frequency Cepstral Coefficients (MFCCs) could reliably predict each AU-derived construct, thereby strengthening the interpretation that these indices reflect genuine affective expressions.

For each construct, AU scores were converted into binary labels by applying threshold estimation. We primarily employed Otsu's method, which identifies the cutoff that maximizes between-class variance in score histograms. Where Otsu failed due to skewed distributions, fallback procedures (such as Gaussian mixture midpoints or percentile cutoffs) were applied. Cheerfulness thresholds were determined from a weighted composite of AU06 (cheek raiser), AU12 (lip corner puller), and AU25 (lips part); confidence from AU05, AU06, AU07, AU12, and AU23; and fatigue from PERCLOS, defined as the proportion of frames with AU45 (eye closure) > 0.5.

On the acoustic side, audio recordings were resampled to 16 kHz, segmented into one-second windows with 50% overlap, and thirteen MFCCs (MFCC0–MFCC12) were extracted using LibROSA. To capture vocal dynamics, the standard deviation across frames was computed for each coefficient within each window. Logistic regression with L2 regularization was then used to predict AU-based labels from MFCC features, and performance was evaluated using cross-validated F1 scores.

The consolidated analysis identified compact, construct-specific MFCC subsets. Cheerfulness was best predicted by the triplet MFCC7, MFCC12, and MFCC6, achieving an F1 score of 0.633. Confidence favored a smaller pair—MFCC7 and MFCC12—with F1 = 0.622. Fatigue, derived from ocular dynamics, aligned only modestly with vocal features, with the quartet MFCC3, MFCC10, MFCC11, and MFCC6 reaching F1 = 0.483 (Table 1).

The interpretation of these combinations reveals distinct acoustic signatures. For cheerfulness, MFCC6 and MFCC7 reflect mid-range spectral modulations linked to prosodic energy, while MFCC12 captures fine spectral detail. Together, they suggest that cheerfulness is expressed through brightening of the spectrum and dynamic articulation, consistent with smiling and lively speech. Confidence was characterized by MFCC7 and MFCC12, emphasizing high-frequency precision and vocal clarity, features associated with resonant and stable articulation. In contrast, fatigue was characterized by a mixed set: MFCC3 indicated reduced spectral energy, MFCC6 reflected mid-range variability, while MFCC10 and MFCC11 indexed high-frequency irregularities. This combination suggests that fatigue is associated with flattened energy and irregular spectral patterns, consistent with monotony or reduced vocal effort.

In summary, these findings demonstrate that cheerfulness and confidence exhibit distinct, replicable acoustic signatures that align with AU-based indices, thereby validating them as affective markers. Fatigue, however, remained primarily facial in nature, with limited but complementary effects on the voice. The joint use of AU-derived thresholds and MFCC-based validation, therefore, provides a stronger multimodal foundation for assessing emotional expression.

Table 1. MFCC subsets associated with AU-derived emotion scores.

Construct	Optimal MFCC subset	F1	Interpretation
Cheerfulness	MFCC7, MFCC12, MFCC6	0.633	Bright spectral detail and dynamic articulation; smiling, energetic prosody
Confidence	MFCC7, MFCC12	0.622	High-frequency clarity and stability; resonant and controlled speech
PERCLOS	MFCC3, MFCC10, MFCC11, MFCC6	0.483	Flattened low-frequency energy and irregular high-frequency patterns; monotony and reduced vocal effort

4 Conclusion

This study demonstrates the value of a multimodal approach to emotion recognition by combining facial Action Units (AUs) with acoustic features such as Mel-Frequency Cepstral Coefficients (MFCCs). Whereas AU-derived indices of cheerfulness, confidence, and fatigue provided theoretically grounded starting points, the additional integration of MFCC analysis enabled a deeper validation of these constructs. Through data-driven thresholding procedures, each AU-based score was transformed into a binary label, which was then tested against MFCC-derived predictors.

The results revealed clear and distinct acoustic signatures for cheerfulness and confidence. Cheerfulness was best captured by MFCC7, MFCC12, and MFCC6, reflecting dynamic articulation and spectral brightening, consistent with the vocal correlates of smiling and lively affect. Confidence was characterized by a smaller set of high-order MFCCs (MFCC7, MFCC12), emphasizing clarity and resonance in the higher-frequency spectrum. By contrast, fatigue—operationalized through PERCLOS—showed weaker but non-trivial convergence with a heterogeneous set of MFCCs (MFCC3, MFCC10, MFCC11, MFCC6), suggesting that fatigue manifests only partially in the voice and remains better anchored in ocular behavior.

Taken together, these findings highlight the necessity of multimodal frameworks for understanding affective states in natural interactions. Facial and vocal modalities offer complementary windows into emotional expression, and their integration provides more reliable and ecologically valid indicators than either channel alone. While cheerfulness and confidence appear to have replicable multimodal signatures, fatigue illustrates the limits of acoustic inference and underscores the importance of face-dominant measures for certain affective states.

Future Directions. Building on the current framework, several avenues for future research can be envisioned. First, incorporating additional acoustic features such as jitter, shimmer, and spectral flux may improve sensitivity to vocal markers of fatigue and stress. Second, multimodal integration could be extended to include physiological signals (e.g., heart rate variability, skin conductance) and contextual cues (e.g., task difficulty, environmental demands), which may enhance the ecological validity of emotion recognition. Third, advanced machine learning architectures, such as deep multimodal fusion networks or Bayesian hierarchical models, could be applied to better capture individual variability and nonlinear dynamics of affective expression. Finally, longitudinal studies across diverse populations and settings are necessary to ensure the robustness and generalizability of the proposed framework, paving the way for practical applications in occupational health monitoring, adaptive human–computer interfaces, and personalized well-being interventions.

By situating AU-based measures within a multimodal validation framework and identifying concrete directions for extension, this study contributes to the ongoing development of affective computing systems that are not only robust and interpretable, but also adaptable to the complexities of real-world emotional communication.

References

1. Pantic, M., Rothkrantz, L.J.M.: Toward an affect-sensitive multimodal human-computer interaction. Proc. IEEE **91**(9), 1370–1390 (2003)
2. Zeng, Z., Pantic, M., Roisman, G.I., Huang, T.S.: A survey of affect recognition methods: Audio, visual, and spontaneous expressions. IEEE Trans. Pattern Anal. Mach. Intell. **31**(1), 39–58 (2009)
3. D'Mello, S., Kory, J.: A review and meta-analysis of multimodal affect detection systems. *ACM Comput. Surv.* **47**(3), 43:1–43:36 (2015)
4. Eyben, F., Wöllmer, M., Schuller, B.: openSMILE – The Munich versatile and fast open-source audio feature extractor. In: *Proc. ACM Int. Conf. Multimedia*, pp. 1459–1462 (2010)
5. El Ayadi, M., Kamel, M.S., Karray, F.: Survey on speech emotion recognition: Features, classification schemes, and databases. Pattern Recogn. **44**(3), 572–587 (2011)
6. Scherer, K.R.: Vocal communication of emotion: A review of research paradigms. Speech Commun. **40**(1–2), 227–256 (2003)
7. Cowie, R., Douglas-Cowie, E., Savvidou, S., McMahon, E., Sawey, M., Schröder, M.: 'FEELTRACE': An instrument for recording perceived emotion in real time. In: *ISCA Workshop on Speech and Emotion*, pp. 19–24 (2000)
8. Ekman, P., Friesen, W.V.: Facial Action Coding System: Investigator's Guide. Consulting Psychologists Press, Palo Alto (1978)

Question-Answering Systems for Tourism: Development of a RAG-Based Prototype for Ecuadorian Places

Yahaira Benítez-Morejón, Janneth Chicaiza(✉), and Priscila Valdiviezo-Diaz

Universidad Técnica Particular de Loja, Loja, Ecuador
{yibenitez,jachicaiza,pmvavldiviezo}@utpl.edu.ec

Abstract. Technological advances have facilitated content creation, dissemination, and global content distribution almost immediately. Advances in large language models (LLMs) facilitate the development of natural language processing (NLP) applications such as Q&A systems. This work aims at developing a Q&A system based on a Retrieval Augmented Generation (RAG) architecture for the domain of tourism in Ecuador. The system design includes items such as user, query, embeddings, retrieval model, generative model, and finally, the answer generated by the system. The retrieval component searches for relevant information in the knowledge base created for the project, providing context to the generative models `Meta-Llama-3-8B-Instruct` and `Gemma-7b-it` to produce reliable and accurate answers. The results indicated a reasonable performance in information retrieval and a superiority of the `Meta-Llama-3-8B-Instruct` model in text prediction. The interface, developed with Streamlit, offers a friendly experience for users, allowing effective consultations about tourism in Ecuador.

Keywords: RAG architecture · question-answering system · LLMs · Ecuador · Tourism

1 Introduction

Technological advances have facilitated content creation, dissemination, and global content distribution almost immediately. In response to the problem of digital information overload, the community needs agile systems that provide reliable answers to specific information needs. To address this issue, this paper proposes the study of generative question-answering (Q&A) systems and the implementation of a Q&A prototype based on a RAG (Retrieval-Augmented Generation) architecture. RAG combines the ability of generative models to understand human language [4] with the efficiency of information retrieval models to find relevant information in large text collections.

Supported by Universidad Técnica Particular de Loja.

Tourism is one of the most dynamic industries worldwide [18] because it constantly evolves in response to changing global trends. Furthermore, thanks to the removal of trade and travel barriers, the tourism and hospitality industry has been playing a key role in economic progress worldwide by creating job sources and reconciling shared interests of tourists, residents, and businesses [10]. In particular, when marginal developing nations are involved, tourism can significantly boost their inhabitants' development. With the increase in the tourism industry and the popularity of digital upgrades in attractions, it is crucial to develop a Q&A system that provides details of the attractions [8].

To contribute to providing answers in demand for information specific to the needs of tourists, this paper analyzes the application of Q&A systems in the tourism domain and presents a specific proposal to provide answers related to Ecuadorian places. In countries like Ecuador, with its vast natural and cultural attractions for visitors, the tourism industry can generate employment and monetary flow for the population [5].

The Q&A prototype was designed using the RAG (Retrieval-Augmented Generation) architecture; therefore, the system pipeline was defined from its basic components: 2) a knowledge base created from texts extracted from Wikipedia; 3) a retriever based on dense representations of texts or embeddings; and 4) a generator based on large language models (LLMs). For this last component, pre-trained language models were tested, like `meta-llama/Meta-Llama-3-8B-Instruct` and `google/gemma-7b-it`. By focusing on tourism as the application domain, the prototype can be scaled to become a product that encourages tourists to choose Ecuadorian destinations, thus promoting the economic development of different types of businesses.

This paper is organized as follows: Sect. 2 describes the research background and related work. Section 3 presents the prototype design and implementation. Section 4 presents the results of the preliminary evaluation. Finally, the conclusion is presented in the last section.

2 Background

2.1 Question-Answering Systems

Question-answering systems are designed to receive questions formulated by humans in natural language, and by using natural language processing (NLP), machine learning (ML), and information retrieval (IR) methods, they can understand the question and answer it appropriately [16].

Currently, there are multiple applications of Q&A systems in various fields such as education [15], medicine [13], agriculture [19], environmental sciences [16], and others. In tourism, Q&A systems provide information on tourist destinations, personalized recommendations, reservations, and travel planning [8].

2.2 RAG Architecture

Retrieval-Augmented Generation (RAG) is an approach that involves retrieving text from a knowledge base to provide as a context to a generative language model, so that the model can create a reliable result for a specific task. Therefore, RAG combines text retrieval and generation techniques [9]. For a Q&A system, the RAG architecture works as follows:

1. Question processing: When a user enters a question into the RAG system, it is analyzed using a language model, which codifies the question like a contextual representation named embedding.
2. Retrieval: Once the system analyzes the question, it searches the knowledge base for relevant information using an information retrieval model. The database is typically composed of documents or data indexed like embeddings.
3. Information integration: Once the system identifies the relevant information in the retrieval, it combines the data with the knowledge available from the selected language model (LLM) to generate an answer.
4. Response generation: Finally, combining the data retrieved from the knowledge base with the general knowledge of the LLM, the system generates a complete answer.

2.3 Related Work

To find proposals for Q&A systems based on the RAG architecture applied to the tourism domain, we searched the Scopus database. In this database, we used keyword combinations to retrieve the most relevant works; however, no RAG systems in tourism were found. Therefore, the search for existing RAG proposals for any RAG-based Q&A system was generalized.

From the search string: TITLE-ABS-KEY (rag) AND TITLE-ABS-KEY (question) AND TITLE-ABS-KEY (answer*) AND PUBYEAR > 2017, 27 documents were retrieved in Scopus. Then, three inclusion criteria were established to identify the closest works: i) Is the proposal based on the RAG architecture? ii) Does the proposal aim to implement a Q&A system? iii) Is the document accessible from the web? As a result of this filtering process, 10 documents met the three criteria. Table 1 shows the application domain or datasets used, the main retrieval methods and language models used, and the performance of each proposal.

From the analysis of related works on Q&A systems based on RAG architecture, we noted that each proposal uses different datasets and information sources because they have been tested in different domains. Furthermore, as seen in Table 1, the most widely used language models are those based on GPT [6,7,9,12,17], along with Orca [7,11]. Also, methods that combine images and text are being developed to improve response generation in Q&A systems [2], and several architectures have been proposed for open-domain Q&A systems such as the RAG-end2end application [14].

Furthermore, as demonstrated in the reviewed papers, the proposed Q&A systems improve their performance through the use of language models and the integration of diverse sources of information, such as text, images, and metadata. Therefore, the RAG architecture improves question processing and provides more accurate and reliable answers.

3 Prototype Implementation

For the implementation of the Q&A prototype for tourism, the following libraries were used: Transformers, datasets, faiss, sentence transformers, bitsandbytes, accelerate, pyngronk, and streamlit. Additionally, for the generative component, two pre-trained language models were tested: `Meta-Llama-3-8B-Instruct` and `Gemma-7b-it`, which are available on the Hugging Face platform. The following describes how these tools were used to create each component of the RAG architecture.

3.1 Knowledge Base Creation

To create the system's knowledge base (KB), Wikipedia pages related to tourism in Ecuador were used as a source of information. This component was automatically created through a module proposed in [3] that identifies a subset of Wikipedia pages related to a specific domain based on a root concept that represents that domain. Then, after identifying the URLs of the pages related to Ecuadorian tourism, the module extracts the textual content of each page, divides it into smaller fragments (chunks), and finally, transforms the preprocessed text into vector representations or embeddings. The module in charge of generating embeddings uses the `all-mpnet-base-v2` model. This model is a powerful tool for converting sentences into high-dimensional vectors, allowing its use in various applications such as semantic search, data clustering, and calculating similarity between sentences. In this project, for the automatic extraction of text from pages related to tourism in Ecuador, the root concept that was set was the DBpedia category, Tourism in Ecuador. DBpedia is an RDF graph that draws on information from Wikipedia.

Table 1. Description of Related Work

Ref.	Domain/dataset(s)	Retrieval methods (R) and LLMs (G)	Performance
[9]	Document Question Answering Dataset (DocuQA) and Stanford Question Answering Dataset (SQuAD).	**R:** Facebook AI Similarity Search (FAISS). **G:** GPT-3.5-turbo.	The proposal outperforms other systems: the accuracy of the proposal reaches 96% for DocuQA and 95.5% for SQuAD vs. the accuracy of other systems: 55% for DocuQA and 85.7% for SQuAD.
[7]	PDF files in the context of the PCI DSS version 4.0 standard.	**R:** FAISS. **G:** Orca 2 (7b y 13b) and other models like Llama-2 (13b & 70b), GPT-3.5-Turbo and GPT-4.	The approach with Orca-2 demonstrates a superior performance in faithfulness, answer relevance, overall score, and inference speed compared to the other models.
[2]	The proposal is used in the domain of multimodal open-ended questions and uses datasets such as LAION, CC, VQA and PAQ.	**R:** Approach based on maximum inner product search (MIPS) to find nearest neighbors. **G** A model based on the T5 decoder.	The proposal outperforms pre-trained models in vision-language (VLP) tasks across several metrics, e.g., in retrieval-F1, MuRAG scores 74.6 versus 70.9 of VLP + VinVL.
[6]	The Qatar Living forum (SemEval-2017-task 3) and Yahoo! Answers.	**R:** Retrieving CQA (Community Question Answering) questions using FAISS. **G:** Compares GPT-4 base with RAG using GPT.	The ITRLMRAG proposal (GPT-4 with RAG) outperforms all the compared approaches, achieving a MAP increase of 4.25 points compared to the best team in SemEval-2017.
[14]	CovidQA, NewsQA dataset, and QAConv.	**R:** The similarity between a question and a passage is calculated by taking the dot product. **G:** BART.	The proposal, RAG-end2end-QA+R, which is fine-tuned with question-answer (QA) pairs and reconstruction signals, improves the results of other architectures.
[12]	The divorce-related QA pairs and legal provisions and cases mentioned in the QA pairs are gathered from two Korean sites.	**R:** The retriever identifies the most relevant document for a query, based on similarity in the embedding space. **G:** GPT-4.	The proposed Eval-RAG demonstrates a better correlation with legal expert evaluations compared to LLM-based evaluators such as FairEval and ChatEval.
[1]	The proposal addresses issues related to petroleum tank cleaning and uses the ELI5 dataset, Wiki Snippets Indexes & information about industrial processes.	**R:** Uses the Structured Activation Vertex Entropy (SAVE) method in combination with Question Answering Machine (QAM) algorithms. **G:** ELI5.	The RAG-based proposal proves to be a good option to complement the traditional ARIZ (Algorithm for Solving Inventive Problems) approach used in these problems.
[17]	The proposal is used in the educational field and uses information from e-books, theses and articles.	**R:** Semantic search based on the similarity metrics. **G:** GPT-4 Turbo.	In the usability tests, the UI obtained positive results in areas such as visual consistency, intuitive navigation, and clarity of information.
[20]	The proposal was evaluated on two open-domain QA benchmarks: CSQA2.0 and StrategyQA.	**R:** For StrategyQA, the retriever was implemented as a sparse BM25 algorithm. For CSQA2.0, the questions were fed into Google Search. **G:** T5-11B and T5-Large.	The proposed IAG-GPT, for the CSQA2.0 dataset, obtained an accuracy of 80.0, while ChatGPT achieved 60.0. And, for StrategyQA, IAG-GPT obtained 74.0 accuracy, while ChatGPT achieved 52.0.
[11]	The proposal is used for smart agriculture with data from: EU Regulation 2018/848 & Climate-smart Agriculture Sourcebook.	**R:** Similarity metrics, such as cosine similarity, are used to identify the most relevant fragments. **G:** Mistral (7b), Llama2 (7b) and Orca2(7b).	Overall, the Mistral-based proposal (7b) proves to be effective in generating contextual and relevant responses within the smart agriculture domain.

Figure 1 presents a path of the steps performed by the existing component to extract text from Wikipedia: Starting from a root concept that represents the domain of interest, the component executes SPARQL queries to traverse the graph and thus obtain the subconcepts associated with the root concept. Then, from the set of subconcepts, the semantic relationship *skos:broader* is used to reach other subconcepts, and so on iteratively until reaching deeper nodes. Finally, from these SKOS concepts, we can reach Wikipedia pages by using the semantic path $\sim dcterms:subject/foaf:isPrimaryTopicOf$.

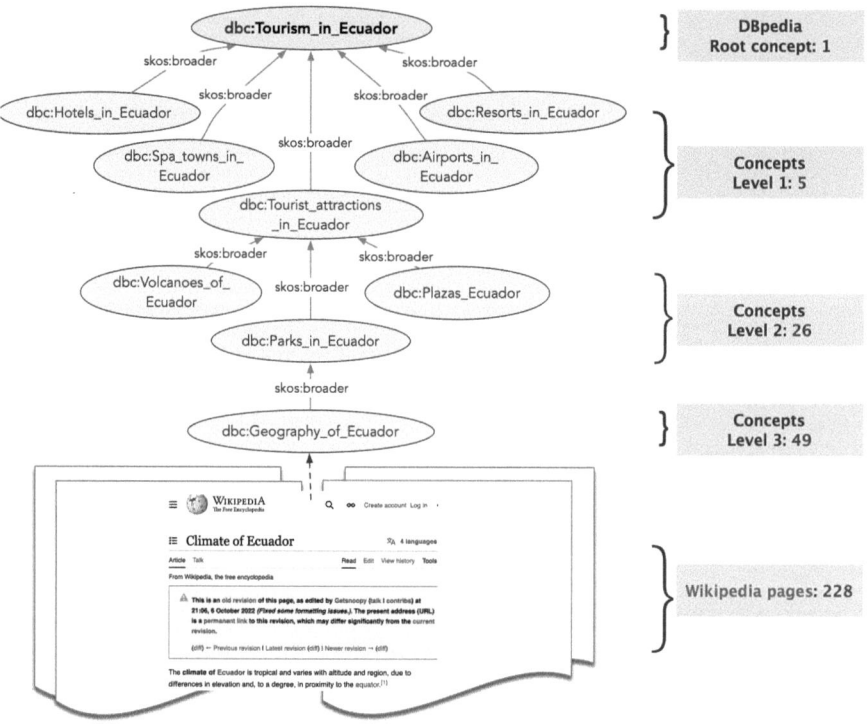

Fig. 1. Traversing DBpedia using SKOS concepts, reaching the URL of Wikipedia pages related to tourism.

After traversing DBpedia, from the root node to their related concepts, we found 228 Wikipedia pages that describe different tourist sites in Ecuador. Figure 1 shows a sample of the SKOS concepts obtained in DBpedia, from which we arrive at Wikipedia pages that describe tourist sites in the country, such as San_Rafael_Falls or Puerto_López, or that group tourist sites from a specific category, such as Monuments_of_Ecuador.

The module used to create the knowledge base completes the creation of the knowledge base once it has vectorized the pre-processed content; this allows the retriever to access this component, facilitating the automatic processing of the extracted information. The resulting database contains 854 entries or documents, and each vector (embedding) is linked to its corresponding metadata, which includes identifiers such as URL, title, content, creation date, and other details relevant to its context. The database was stored in a Hugging Face repository, a platform comprised of more than 900,000 models, 200,000 datasets, and 300,000 demo applications.

3.2 System Design

This section presents the logical design of the system based on the RAG architecture and the prototype implementation, highlighting the key components of the architecture, such as the extractive model for retrieving relevant information and the generative model for generating coherent responses, as well as the integration process between the two. Figure 2 identifies the main flow of the system and its main components.

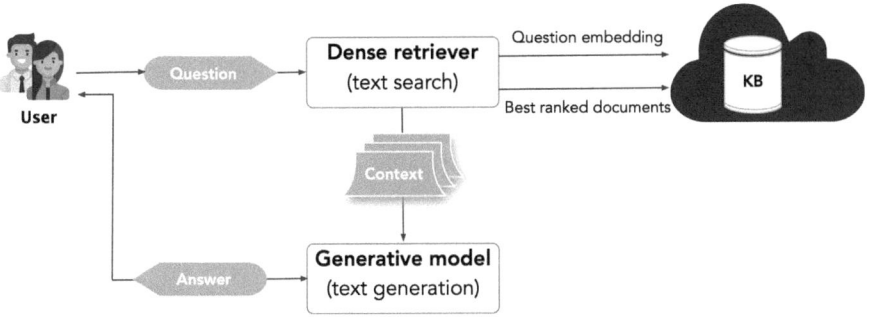

Fig. 2. Logical design of the system.

Each item identified in Fig. 2 is described below:

- User: Is the person who asks the Q&A system a question. In the specific context of tourism, the user can be a travel agent, tourist, student, or anyone interested in learning about tourism in Ecuador.
- Question: This is the question posed by the user to the system. Once the query has been entered, the system creates the embedding (dense vectors of real numbers) so that the search component (retriever) can process it. The question embedding represents the user's question in the form of dense vectors of real numbers.
- Dense Retriever: This is the extractive component of the architecture that is responsible for searching the knowledge base for relevant information. To find the highest-ranked documents, it uses similarity metrics to compare the embedding of the user's question with the embeddings of each document in the knowledge base.

- Knowledge base (KB): It refers to the set of documents that were obtained from the tourism domain and converted into embeddings to be stored in a vector database.
- Context: It is the information extracted from the database by the retriever and sent to the generative model to create the answer.
- Generative model: It is the LLM responsible for generating coherent and accurate answers based on both the relevant documents retrieved and the user's question.
- Answer: It is the answer generated by the generative model and delivered to the user.

Relevant details of the two main processing components of the described architecture are provided below:

- **Dense Retriever**: In the RAG architecture, the retriever is responsible for searching for relevant information in the database or corpus. The retrieval model provides the generative model with the relevant and specific context to accurately and coherently answer questions posed by users on the topic of tourism in Ecuador. Once the user enters a question, the system (Q&A) encodes it using the "sentence-transformers/all-mpnet-base-v2" model, which is an optimized variant of the MPNet model. The resulting vector captures the semantic information of the knowledge base previously loaded into the system. In the prototype, to search for documents in the knowledge base, the *get_nearest_examples()* method of the *Dataset* class available in the Python *dataset* library was used. This performs a nearest neighbor lookup, comparing the query embedding with the documents' embeddings stored in the KB; as a result, it returns the most similar documents.
- **Generative model**: In a RAG system, the generative model plays a fundamental role in generating coherent and accurate answers. For this project, two LLMs were analyzed: `Meta-Llama-3- 8B-Instruct` and `Gemma-7b-it`. These models were implemented and executed in a Google Colab notebook due to the high computational resources they require. One of the key elements for interacting with the LLM is the prompt. The prompt indicates to the model the instructions it must follow when responding to the query generated by the user. The prompt must be clear and precise to guide the model toward a coherent and appropriate response.

3.3 Interface of the Prototype

The prototype system features a user-friendly interface, allowing the user to ask questions about the tourism domain. The Streamlit library was used to create the user interface. One advantage of using Streamlit is that it allows you to create interfaces with just a few lines of code. Figure 3 shows the created interface.

4 Preliminary Results

4.1 Experimental Results

Once the prototype interface was created, testing of the entire system was performed, considering the two components of the architecture (retrieval and generative).

Analyzing the answers generated by the Meta-Llama-3-8B-Instruct model, we noted that the texts were shorter about the content found in the KG; additionally, some answers were incoherent. For example, to the question: *What is the main attraction in Loja?*, Llama generated the response: *The main attraction in Loja is the Great Parade of Mariscal, which is held during the Fiestas de Quito....* Since Loja is a city located in the south of the country, the response generated by the model refers to the city of Quito, the country's capital, located in the north-central region.

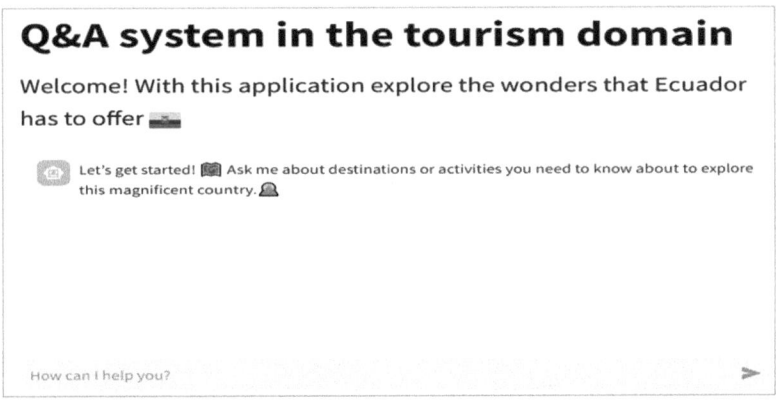

Fig. 3. User interface of the prototype.

On the other hand, using the Gemma-7b-it model, answers were coherent with the context in the knowledge base. However, there are also outputs in which the model simply responds that it doesn't know what it's being asked about, even though the answer is in the context.

Depending on the quality of the answers obtained during experiments, the configuration of the generative model was reviewed, and adjustments were made to the prompt and model parameters to improve the responses of the Q&A system, that is, to obtain contextually appropriate answers in the tourism domain. Specifically, the following changes were made:

- *temperature*: controls the level of randomness with which the model generates responses. A temperature value close to 0 generates more conservative and coherent responses. Conversely, a high value, close to 1, increases the model's creativity in generating responses, but the generated text may be less coherent. The final value for the prototype was 0.5.

- top_p: selects the most likely tokens until the sum of all their probabilities reaches a value of p. With low values, the options are limited to the most likely and consistent ones. Higher values, however, allow for a variety of generated options. To improve the quality and diversity of the generated text, a value of 0.9 was set.
- prompt: defines the instruction for the system to generate text according to the provided specification. Figure 4 shows the final prompt used to generate clear, concise answers based solely on the provided context. Preliminary testing showed that the system generates better responses when a range of 100–150 words is specified.

```
SYS_PROMPT = """
You are an assistant for answering questions.
You are given a question and context. Provide a brief, clear, and conversational answer based only on the context.
Ensure the answer contains between 100 and 150 words.
If the context does not answer the question, simply say "I do not know."
Do not include additional information such as notes or extra comments.
"""
```

Fig. 4. Prompt used for the answer generation.

Once the aforementioned adjustments were made, more exhaustive tests were carried out, the results of which are indicated in the following subsection.

4.2 Evaluation by Component

The goal of this kind of evaluation was to determine the system's accuracy in retrieving relevant text and measure its ability to generate coherent answers in the domain of tourism in Ecuador.

To evaluate the system's effectiveness and performance, a set of 30 questions related to the tourism domain were used, such as 1) *What makes Baños de Agua Santa an important tourist attraction of Ecuador?*, 2) *Where is the Cotopaxi National Park?*, and 3) *What are some museums to visit in Quito?*. Of the total questions, 50% were formulated based on content found in the KB (*Q-Type1*), and the other 50% were formulated based on content not found in the database but related to the domain of tourism in Ecuador (*Q-Type2*).

For the performance evaluation of the Q&A system, specific metrics were used to evaluate each component, such as cosine similarity and precision for the retrieval component and perplexity for the generative component:

- Cosine similarity: Measures the average similarity between the query embedding and the embeddings of the documents selected by the retriever.
- Precision: Measures what proportion of the positive predictions that were made were correct.
- Perplexity: Measures how well a language model predicts a sequence of text.

Regarding the first metric, Fig. 5 presents the average cosine similarity value obtained for the 30 test questions. The 15 questions formulated based on knowledge base content (*Q-type1*) are represented in blue, while the 15 questions formulated from content not present in the knowledge base (*Q-type2*) are shown in green. As expected, the lowest values are found in the second group of questions, as there is no associated content in the KB.

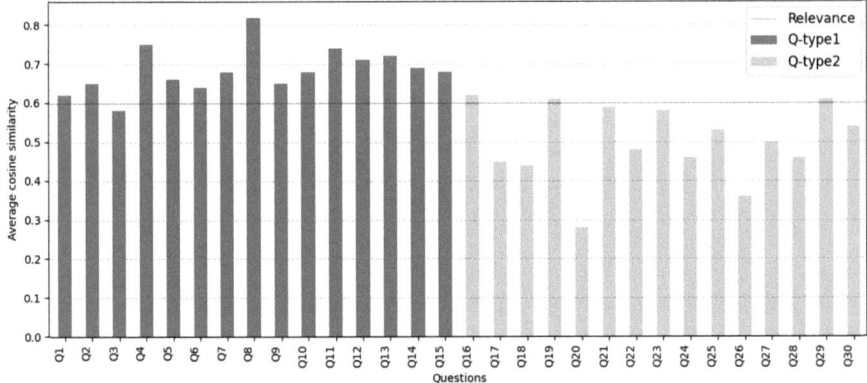

Fig. 5. Cosine similarity values.

Likewise, to determine the system's precision for generating relevant answers, the 30 test questions and a similarity threshold of 0.6 were considered to determine whether an answer is relevant or not (see Fig. 5). In addition, a manual assessment (user's relevance) of the relevance of the answers generated by each model was performed. Finally, automatic relevance, considering the similarity score, was compared with human relevance to establish the overall precision of the system. Table 2 presents the system's performance for generating answers based on each question category.

Table 2. Precision of the Q&A system

Question type	Automatic relevance	User's relevance	Precision
Q-Type1	93.33%	100%	93.33%
Q-Type2	20%	73.33%	46.67%

Finally, to evaluate the generative component, the perplexity metric was applied. During the development of the project, perplexity was calculated based on the average loss per token in the output generated by the model. Figure 6 presents the perplexity values for each model.

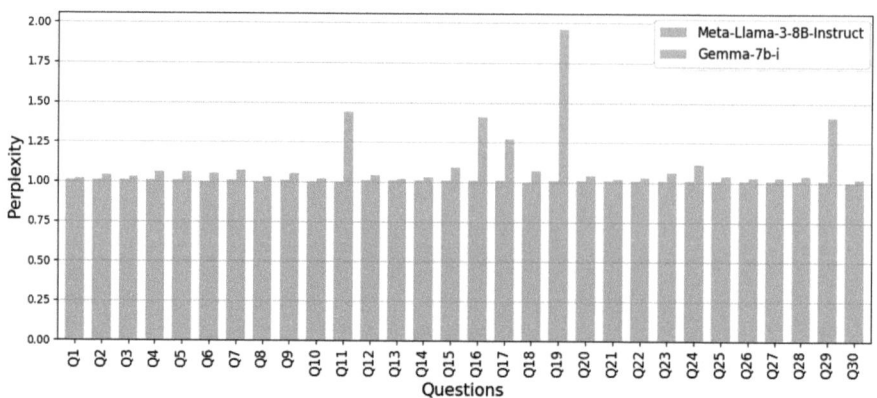

Fig. 6. Perplexity of the generative models.

Once the preliminary tests have been carried out and based on the characteristics of the models considered, we concluded that the Meta-Llama-3-8B-Instruct model offers a stable capacity to generate coherent responses.

4.3 Assessment of the Answers' Reliability

In this part, the results of the reliability evaluation of the answers generated by the models are presented. Through a manual process, each answer was classified into one of 3 possible values, according to the Likert scale: High, Medium, or Low. Figure 7 presents the reliability ratio of the responses generated by each model. In general, the performance of both models is similar, although Meta-Llama-3-8B-Instruct provides a higher rate of average response quality than Gemma-7B-it (23.22% vs. 16.67%).

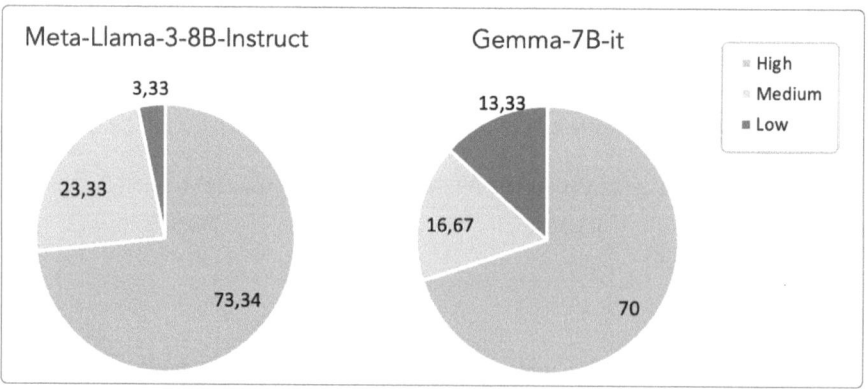

Fig. 7. Reliability assessment of the answers generated by each LLM.

4.4 Discussion

From the results obtained for the test questions, it was observed that each model effectively uses the information available in the KG to generate relevant answers. This finding is confirmed in Fig. 5, because the relevance of the answers generated by the system is higher for the *Q-type1* questions (questions whose answers are found in the KB) than for the questions in the second group *Q-type2* (questions for which there is no information in the KB). As expected, in the second group, the relevance of the answers is lower because the LLM considers only its current and local knowledge to generate the answers. In other words, the average similarity between user questions and documents retrieved from the KG exceeds the relevance threshold (set at 0.6), with the average value being 0.7. Meanwhile, an average similarity value of 0.48 was obtained for the 15 questions formulated based on content not found in the knowledge base but related to the domain of tourism in Ecuador.

On the other hand, based on the results obtained through the precision metric, a clear difference in system performance is observed between the two groups of questions. In the first category, *Q-type1*, the system generated answers with a precision of 93.33%; this indicates that the kB contains relevant information and the model retrieved it effectively. For the second category, *Q-type2*, the system performance was lower, with a precision of 46.67%; that is, the model has difficulty generating relevant answers when it does not find directly related information in the knowledge base. However, the fact that the answers were considered relevant by the user indicates that, in some cases, the system was able to generate valid answers. It is likely that for *Q-type2* questions, even if there is no direct information in the KB, the LLM generated valid answers based on its general knowledge acquired during its training.

Likewise, applying the perplexity metric showed that the `Meta-Llama-3-8B-Instruct` model showed a better performance, with values between 1.00 and 1.01 in predicting a text sequence, which indicates that it is more suitable for this type of response generation task. In contrast, `Gemma-7b-it` presents slightly higher values, in the range of 1.02 to 1.96, which would indicate that, in certain answers, the Google model has more difficulty predicting the next word.

Finally, regarding response reliability, it is worth highlighting that 73.33% of the responses generated by the Llama-3 model were rated as highly reliable, while the medium and low reliability questions represent 26.66%, which could be improved by adding more information to the KB and adjusting parameters. Regarding the responses generated by the Gemma-7b model, 70% of the responses are reliable; that is, two out of three questions received a correct answer. The remaining 30% of the responses correspond to medium and low reliability; this percentage indicates that the model may have somewhat greater limitations in generating certain types of responses.

5 Conclusion

During the development and validation of the Q&A prototype, the effectiveness of the RAG architecture was verified for resolving queries posed in the tourism domain in Ecuador. The system combined a retrieval component, which used the `all-mpnet-base-v2` model to retrieve relevant information from the knowledge base, and a generative component that used the `Llama-3-8B` and `Gemma-7b` language models to generate answers. This strategy was effective because the system can provide accurate and relevant answers related to tourism.

To optimize the system's performance, several tests were performed by varying the hyperparameters and the prompt used by the generative component; these tests were essential for improving the quality of the generated responses.

According to the results, applying the cosine similarity metric showed that the system exhibited reasonable performance in information retrieval. Regarding the application of the perplexity metric, the `Llama-3-8B` model performed slightly better in predicting a text sequence compared to the `Gemma-7b` model.

The Streamlit Python library was used to implement the system interface, and web access was provided for users to access the prototype via a public URL. The designed interface is user-friendly, allowing users to ask questions about the tourism domain in Ecuador and providing consistent answers generated by the trained model.

The Streamlit Python library was used to implement the system interface, and a tunnel with Ngrok enabled users to access the prototype through a public URL. The designed interface is user-friendly, allowing users to ask questions about tourism in Ecuador and providing consistent answers generated by the trained model.

As a future project, we are collecting data from various sources, such as reviews, tourism management entity websites, images, videos, and more, to enrich the knowledge base and thus offer tourists a more realistic experience of what awaits them when they visit Ecuador.

References

1. Brad, S., tetco, E.: An interactive artificial intelligence system for inventive problem-solving. In: Nowak, R., Chrząszcz, J., Brad, S. (eds.) Systematic Innovation Partnerships with Artificial Intelligence and Information Technology, pp. 165–177. Springer (2022)
2. Chen, W., Hu, H., Chen, X., Verga, P., Cohen, W.: MuRAG: multimodal retrieval-augmented generator for open question answering over images and text. In: Goldberg, Y., Kozareva, Z., Zhang, Y. (eds.) Proceedings of the 2022 Conference on Empirical Methods in Natural Language Processing, pp. 5558–5570. Association for Computational Linguistics, Abu Dhabi, United Arab Emirates (2022)
3. Chicaiza, J., Martínez-Velásquez, M., Soto-Coronel, F., Bouayad-Agha, N.: Creating textual corpora based on wikipedia and knowledge graphs. In: Rocha, Á., et al. (eds.) Good Practices and New Perspectives in Information Systems and Technologies, pp. 325–337. Springer (2024)

4. Fan, W., et al.: A survey on rag meeting LLMs: towards retrieval-augmented large language models, pp. 6491–6501 (2024)
5. Gavilanes Montoya, A., et al.: Current situation of tourism in Ecuador: challenges and opportunities. Green World J. **03**, 1–11 (2020)
6. Ghasemi, S., Shakery, A.: Harnessing the power of metadata for enhanced question retrieval in community question answering. IEEE Access **12**, 65768–65779 (2024). https://doi.org/10.1109/ACCESS.2024.3395449
7. Huang, D., Wang, Z.: Evaluation of orca 2 against other LLMs for retrieval augmented generation. In: Wang, Z., Tan, C.W. (eds.) Trends and Applications in Knowledge Discovery and Data Mining, pp. 3–19. Springer, Singapore (2024)
8. Huang, W., Xu, S., Yuhan, W., Fan, J., Chang, Q.: Attractiondetailsqa: an attraction details focused on Chinese question answering dataset. IEEE Access **10**, 86215–86221 (2022). https://doi.org/10.1109/ACCESS.2022.3181188
9. Muludi, K., Fitria, K., Triloka, M., Sutedi, J.: Retrieval-augmented generation approach: document question answering using large language model. Int. J. Adv. Comput. Sci. Appl. **15**(3), 776–785 (2024). https://doi.org/10.14569/ijacsa.2024.0150379
10. OECD: OECD Tourism Trends and Policies 2024. OECD Publishing (2024). https://doi.org/10.1787/80885d8b-en
11. Radeva, I., Popchev, I., Doukovska, L., Dimitrova, M.: Web application for retrieval-augmented generation: implementation and testing. Electronics **13**(7) (2024). https://doi.org/10.3390/electronics13071361
12. Ryu, C., et al.: Retrieval-based evaluation for LLMs: a case study in Korean legal QA. In: Preoțiuc-Pietro, D., et.al. (eds.) Proceedings of the Natural Legal Language Processing Workshop 2023, pp. 132–137. Association for Computational Linguistics, Singapore (2023)
13. Singh, S., Susan, S.: Healthcare question–answering system: trends and perspectives. In: Jain, S., Groppe, S., Mihindukulasooriya, N. (eds.) Proceedings of the International Health Informatics Conference, pp. 239–249. Springer, Singapore (2023)
14. Siriwardhana, S., Weerasekera, R., Wen, E., Kaluarachchi, T., Rana, R., Nanayakkara, S.: Improving the domain adaptation of retrieval augmented generation (RAG) models for open domain question answering. Trans. Assoc. Comput. Linguist. **11**, 1–17 (2023). https://doi.org/10.1162/tacl_a_00530
15. Soares, T.G., Azhari, A., Rokhman, N.: Survey categorizing paper on education question answering systems. In: Ao, S.I., Castillo, O., Katagiri, H., Chan, A., Amouzegar, M.A. (eds.) Transactions on Engineering Technologies, pp. 77–91. Springer, Singapore (2023)
16. Soto-Jiménez, F., Martínez-Velásquez, M., Chicaiza, J., Vinueza-Naranjo, P., Bouayad-Agha, N.: Rag-based question-answering systems for closed-domains: development of a prototype for the pollution domain. In: Arai, K. (ed.) Intelligent Systems and Applications, pp. 573–589. Springer, Cham (2024)
17. Triwicaksana S, M.B., Oktavia, T.: Building a retrieval-augmented generation system for enhanced student learning: case study at private university. J. Theor. Appl. Inf. Technol. **101**(22), 7381–7393 (2023)
18. Truyols, M.: Complete overview of the 5 sectors in the tourism industry. Technical report, MIZE (2023). https://mize.tech/blog/complete-overview-of-the-5-sectors-in-the-tourism-industry/

19. Yang, T., Mei, Y., Xu, L., Yu, H., Chen, Y.: Application of question answering systems for intelligent agriculture production and sustainable management: a review. Resour. Conserv. Recycl. **204** (2024). https://doi.org/10.1016/j.resconrec.2024.107497
20. Zhang, Z., et al.: IAG: induction-augmented generation framework for answering reasoning questions. In: The 2023 Conference on Empirical Methods in Natural Language Processing (2023)

Key Approaches to Neuroleadership: Perspective from the Public Educational Institutions of Montería

Diana Patricia Eljach Hernández[1(✉)], Leonardo Antonio Díaz Pertuz[1], Helmer Muñoz Hernández[1], Kavir Ala Oviedo Prioló[2], and Yamid Fabian Hernández Julio[1]

[1] Universidad del Sinú, Monteria, Colombia
dianaeljach@unisinu.edu.co
[2] Corporación Unificada Nacional de Educación Superior CUN, Bogota, Colombia

Abstract. The purpose of this study was to identify the main domains of neuroleadership in formal educational institutions in Montería, based on contemporary contributions such as those of Atencio Bravo et al. (2020), Avellán and Chávez (2022), Mieses Onofre (2024) and Zambrano and Meza (2024), who have highlighted the importance of integrating the principles of neuroleadership to strengthen educational management, organizational climate and the development of emotional and cognitive competencies in school leaders, among others. It was carried out with a quantitative approach within the positivist paradigm. The study was descriptive-inferential, and the design was non-experimental, cross-sectional. The population consisted of 25 subjects working in educational institutions in Montería. A 12-item questionnaire was designed and approved by five specialists. Cronbach's alpha coefficient was used to determine reliability, yielding $r = 0.86$ (very high reliability). Descriptive statistics and measures of central tendency (arithmetic mean and standard deviation) were used to interpret the data. In summary, the study showed that the neuroleadership focus regions were fairly well represented, with a moderate standard deviation compared to the population mean and moderate reliability. Similarly, decision-making was observed to belong to the present group, although problem-solving, teamwork, and emotional control showed some presence.

Keywords: Official Educational Institutions · Neuroleadership · Problem-solving · Decision-making

1 Introduction

To begin with, it is interesting to note that most organizations in today's world have been pressured by the great changes brought about by the global economy, which has led to consider human talent as the main production factor to break the barriers of competitiveness, which involves taking into account the expectations, needs and competencies of all those people who make up the organization or live in it. It should also be noted that global interest has shifted towards the concentration of efficient approaches aimed at efficiently leading the management of human talent in companies classified as large, medium and small, placing in the first place the collaborators who in recent decades have been recognized as the primary factor for the evolution of society, considering that this evolution could be thought of as a challenge, since any organization must adapt to the environment, motivated by the opportunity to reflect to improve and as a threat because the same change may mean starting from scratch to learn to tolerate. It also indicates that the first studies on leadership were focused on finding the psychological traits inherent to effective leaders; characteristics such as intelligence, willingness, sociability and authority were some of the most widely accepted, but their validation over time in various organizations proved unsuccessful. It is also pointed out that success in leadership was, in many cases, independent of the predominance of these traits. In general terms, the main leadership trends that have emerged in response to changes in the organizational environment have been analyzed, starting initially from the first theories. It is necessary to mention that within any organization, the performance of the leader plays a fundamental role; it is he who guides and motivates the personnel so that, when the work functions are carried out, each of the proposed objectives are achieved. Thus, the leader within the labor field must be a leader who continues to fulfill his role: to guide; the work he performs on his personnel, through actions that allow him to make known the vision and mission of the department within the organization. It should be noted that the organizations of the present millennium require leaders with the ability to make morally accepted decisions, as stated by Mieses Onofre (2024), who argues that the ability to make ethical choices is directly related to the level of moral development of the individual. In this sense, and in a context where organizations seek to adapt to new trends, contemporary leadership requires a capacity for agile diagnosis and conscious action in the face of changing situations. Within this scenario, neuroleadership acquires relevance, defined as an approach that integrates the principles of neuroscience to improve management and leadership practices. According to Avellán and Chávez (2022), neuroleadership strengthens educational management through the development of emotional and cognitive competencies that have a direct impact on the quality of institutional interaction and decision-making. Along the same lines, Mieses Onofre (2024) proposes a pedagogical management model focused on neuroleadership, highlighting that this approach facilitates the development of soft skills in teachers, promotes healthy work environments and enhances organizational performance through the conscious management of emotions and the reduction of work stress. Zambrano and Meza (2024) emphasize that neuroleadership is a determining factor for the transformation of educational management, since it optimizes communication processes, collaborative leadership, emotional self-regulation and organizational learning. Thus, institutions can build organizational cultures that are more resilient, creative and oriented to the well-being of their members. Consequently, it is reaffirmed that neuroleadership represents a key strategic tool for strengthening educational institutions in increasingly dynamic and complex environments, integrating the moral dimension of leadership with the knowledge of the neural processes that underlie human behavior in organizations.

2 Literature Review

2.1 Neuroleadership

In the same context, recent research argues that neuroleadership is an emerging field that examines how physiological processes in the brains of leaders influence their behaviors, interactions and organizational decisions. According to Zambrano and Meza (2024), neuroleadership seeks to optimize leadership effectiveness by understanding the neural mechanisms related to emotional intelligence, empathy and strategic decision making. This discipline integrates advances in cognitive neuroscience to enhance management and adaptation skills in dynamic work environments.

Recent research highlights that neuroleadership seeks to define the neural bases underpinning leadership and organizational management processes. According to Zambrano and Meza (2024), this discipline studies the brain mechanisms underlying individual behavior, motivation, strategic decision-making, emotional regulation, social interaction and continuous learning, considering these aspects essential to strengthen the performance of leaders in changing environments.

In a complementary manner, Avellán and Chávez (2022) argue that neuroleadership integrates emotional and cognitive knowledge in decision-making processes, favoring conflict resolution, the management of high-performance teams, the promotion of self-learning and managerial innovation, thus contributing to the development of creative and motivational competencies that are fundamental for effective leadership.

For the purposes of this research work, the author assumes a theoretical position based on the recent approaches of Zambrano and Meza (2024), who emphasize that neuroleadership is a scientific discipline that bases its postulates on the neural bases of leadership and organizational management. This approach allows understanding the brain processes that determine individual behavior in work performance, as well as fundamental aspects such as motivation, strategic decision making, emotional intelligence and social skills, consolidating itself as a key theoretical support in the development of effective leaders in complex environments.

It should be noted that the author points out aspects such as intelligence and individual learning, among other aspects linked to the institutional or organizational world, as well as leadership itself, as it could be present in the public educational institutions of Colombia, specifically those of the municipality of Monteria, objects of the current research, where the staff must understand the actions, capabilities and emotions of a leader within an institution.

2.2 Areas of the Neuroleadership Approach

The focus areas of neuroleadership interpret and apply neuroscience findings to organizational management. According to Avellán and Chávez (2022), neuroleadership enables leaders to assume their roles more consciously, enhancing managerial skills and strengthening interpersonal relationships to achieve collective goals. Similarly, Zambrano and Meza (2024) emphasize that neuroleadership examines the biological foundations of human behavior, particularly the rational and emotional components that influence strategic decision-making and effective performance within organizations.

Recent studies highlight that the core areas of neuroleadership—strategic decision-making, problem-solving, teamwork, and emotional management—are closely linked to organizational efficiency. Mieses Onofre (2024) states that integrating these elements increases individual satisfaction at work and promotes an organizational commitment aimed at rational and conscious problem-solving.

Furthermore, Zambrano and Meza (2024) argue that the focus areas of neuroleadership define essential leadership competencies, helping leaders remain dynamic and adaptable in complex environments. These areas guide institutional practices toward meaningful change and foster both personal and organizational growth. Research by Avellán and Chávez (2022), as well as Mieses Onofre (2024), supports the idea that neuroleadership encourages a shift from traditional leadership models to more emotionally intelligent and development-focused approaches.

This study adopts the theoretical framework of Zambrano and Meza (2024), asserting that neuroleadership transforms neuroscientific findings into practical managerial competencies, observable in the behavior and performance of staff in public educational institutions, particularly through attention to the biological and rational bases of leadership and management.

2.2.1 Decision Making

According to Zambrano and Meza (2024), the area of focus related to decision making constitutes one of the fundamental pillars that determine the success or failure of a leader in his or her organizational performance. This decisional process represents one of the most frequent and critical realities faced by a rector within an institution, requiring not only advanced cognitive skills, but also a high level of motivation and emotional self-regulation to make strategic decisions that promote the fulfillment of institutional objectives. Likewise, Avellán and Chávez (2022) emphasize that effective leadership is strengthened through continuous improvement of the ability to make sound decisions in contexts of uncertainty and change.

In this same context, research such as that of Onofre (2024) highlights that self determination in leadership is manifested in the ability to make conscious decisions based on the recognition of one's own needs, possibilities and aspirations. Self-determination is configured from the integration of multiple dimensions: a functional dimension, evidenced in the construction of sustainable and strategic professional projects; and a motivational dimension, reflected in intrinsic motivation and satisfaction derived from professional choice and performance. Thus, self-determination is consolidated as a key element to enhance effective leadership and organizational commitment in dynamic environments.

According to Zambrano and Meza (2024), these neurobiological processes allow individuals to perform their functions with greater satisfaction, effectiveness and benefit, by optimizing their cognitive and emotional performance in complex organizational contexts. Likewise, Mieses Onofre (2024) emphasizes that strengthening these neural connections through neuroleadership practices has a direct impact on the quality of learning and strategic decision making, considering that in every brain process there is a system of communication and interrelation of neurons, through which it is possible to learn and therefore produce knowledge and emotions that lead to the triggering of particular actions.

For the present research work, the author adopts as a theoretical position the approaches of Zambrano and Meza (2024), who emphasize that decision making represents one of the essential aspects through which the success of a leader can be assessed. This process is one of the most frequent realities faced by managers in official educational institutions, where motivation and analytical skills are essential to make strategic decisions that have a positive impact on the organization. In this way, making the right decisions is consolidated as a key element to achieve institutional objectives and strengthen managerial leadership.

2.2.2 Troubleshooting

In the same context, Guzmán Gallardo & Seminario Flores (2024) validate the idea that a leader's skills are mediated and built through a complex and systematic process that takes place precisely in action, that is, in the context itself where he/she develops his/her abilities and problem solving allows him/her to somehow face a problematic situation within an organization.

For Humánez (2024), problem solving itself implies the rational search for a solution or solutions through a series of strategies that help to solve or confront a problematic situation. Thus, four basic problem-solving skills are considered: (1) Definition and formulation of the problem; (2) Generation of alternative solutions; (3) Decision-making; (4) Application of the solution and verification of its usefulness.

Or Castro Arce (2023), this approach corresponds to the ability to establish successful processes for conflict resolution, one of the most important areas of focus of neuroleadership, given that a certain relationship is established between decision-making and the ability to negotiate, where, from neuroleadership, it is explained how people make decisions, solve problems, manage change, collaborate with others more efficiently and manage their emotions in a successful way, leading to understand their reactions, becoming a sine qua non condition to direct resources and human energy in a successful way to solve problems.

From the above mentioned, it is indicated that the theoretical postulates agree in pointing out that problem solving highlights the variation of basic skills that coexist in an articulated manner, with the objective of empowering decision making and motivation for conflict resolution in the organizational environment. This aspect must transcend to the human sphere and understand the situations surrounding the leader's functions in order to establish an effective negotiation and to solve and improve processes that pose a challenge and require proposals for continuous improvement.

However, for the purpose of the current research work, the theoretical postulate of (Aragón and Carrillo, 2025), who point out that problem solving implies the search for one or several solutions, by means of strategies that help to solve or confront a problem; a situation that could occur within official educational institutions, where their personnel, based on this understanding, can manage to solve and improve the processes within them. It should be noted that in order for these institutions to solve problem situations, they must define and formulate the problem, generate alternative solutions, make decisions and apply the solution.

2.2.3 Teamwork

On the other hand, (Cedillo Sánchez and Guerrero Collaguazo, 2025) indicate that it can be said that teamwork requires empathetic qualities because it is evident that no one likes to work with irritable, domineering or cold leaders. However, those who maintain an optimistic and enthusiastic leadership tend to retain the attention and interest of their employees for much longer. The above corresponds to consider that an adequate emotional management is required to lead other people within an organization, as a necessary skill to achieve the desired achievements.

For (Toledo Solórzano, 2023), teamwork represents more than a process of actions developed collaboratively, the possibility of establishing sustainable working relationships, through which those subjects with whom a common goal or objective is shared are recognized as members; that is why one of the areas of focus of neuroleadership is centered on understanding how a leader manages to motivate and establish relationships between subjects who actively contribute to achieve a collective purpose.

According to Eras E. Vicente & Coronado M. Delia (2024), the members of the group assign to a member the role of spokesperson, scapegoat, leader, among others. It will be necessary for the member to assume the role, to "take charge". This game of allocation and assumption of roles takes place in the interaction among the members of the institution. In other words, the leader is the person who understands what the group needs and proposes a credible way to solve that need. If the situation or the members of an organization change, the leader may then cease to be a leader at another time, just as he or she may not be a leader in another group.

It should be noted that the theories presented coincide in pointing out that teamwork refers to the importance of establishing roles and commitments in the middle of the processes, as a way not to condition obligations, but to establish an organized and committed work, attributing to others ethical elements such as responsibility and sense of belonging, since what is presented rather than defining a leader to lead, what is sought is a reference that in addition to motivating achieves concrete actions and plans to follow.

However, the researcher takes a position with the theory of the author Zapata & Atencio (2020), who point out that teamwork, apart from being a process of actions developed in a collaborative manner, represents the possibility of establishing working relationships, through which the members of the institutions, in this specific case, the educational institutions under study, are recognized as subjects with whom a common goal or objective is shared. Therefore, neuroleadership focuses on understanding how a leader manages to motivate and establish relationships between subjects.

2.2.4 Emotional Management

For Castro Arce (2023), emotional management can be conceived as the way through which the leader assumes a thought-out conduct oriented towards the development of previously established actions, and from which he manages to obtain benefits for himself and for others, in this aspect, the ability to recognize himself as a subject endowed with capabilities that make him different, with merits to achieve objectives and to make others achieve them as well.

Some authors have made contributions regarding certain theories associated with emotions, among them García Andrade (2020), who, based on the theory of expectation states, states that the valuation of the same object, fact or person will depend on the previous expectations of the subject, which can modify the resulting emotional experience. It should be noted that, in the field of social interaction, a key factor that exists is whether individuals live up to the expectations they arouse in others, depending on their position of power or status and the leadership they exercise for the management of the work team.

Theoretical positions such as those of Yepes and Ospina (2022, p. 84), state that "emotions are dynamic bodily dispositions that define the different domains of action in which we move". Hence, when we change emotion, we change the domain of action. That is to say, if the emotional circumstances change, the way of reasoning changes. Therefore, it is considered that the rationality of the human being is based on the realization of operations conditioned by the emotional dimension. In a certain way it is considered that consciousness and thought originate from the emotional disposition that filters reality.

It is also indicated that the referenced theories coincide in pointing out that it is possible to specify that emotional management corresponds in a certain way to the interaction between emotional and cognitive elements that lead the subject to experience feelings that categorically condition his actions. That is why it could be said that a leader recognizes the priority of his functions and actions and therefore manages and focuses his emotions towards his objectives, this being a fundamental characteristic through which it is possible for the leader to achieve everything proposed.

It should be noted that the researcher takes a position with the theoretical postulate of the author Castro Arce (2023), who states that emotional management is defined as the way through which the leader assumes a thoughtful and oriented behavior towards the development of previously established actions, by means of which he achieves benefits for him and the other members of an organization. Such a situation could arise in official educational institutions, where the staff could have the ability to recognize themselves as a subject with capabilities and merits to achieve objectives.

2.3 Methodology

According to Córdoba Natali (2023), methodological support encompasses the set of activities and procedures that a researcher undertakes to answer a research question. Methodology refers to how a specific activity is conducted. This study follows a positivist approach, which, as Castañeda Mota (2022) states, holds that scientific knowledge reflects objective reality and must be discovered through empirical and experimental methods without the observer's interference. Consequently, the study applies a quantitative method, using pre-established rules, empirical operations, and descriptive and parametric statistical techniques.

The research focuses on the Neuroleadership Focus Areas, using quantitative techniques and statistical procedures to analyze the participants' responses based on predefined categories. It is classified as explanatory research, which, according to Pomalaya Villalobos (2024), seeks to identify the causes of a social phenomenon and the conditions under which it occurs. Pérez and Campo (2024) also highlight its structured nature, combining exploration, description, and correlation for better understanding.

The research is also field-based, conducted directly in the context where the studied variables appear—official educational institutions in Montería, Córdoba, Colombia. Aragón and Navas (2023) describe field research as data collection from real-world subjects without manipulating variables. As non-experimental research, according to Castañeda Mota (2022), it observes and describes without altering the phenomena.

The study used a cross-sectional design, collecting data at a single point in time. Quintero, Barón, and Acosta (2021) describe this design as ideal for capturing variable behavior across groups at a specific time. The universe included 8,200 institutions, of which 61 are in Montería. A purposive sample of five institutions—urban and rural—was selected based on classification by the Secretary of Education and criteria such as proximity, data accessibility, and leadership seniority.

Authors like Gallegos (2021), Vizcaíno & Cedeño (2023) support purposive, non-probabilistic sampling for targeting relevant subgroups. The selection used the $Z_{0.05}^2$ test statistic with a 95% confidence level. Institutions selected met criteria such as institutional classification (A +, A, B, C, D), proximity, and having rectors with more than four years of service to ensure depth of data.

Table 1. Purposive sample of the research

EDUCATIONAL INSTITUTIONS OFFICIAL	CLASSIFICATION	REPORTING UNITS			TOTAL
		RECTOR	COORDINATOR	SCHOOL COUNSELOR	
I.E. Antonio Mariño	A + Plus	01	03	01	05
I.E. Normal Superior de Montería	A	01	03	01	05
I.E. Cristóbal Colón	B	01	03	01	05
I.E. Camilo Torres Mocarí	C	01	03	01	05
I.E. Guateque	D	01	03	01	05
TOTAL		05	15	05	25

Source: Eljach (2020)

Table 1 shows the 05 official educational institutions classified in 2019 as A +, A, B, C and D, selected as an intentional sample for the purposes of the current study, together with the reasoned informant units such as: 05 Rectors of the institutions to be studied, 15 Coordinators (without any specification), as well as 05 School counselors, which totaled 25 informant units selected for this research.

Gallego (2021), states that the survey is the collection of specific data within a specific topic and through the use of questionnaires or interviews with precise questions or answers that allow a quick tabulation and analysis of the information, which will make possible the process of data collection in a detailed and relevant way with the objectives, as well as the research design. Aragón and Navas (2023), for their part, point out that data collection techniques are the different ways and means of obtaining information, and that instruments are the material means used to collect and store data. Tambo Vera (2021) also indicates that, in general, a data collection instrument is the means by which it is possible to apply a particular data collection technique. It should be noted that for the purpose of the current research, a 12-item questionnaire was used as an instrument to measure the Neuroleadership Focus Areas. The questionnaire was designed with multiple response options of 05 alternatives, such as Strongly Agree-5, Agree-4, Neither Agree Nor Disagree-3, Disagree-2 and Strongly Disagree-1. It should be noted that the validity of the instrument was obtained by means of 05 experts, who will issue their judgments as specialists on the construction of the preliminary version of the survey and the questionnaire, specifying its congruence with the research objectives, the clarity of the questions, their correct wording, among other considerations that could arise. R. Pérez, L. Torres, others (2022) point out that the validity of an instrument is defined as the degree to which a test measures what is proposed in an investigation. In the same way, Gallegos (2021) points out that validity determines the level of strength, firmness and quality of the instrument, inferring that the instrument fulfills the objective for which it was elaborated, as well as for the judgment. It is important to point out that all research should show or identify its degree of reliability, in such a way that this will give greater confidence to the participant interested in the research. In view of this, it is significant to point out that the reliability marked by the authors Velasco, Manzano and Pérez (2022), refers to the degree to which its repeated application on the same subject or object produces the same results, therefore it is determined by means of multiple techniques. In this case, Cronbach's Alpha coefficient was used to quantify the level of reliability of a measurement scale for the unobservable magnitude constructed from the observed variables. In view of these considerations, we proceeded to tabulate the responses in the statistical program Microsoft Excel - Version 10, in Spanish, obtaining a significant value for the questionnaire. In view of the above, it is indicated that for the variable under study a value of $r = 0.91$ was obtained, which according to the scale is considered very high reliability. In this sense, the data will be tabulated in a double-entry matrix using the Microsoft Excel program, which will make it possible to organize and systematize information from vertical columns representing the items and horizontal columns representing the subjects, which concentrate and relate the information to be obtained. Regarding tabulation, Velasco, Manzano and Pérez (2022), mention that it is the systematization of the information which when processed and quantified allows the presentation of tables, to obtain an overall view of the whole set of data grouped in the calculation of absolute frequencies (AF) and relative frequencies (FR%) of the responses given by the units of analysis, which were represented in a table, indicating the arithmetic means. Finally, the results were compared and verified with the assumptions established in the theoretical framework, in accordance with the postulates of the authors versed in the research topic, by means of the interpretation of the corresponding scale, which served as the basis for the general conclusions and recommendations. The interpretation scale was constructed.

Table 2. Scale of interpretation of the arithmetic average

CATEGORIES	RANGE	ALTERNATIVES
Very present	4,21–5,00	Totally agree
Present	3,41–4,20	Agreed
Moderately present	2,61–3,40	Neither agree nor disagree
Little present	1,81–2,60	Disagree
Not present	1,00–1,80	Strongly disagree

Source: Eljach (2020)

Table 1 shows the alternative responses to be considered to measure the areas of focus of neuroleadership within the official educational institutions of Monteria, assuming the Attitude scale, together with its ranges that go from one (1) to five (5), as well as the categories to measure the average or arithmetic mean according to the total percentage of the responses to the indicators.

2.4 Results

The following is the analysis and interpretation of the results of the research entitled: Areas of focus of neuroleadership: a determination from the educational institutions of Monteria, which were reached once the instrument was applied, which emanates from the field work carried out and which could be a revealing contribution.

Next, the analysis and discussion of Table 3 is presented, referring to the Areas of focus of neuroleadership, beginning with the Decision-making indicator, the results of which showed that out of 25 respondents, 11, representing 44%, leaned towards the option neither agree nor disagree when indicating that decision making is fundamental for the success of the educational institutions studied, where the fulfillment of institutional objectives depends on the leader's decision making and the leader's task is to face a problematic situation in order to solve it successfully.

On the other hand, 7 of the subjects, reflecting 28% were inclined to agree with the option indicated in the previous paragraph, 5 workers with 20% assumed the option totally agree as described above, 2 collaborators assuming 8% marked the option in disagreement and no subject was inclined to totally disagree. The average of the Decision-making indicator was 3.60, which when compared with the arithmetic mean scale was positioned in the present category, with a standard deviation of 0.56, indicating a moderate dispersion of the data with respect to the mean and a moderate reliability. Continuing with the analysis and discussion, the problem solving indicator is presented, indicating that of the 25 respondents, 8, representing 32%, leaned towards the option of agreement, that is, the leader thinks that problem solving implies the application of strategies, also faces problematic situations and solves them successfully, and also has the ability to generate alternative solutions to a problem. On the other hand, 7 subjects with 28% indicated the option neither agree nor disagree with respect to the comments made previously, 5 collaborators reflected with 20% stated that they disagreed with the opinions issued previously, 3 workers represented with 12% indicated that they totally agreed

Table 3. Areas of focus of neuroleadership

Response alternatives		TD		DA		NDA/NED		ED		TED		TOTAL		AVG	STD DEV
Indicators	Items	Fa	Fr	Fa	Fr	Fa	Fr	Fa	Fr	Fa	Fr	Fa	Fr		
Decision Making	13–15	5	20.00	7	28.00	11	44.00	2	8.00	0	0.00	25	100	3.60	0.56
Problem Solving	16–18	3	12.00	8	32.00	7	28.00	5	20.00	2	8.00	25	100	3.24	0.57
Teamwork	19–21	4	16.00	8	32.00	6	24.00	5	20.00	2	8.00	25	100	3.33	0.60
Emotional Management	22–24	3	12.00	8	32.00	7	28.00	5	20.00	2	8.00	25	100	3.20	0.65
Dimension Total		4	15.31	8	31.63	7	28.57	5	20.41	1	04.08	25	100	3.34	0.59
ARITHMETIC MEAN	3.34														
CATEGORY	MODERATELY PRESENT														
STANDARD DEVIATION	0.59														

Interpretation Scale for the Mean or Average

Categories	Ranges	Alternatives	Total Average of the Dimension
Very Present	$4.21 \geq X \leq 5.00$	Totally Agree	3.34
Present	$3.41 \geq X \leq 4.20$	Agree	
Moderately Present	$2.61 \geq X \leq 3.40$	Neither Agree Nor Disagree	
Slightly Present	$1.81 \geq X \leq 2.60$	Disagree	
Not Present	$1.00 \geq X \leq 1.80$	Totally Disagree	CATEGORY: MODERATELY PRESENT

Source: Eljach (2020)

with the approaches indicated previously and 2 individuals with a percentage of 8% were inclined to state that they did not agree with the approaches referred to previously. It should be noted that the average of the Problem solving indicator was 3.24, which when compared with the arithmetic mean scale was positioned in the moderately present category, with a standard deviation of 0.57, indicating a moderate dispersion of the data with respect to the mean and moderate reliability. Next, the Teamwork indicator is presented. Of the 25 respondents, 8, representing 32%, agreed that the leader considers that teamwork represents a process of collaboration among the personnel, sharing common goals among the members of the organization and focusing on how to motivate the personnel to achieve a collective purpose that is in line with the program established by the institution. Next, it is indicated that 6 of the respondents with a percentage of 24% were inclined to answer neither agree nor disagree with respect to the aforementioned approach, 5 subjects with 20% responded to disagree with the comments made above, while 4 collaborators with 16% responded to agree with what was explained above, finally it is shown that only 2 informant units that assume 8% stated that they totally disagreed with the aforementioned approaches. It should be noted that the average of the Teamwork indicator was 3.33, which when compared with the average scale is located in the moderately present category, with a standard deviation of 0.60, indicating a moderate dispersion of the population with respect to the average and a moderate reliability, which can be observed in the study analysis table. Then, the Emotional Management indicator is presented, for which of the 25 subjects surveyed, 8, representing 32%, stated that they agreed that the Institution assumes a conduct oriented to the development of established actions, where the leader reasons the oriented conducts achieving benefits for the members of the organization, having the capacity to recognize himself as a subject with merits for the achievement of the institutional objectives. On the other hand, 7 subjects of those surveyed, representing 28%, answered neither agreeing nor disagreeing with the approaches indicated above, 5 informant units reflecting 20% answered disagreeing with the approaches indicated, 3 workers with a percentage of 12% answered agreeing with the approaches indicated and only 2 people with 8% were inclined to the option totally disagreeing with the approaches indicated above. It should be noted that the average of the Emotional Management indicator was 3.20, which when compared with the scale of interpretation of the mean is in the category moderately present, with a standard deviation of 0.65, which is considered a high dispersion with respect to the mean of the population and a high reliability. In view of the above, it is indicated that the mean of the dimension Areas of the neuroleadership approach was 3.34, which when compared with the scale of interpretation of the arithmetic mean is located in the category moderately present, with a standard deviation of 0.59, which is considered moderate with respect to the mean of the population under study and moderate reliability. Table 2 shows the results for the ANOVA test, the basic operation of which consists of calculating the mean of each of the groups, allowing the comparison of multiple means, but it does so by using variances in order to measure significant differences between the indicators of each dimension (Table. 4).

Table 4. One-factor ANOVA: Neuroleadership focus areas.

Sum of squares	gl	Root mean square	F	Sig.
0,132	2	0,066	7,222	0,062
0,357	39	0,009		
0,489	41			

Source: Results of questionnaire application. Eljach (2020)

For the test, the following hypotheses are formulated:

H0: $\mu1 = \mu2 = \mu3$ (There are no significant differences between the areas of focus of Neuroleadership) Hl: $\mu1 \neq \mu2 \neq \mu3$ (There are significant differences between the areas of focus of Neuroleadership) ($\mu1, \mu2, \mu3$) Average of responses for each indicator.

In the same it is observed that Sig. $0.062 > 0.05$, to which H0 is not rejected, presenting that there are no significant differences between the indicators of the dimension Areas of focus of the Neuroleadership, results that reiterate the descriptive ones as for the interpretation of the mean of answers when being located in the category moderately present in these institutions, when manifesting the test that there are no significant differences.

3 Discussion

The following is a discussion of the results according to the data shown above, starting with the results of this indicator Decision-making, which coincide with the theory of Avellán and Chávez (2022), who state that this area of focus corresponds to one of the fundamental aspects through which a leader can be considered successful or not; It should be noted that this is one of the realities that a manager continuously faces, and therefore requires a degree of motivation to make decisions that are considered decisive in his or her managerial exercise and allow the fulfillment of organizational objectives.

Likewise, the discussion of the Problem Solving indicator is presented, pointing out that the results obtained coincide moderately with the postulate of Humánez (2024), for whom problem solving itself implies the rational search for a solution or solutions through a series of strategies that help to solve or confront a problematic situation. Four basic skills are considered: 1) definition and formulation of the problem, 2) generation of alternative solutions, 3) decision making, and 4) application of the solution and verification of its usefulness.

Within the same context, the Teamwork indicator is shown, whose results are moderately consistent with Castro Arce (2023), who states that teamwork represents more than a process of actions developed collaboratively, the possibility of establishing sustainable working relationships, through which those subjects with whom a common goal or objective is shared are recognized as members; that is why one of the areas of focus of neuroleadership is centered on understanding how a leader manages to motivate and establish relationships between subjects who actively contribute to achieve a collective purpose.

Next, the Emotional Management indicator is presented, for which the results coincide moderately with the theory of Castro Arce (2023), who states that emotional management can be conceived as the way through which the leader assumes a thoughtful and oriented behavior towards the development of previously established actions, and from which he manages to obtain benefits for himself and for others, in this aspect, the ability to recognize himself as a subject endowed with capabilities that make him different, with merits to achieve objectives and make others also achieve them is fundamentally pointed out.

Finally, the discussion of the results obtained for the dimension under study is shown, for which it is indicated that these coincide moderately with the theoretical postulate of Avellán and Chávez (2022), who state that the Areas of focus of neuroleadership, interpret and translate the discoveries of neuroscience, where leadership is assumed and the managerial skills performed are acceptable; it is where relationships are established to achieve the objectives and make others achieve them.

4 Conclusion and Recommendations

As final considerations, it is concluded that the Neuroleadership Focus Areas were found to be moderately present, with a moderate standard deviation with respect to the mean of the population under study and moderate reliability. In view of the above, it is indicated that the Decision-making indicator was found to be in the present category, while the indicators Problem solving, Teamwork and Emotional management were found to be moderately present. In view of the above, it is indicated that the personnel surveyed expressed neither agreeing nor disagreeing that decision making is fundamental for the success of the educational institutions studied, where the fulfillment of the institutional objectives depends on the leader's decision making and the leader's task is to face a problematic situation in order to solve it successfully.

The staff also agreed that the leader believes that problem solving involves the application of strategies, confronting uncertain situations and successfully resolving them, with the ability to generate alternative solutions to a difficulty. Respondents also agreed that the leader considers that teamwork represents a process of collaboration among the staff, sharing common goals among the members of the institution and focusing on how to motivate the staff to achieve a collective purpose.

Within the same context, it is also concluded that the personnel of the official educational institutions, expressed their agreement that these organizations assume a conduct oriented to the development of established actions, where the leader reasons the oriented conducts, thus achieving benefits for the members of the same, with the leader having the capacity to recognize himself as a subject with merits for the achievement of the institutional objectives.

Therefore, it is considered that the objective of the courses would be to make managers and institutional leaders aware of the strategies to solve problems and the variables that influence the process of selecting effective alternatives in the work and/or personal context, impacting their work team and the institution.

It should be noted that the participants (Rectors, Coordinators and School Counselors) will be able to carry out group exercises that will allow them to evaluate their own decision-making style and learn the best practices for their managerial performance and thus ensure that the work team achieves the proposed objectives.

Likewise, it is indicated that the courses would be developed in a semester period, where the responsible for the same would be the Human Capital management of each official educational institution studied, considering for its execution from the expert personnel in the area (facilitator), financial resources, place of development of the courses, among others. It is also indicated that the benefit that such courses would offer would be the development and improvement of the personal and professional competences of the participants, in the focus areas of neuroleadership.

References

Quintero, A.F.S., Barón, W.S., Acosta, C.A.G.: Pro-environmental behavior and environmental knowledge in university students: does the area of knowledge make a difference? SIELO 14(1), 14(1), 64–82 (November 10, 2021). https://doi.org/10.21615/cesp.14.1.6

Álvaro, T.V.: a Responsabilidad Social Empresarial y el desarrollo sostenible por la constructora Corporación de Servicios y Soluciones Integrales SAC. (P. 15–20, Ed.) Lima - Peru: Universidad Cesar Vallejo (2021). chrome-extension://efaidnbmnnnibpcajpcglclefindmkaj/https://core.ac.uk/download/pdf/483440926.pdf

Ana Manuela Palma Avellán, C.O.: Educational Management and Neuroleadership at the Secondary Education level in Manabí. Polo del Conocimiento (CASEDELPO), 7(11) (November 11, 2022). https://www.polodelconocimiento.com/ojs/index.php/es/article/view/4872/11782

Andrade, A.G.: Emotional perception: sociology and affective neuroscience. Mexicana de Sociologia **82**(4), 82(4), 841–870 (October 2020). https://doi.org/10.22201/iis.01882503p.2020.4.59209

Andrea Álvarez Yepes, D.M.: Network of conversations and emotions: interweaving meanings between systemic family therapy and Humberto Maturana's thought. Universidad Luis Amigó, **9**(1), 77–94 (2022). file:///C:/Users/Ana%20Figueroa%20Sibaja/Desktop/HELMER%20WORK/TASKS%20ARTICLES%20ESPA%C3%91A/journalsauthors,+Network.pdf

Arce, M.M.: Educational management from neuroleadership: a proposal for strengthening the organizational climate of the Colegio Humanístico Costarricense, Campus UNA. Universidad Nacional de Costa Rica, 44–49 (October 2023). chrome-extension://efaidnbmnnnibpcajpcglclefindmkaj/https://repositorio.una.ac.cr/server/api/core/bitstreams/a57eed5f-8392-4f7c-a531-94aa3b2bdd6d/content

Atencio Bravo E.A, Lora, L.J.R., Berastegui, Y.B.Z.: Neuroliderazgo como estrategia para el fortalecimiento de la gestión directiva en instituciones educativas, Actual. Investig. Educ vol.20 n.1 San Jose Jan (April 2020). https://www.scielo.sa.cr/scielo.php?pid=S1409-47032020000100416&script=sci_arttext

Córdoba, N.S.: Research methodology I. Rosario: Universidad Nacional de Rosario Facultad de Ciencias Médicas Escuela de Enfermería (2023). chrome-extension://efaidnbmnnnibpcajpcglclefindmkaj/https://rephip.unr.edu.ar/server/api/core/bitstreams/20a602cf-3b46-4b56-a3de-f52ab0e0fe7c/content

Dora Isabel Cardenas Ramos, A.C.: Business plan for the technological platform for psychological and emotional wellness services "VitalMente", period 2020–2025. USIL (2022). chrome-extension://efaidnbmnnnibpcajpcglclefindmkaj/https://repositorio.usil.edu.pe/server/api/core/bitstreams/594ad9ef-6fa4-416c-b259-15c46d057f62/content

Aragón, E.E.P., Carrillo, C.: Leadership styles and conflict resolution in educational institutions in Guajira-Colombia. Andean Journal of Pedagogical Sciences (2025). file:///C:/Users/Ana%20Figueroa%20Sibaja/Desktop/HELMER%20WORK/TASKS%20ARTICLES%20ESPA%C3%91A/elvis-eliana-pinto-aragon-and-others.pdf

Eliana Gallardo Echenique, M.: Research methodology: interactive self-training manuals. Universidad Continental. Huancayo: Universidad Continental Av. San Carlos 1980, Huancayo-Peru (2022). chrome-extension://efaidnbmnnnibpcajpcglclefindmkaj/https://repositorio.continental.edu.pe/bitstream/20.500.12394/4278/1/DO_UC_EG_MAI_UC0584_2018.pdf

Evelio Antonio Sará Pérez, M.A. Polya's Method in solving linear equation problems (2024). Obtenido de file:///C:/Users/Ana%20Figueroa%20Sibaja/Desktop/HELMER%20TRABAJO/TAREAS%20ARTICULOS%20ESPA%C3%91A/249-Texto%20del%20art%C3%ADculo-566-1-10-20240710.pdf

Gallegos, W.L. (2021). Effects of a prosocial behavior program in college students. Education **27**, 137–154 (140). https://revistas.unife.edu.pe/index.php/educacion/article/view/2430

Guzmán Gallardo, D.D.: Case analysis in a financial institution case Susana Cueva. PIRHUA. (2023). https://pirhua.udep.edu.pe/item/184a25a3-d3df-4e3d-806f-ed72cab72d32

Humánez, J.: Strategies based on neuroleadership and its influence on the organizational climate of educational institutions. Social Innova Sciences (2024). http://socialinnovasciences.org/ojs/index.php/sis/article/download/154/210

José Felipe Ramírez-Pérez, V.G.: Reliability and validity of an instrument to measure the competitive performance of health institutions using information and communication technologies. CienciaUAT **16**(2), 97–115 (May 16, 2022). https://doi.org/10.29059/cienciauat.v16i2.1539

Cedillo Sánchez, K.D., Guerrero Collaguazo, D.M.. niversidad de Cuenca, Ecuador (April 03, 2025). chrome-extension://efaidnbmnnnibpcajpcglclefindmkaj/https://rest-dspace.ucuenca.edu.ec/server/api/core/bitstreams/5432253a-df5a-405f-88fa-36edd543131a/content

María de los Ángeles Cienfuegos Velasco, P. J.: The quantitative and qualitative from a statistical treatment. Int. Soc. Sci. Human. **11**(21), 12(1) 48–63 (January 2022). https://www.ricsh.org.mx/index.php/RICSH/article/view/275

Mina, J.D.: Feasibility study for the creation of a manufacturing company of cleaning products in the municipality of Santiago de Cali, Valle del Cauca. Universidad Nacional Abierta y a Distancia - UNAD, 100 – 135 (2021). chrome-extension://efaidnbmnnnibpcajpcajcglclefindmkaj/https://core.ac.uk/download/pdf/481481405.pdf

Mota, M.M.: The scientificity of quantitative, qualitative, and emergent methodologies. Digital de Investigación en Docencia Universitaria **16**(1), 6 (2022). https://doi.org/10.19083/ridu.2022.1555

Navas, E.E.: Chapter 2: Work teams at the University of La Guajira. In Entrepreneurship and Social Economy in Latin America. Product of the International Congress of knowledge: Society - Enterprise - State: pragmatika editions (2023). https://www.researchgate.net/publication/377761773_CIC-SEE-2023-Libro1-EJE1-Emprendimiento_y_Economia_Social_en_America_Latina#page=48

Structural Analysis as a Tool to Establish the Influence of ICT in the Educational Processes of Rural Institutions in San Antero, Cordoba, Colombia

Leonardo Antonio Díaz Pertuz[1(✉)], Diana Patricia Eljach Hernández[1], Yamid Fabian Hernández Julio[1], Helmer Muñoz Hernández[1], and Kavir Ala Oviedo Prioló[2]

[1] University of Sinú, Cordoba, Colombia
leonardodiazp@unisinu.edu.co
[2] National Unified Corporation of Higher Education CUN, Bogota, Colombia

Abstract. The present article describes a study conducted in rural educational institutions in the municipality of San Antero, Colombia, using the MICMAC and MACTOR software to analyze the influence of ICT on educational processes. The objective is to identify the actors and factors that can positively or negatively impact educational quality, aiming to contribute to its improvement. Various aspects are examined, such as structural barriers and limited perceptions of ICT in rural contexts, with the purpose of promoting educational innovation and benefiting the student as the central axis of the process. The importance of comprehensively addressing these aspects is emphasized, involving multiple actors and focusing on both technological infrastructure and teacher training. This holistic approach is essential to achieve effective integration of ICT in rural educational institutions, thereby enhancing the quality of education in these areas.

Keywords: Information and Communication Technologies (ICT) · innovation · educational quality · Resources · academic performance · educational processes

1 Introduction

Today, educational processes must keep pace with technological advances to achieve the desired change and transformation. This poses a significant challenge, especially considering the existing technological gaps, poverty levels, and educational disparities. Consequently, contemporary education faces the task of optimizing all processes to overcome the low educational standards prevalent in most educational institutions located in rural or remote areas.

Educational processes encompass a variety of procedures, methods, and actions that involve all members of the educational community to ensure the comprehensive development of students. Strategic planning of these processes involves considering various aspects, including needs assessment, goal setting, design, and evaluation. As Henao et al.

(2006) highlight, planning in education aims to contribute to the improvement of the educational process. Similarly, Rodríguez et al. (2022) emphasize the pedagogical process as the educational process that manifests the relationship between education, instruction, teaching, and learning, with the ultimate goal of developing students' personalities to prepare them for life as social beings.

With the integration of Information and Communication Technologies (ICT) into educational processes, motivation becomes a fundamental component for achieving educational success. Therefore, institutional planning and teaching programming must incorporate strategies to spark students' interest in learning.

In this context, this article presents a study carried out to determine the influence of ICTs in educational processes in rural public institutions in the municipality of San Antero, Córdoba-Colombia, through the implementation of the structural analysis carried out by the MICMAC and MACTOR tools, whose objective is to identify actors and factors that may positively or negatively influence the processes in these rural educational institutions, the findings show in detail the direct influence of six variables that intervene in the educational processes of institutions located in rural areas of the municipality of San Antero, department of Córdoba-Colombia, such as: (POLED) Educational Policies, (TICS) Information and Communication Technologies, (REDUC) Educational Resources, (RENAC) Academic Performance, (INMET) Methodological Innovation and (ACTCU) Curricular Update, allowing more relevant, assertive and less exhausting processes that contribute to improving educational quality and promoting educational innovation in all educational contexts, with the main focus of benefiting the students, on whom these processes revolve.

2 Influence of Tics in Educational Processes

Every educational process includes a set of procedures, methods, and actions involving all members of the educational community, all with the goal of achieving the comprehensive development of students. This entire process must be planned taking into account various aspects such as available technological capabilities, current educational trends, and labor market demands, but, above all, the needs of the context and the students.

Henao et al. (2006) understand planning as "an act that includes needs analysis, objective setting, goals, design, and evaluation; its central purpose is to contribute to the improvement of the educational process." (p. 215). For Rodríguez et al. (2022), the pedagogical process is identified as the educational process where the relationship between education, instruction, teaching and learning is manifested, aimed at the development of the student's personality for their preparation for life as a social being.

ICTs energize learning environments by generating spaces where motivation becomes a fundamental part of achieving success in educational processes. Therefore, institutional planning and teaching programming must consider this, establishing strategies that capture students' interest in learning. In this sense, authors such as Deci and Ryan (2000) have highlighted the importance of intrinsic motivation in the educational context, which arises when students feel autonomous, competent, and connected to the content and learning activities.

Likewise, according to Csikszentmihalyi (1990), the state of flow, characterized by deep engagement and enjoyment in completing a challenging task, can be promoted through the effective integration of ICTs into the educational process. Therefore, it is essential for teachers to design learning experiences that leverage the potential of ICTs to foster autonomy, competence, and connection to the content, thereby creating a motivating environment that drives students' educational success.

The integration of Information and Communication Technologies (ICTs) into educational processes has transformed the way we teach and learn. Authors such as Prensky (2001) have highlighted the importance of understanding students in the digital age, who are digital natives and possess innate skills in using technology. In this sense, ICTs offer tools and resources that can be adapted to different learning styles, allowing for greater personalization and flexibility in teaching (Johnson et al., 2016). The implementation of educational platforms, mobile applications, and multimedia resources in the classroom can enrich the teaching-learning process, facilitating the understanding of complex concepts and stimulating active student participation (Almerich et al., 2016).

However, it is important to recognize that the mere presence of ICTs does not automatically guarantee an improvement in educational quality. Authors such as Cuban (2001) warn about the risks of adopting technologies without adequate planning, which can lead to superficial or ineffective use of these technologies in the classroom. Therefore, it is essential that teachers receive ongoing training in the pedagogical use of ICTs and that there is institutional support that promotes their effective integration into the school curriculum (Ertmer, 2005).

Furthermore, the digital divide that persists in many educational contexts must be taken into account, where not all students have equal access to ICTs and a quality internet connection (Warschauer, 2003). This inequality can exacerbate educational and social disparities, deepening the exclusion and marginalization of certain groups of students (Selwyn, 2010). Therefore, it is necessary to implement public policies that promote equity in access to technology and ensure that all students can benefit from the opportunities offered by ICTs in the educational field.

From this research the following question arises: Is there a significant relationship between the use of ICT, institutional educational processes and the levels of academic motivation of public-school students in the rural area in the municipality of San Antero, Córdoba-Colombia?

3 Methodology

The method used for this research is based on strategic foresight, which, according to Goyeneche & Parodi. (2017) they describe as:

A tool for mobilizing and articulating collective knowledge; that which is embedded in society, dispersed in the perspectives of different experts, involved and interested in different topics. It is a way of structuring an informed discussion about the possible forms the future will take; not in terms of prediction or prognosis, but rather as different possible scenarios, which allow us to identify the risks and opportunities in each and help us devise agreed-upon courses of action to choose a desired situation from among the possible ones and define a strategy to move toward it. (p. 5).

Through this method, a course of action can be established to take the necessary measures to provide answers, solutions, and solutions to various difficulties that have arisen or may arise in the near future.

On the other hand, the authors express that the concepts related to foresight, strategy and planning are closely related and state that:

These elements are constantly undergoing feedback, and their dynamic transformation is a challenge for implementing action. Thus, strategic foresight is essential for building agreements, generating action-oriented alliances, and consolidating strategic directions that enable policy implementation (p. 9).

4 Procedure

This research is carried out at the Julio C. Miranda Educational Institution, located in the rural area of the municipality of San Antero, Córdoba (Colombia), selected for its proximity to the researcher's residence. According to the Departmental Secretariat of Education (2023), this municipality has one rural educational institution and two primary schools. In addition, a review of the relevant literature was conducted, in which the fundamental aspects related to the objective of the topic addressed were defined. To this end, various relevant studies in the area were taken into account. For example, Almerich, Suárez-Rodríguez, and Díaz-García (2016) examined the challenges and risks for teachers in the integration of ICTs in education. In addition, Cuban (2001) provided a critical view on the use of computers in the classroom, and studies on the influence of ICTs on academic performance were also reviewed, such as the work by Johnson, Adams, and Cummins (2016), which explored emerging trends in educational technology.

Research directly linked to the use of ICTs in rural education was used, as well as research related to the prospective method. These were taken primarily from Google Scholar, some important journals such as SciELO, and a research support tool such as Scopus.

The aspects considered to initiate the review were the definition and importance of educational processes, ICTs, and their influence on educational processes. Furthermore, the variables that have the greatest influence on these processes, either positively or negatively, were analyzed. Therefore, Michel Gode's strategic foresight technique was taken into account, as well as the information obtained through the implementation of cross-impact matrix foresight tools, such as MICMAC and MACTOR software. These tools allowed for cross-referencing or correlation of variables, actors, and others, assigning a score, either from 0 to 3 or 0 to 4, depending on the case, where:

0 = No influence
1 = Weak / Processes
2 = Moderate influence / Projects
3 = Strong influence / Mission
4 = Potential influence / Existence

First, the tool was executed

4.1 Application of the MICMAC Tool

The MICMAC tool in its parameterization of variables requires as a first measure, that these are presented with the name that is used to establish their management (long name), as a second, the representative acronym of the variable (short title) and as a third a brief description of its function or purpose (description), aspects that are shown in Table 1.

4.2 List of Variables

The variables worked on are:

1. Plan for the acquisition and supply of inputs (PASI)
2. Curriculum Update (ACTCU)
3. Methodological Innovation (INMET)
4. Cross-curricular projects (PTRES)
5. Academic performance (RENAC)
6. Educational Coverage (COED)
7. Infrastructure (INFRE)
8. Preventive maintenance of equipment (PMRE)
9. Educational inclusion (INED)
10. Educational Policies (POLED)
11. Information and Communication Technologies (ICT)
12. Economic support (APEC)
13. Economic growth (CRECO)
14. Internet (INTR)
15. Science, Technology and Innovation Plan (PCTI)
16. Budget (PRES)
17. Illiteracy (ANFB)
18. Standards (EST)
19. Educational system (SISED)
20. Human Talent (TALHU)
21. Inequality (DESG)
22. Social conflict (CONSOC)
23. Technological Advances (AVTEC)
24. Educational community (COMED)
25. Diversity (DIVD)
26. Technological Resources (CHALLENGE)
27. Traditional Education (EDTRA)
28. Educational Resources (REDUC)
29. Digital Resources (REDIG)
30. Student Enrollment System (SIMATE)
31. Connectivity (CNTVD)

Table 1. Description of Variables

No.	Long Title	Short Title	Description
1	Inclusive Institutional Educational Project	PEIDI	An instrument that organizes and gives meaning to the management of an educational institution, with its principles, goals, strategies, etc., including care for children with disabilities.

(continued)

Table 1. (*continued*)

No.	Long Title	Short Title	Description
2	Plan for the acquisition and supply of inputs	PASI	Planning and management of acquisitions and supplies necessary to identify the resources and products to be acquired annually
3	Curriculum update	ACTCU	Adjustment of curricular guidelines, taking into account the functions of the Institution and the needs of society, to seek educational quality
4	Methodological innovation	INMET	Changes concerning the teaching process to achieve better student learning
5	Cross-curricular projects	PTRES	They are pedagogical strategies that allow planning, developing and evaluating the curriculum in the educational establishment, thereby making it possible to improve the quality of the teaching process and the integral development of the student, which facilitates acquiring learning for life.
6	Academic performance	RENAC	It is the way to evaluate the ability to learn, study and the way in which one learns during training.
7	Educational coverage	COED	The capacity of the education system to serve the population that demands services in a given grade level, which can be estimated from different concepts and statistical sources. (DANE)
8	Infrastructure	INFRE	Set of facilities, services and technical means that support the development of activities.
9	Preventive maintenance of equipment	MPRE	Regularly scheduled maintenance work to prevent future anomalies and unforeseen events and to repair devices before they fail.

(*continued*)

Table 1. (*continued*)

No.	Long Title	Short Title	Description
10	Educational inclusion	INED	Process aimed at guaranteeing the right to a quality education for all students under equal conditions, paying special attention to those most excluded. (OEI)
11	Educational policies	POLED	Guidelines for action regarding education involve laws, resolutions, and regulations that determine the nation's pedagogical doctrine and set its objectives.
12	Information and Communication Technologies	ICT	Set of resources and technological solutions to collect, process, store and transmit information.
13	Financial support	APEC	Economic or financial aid to support the educational crisis
14	Economic growth	CRECO	Increase in income, values and goods produced by a country's economy
15	Internet	INTR	Global computer network, which facilitates direct connection between technological devices
16	Science, Technology and Innovation Plan	PCTI	Document that promotes sustainable economic and social development, whose main drivers are science, technology, and innovation.
17	Budget	PRES	Plan of the operations and resources of a company, which is formulated to achieve the proposed objectives in a certain period and is expressed in monetary terms
18	Illiteracy	ANFB	Lack of basic knowledge that prevents reading or writing a simple sentence
19	Standards	EST	Documented agreements containing precise technical specifications or criteria that are used consistently
20	Educational system	SISED	It includes everything involved in educating students in public or private schools, and includes the levels of education provided.

(*continued*)

Table 1. (*continued*)

No.	Long Title	Short Title	Description
21	Human talent	TALHU	It is mainly made up of the knowledge, information, skills and abilities that individuals possess.
22	Inequality	DESG	Lack of balance between people, conditions or access to resources, services and positions in society
23	Social conflict	CONSOC	Actions of different interdependent sectors whose objectives, interests, values or needs conflict
24	Technological advances	AVTEC	It relates to new technologies, ways of using them and all those new products derived from technology that have been incorporated into current life.
25	Educational community	COMED	It is made up of students, educators, parents, graduates, teaching directors, and school administrators. (MEN)
26	Diversity	DIVD	People with different physical, social and personal characteristics, in a group or organization
27	Technological Resources	CHALLENGE	They are elements that come from scientific-technical innovation to enable and facilitate some type of work.
28	Traditional Education	EDTRA	Transmissionist teaching method, which educates through repetition, memory and discipline.
29	Educational Resources	REDUC	These are the materials, spaces, tools, or equipment necessary to support learning and contribute to developing educational processes in the best possible way.
30	Digital Resources	REDIG	An element in digital format used to display and store on a direct electronic device or with network access.
31	Student Enrollment System	SIMATE	Tool to organize and control the enrollment process, keeping it updated from all educational institutions nationwide

(*continued*)

Table 1. (*continued*)

NO.	LONG TITLE	SHORT TITLE	DESCRIPTION
32	Connectivity	CNTVD	Ability to maintain connections between various devices, networks or systems, to transfer data

Fountain: Own elaboration

4.3 Direct Influence Matrix (DIM)

An analysis is performed of each of the aspects that make up the Direct Influence Matrix, taking a horizontal reading of each one and assigning a score from 0 to 3, according to the influence each has on the other; as an evaluation (0 = no influence and 3 = very influential).

This can be seen in Table 2, which identifies the main factors, their status, and their relationship to each other, thus allowing us to identify the variables with the greatest impact.

Table 2. Direct Influence Matrix

	1: PEIDI	2: PASI	3: ACTCU	4: INMET	5: PTRs	6: RENAC	7: COED	8: INFRE	9: APRE	10: INED	11: POLED	12: TICS	13: APEC	14: CRECO	15: INTR	16: PCTI	17: PRES	18: ANFB	19: EST	20: SISED	21: TALHU	22: DESG	23: CONSOC	24: AVTEC	25: COMED	26: DIVD	27: RETO	28: EDTRA	29: REDUC	30: REDIG	31: SIMATE	32: CNTVD
1: PEIDI	0	2	3	3	3	2	2	1	1	3	3	3	3	1	1	2	2	2	3	3	2	1	1	2	3	3	2	3	2	2	2	1
2: PASI	2	0	0	2	2	2	2	2	3	0	1	3	3	3	2	2	3	0	0	0	0	0	1	0	0	2	0	3	3	0	2
3: ACTCU	3	0	0	3	3	3	2	0	0	2	3	2	1	2	1	1	3	3	3	3	3	1	0	2	2	2	3	5	3	1	1
4: INMET	3	1	3	0	3	3	2	0	0	2	3	3	3	2	2	2	2	2	3	3	3	1	0	3	2	2	3	3	2	2	1	1
5: PTRES	3	1	3	3	0	3	2	1	0	3	3	2	2	0	1	0	1	3	1	3	1	3	3	2	3	3	2	3	3	3	3	0
6: RENAC	3	1	2	2	1	0	0	0	2	3	2	2	1	1	1	1	3	3	3	3	3	1	2	3	3	3	3	3	3	3	1
7: COED	3	1	2	2	1	3	0	2	0	2	3	2	1	1	1	2	2	1	1	1	3	1	0	2	2	0	0	0	0	3	0
8: INFRE	1	1	0	0	0	2	2	0	3	2	0	1	3	2	1	1	3	0	0	1	0	1	2	0	1	0	2	0	1	0	0	0
9: MPRE	1	3	0	0	1	1	0	2	0	0	1	3	2	2	0	0	3	0	0	1	0	0	0	1	0	0	3	0	3	3	0	1
10: INED	3	1	3	3	3	3	3	1	0	0	3	2	0	0	0	1	2	2	3	3	3	1	1	3	3	1	1	1	1	1	3
11: POLED	3	3	3	3	3	3	3	3	3	3	0	3	3	1	1	1	3	2	2	3	3	3	1	1	3	3	3	3	3	3	3
12: TICS	1	3	3	3	3	2	2	1	3	2	3	0	2	1	3	1	1	2	0	2	2	2	1	3	1	2	3	3	3	2	3
13: APEC	2	3	1	2	2	1	2	2	3	0	3	3	0	3	3	0	2	0	1	1	0	0	0	3	1	0	3	2	3	3	1	3
14: CRECO	0	2	1	2	2	1	2	3	3	0	1	3	3	0	1	3	3	1	0	1	2	2	3	1	0	3	1	3	1	3	1	2
15: INTR	1	0	1	1	1	1	0	0	0	2	1	2	0	0	0	1	1	0	0	0	2	1	3	0	1	2	3	3	1	1	3
16: PCTI	1	1	1	1	1	0	0	0	0	1	1	2	1	1	2	0	1	1	0	0	0	2	1	3	0	1	2	2	1	2	1	3
17: PRES	1	3	1	2	2	1	1	3	3	1	1	3	3	0	3	0	0	0	0	1	1	0	1	3	3	3	1	3
18: ANFB	2	0	0	0	2	3	3	0	0	3	3	0	0	0	0	0	0	0	2	2	3	3	0	3	0	0	0	0	3	0
19: EST	3	0	3	3	3	2	3	3	0	3	3	3	1	0	1	1	0	3	0	3	2	3	3	3	3	3	3	3	3	3	3
20: SISED	3	3	3	3	3	3	3	2	1	3	3	2	1	0	0	1	3	0	0	0	1	3	3	0	3	3	3	3	3	3	1
21: TALHU	2	0	3	3	3	3	2	0	1	3	3	0	0	0	0	3	0	3	0	3	3	0	3	3	0	3	2	2	2	2	0
22: DESG	3	2	3	3	3	3	1	0	0	3	3	0	0	0	0	0	3	2	3	2	0	3	0	3	3	0	1	2	2	3	2
23: CONSOC	3	2	2	2	2	3	2	2	3	3	1	1	0	0	0	3	2	2	3	0	0	3	3	0	0	2	2	2	2
24: AVTEC	1	2	2	2	2	2	1	1	2	1	2	3	1	0	2	3	1	2	1	2	1	2	2	0	2	2	3	2	3	3	1	2
25: COMED	3	3	3	3	3	3	3	1	0	2	3	1	1	1	0	1	0	3	0	2	2	3	3	0	2	0	0	0	0	2	0
26: DIVD	2	2	2	2	2	3	2	0	3	3	0	2	1	0	1	1	2	2	2	2	3	0	2	0	0	3	3	3	3	1
27: RETO	1	2	2	2	2	2	2	0	2	3	3	0	0	3	2	1	2	1	2	2	3	2	2	0	2	3	3	2	3
28: EDTRA	3	2	3	3	3	3	3	2	0	3	3	3	2	1	2	2	2	3	3	3	3	3	3	3	3	3	3	0	3	3	1	1
29: REDUC	3	2	3	3	3	3	3	1	1	2	2	1	1	1	0	2	2	0	3	2	3	1	2	2	3	3	3	0	1	1
30: REDIG	1	2	3	3	3	3	2	1	0	3	2	3	1	0	2	2	1	2	1	2	1	1	2	3	3	0	1	1
31: SIMATE	2	1	1	1	1	0	3	2	2	3	0	3	3	0	0	3	3	1	2	1	3	0	2	3	0	0	0	0	0	0
32: CNTVD	0	0	0	2	2	1	0	0	1	1	2	2	0	0	3	1	1	2	0	1	1	2	1	2	1	2	2	3	3	1	2	0

Fountain: Own elaboration

Table 2 shows the rating assigned to each factor. In the case of purely institutional factors related to management and finance, they were given a high rating because the proper functioning, management, and use of ICT in institutional processes largely depend on them.

That is, the Institutional Educational Project influences the institution's acquisition plan, curriculum updates, methodological innovation, academic performance, preventive equipment maintenance, educational policies, technological resources, the educational system, educational and digital resources, among others. Therefore, it has less influence and is given a low rating compared to external aspects or those related to government policies such as educational coverage, infrastructure, the Science and Technology Plan, inequality, social conflict, diversity, and the national enrollment system.

But when analyzing the influence of these others on institutional aspects, it is clear that, on the contrary, they do have an impact, since the proper functioning of the financial and, obviously, the academic aspects of educational institutions largely depend on the support of external entities, whether national or governmental.

Likewise, the Potential Direct Influence Matrix (PDIM) represents the potential influences and dependencies present between variables. As shown in Figs. 1 and 2.

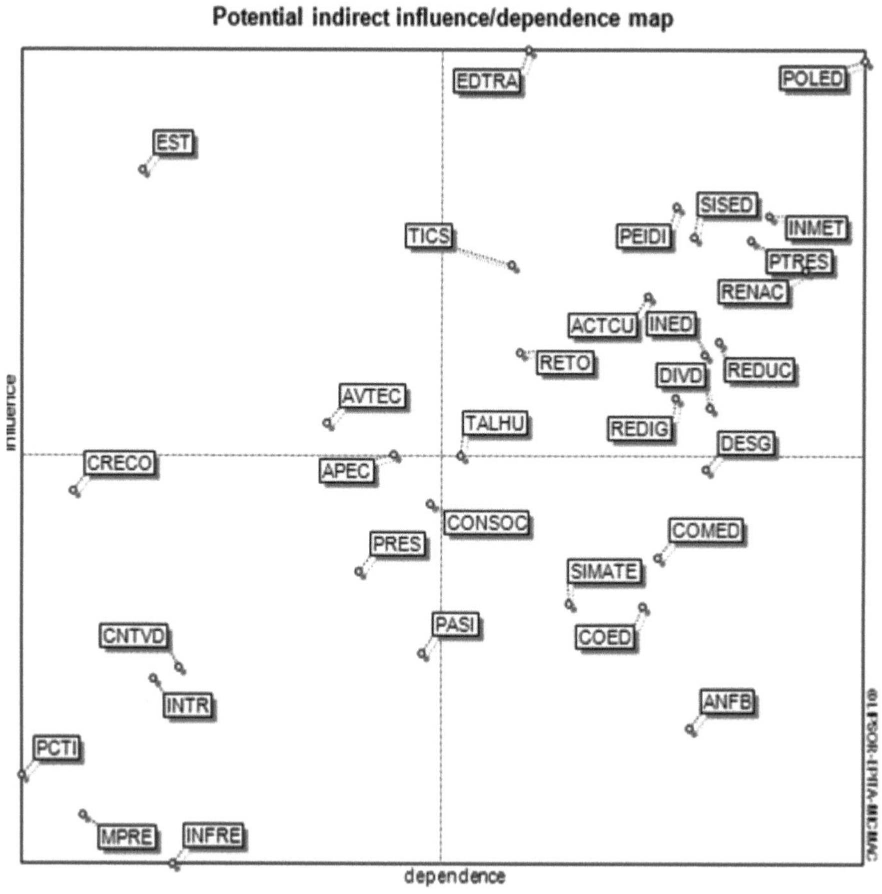

Fig. 1. Map of direct influence/dependence, **Source:** Own elaboration

Figure 1 shows the interpretation of the MICMAC tool in the classification of the variables with the greatest *influence/dependence* of the 32 that were taken into account as the object of study. To do this, it is established that the variables with these characteristics are located in the upper right quadrant.

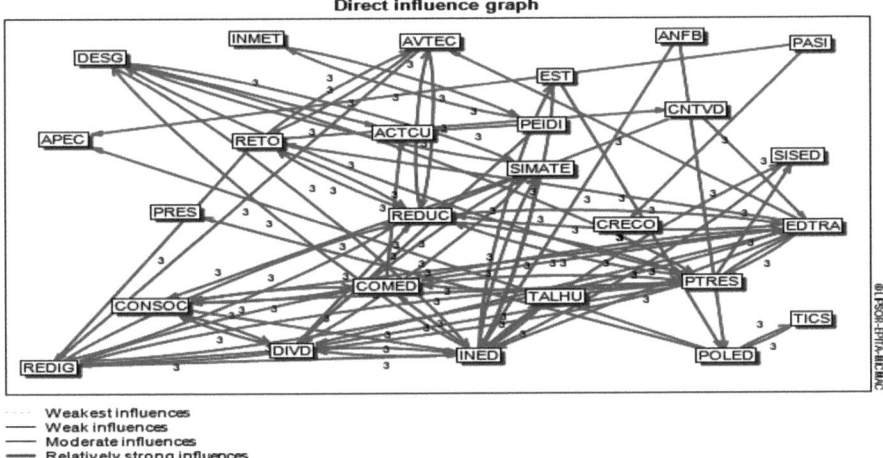

Fig. 2. Direct influence graph. **Source:** Prepared by the authors.

5 Application of the MACTOR Tool

5.1 List of Actors

32. Ministry of National Education (MEN)
33. Certified Territorial Entity (ETC)
34. Teaching Directors (DIRDOC)
35. Teaching Team (EDOC)
36. Parents of Families (PADFAM)
37. Educational Community (COMED)
38. Students (STUD)
39. Ministry of Finance (MINHAC)
40. Congress of the Republic (CONREP)
41. Technology Companies (EMPTEC)
42. Technology and Computer Science Teachers (DTECINF)
43. Resource Evaluation Committee (COMEVREC)
44. Telecommunications Companies (EMPTEL)
45. Innovative Teachers (DOCINNOV)
46. Institutional Academic Council (CONACIN)
47. Traditionalist Teachers (DOCTRAD)

5.2 Battlefield Construction: Introducing Variables

Below are the challenges to be achieved in each of the (6) strategic variables presented. Where the actors that influence either positively or negatively are indicated

5.2.1 List of Variables

1. (POLED) Educational Policies
2. (TICS) Information and Communication Technologies
3. (REDUC) Educational Resources
4. (RENAC) Academic Performance
5. (INMET) Methodological innovation
6. (ACTCU) Curriculum update

Table 3. Battlefields

VARIABLE	CHALLENGE	ACTORS IN FAVOR	ACTORS AGAINST
(POLED) Educational Policies	Support the implementation of educational policies in rural institutions.	Ministry of National Education (MEN) Departmental and Municipal Education Secretariats Parents' associations Educational community	Congress of the Republic Treasury Primary School Teachers
(TICS) Information and Communication Technologies	Transforming education at rural educational institutions in San Antero into an environment more aligned with contemporary technological demands.	Ministry of National Education Local technology companies	Educational community misinformed about ICTs
(REDUC) Educational Resources	Improve the availability and quality of digital educational resources to support the effective integration of ICTs into the institutional processes of rural education in San Antero.	Department of Educational Technology Educational Resources Evaluation Committee Technology and telecommunications companies	Treasury Primary school teachers with resistance to change

(*continued*)

Table 3. (*continued*)

VARIABLE	CHALLENGE	ACTORS IN FAVOR	ACTORS AGAINST
(RENAC) Academic Performance	Improve the incorporation of ICTs into institutional processes in the rural area of San Antero, strengthening transversality and optimizing the use of technological resources.	Management team and pedagogical coordinators Teachers committed to educational innovation Students motivated by the use of technology	Teachers with resistance to change Students with limited access to technology
(INMET) Methodological Innovation	Implement new teaching methodologies that effectively integrate ICTs into rural educational processes in San Antero, thus promoting creativity, active student participation, and the development of digital skills relevant to the 21st century.	Management team and pedagogical coordinators Teachers enthusiastic about educational innovation Students motivated by active learning	Students motivated by active learning Institutional and cultural barriers Administrative sector disconnected from the pedagogical field
(ACTCU) Curriculum Update	Transform institutions in the rural areas of the municipality of San Antero into an educational environment more aligned with contemporary technological demands.	Institutional Academic Council Ministry of National Education	Traditionalist teachers

Source: Own elaboration

Table 3 presents a detailed analysis of the challenges and actors involved in the integration of ICTs in educational processes in rural areas of the municipality of San Antero. Each variable identifies a specific area for improvement and shows the actors for and against each challenge.

Looking at the challenges posed, the need for political and financial support to bolster the implementation of educational policies that promote the integration of ICTs in rural institutions stands out; however, opposing actors, such as the Congress of the Republic and the Ministry of Finance, have been identified, which can represent bureaucratic and financial barriers.

Regarding ICTs, their potential to transform education in rural settings is recognized, but the lack of information and training among the educational community is pointed out as a significant barrier.

Improving digital educational resources is another crucial aspect, with supporting actors such as the Department of Educational Technology and technology companies, but with resistance to change among some teachers being a significant barrier.

Academic performance and methodological innovation are influenced by both enabling actors and institutional and cultural barriers, highlighting the complexity of successful ICT implementation in rural educational settings.

Finally, curricular updating is posed as a key challenge, with the need to align the educational curriculum with current technological demands.

Table 4. List of Actors and Description of Objectives

No.	LONG TITLE	SHORT TITLE	DESCRIPTION
1	Educational Policies	POLED	Support the implementation of educational policies in rural institutions.
2	Information and Communication Technologies	ICT	Transforming education at rural educational institutions in San Antero into an environment more aligned with contemporary technological demands.
3	Educational Resources	REDUC	Improve the availability and quality of digital educational resources to support the effective integration of ICTs into the institutional processes of rural education in San Antero.
4	Academic Performance	RENAC	Improve the incorporation of ICTs into institutional processes in the rural area of San Antero, strengthening transversality and optimizing the use of technological resources.
5	Methodological Innovation	INMET	Implement new teaching methodologies that effectively integrate ICTs into rural educational processes in San Antero, thus promoting creativity, active student participation, and the development of digital skills relevant to the 21st century.
6	Curriculum Update	ACTCU	Transform institutions in the rural areas of the municipality of San Antero into an educational environment more aligned with contemporary technological demands.

Source: Own elaboration

Table 4. as a whole shows the diversity of actors involved and the specific objectives they pursue in the process of integrating ICTs into rural educational institutions in San Antero. Collaboration among these actors is highlighted as crucial to successfully achieving these objectives and improving the quality of education in these communities.

6 Direct Influence Matrix (DIM)

In this Direct Influence Matrix (DIM), Actor is related to Actor by Actor through the table of actor strategies, that is, it is about describing the direct influence that actors have on each other, as shown in Fig. 3.

Direct Influence Matrix (DIM)

MID	MEN	ETC	DIRDOC	EDOC	PADFAM	COMED	ESTUD	MINHAC	CONREP	EMPTEC	DTECINF	COMEVREC	EMPTEL	DOCINNOV	CONACIN	DOCTRAD
MEN	0	4	4	4	3	2	3	0	0	1	4	4	1	4	4	4
ETC	0	0	4	4	3	4	4	0	0	1	4	4	1	4	4	4
DIRDOC	0	0	0	4	3	3	4	0	0	0	4	4	0	4	4	4
EDOC	1	2	2	0	2	2	4	0	0	0	3	3	0	4	4	4
PADFAM	0	0	1	2	0	2	4	0	0	0	2	3	0	3	3	3
COMED	1	1	3	3	4	0	4	0	0	0	3	3	0	3	3	3
ESTUD	3	3	3	4	3	4	0	0	0	1	4	4	1	4	4	4
MINHAC	1	3	3	3	2	3	3	0	2	4	2	3	4	3	2	2
CONREP	3	4	1	1	0	3	3	3	0	2	1	1	1	1	1	1
EMPTEC	0	1	1	1	1	1	1	0	0	0	3	3	3	3	0	0
DTECINF	0	0	1	2	2	1	4	0	0	2	0	2	3	3	1	2
COMEVREC	1	1	1	2	0	3	3	0	0	1	3	0	1	3	2	3
EMPTEL	0	1	1	1	1	1	1	0	0	3	3	3	0	3	0	0
DOCINNOV	2	2	2	3	1	3	4	0	0	2	4	1	2	0	1	4
CONACIN	0	0	2	3	1	2	4	0	0	0	4	3	0	4	0	4
DOCTRAD	0	0	1	3	1	3	4	0	0	0	3	3	1	2	3	0

Fig. 3. Direct Influence Matrix (MID) Actor X Actor. **Source:** Own elaboration

In Fig. 3, the degree of influence between the actors involved in the integration of ICTs shown in Table 3 is determined. It can be seen that the influence is assigned a score of 0 to 4, depending on the effect that one has on the other. The assessment is taken from the influence of the horizontal variable on the vertical variable, that is, for example, the influence value of the variable MEN on DTECINE is 4, but this one on the MEN is 0, where:

0: No influence
1: Processes
2: Projects
3: Mission
4: Existence

Valued Positions Matrix

2MAO	POLED	TICS	REDUC	RENAC	INMET	ACTCU
MEN	4	3	4	3	4	4
ETC	4	3	2	3	4	4
DIRDOC	4	2	4	3	4	4
EDOC	4	4	4	3	4	4
PADFAM	1	0	0	3	0	1
COMED	3	1	1	4	2	4
ESTUD	2	1	0	4	1	2
MINHAC	3	4	4	2	3	2
CONREP	1	2	3	0	1	0
EMPTEC	1	4	4	3	4	2
DTECINF	3	4	4	3	4	4
COMEVREC	4	3	4	4	4	4
EMPTEL	1	4	3	3	4	2
DOCINNOV	4	4	4	4	4	4
CONACIN	4	3	3	4	4	4
DOCTRAD	2	1	1	4	1	1

© LIPSOR-EPITA-MACTOR

Fig. 4. Valued Position Matrix (2MAO). **Source:** Prepared by the authors.

Figure 4 shows vertically the actors that make up the direct influence matrix of Fig. 3 and how these, through a valuation assigned from 0 to 4 (described above), influence the six most relevant variables that intervene in the educational processes of the institutions located in rural areas of the municipality of San Antero, department of Córdoba-Colombia, obtained through the implementation of the MICMAC tool.

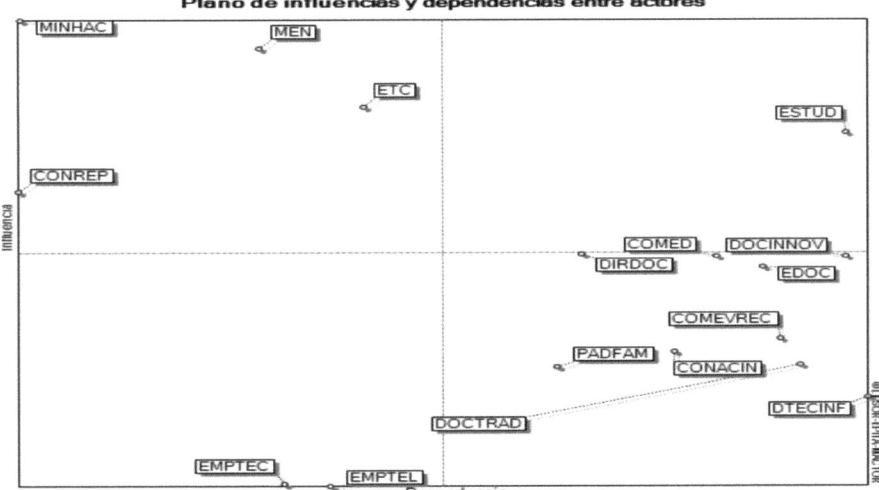

Fig. 5. Map of influences and dependencies between actors. **Source:** Prepared by the authors.

Figure 5 shows the plane of influence and dependence between actors, according to their positions with respect to the influence and/or dependence presented, whether direct or indirect.

7 Conclusions

The structural analysis carried out using the MICMAC and MACTOR tools has allowed us to identify in detail the influence of ICTs on the educational processes of institutions located in rural areas of the municipality of San Antero, department of Córdoba, Colombia.

First, it was evident that educational policies, particularly those related to the implementation of technology in rural areas, play a crucial role in promoting the integration of ICTs in educational institutions. Supporters of this integration include the Ministry of National Education and parent-teacher associations, while the most restrictive factors come from the Congress of the Republic and the Ministry of Finance.

Secondly, the lack of awareness and support from the educational community regarding ICTs was found to be a significant obstacle to their successful implementation; this highlights the need for training and awareness programs to promote greater technology adoption in rural educational settings.

Furthermore, the importance of improving the availability and quality of digital educational resources to support the effective integration of ICTs into educational processes was identified. This requires collaboration between the educational technology department, technology and telecommunications companies, and the educational resources evaluation committee to overcome financial and logistical barriers.

Taking the above into account, this research could be further enriched by considering the long-term implications of the integration of ICTs in the development of educational institutions in rural areas, evaluating their impact on the quality of learning, the decrease in school dropout rates, and the reduction of the very marked social gap between rural and urban areas. Ávila (2022) points out that the implementation and appropriation of ICTs in rural contexts generate pedagogical transformations when contextualized with the real needs of communities. Similarly, Lugo and Ithurburu (2019) highlight in their studies that digital public policies must consider differentiated territorial approaches in their planning and implementation so that the processes are relevant to rural areas. Likewise, comparisons should be made with other contexts in rural areas of the country and in Latin America, in order to identify common patterns where the process can be replicated, as suggested by Bonilla and Muñoz (2022) in their study on technology-mediated education in rural regions during the pandemic. This observation argues for the contextualization of the findings, but would also provide strategies for the formulation of new public policies aimed at closing the educational digital divide in rural areas, as proposed by Sáenz et al. (2023) in their review of the effects of ICTs in rural communities in times of crisis.

References

Almerich, G., Suárez-Rodríguez, J.M., Díaz-García, I.: Information and communication technologies (ICTs) in education: Challenges and risks for teachers. Journal of Education **371**, 41–65 (2016)
Ávila, N.P.: Using technology in rural school teaching as a strategy for educational transformation. Voices and Educational Realities (2022)
Booth, T., Ainscow, M.: Index for Inclusion: Developing Learning and Participation in Schools (2011)
Bonilla, O.P., Muñoz, D.E.: Rural education mediated by traditional technology in times of pandemic. Journal Between Science and Engineering (2022)
Csikszentmihalyi, M.: Flow: The psychology of optimal experience. Harper & Row, (1990)
Cuban, L.: Oversold and underused: Computers in the classroom. Harvard University Press (2001)
Deci, E.L., Ryan, R.M.: The "what" and "why" of goal pursuits: Human needs and the self-determination of behavior. Psychol. Inq. **11**(4), 227–268 (2000)
Ertmer, P.A.: Teacher pedagogical beliefs: The final frontier in our quest for technology integration? Education Tech. Research Dev. **53**(4), 25–39 (2005)
Freire, P.: Pedagogy of the Oppressed (1970)
Goyeneche, G., Parodi, T.: Introduction to Prospective–Methodological synthesis. Planning Directorate, 20–27 (2017)
Henao, G., Ramirez, L., Ramirez, C.: What is psychopedagogical intervention: Definition, principles and components. The Medellin-Colombia USB Agora **6**(2), 215–226 (2006)
Johnson, L., Adams, S., Cummins, M.: NMC horizon report: 2016 K-12 edition. The New Media Consortium (2016)
Lave, J., Wenger, E.: Situated Learning: Legitimate Peripheral Participation (1991)
Papert, S.: The Children's Machine: Rethinking School in the Age of the Computer (1993). https://doi.org/10.5555/139395
Prensky, M.: Digital natives, digital immigrants part 1. On the Horizon **9**(5), 1–6 (2001)
Rodríguez, M., Pérez, J., Gómez, A.: Pedagogy in the digital age. Technological Publishing House of Antioquia (2022)
Rodríguez, Z., Delvaty, E., Deulofeu, B., Pérez, Z.: The pedagogical process and formative objectives in education. Edumecentro **14**, 2120 (2022)
Sáenz, M.L.S., Jácome, R.T., Caraballo, L.H. Information and communications technologies and rural education in times of pandemic. Unimar Journal (2023)

Secretaria, de Educación Departamental.: Boletín Estadístico del Sector Educativo en el Departamento de Córdoba (2023) chrome-extension://efaidnbmnnnibpcajpcglclefindmkaj/https://seducacion.cordoba.gov.co

Selwyn, N.: Schools and schooling in the digital age: A critical analysis. Routledge (2010)

Thomas, J.: A Review of Research on Project-Based Learning (2000)

Warschauer, M.: Technology and social inclusion: Rethinking the digital divide. MIT Press (2003). https://www.researchgate.net/publication/31775276_Technology_and_Social_Inclusion_Rethinking_the_Digital_Divide_M_Warschauer

An Efficient Federated Utility-Mining Algorithm

Tzung-Pei Hong[1,2(✉)], Jing-Chi Yang[2], Yu-Chuan Tsai[3], and Chun-Hao Chen[4]

[1] Department of Computer Science and Information Engineering, National University of Kaohsiung, Kaohsiung, Taiwan, R.O.C.
`tphong@nuk.edu.tw`
[2] Department of Computer Science and Engineering, National Sun Yat-Sen University, Kaohsiung, Taiwan, R.O.C.
`charliejcyang@icloud.com`
[3] Library and Information Center, National University of Kaohsiung, Kaohsiung, Taiwan, R.O.C.
`yjtsai@nuk.edu.tw`
[4] Department of Computer Science and Information Engineering, National Kaohsiung University of Science and Technology, Kaohsiung, Taiwan, R.O.C.
`chench@nkust.edu.tw`

Abstract. Mining data securely and effectively from multiple sources has become a critical issue in the era of rapid information growth. This paper thus addresses the federated utility mining problem within a horizontal framework, in which items are nearly identical across clients, but users differ. We propose a federated utility-mining algorithm that transmits local high-utility itemsets, rather than transaction data, from clients to a server, thereby increasing data privacy. Clients independently mine high-utility itemsets, and the server integrates these itemsets to produce final results. Although the proposed algorithm discovers fewer high-utility itemsets than the traditional utility mining algorithm, it can save significant integration time and avoid transmitting the original datasets. We then conduct experiments on two real datasets by allocating the data among three clients, with the results showing that the proposed algorithm achieves good performance.

Keywords: Utility Mining · High-utility Itemset · Federated Mining · Horizontal Framework · Data Privacy

1 Introduction

Rapid advancements in information technology have led to the era of big data, characterized by the generation of massive data, prompting the development of efficient mining techniques [1, 2]. Concurrently, rising concerns about data privacy, exemplified by incidents such as the Facebook-Cambridge Analytica scandal [3], have prompted strict regulations like GDPR [4] and CCPA [6], posing new challenges for data mining and machine learning.

Federated Learning (FL), proposed by Google in 2017, enables decentralized collaborative training without directly sharing private data, preserving privacy by exchanging only model parameters [12, 13]. However, existing FL research mainly focuses on machine learning tasks, with limited exploration of federated data mining. In recent years, several studies have started exploring the integration of FL with data mining methods [20, 21]. Wu et al. combined FL with frequent itemset mining, fulfilling the need for collaborative mining from multiple data sources [32]. However, existing FL research mainly focuses on machine learning tasks, with limited exploration of federated data mining, particularly high-utility itemset mining (HUIM) [10, 11]. In high-utility itemset mining, the external profits of items need to be considered when discovering interesting patterns in a transaction database. Many researchers have devoted themselves to the problem of mining high-utility itemsets [5, 9, 10, 11 and 33].

This paper proposes a Fast Federated High-Utility Itemset Mining (FFHUIM) framework, integrating the traditional utility mining algorithms with federated mining to produce rapid yet accurate outcomes suitable for real scenarios. FFHUIM is compared with a traditional approach using the Retail and the Foodmart datasets provided by SPMF [34] to evaluate its effectiveness. It contributes significantly to federated mining by demonstrating how advanced mining techniques adapt to decentralized environments while preserving data privacy and mining accuracy.

2 Related Work

2.1 High-Utility Itemset Mining

High-utility itemset mining (HUIM) represents a significant development beyond traditional frequent itemset mining (FIM) [33]. HUIM emphasizes the utility of itemsets, typically measured in profit, cost, or other quantitative metrics. Given a transaction database and a user-defined minimum utility threshold, an itemset is defined as a high-utility itemset (*HUI*) if its total utility across all transactions is no less than the threshold. This shift highlights the importance of item value over mere occurrence, making HUIM particularly relevant in contexts where the economic or practical significance of itemsets outweighs their ubiquity [9]. A critical challenge in HUIM is identifying itemsets that yield high utility without necessarily being frequent. This challenge arises because the utility of an itemset does not exhibit the downward closure property, a key trait used in frequent itemset mining to effectively reduce the search space [10, 11]. Integrating a traditional utility mining algorithm within federated mining (FM) systems represents a substantial advancement in this field [20–22]. FM enables the decentralized processing of data, which helps maintain privacy and mitigates the risks associated with centralized data storage and analysis. By adapting the utility mining algorithm to operate within FM frameworks, this research addresses the scalability and privacy challenges that traditional HUIM methods encounter when applied to distributed data sources.

2.2 Federated Learning

Federated learning enables decentralized training of machine learning models, where raw data remains on local devices, and only model updates are shared with a central server [8]. As shown in Fig. 1, the process of federated learning mainly involves four steps:

(1) A central server initializes and distributes a global model to a set of selected clients.
(2) Each client independently trains the model using its local dataset.
(3) After training, the locally updated model parameters are sent to the server.
(4) The server aggregates these updates to refine the global model.

If the aggregated model does not achieve satisfactory performance, the process repeats until convergence.

Depending on the structure and overlap of data among participants, FL can be categorized into three types: Horizontal Federated Learning (HFL), Vertical Federated Learning (VFL), and Federated Transfer Learning (FTL) [24]. HFL is suited for scenarios where participants share similar feature spaces but have distinct user groups, such as multiple retail companies collecting the same customer behavior attributes but serving different customer bases. In contrast, VFL applies to cases where different organizations hold complementary attributes about the same set of users, for instance, a bank and an e-commerce platform that share a common user base but maintain separate financial and behavioral datasets. FTL, on the other hand, addresses cases where both the user groups and feature spaces differ significantly.

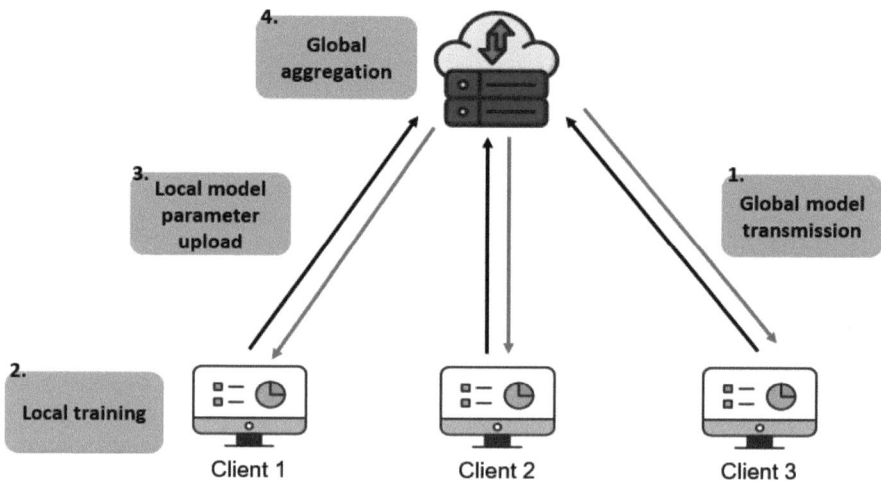

Fig. 1. The process of federated learning.

2.3 Federated Data Mining

While federated learning (FL) has garnered considerable attention in machine learning applications [8], relatively few studies have explored its potential in data mining [14], particularly in high-utility itemset mining (HUIM) tasks [10, 11]. Given the promising privacy-preserving characteristics of FL [13, 15], such as maintaining data locality and ensuring differential privacy [8], this paper aims to extend the FL concept to data mining applications.

Federated data mining (FDM) represents a significant shift from centralized to distributed data processing paradigms [22]. In the FDM approach, data remains in local environments, such as users' mobile devices or geographically dispersed servers, ensuring compliance with data locality and privacy regulations. Unlike traditional data mining, where datasets are aggregated centrally, FDM allows data holders to collaboratively mine valuable knowledge without sharing raw data. This approach aligns with modern data privacy concerns and regulatory frameworks such as GDPR and CCPA, which impose strict guidelines on data sharing and privacy. FDM is particularly well-suited for scenarios where centralized data storage and analysis are not feasible due to legal, ethical, or operational limitations. It has emerged as a promising solution for privacy-preserving data mining in various fields, including retail, healthcare, and finance [32].

Previous studies that combine federated learning with frequent itemset mining have demonstrated great potential. Hong et al. [22] introduced a federated mining framework for complete frequent itemsets, which optimizes both the efficiency and accuracy of the mining process in decentralized environments. Furthermore, Hong et al. [20] proposed a federated approach for erasable-based itemset mining in manufacturing, enabling the identification of low-profit materials without sharing raw data. These strategies represent significant advances in privacy-preserving mining techniques. However, such approaches have not yet been widely applied to high-utility itemset mining in real-world scenarios.

3 Proposed Algorithm

3.1 Notation

Table 1 summarizes notations used in this work.

Table 1. Notation.

Symbol	Description
N	The number of clients participating in the federated mining
S	The aggregation server
C	The set of clients, represented as $C = \{C_1, C_2, ..., C_N\}$
X	An itemset
λ	The minimum utility ratio threshold
$C_i.TD$	The local transaction database of the i-th client
$C_i.|TD|$	The number of transactions in $C_i.TD$
$C_i.HUI$	A locally high-utility itemset of the i-th client
$C_i.DU$	The local database utility value of all transactions in the i-th client
$S.DU$	The sum of $C_i.DU$s used in the aggregation server
$C_i.U(X)$	The local utility value of X in the i-th client
$S.U(X)$	The aggregated utility of X in the server
$S.HUI$	An aggregated high-utility itemset

3.2 The Proposed Algorithm

The flowchart of the proposed algorithm, FFHUIM, is depicted in Fig. 2. It is divided into two parts: client side and server side. The client side will handle mining from its returned mined knowledge to form and adjust their business policies. The proposed FFHUIM algorithm is described as follows.

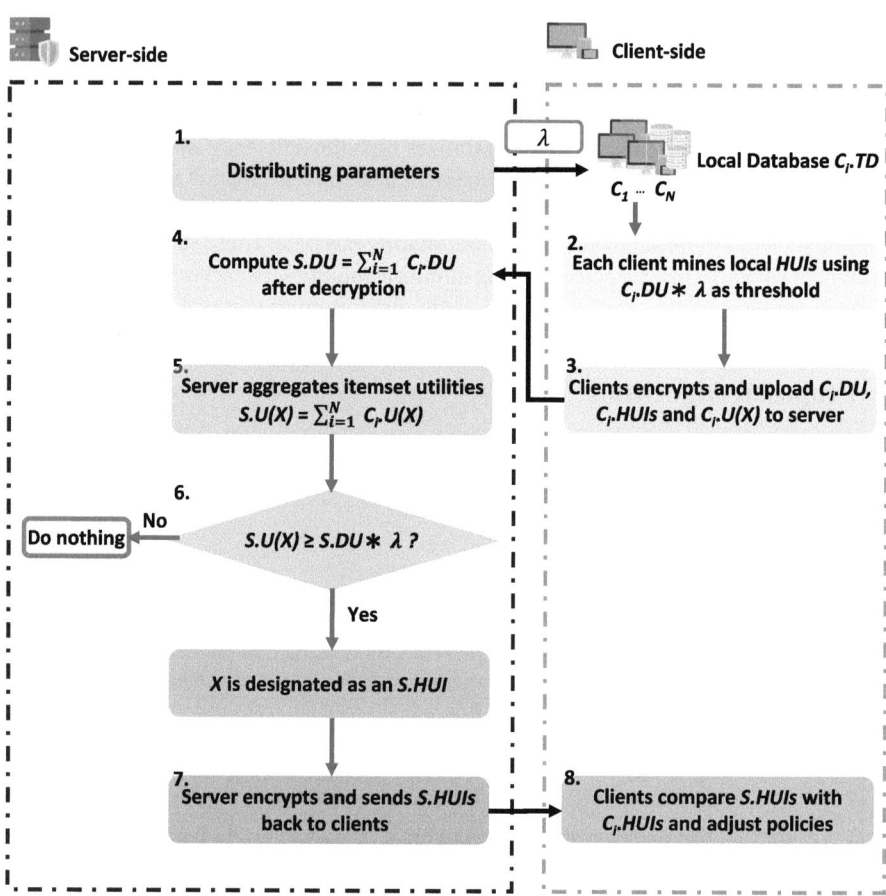

Fig. 2. The flowchart of the proposed FFHUIM algorithm.

Algorithm: The FFHUIM algorithm

Input: The client sources (transaction databases) and the minimum utility ratio threshold λ

Output: The aggregated high-utility itemsets.

Step 1: The server S transmits the minimum utility ratio threshold λ to all the clients.

Step 2: Each client C_i mines its own high-utility itemsets (HUI_i) from its local database ($C_i.TD$) by a traditional high-utility mining algorithm based on the local utility threshold value of $C_i.DU * \lambda$, where $C_i.DU$ represents the locally total utility value of all transactions in C_i.

Step 3: Each client C_i encrypts and uploads its $C_i.DU$, and it's each high-utility itemset X in $C_i.HUI$s with the local utility value $C_i.U(X)$ to the server.

Step 4: The server calculates $S.DU$ after it receives $C_i.DU$ uploaded from all the clients as follows:

$$S.DU = \sum_{i=1}^{N} C_i.DU.$$

Step 5: The server calculates $S.U(X)$ for each X sent from a client to it as follows:

$$S.U(X) = \sum_{i=1}^{N} C_i.U(X).$$

Step 6: The server designates each itemset X in **Step 5** as a globally high-utility itemset $S.HUI$ if $S.U(X) \geq S.DU * \lambda$.

Step 7: The server encrypts and returns all the $S.HUI$s to the clients.

Step 8: Each client C_i can compare the returned $S.HUI$s with $C_i.HUI$s to form and adjust their business policy.

3.3 An Example of the Proposed Algorithm

This section demonstrates an example of the proposed FFHUIM algorithm. In this example, we only demonstrate the feasibility of our proposed algorithm, so we ignore the encryption and decryption steps in the communication between each client and server. This example involves three clients, with their transaction datasets depicted in Tables 2, 3, and 4, respectively. Their transaction numbers are 3, 4, and 5, respectively. There are five distinct items in the datasets. The external utility table for these five items is presented in Table 5.

Table 2. C_1's transaction database.

TID \ ITEM	A	B	C	D	E	TU
T_1	1	7	0	1	0	15
T_2	1	1	0	2	0	12
T_3	0	0	3	1	1	13

Table 3. C_2's transaction database.

TID \ ITEM	A	B	C	D	E	TU
T_1	0	0	4	1	1	15
T_2	2	4	2	2	0	24
T_3	0	0	1	0	1	6
T_4	0	5	5	0	0	15

Table 4. C_3's transaction database.

TID \ ITEM	A	B	C	D	E	TU
T_1	4	7	0	1	0	30
T_2	1	8	0	0	0	13
T_3	2	0	0	1	1	17
T_4	0	0	1	2	3	20
T_5	0	4	0	0	4	20

Table 5. The external utility table.

ITEM	PROFIT($) (per unit)
A	5
B	1
C	2
D	3
E	4

Step 1: Assume the minimum utility ratio threshold $\lambda = 0.2$. For the three data sources, the FFHUIM algorithm proceeds as follows.

Step 2: Each client C_i finds its high-utility itemsets (HUI_i). An itemset X is included in HUI_i if:

$$C_i.U(X) \geq C_i.DU * \lambda$$

Take the itemset $\{A\}$ in Client 1 as an example. We have $C_1.U(\{A\}) = 10$ and $C_1.DU * \lambda = 40 * 0.2 = 8$. Since $C_1.U(\{A\}) \geq C_1.DU * \lambda$, i.e. $10 \geq 8$. It follows that $\{A\}$ satisfies the high-utility criterion and therefore belongs to $C_1.HUIs$. The result of C_1's $HUIs$ is shown in Table 6 using any traditional utility mining algorithm. We may get the $HUIs$ for the other two clients, as shown in Tables 7 and 8.

Table 6. C_1's high-utility itemsets, $C_1.HUIs$.

Level	Pair of an itemset and its actual utility
L1	({A}, 10), ({B}, 8), ({D}, 12)
L2	({AB}, 18), ({AD}, 19), ({BD}, 17), ({CD}, 9), ({CE}, 10)
L3	({ABD}, 2), ({CDE}, 13)

Table 7. C_2's high-utility itemsets, $C_2.HUIs$.

Level	Pair of an itemset and its actual utility
L1	({C}, 24)
L2	({AB}, 14), ({AC}, 14), ({AD}, 16), ({BC}, 23), ({DE}, 21), ({CE}, 18)
L3	({ABC}, 18), ({ABD}, 20), ({ACD}, 20), ({BCD}, 14), ({CDE}, 15)
L4	({ABCD}, 24)

Table 8. C_3's high-utility itemsets, $C_3.HUIs$.

Level	Pair of an itemset and its actual utility
L1	({A}, 35), ({E}, 32)
L2	({AB}, 40), ({AD}, 36), ({BE}, 20), ({DE}, 25)
L3	({ABD}, 30), ({CDE}, 20)

Step 3: Each client C_i encrypts and uploads its $C_i.DU$, and it's each high-utility itemset X in $C_i.HUIs$ with the local utility value $C_i.U(X)$ to the server.

Step 4: After receiving each client's $C_i.DU$, the server computes $S.DU = C_1.DU + C_2.DU + C_3.DU = > 200 = 40 + 60 + 100$.

Step 5: The server aggregates the actual utilities of each itemset X received from the clients. For example, $S.U(\{A\}) = 10 + 0 + 35 = 45$. The aggregate results are displayed in Table 9.

Table 9. The server aggregates results.

Level	Pair of an itemset and its actual utility
L1	{A} - 45, {B} - 8, {C} – 24, {D} - 12, {E} - 32
L2	{AB} - 72, {AC} - 14, {AD} - 71, {BC} - 23, {BD} - 17, {BE} - 20, {CD} - 9, {CE} - 28, {DE} - 46
L3	{ABC} - 18, {ABD} - 52, {ACD} - 20, {BCD} - 14, {CDE} - 48
L4	{ABCD} - 24

Step 6: The server designates each itemset X in Step 5 as an *S.HUI* if $S.U(X) \geq S.DU * \lambda$. For example, $S.U(\{A\}) = 45$ and $S.DU * \lambda = 200 * 0.2 = 40$. Since $S.U(\{A\}) \geq S.DU * \lambda$, i.e. $45 \geq 40$. It follows that $\{A\}$ satisfies the high-utility criterion and therefore belongs to *S.HUIs*. The result of *S.HUIs* is shown in Table 10.

Step 7: The server encrypts and returns all the *S.HUIs* to the clients.

Step 8: Each client C_i can compare the returned *S.HUIs* with $C_i.HUIs$ to form and adjust their business policy.

Table 10. The server high-utility itemsets, *S.HUIs*.

Level	Pair of an itemset and its actual utility
L1	{A} - 45
L2	{AB} - 72, {AD} - 71
L3	{ABD} - 77, {CDE} - 48

4 Experimental Results

In this section, we compared the performance of a traditional utility mining algorithm (the two-phase utility mining algorithm, proposed by Liu [5], referred to as two-phase) and the FFHUIM algorithm under different minimum utility ratio thresholds. Note that any utility mining algorithm can act as the local mining approach in clients. Two open datasets, Retail and Foodmart, obtained from SPMF [34], which is an open-source software and data mining library, were considered for the experiments. The experimental setup includes an AMD Ryzen 7 3700X 8-Core Processor (3.59 GHz) and 16 GB of RAM, running the Microsoft Windows 10 operating system.

We first introduce the experimental results for the Retail dataset. The threshold λ was varied from 0.0006 to 0.001 to evaluate scalability and efficiency under varying utility thresholds. As illustrated in Fig. 3, the FFHUIM algorithm may discover fewer HUIs compared to the traditional utility mining algorithm. Because each client uploads only its local high-utility itemsets to the server, the integration in the server may miss some global high-utility itemsets if they are not locally high in all the clients. Nevertheless, it effectively controls the number of missing HUIs within an acceptable range. Moreover, FFHUIM can benefit from simple and approximate server aggregation and data decentralization in certain scenarios, reducing execution time.

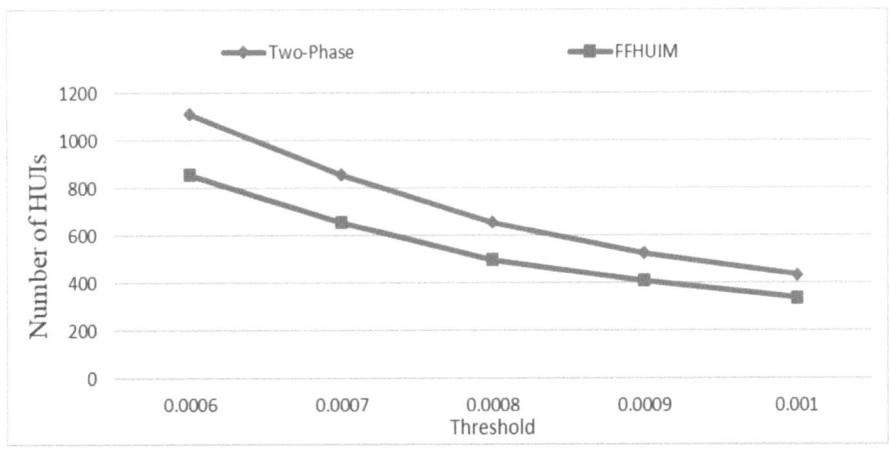

Fig. 3. Impact of threshold variations on the number of HUIs for Retail.

Regarding execution efficiency, Fig. 4 demonstrates that FFHUIM maintains competitive runtime performance. The execution time is recorded as the maximum local execution time among the clients plus the server aggregation time. Because the server aggregation adopts an approximate mechanism, it can reduce the rescanning time for acquiring totally accurate mining results. Particularly under lower threshold settings, where the traditional algorithm suffers from an explosion in candidate itemsets and prolonged processing time, FFHUIM demonstrates stable and scalable behavior due to its ability to process data locally before aggregation. These results validate the practical feasibility of FFHUIM, showing that it achieves a favorable balance between utility mining accuracy and computational efficiency while adhering to federated constraints such as privacy preservation and decentralized computation.

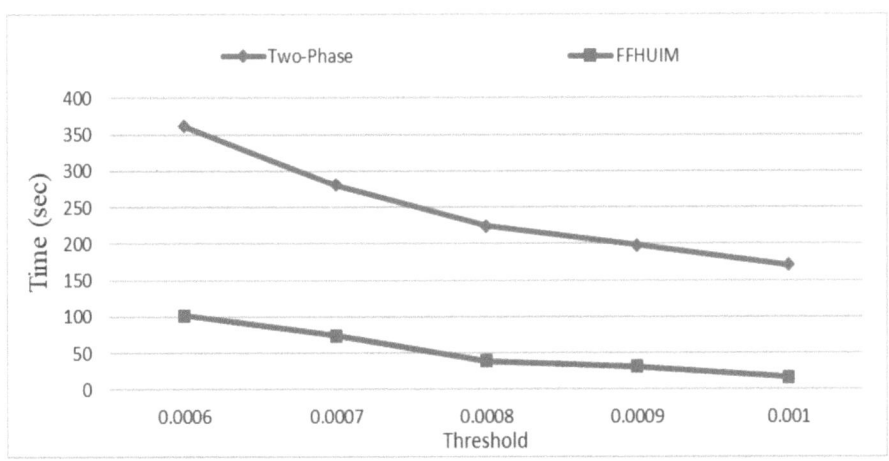

Fig. 4. Impact of threshold variations on execution time for Retail.

We then introduce the experimental results for the Foodmart dataset. The threshold λ was varied from 0.0006 to 0.0008. The comparison of the obtained high-utility itemsets by the two algorithms is illustrated in Fig. 5. Similar to the results for the Retail dataset, the FFHUIM algorithm may still produce a slightly different number of discovered *HUIs* compared to the traditional utility mining algorithm for Foodmart. The comparison of the execution efficiency is shown in Fig. 6.

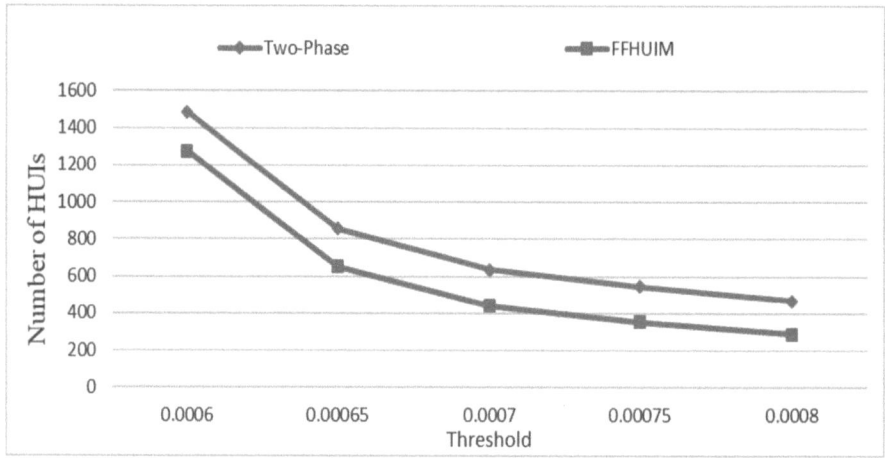

Fig. 5. Impact of threshold variations on the number of *HUIs* for Foodmart.

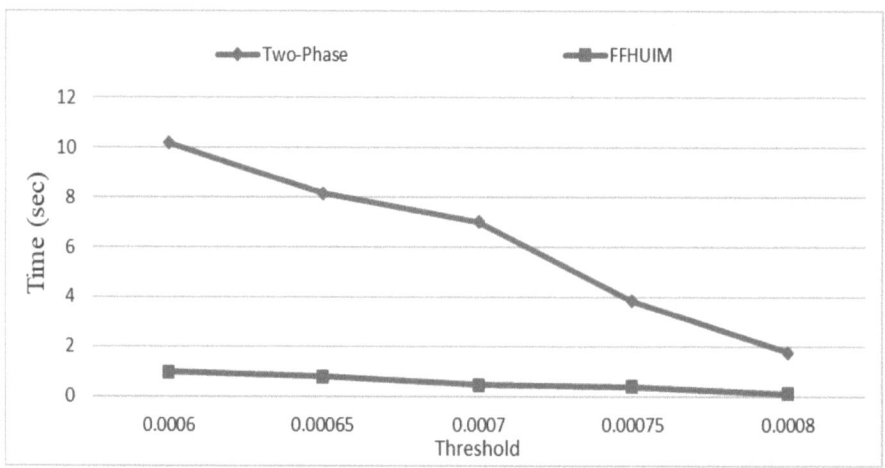

Fig. 6. Impact of threshold variations on execution time for Foodmart.

5 Conclusion and Future Work

This paper successfully demonstrates the comparative application of the traditional utility mining and FFHUIM algorithms, exploring their performance and effectiveness in decentralized data environments. The research highlights the advantages of the FFHUIM algorithm in terms of scalability, privacy preservation, and efficiency when applied to distributed datasets. Through various experimental setups and the analysis of real-world datasets, this paper demonstrates that the FFHUIM algorithm offers significant enhancements to the efficiency and effectiveness of utility mining in distributed data environments. Future research could explore integrating the concept of a pre-large [20] condition to reduce the *HUI* loss. This adjustment would allow for a more robust capture of itemsets that narrowly miss the high utility threshold during the server-side aggregation phase. Additionally, refining this mechanism could significantly enhance the adaptability and efficiency of the mining process, particularly in dynamic environments where data is constantly evolving.

Acknowledgements. This work was supported by the grant NSTC 112-2221-E-390-014-MY3, the National Science and Technology Council, Taiwan.

Disclosure of Interests. The authors have no competing interests to declare that are relevant to the content of this article.

References

1. Agrawal, R., Imieliński, T., Swami, A.: Mining association rules between sets of items in large databases. The 1993 ACM Sigmod International Conference on Management of Data, 207–216 (1993)
2. Agrawal, R., Srikant, R.: Fast algorithms for mining association rules. The 20th International Conference on Very Large Data Bases, 487–499 (1994)
3. Sushama, C., Kumar, M.S., Neelima, P.: Privacy and security issues in the future: A social media. Material Today: Proceedings (2021)
4. Voigt, P., Von Dem Bussche, A.: The EU General Data Protection Regulation (GDPR). Springer International Publishing **10**, 10–5555 (2017)
5. Liu, Y., Liao, W.K., Choudhary, A.N.: Two-phase algorithm for fast discovery of high-utility itemsets. The 9th Pacific-Asia Conference on Knowledge Discovery and Data Mining, 689–695 (2005)
6. Pardau, S.L.: The California consumer privacy act: Towards a European-style privacy regime in the United States. Technology Law and Policy **23**, 68 (2018)
7. Molina, C., Prados-Suarez, B., Martinez-Sanchez, B.: Federated mining of interesting association rules over EHRs. Applying the FAIR Principles to Accelerate Health Research in Europe in the Post COVID-19 Era, **287**, 3–7 (2021)
8. Wang, Z., Zhu, Y., Wang, D., Han, Z.: FedFPM: A unified federated analytics framework for collaborative frequent pattern mining. The 2022 IEEE Conference on Computer Communications, 61–70 (2022)
9. Chan, R., Yang, Q., Shen, Y.D.: Mining high-utility itemsets. The Third IEEE International Conference on Data Mining, 19–26 (2003)
10. Han, M., Gao, Z., Li, A., Liu, S., Mu, D.: An overview of high-utility itemsets mining methods based on intelligent optimization algorithms. Knowl. Inf. Syst. **64**, 2945–2984 (2022)
11. Fournier-Viger, P., Lin, J.C.W., Truong, T., Nkambou, R.: A survey of high-utility itemset mining. High-utility Pattern Mining. Stud. Big Data, **51**, 1–46 (2019)
12. Chen, Y., Gan, W., Wu, Y., Yu, P.S.: Privacy-preserving federated mining of frequent itemsets. Inf. Sci. **625**, 504–520 (2023)

13. El Ouadrhiri, A., Abdelhadi, A.: Differential privacy for deep and federated learning: A survey. IEEE Access **10**, 22359–22380 (2022)
14. Yu, B., Mao, W., Lv, Y., Zhang, C., Xie, Y.: A survey on federated learning in data mining. WIREs Data Min. Knowl. Discovery, **12**(1) (2022)
15. Wei, X.: A multi-modal heterogeneous data mining algorithm using federated learning. J. Eng., 458–466 (2021)
16. Cheng, Z., Fang, W., Shen, W., Lin, J.C.W., Yuan, B.: An efficient utility-list based high-utility itemset mining algorithm. Appl. Intell. **53**, 6992–7006 (2023)
17. Yan, Y., Niu, X., Zhang, Z., Fournier-Viger, P., Ye, L., Min, F.: Efficient high-utility itemset mining without the join operation. Inf. Sci. **681** (2024)
18. Chen, J., Yang, S., Ding, W., Li, P., Liu, A., Zhang, H., Li, T.: Incremental high average-utility itemset mining: Survey and challenges. Sci. Rep. **14**, Article no. 9924 (2024)
19. Kumar, R., Singh, K.: High-utility itemsets mining from transactional databases: A survey. Appl. Intell. **53**, 27655–27703 (2023)
20. Hong, T. P., Kuo, M. J., Chen, C. H., Li, K. S. M.: Federated erasable-itemset mining with quasi-erasable itemsets. The 16th Asian Conference on Intelligent Information and Database Systems (ACIIDS 2024), 299–307 (2024)
21. Hong, T. P., Chen, C. C., Chen, C. H., Li, K. S. M.: Privacy preserving based on SHA encryption and cluster analysis in federated frequent itemset mining. The 17th Asian Conference on Intelligent Information and Database Systems (ACIIDS 2025), 15683, 83–94 (2025)
22. Hong, T. P., Hsu, Y. P., Chen, C. H., Wu, M. T.: A federated mining framework for complete frequent itemsets. The 2023 IEEE International Conference on Systems, Man, and Cybernetics, 2483–2488 (2023)
23. Kairouz, P., et al.: Advances and open problems in federated learning. Foundations and Trends in Machine Learning **14**, 1–210 (2021)
24. Li, L., Fan, Y., Tse, M., Lin, K.Y.: A review of applications in federated learning. Comput. Ind. Eng. **149**, 106854 (2020)
25. Peng, A. X., Koh, Y. S., Riddle, P.: mHUIMiner: A fast high-utility itemset mining algorithm for sparse datasets. Pacific-Asia Conference on Knowledge Discovery and Data Mining, 196–207 (2017)
26. Qu, J. F., Liu, M., Fournier-Viger, P.: Efficient algorithms for high-utility itemset mining without candidate generation. High-utility Pattern Mining: Theory, Algorithms and Applications. Springer (2018)
27. Ryang, H., Yun, U.: Top-k high-utility pattern mining with effective threshold raising strategies. Knowl.-Based Syst. **76**, 109–126 (2015)
28. Ryang, H., Yun, U.: High-utility pattern mining over data streams with sliding window technique. Expert Syst. Appl. **57**, 214–231 (2016)
29. Ryang, H., Yun, U., Ryu, K.: Discovering high-utility itemsets with multiple minimum supports. Intelligent Data Analysis **18**(6), 1027–1047 (2014)
30. Shie, B.E., Yu, P.S., Tseng, V.S.: Efficient algorithms for mining maximal high-utility itemsets from data streams with different models. Expert Syst. Appl. **39**(17), 12947–12960 (2012)
31. Song, W., Huang, C.: Discovering high-utility itemsets based on the artificial bee colony algorithm. The 22nd Pacific-Asia Conference on Knowledge Discovery and Data Mining, 3–14 (2018)
32. Wu, J. M. T., Teng, Q., Huda, S., Chen, Y. C., Chen, C. M.: A privacy frequent itemsets mining framework for collaboration in IoT using federated learning. ACM Transactions on Sensor Networks, 18 (2022)
33. Luna, J. M., Fournier-Viger, P., Ventura, S.: Frequent itemset mining: A 25-year review. Data Min. Knowl. Discovery **9** (2019)
34. SPMF: An Open Source Data Mining Library, Available: https://www.philippe-fournier-viger.com/spmf/

The Study of Value Creation and AI Agent in Social Network: Governance and Ethical Risks Aspects

Chian-Hsueng Chao[✉], Pei-Chen Hsu, Xiu-Hua Wei, and Tian-Yi See

National University of Kaohsiung, Kaohsiung, Taiwan
cchao@nuk.edu.tw, {a1101224,a1113324,a1093328}@mail.nuk.edu.tw

Abstract. As artificial intelligence continues to grow in many areas, AI agents have emerged as key players in social media platforms. Through the capabilities, such as reasoning and cognitive decision-making process, the self-sufficient and autonomous agents are transforming the user-generated contents (UGC) into AI Generated Content (AIGC). However, those self-decision-making features challenge the user trust, safety, and engagement. Our observations show that users often feel both interest and concern toward those AI-driven features, especially in transparency and fairness issues. This research analyzes the balance of human versus AI agent control of decisions in value creation for social network. There are three primary concerns: user individual security, transparency of decision making, and fair data access, along the dimensions of governance, ethical and accountability. Our purpose is to propose an original, compliant, cross-jurisdictional AI agent governance model that offers policy guidance and pragmatic strategies for the AI-enabled value creation in the Social Network.

Keywords: AI Agent Governance · Autonomous Decision-Making · Ethics and Accountability

1 Introduction

1.1 Research Background

Artificial intelligence (AI) agents—autonomous computational entities capable of perceiving their environment, making decisions, and executing actions (Russell & Norvig, 2021)—are reshaping digital interactions.

The evolution of artificial intelligence (AI) has brought a paradigm shift in online communities, especially in the field of online social networks and media (OSNEM). AI is increasingly undertaking numerous communication works that were previously carried out by human, including post management, sentiment analysis, information filtering, and content moderation. By leveraging the natural language processing (NLP), recommendation system, cognitive decision-making process, and personalized services, the self-sufficient and autonomous agents are transforming the user-generated contents (UGC) into AI Generated Content (AIGC). The AI agent opens up a new era of community engagement and value creation.

1.2 Motivation of the Study

AI agents provide new possibilities for value creation in the social media context. With intelligent recommendations, real-time responses, and content creation, they are able to enhance users' involvement, enlighten insights, speed up knowledge sharing, and co-create value between users and platforms. For instance, an AI agent on an online forum can pre-categorize discussion points and greet new members proactively. This is a paradigm of value co-creation through interaction across various stakeholders.

However, the opportunities come with new challenges: algorithmic bias, privacy violations, the spread of disinformation, and blurred ethical lines. One must learn how AI agents impact the social network, regulation, and platforms operation. This study focuses on how AI agents promote co-creation, and how they influence social network. The balance between human and AI agent in the decision-making and value creation in social network. Issues, such as ethical and regulation/regulatory are also included to ensure risk control and maximum advantages when applying AI agent in social network. The purpose is to propose an original, compliant, cross-jurisdictional AI agent governance model that offers policy guidance and pragmatic strategies for the AI-enabled value creation in the Social Network.

2 Literature Review

2.1 The AI Agent

AI agents are computer programs that have the autonomy to sense their environment, make decisions, and take actions to accomplish assigned objectives with minimal human intervention (Russell & Norvig, 2021). This autonomy allows AI agents to operate in online societies and help managing complex multi-party interactions. The latest development of large language models (LLMs) and generative AI has expanded agents the abilities in community tasks, such as sentiment analysis, content moderation, recommendation, and real-time interaction facilitation (Zhang et al., 2024). Recent advancements have enabled AI agents to simulate human-like communication, generate relevant responses, and dynamically adjust behavior based on social feedback. For example, the AI-powered recommendation systems can map users' behavior and interests, and then suggests with customized content that enables greater participation and co-creation in online communities. Because artificial intelligence and autonomous AI-agent have gradually moved away from the manual intervention in decision-making, they can interact with us or handle things on our behalf, so people need value requirements such as ethics, responsibility, and law for the AI equipped autonomous machines.

2.2 The Value Co-Creation

According to Vargo and Lusch (2016), value is co-created through the interaction and integration of resources by multiple actors within a service ecosystem. The same goes for online communities, where value is not created through one-way content streaming but through interactive and ongoing exchange among multiple participants. AI agents are the new participants here—intermediaries and facilitators in the co-creation process. Their participation is on behalf of the Service-Dominant Logic understanding that value is not inherent in outputs but instead is actualized through shared, multi-actor interactions.

The Service-Dominant Logic (S-D Logic) sees "service" (not goods exchange) as the basis of all economic exchange. In their framework, value is created through two-way processes between firms, consumers, and stakeholders, who are seen as active resources. Unlike one-way delivery of goods, S-D Logic focuses on interactive "service-for-service" exchange. From a social value co-creation perspective, companies act as platforms that let stakeholders exchange knowledge, ability, and experience, so they co-create social and personal value. This idea supports customer trust and emotional bonds. S-D Logic concepts such as value-in-use, resource integration, and ecosystem-based interaction provide a foundation for understanding how AI agents can join value co-creation in digital social platforms.

Researchers tend to have broad agreement that value co-creation is not generic, it is a multi-level, context-dependent process influenced by interaction between multiple actors. Prahalad and Ramaswamy (2004) proposed the DART framework—Dialogue, Access, Risk Assessment, and Transparency—to stress the experiential and participative nature of co-creation. Normann and Ramirez (1993) added the concept of "value constellations," where value is created through redefining roles within networks, not linear supply chains.

For the public sphere, Osborne (2018) and Nambisan & Nambisan (2008) generalized co-creation into public service logic with emphasis on its ecosystemicness and positing active citizen roles as an explorer, ideator, designer, and diffuser. Digital platforms further facilitate this activity by allowing shared experiences and deep interactions between several actors.

In addition to the above philosophies, there are some aspects regarding to the AI co-creation practices which are summarized as follows:

1. Technical Aspects

    AI models use NLP, machine learning, and deep language models to control content moderation, recommendation, sentiment analysis, and analytics. AI models can perform numerous tasks with little human involvement. It is believed that while these models are efficient, there are still challenges in handling more nuanced or context-dependent cases.

2. Interaction Aspects

    AI agents use NLP, deep language models, and machine learning to perform content moderation, recommendations, sentiment analysis, and analytics. They support many operations with little human intervention.

3. Governance Aspects

    The growth of AI agents creates novel dynamics in content regulation and community governance. AI agents are capable of detecting and filtering harmful or inappropriate content (e.g., hate speech, violent images, fake news) automatically and continuously, greatly diminishing manual moderation. This feature enhances the safety of platforms and rule enforcement consistency, but it also challenges the ethical boundaries and accountability of AI-powered governance.

4. Critical Perspectives and Challenges

    While co-creation is the buzz, there was a warning by some scholars against the overuse of the term co-creation as a buzzword. Jansen and Pieters (2013) believe that successful co-creation entails active, intensive, and central user participation, as opposed to tokenistic participation. Ritual, routine practice, and inclusion are what will maintain co-creation, particularly in online contexts. These perspectives highlight the importance of websites taking action in an attempt to gain genuine, meaningful user involvement and pre-empt shallow or tokenistic co-creation. This study recognize that making co-creation truly meaningful is still a real challenge for many platforms.

Furthermore, the structural, procedural, and relational practices model proposed by Papagiannidis et al. (2025) further enrich the institutional design of AI governance in multi-stakeholder environments. The technical dimension of this concept is characterised by its emphasis on data and model management, interpretability, and information security. The purpose of these elements is to support the stable operation of AI on community platforms. The interactive dimension focuses on transparency in human-machine interactions, user autonomy, and ultimate human oversight. All these elements address challenges of trust and black-box opacity. And the governance dimension encompasses accountability, privacy, diversity, and the configuration of internal governance structures. These elements provide more robust institutional support.

The focus on relational practices can be conceptualised as a response to the S-D Logic's concept of resource integration through actor-to-actor networks. This suggests that when AI agents engage in community interactions, their governance behaviours must align with users' social role expectations and ethical trust. Therefore, it is erroneous to consider AI as merely a technical apparatus; it must also be regarded as an ethical agent within the value network that bearing the implications of institutional responsibility.

This study also found that the majority of current AI ethics and governance principles remain at a declarative level, lacking concrete implementation mechanisms. Furthermore, an overemphasis on participation and inclusion, although it can enhance trust that may also lead to decreased decision-making efficiency. This practice-oriented model assists in elucidating the ethical boundaries and governance pathways for AI agents' co-creative participation in community ecosystems, and thus facilitating a transition from principles to institutional action.

2.3 The Role of the AI Agent in Community Co-Creation

2.3.1 Practical Roles of AI Agents in Online Communities

Due to the rapid development of generative artificial intelligence and multi-agent systems, AI agents have transformed from passive tools into active and versatile participants in community co-creation. Drawing on the perspective of service-dominant logic, AI agents now take on multiple roles such as curation, collaboration, and motivation to facilitate the discovery of issues, the integration of resources, and the creation of new value within online communities, as summarized in Table 1.

Table 1. Roles and Contributions of AI Agents in Community Co-Creation.

Role	Real-world Practice/ Example	Observations/ Human Insights
Curator	Generative AI on platforms like Netflix creates tailored recommendations for users; AI tools on social forums help community managers spot trending topics or moderate content flow.	Personalized feeds boost convenience, but sometimes result in repeated suggestions and less variety. Many communities introduce human curation to highlight fresh or niche content.
Collaborator	AI agents routinely screen new posts, flag potential problems, and handle repetitive moderation tasks; Wikipedia and YouTube use bots for these roles, letting moderators focus on planning and creative work.	Automated screening increases efficiency, but challenging cases or subtle context often need people to make the final decision. Combining AI and human judgment usually leads to better outcomes.
Motivator	AI agents deliver personalized encouragement and recommend virtual badges or challenges based on user activity. Duolingo use these methods to keep users engaged and reward contributions.	Digital rewards are effective in the short term, but sustained engagement typically comes from meaningful interaction and authentic community feedback.

2.3.2 Ethical and Institutional Foundations of Co-Creation

If AI agents are to serve as value co-creators on community platforms, beyond their functional roles, they must also assume corresponding ethical responsibilities and institutional obligations. Floridi et al. (2021) proposed five core principles of AI ethics, including non-maleficence (harm avoidance), fairness and justice, respect for autonomy, explicability, and privacy protection that provide a philosophical foundation for AI behavior. UNESCO (2021), in its Recommendation on the Ethics of Artificial Intelligence, notes similarly that regulatory and evaluation mechanisms should be established by governments, replacing corporate self-regulation and thereby laying the groundwork for a global governance consensus. However, without transparent mechanism and institutional support, ethical principles will be difficult to implement. Diakopoulos (2020) points out that algorithms on community platforms do not merely perform tasks such as classification, ranking, and filtering; they also directly shape information architectures and values. If the decision-making process lacks the explainability and error-correction capabilities, information asymmetries and trust erosion will occur. To this end, he proposes five standards for information disclosure: explanations of human involvement, disclosure of data sources, articulation of model assumptions, indicators of inference error, and labeling of application scope. All of which help establish accountability mechanisms on platforms and a foundation for user understanding.

At the institutional level, Navdarashvili (2023) observes that current AI development is overly driven by corporate interests, leading to ethical washing and the marginalization of community values. He proposes a Multi-Stakeholder model as an alternative governance approach, arguing that users, developers, platforms, and the public sector should jointly participate in AI governance to achieve decision-making transparency and value alignment. In practice, this model emphasizes five key benefits: (1) fostering transparent dialogue between AI and the community, (2) ensuring users' rights to information and choice, (3) establishing ethical oversight mechanisms, (4) preventing power imbalances and breakdowns of trust, and (5) addressing demands for fairness and diversity. These design principles align with the relational orientation, interactive participation, and resource reconfiguration spirit advocated by S-D Logic. In summary, ethical principles, accountability mechanisms, and multi-stakeholder governance models laid the institutional foundation for AI agents' co-creative engagement in community ecosystems, which not only respond to users' value expectations but also help build a sustainable, trustworthy co-creation ecosystem.

3 Analysis

3.1 AI-Enabled Co-Creation Processes

3.1.1 Adaptive Workflow

The AI agents sense and react to user motives like information search, self-expression, and social reward, which facilitates accurate and adaptive engagement (Kaartemo & Helkkula, 2018). Calibration is continuous since user motives change over time along with community dynamics. In Fig. 1, There are three steps in the process: AI content creation, human editing, and community feedback. The circularity facilitates ongoing calibration to changing user motivations and social dynamics.

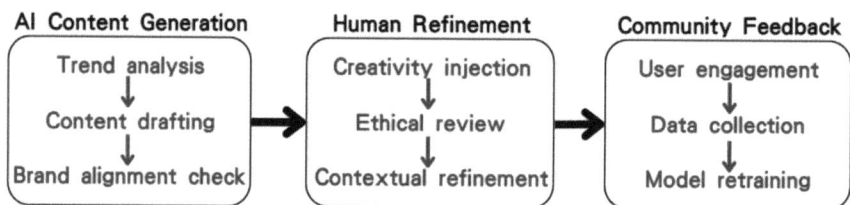

Fig. 1. Adaptive co-creation workflow

AI-driven personalization strives to balance broad relevance with individual preference, bundling similar topics for each user (Zhang et al., 2024). In Fig. 2, the four-step path from profiling to co-creation illustrates this process—profiling, adaptive bundling, targeted recommendation, and feedback-driven refinement. When personalization is pushed too far, users see less-varied content; continuous calibration is therefore required to preserve serendipity and discovery.

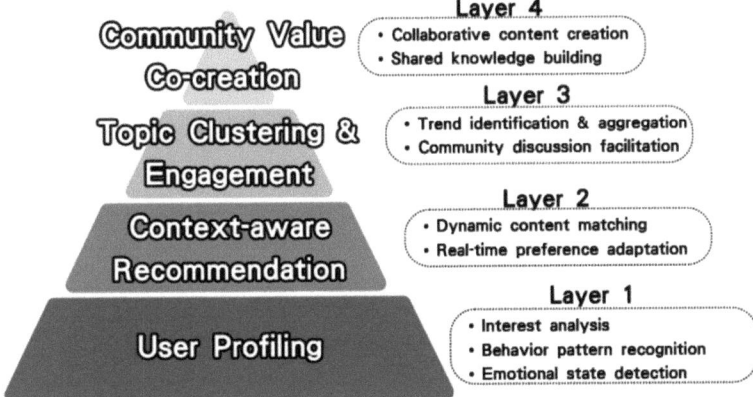

Fig.2. Four-step path from profiling to co-creation

3.1.2 Motivation Recognition

AI agents infer what drives community members, such as information seeking, self-expression, and social recognition that can be analyzing behaviour logs and sentiment signals (Kaartemo & Helkkula, 2018). In Fig. 3, the model iterates through three stages: (1) ingesting user data, (2) clustering motivational patterns with adaptive learning, and (3) updating its parameters with engagement feedback. This continuous calibration keeps motivational tactics aligned with the community's evolution, sustaining both involvement and brand loyalty.

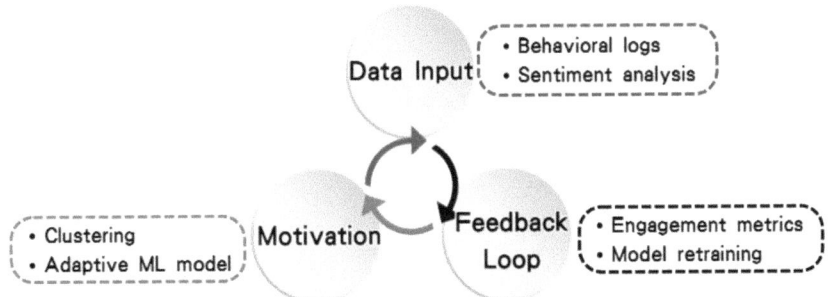

Fig. 3. Motivation-recognition loop

3.2 Thematic Perspective

From the thematic perspective, based on literature review, the AI agents support community co-creation can be grouped into four recurring themes, as summarized in Table 2. All the key findings are solid research foundation of this study.

Table 2. Thematic Areas of AI Agent Contribution to Community Co-Creation.

Thematic Area	Key Functions	Practical Examples / Implications
Personalization & User Experience	1. Tailors content and interactions to individual preferences (Saheb et al., 2024; Bassano et al., 2020)	Netflix's recommender system reduces churn and boosts satisfaction.
	2. Increases engagement and satisfaction by learning from user data (Vargo & Lusch, 2016)	
Efficiency & Process Automation	1. Automates content filtering, tagging, and support (Kupershtein et al., 2025; Nedungadi et al., 2025)	Meta's automated moderation handles millions of posts daily.
	2. Reduces labor costs and response times, enabling scalable moderation (Gupta, 2025)	
Content Creation & Optimization	1. Generates and optimizes content for engagement (Bail, 2024)	Coca-Cola uses AI for brand-consistent content generation.
	2. Enables real-time feedback and strategy adjustment; risk of content flooding if unmanaged	
Community Monitoring & Intelligence	1. Real-time monitoring of behavior and sentiment (Diakopoulos, 2025; Mansfield et al., 2025)	Twitter detects coordinated bot activity; Wikipedia flags disputes.
	2. Detects trends, issues, and supports proactive moderation (Jaiswal et al., 2024)	

3.3 The Governance Framework

3.3.1 Ethical Risks

These recurring co-creation patterns fit the service-dominant logic view that value emerges from fluid, multi-actor interactions. As AI agents shift from passive tools to active service partners, their growing role on social platforms also brings a new set of ethical and governance challenges. The risks are described as follows: AI-powered systems in online communities face five primary risks: (1) unconsent-ed data use violating user rights (GDPR, Art.6); (2) surveillance capitalism enabled by opaque data ecosystems; (3) algorithmic bias requiring routine audits; (4) misinformation challenges from generative content, and (5) lack of transparency, calling for explainable AI (XAI) and ethical oversight. Mitigating risk of AI agents associated with the development, deployment. The goal is to minimize potential harms and maximize the benefits of AI.

3.3.2 Principles and Governance Cycle

To ensure ethical and sustainable AI-supported co-creation, this framework uses four core principles: transparency, user participation, fairness, and norms-based alignment. These principles guide a five-phase governance cycle: information disclosure, user engagement, data collection, continuous adjustment, and iterative refinement. As shown in Fig. 4, this approach helps platforms stay accountable, inclusive, and responsive to changing community needs and ethical standards.

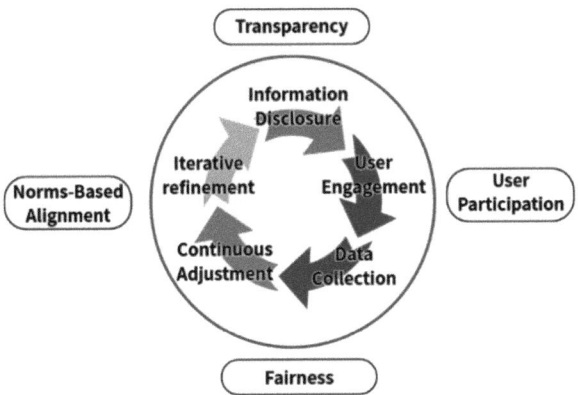

Fig. 4. Four principles and five-phase governance cycle

3.4 The Model of Governance

As shown in Fig. 5, this concentric governance model visually summarizes the framework developed in this study. At the center is sustainable co-creation, which aligns with the principles of service-dominant logic. This core objective is supported by three governance strategies: transparency and participation, inclusive incentive design, and clearly defined roles and responsibilities. These strategies are part of a broader structure built on two foundational conditions. The first is adaptive feedback, which keeps the system responsive to changes in community dynamics. The second is institutional trust, which gives the governance system legitimacy and encourages meaningful user engagement in AI-supported environments. Together, these elements create a coherent and practical model for ethical, inclusive, and sustainable AI governance.

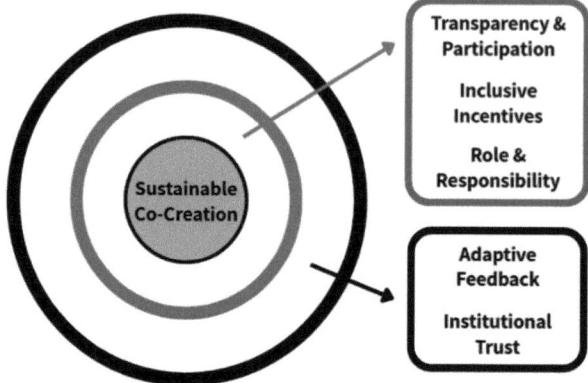

Fig. 5. Concentric governance framework for AI-enabled sustainable co-creation

3.5 Collaborative Moderation and Inclusive Incentive Design

To be more effective and inclusive in AI governance, this framework uses a multi-tier moderation system and recognition-based incentives. AI agents handle routine tasks, while human moderators and third-party reviewers deal with complex cases and ensure fairness (GPAI, 2023). At the same time, badges, ratings, and peer feedback help encourage participation. To prevent exclusion or shallow engagement, incentives should balance visibility with meaningful contribution (Jiao & Fan, 2019).

4 Conclusion

This study aimed to tackle the ethical, participatory, and structural issues raised by the growing role of AI agents in online communities. Using a governance framework based on service-dominant logic (S-D Logic), the study showed that sustainable co-creation can be supported through transparent decisions, inclusive design, and clear human-AI role division. Under the S-D Logic paradigm, AI agents are not just tools but active participants in the co-creation of service value within digital ecosystems.

As stated by Zawacki-Richter et al. (2022), while most countries emphasize the development of AI technologies and talent in policy documents, discussions related to ethical practices, transparency mechanisms, and user participation are extremely scarce. This is particularly true in areas like AI for communities and AI for public communication, where many decision-making processes fail to reflect users' real experiences and ethical concerns. As a result, a top-down governance model dominated by enterprises has emerged, which struggles to respond to the value differences and rights demands of actual community members. This governance gap highlights the lack of continuous, reciprocal, and equal interaction between designers and users, resulting in a shortage of ethical risk alerts and institutionalized co-creation. To address this issue, AI agents can be designed as mediators of participatory governance, enhancing the voices of various stakeholders through three key mechanisms: information transparency, participatory feedback, and ethical risk alerts.

First, AI agents can offer information disclosure and explanation functions by automatically providing their decision logic, data sources, and prediction basis. This helps users understand recommendation outcomes, ranking logic, and behavioral suggestions, thereby demystifying AI's black-box nature. Second, AI agents can record and analyze user feedback (such as inappropriate recommendation flags or disagreements), serving as triggers for bias detection and fine-tuning. They can also report community concerns to developers and governance bodies, informing future system redesigns. Third, AI agents can assist in identifying ethical risks, such as discriminatory results, unfair content promotion, or user marginalization, and encourage users to express concerns that can activate consensus-building or governance processes.

The study presented four key findings. (1) AI agents enable personalized interactions and foster community topic development. By leveraging behavioral data and sentiment analysis, they suggest relevant content, enhance engagement, and simplify information flow. (2) Collaborative governance improves the efficiency and fairness of content moderation. A multi-level review system—integrating AI pre-screening, user reporting, human review, and third-party verification—ensures both operational effectiveness and ethical discernment. Human-AI collaboration remains essential for managing sensitive cases. (3) Gamification and digital incentives promote co-creation and social recognition by supporting peer acknowledgment and user-led content management. This encourages community cohesion and activity. (4) New challenges under AI participation must be proactively addressed, particularly the development of stable human-AI collaboration models where human control is maintained while maximizing technological benefits. Future research should focus on enhancing human-AI collaboration to generate greater value across social networks.

References

Bail, C.A.: Can generative AI improve social science? Proc. Natl. Acad. Sci. **121**(21), e2314021121 (2024)

Bassano, C., Barile, S., Saviano, M.L., Cosimato, S., Pietronudo, M.C.: AI technologies and value co-creation in a luxury-goods context. In: Proceedings of the 53rd Hawaii International Conference on System Sciences (HICSS), 1–13 (2020)

Diakopoulos, N.: Prospective algorithmic accountability and the role of the news media. In: Noorman, M., Verdicchio, M. (eds.) Computer Ethics Across Disciplines, pp. 75–90. Springer (2025)

Donato, H.C., Farina, M.C., Donaire, D., dos Santos, I.C.: Value co-creation and social-network analysis on a network-engagement platform. Revista de Administração Mackenzie **18**(5), e20231805 (2017)

Diakopoulos, N.: Accountability in algorithmic decision making. Commun. ACM **59**(2), 56–62 (2016)

García-Huete, E., Ignacio-Cerrato, S., Pacios, D., Vázquez-Poletti, J.L., et al.: Generative AI, colour patterns and war-imagery diffusion in social media. Frontiers in Artificial Intelligence **7**, 127 (2025)

Global Partnership on Artificial Intelligence (GPAI).: GPAI Ministerial Declaration 2023. Approved at the GPAI Ministerial Council Meeting, New Delhi, 13 December 2023

Jaiswal, A., Shah, A., Harjadi, C., Windgassen, E., Washington, P.: Ethics of using social-media data to train AI for digital phenotyping. JMIR Formative Research **8**, e51814 (2024)

Jansen, S., Pieters, W.: The challenge of co-creation—A critical approach. In: Proceedings of the 21st European Conference on Information Systems (ECIS), 1–12 (2013)

Jiao, J., Fan, J.: Customer-enterprise social value co-creation: A literature review and prospects. Foreign Economics & Management **41**(2), 40–54 (2019)

Johansson, F.: All aboard the AI Express: An exploratory study on AI implementation for enhanced digital servitization from an S-D Logic perspective [Master's thesis, Jönköping University] (2023)

Kaartemo, V., Helkkula, A.: A systematic review of artificial intelligence and robots in value co-creation: Current status and future research avenues. Journal of Creating Value **4**(2), 211–228 (2018)

Kupershtein, L., Zalepa, O., Sorokolit, V., Prokopenko, S.: An AI-agent system for fact-checking with large language models. CEUR Workshop Proceedings **3646**, 30–41 (2025)

Mansfield, K.L., Ghai, S., Hakman, T., et al.: Digital harms in youth: A social-media–AI perspective. The Lancet Child & Adolescent Health **9**(3), 210–220 (2025)

Nedungadi, P., Veena, G., Tang, K.Y., et al.: AI techniques for social-media BERTopic modelling. IEEE Access **13**, 107406–107421 (2025)

Nambisan, S., Nambisan, P.: How to profit from a better virtual customer environment. MIT Sloan Manag. Rev. **49**(3), 53–61 (2008)

Normann, R., Ramírez, R.: From value chain to value constellation: Designing interactive strategy. Harv. Bus. Rev. **71**(4), 65–77 (1993)

Navdarashvili, T.: Multi-stakeholder model as a viable option for a pro-social AI governance. University of Milan (2023)

Osborne, S.P.: Co-production and the co-creation of value in public services. In: Brandsen, T., Steen, T., Verschuere, B. (eds.) Co-Production and Co-Creation: Engaging Citizens in Public Services, pp. 18–26. Routledge (2018)

Pan, L., Wang, C.Y., Zhou, F., Lü, L.: Social-media complexity in the generative-AI era. Nat. Sci. Rev. **12**(1), nwad302 (2025)

Prahalad, C.K., Ramaswamy, V.: Co-creating unique value with customers. Strategy & Leadership **32**(3), 4–9 (2004)

Papagiannidis, E., Vafeas, M., Bourlakis, M.: Responsible artificial intelligence governance: A review and research framework. J. Strateg. Inf. Syst. **34**(1), 101885 (2025)

Russell, S., Norvig, P.: Artificial intelligence: A modern approach, 4th edn. Pearson (2021)

Saheb, T.S., Sidaoui, M., Schmarzo, B.: AI–social-media convergence: A bibliometric review. Telematics and Informatics Reports **14**, 100035 (2024)

Sethupathi, G.M., Akshay, U., Surya, C.M., Nath, G.M., Venkat, S.: AI-driven social-media marketing. International Journal of Engineering and Advanced Technology **9**(1S), 54–58 (2019)

Vargo, S.L., Lusch, R.F.: Institutions and axioms: An extension and update of service-dominant logic. J. Acad. Mark. Sci. **44**(1), 5–23 (2016)

Wang, Y., Smith, A., Zhang, X.: Human-AI collaboration in online communities: Opportunities and challenges. Comput. Hum. Behav. **120**, 106765 (2022)

Zhang, X., Li, Y., Wang, S.: AI-powered recommendation systems in online communities: A review. Inf. Process. Manage. **61**(1), 102456 (2024)

Zawacki-Richter, O., Marín, V.I., Bond, M., Gouverneur, F.: Education for AI, not AI for education: The role of education and ethics in national AI policy strategies. Learn. Media Technol. **47**(1), 146–162 (2022)

Gupta, M.: AI agents in 2025: Transforming business, redefining leadership and ac-celerating digital transformation. Forbes Technology Council (2025, March 7). https://www.forbes.com/sites/forbestechcouncil/2025/03/07/ai-agents-in-2025-transforming-business-redefining-leadership-and-accelerating-digital-transformation/,last. Accessed 5 May 2025

SmythOS.: AI agent ethics: Understanding the ethical considerations (2024, February 15). https://smythos.com/ai-agents/ai-agent-development/ai-agent-ethics/,last. Accessed 5 May 2025

European Union: Digital Services Act (Regulation EU 2022/2065). Official Journal of the European Union (2022). https://eur-lex.europa.eu/eli/reg/2022/2065/oj/eng,last. Accessed 5 May 2025

Enhancing Sustainability Education Through Social Media Platforms

Dario Liberona(✉)

Seinajoki University of Applied Sciences, Seinajoki, Finland
Dario.liberona@seamk.fi

Abstract. Social networks have emerged as powerful platforms for communication, Universities are now educating the Z generation and in around five years they will start teaching the Alpha generation, the youngest and more digital generation ever and probably the most influenced by social networks. This article explores the potential of social media networks in promoting and teaching sustainability in Higher Education Institutions. It delves into various strategies, content formats, and engagement techniques that can effectively convey complex environmental and social concepts, inspire action, and foster a global community dedicated to a more sustainable future. Drawing upon real-world examples and academic research, this article provides a comprehensive guide for individuals, organizations, and educators seeking to harness the reach and interactivity of social networks for sustainability education and advocacy.

Keywords: Social media networks · Sustainability · Circular economy teaching · Social media content

1 Introduction

1.1 About Social Media Networks

Social media networks have become very important part of our daily lives in the XXI century. Today it is difficult to visualise the modern life without social media, specially for younger generations, if we consider the definitions and popularization of marketing concept of generations X, Y (millennials), Z and soon Apha (A.M.A.org 2022 and Seriac 2023), we can relate them to different educational generations in Higher Education and to Social Media (McCrindle 2018) we can see and increasing use of social media in younger generations (Table 1).

Social media has become an important tool for marketing campaign and practices, is an important part of promotion mix of any retail or massive customers' companies. Starting from basic means of exchanging information, it grew to development of means of communication whose purpose was manifold. With the advent of social media customers and businesses have become well connected with each other. During these past two decades, many social media platforms have evolved rapidly, they have become more sophisticated in terms of technology, more massive in terms of users, and with a higher influence on customers and general media.

During the past 25 years, we have witness how social media networks have evolved from niche online communities to ubiquitous platforms with important influence in global communication, culture, and commerce. The early 2000s to 2009 there was the rise of desktop-centric networking sites like Friendster and MySpace mostly in United States and not very global, with hundreds of thousands of users, mostly connecting existing friends with basic profiles, also during these years LinkedIn was launched. The latter half of that decade, was marked by the launch of Facebook, YouTube, and Twitter, initiated the shift towards broader public engagement and user-generated content, setting the stage for the mobile revolution (Sheikh, 2023). The 2010s were defined by the explosion of visual content (Instagram, Snapchat) and the mass adoption of smartphones, making social media truly mobile-first. Entering the 2020s, short-form video platforms like TikTok founded in 2016 ascended and we witness some changes, the growth of Linkedin, the switch form facebook to Meta and Twitter to X. Looking ahead to 2025–2030, the landscape is poised for deeper integration of AI for personalization and content creation, the maturation of social commerce, and the potential emergence of decentralized social platforms, all while navigating an ever-evolving regulatory environment. There are multiple new social media platforms being launched every year like Noplace, Partiful, Quest,

Table 1. Generations segmentation and social media (own elaboration base on multiple articles)

Generation	Approximate Birth Years	General Characteristics	Typical Usage of Social Media	Mostly Educated Higher Education (Years)	Working Status
Baby Boomers	1946 – 1964	- Grew up in a post-WWII era of economic prosperity and social change.	- Late Adopters/Minimal Use for Many: Many are not active social media users.	1965 - 1984	Mostley Retired
		- Influenced by Civil Rights Movement, Vietnam War, counter-culture.	- Facebook Dominance: If they use social media, Facebook is by far the most popular platform, often used to connect with family and friends.	Early computers, punching cards Main Frames	
		- Optimistic, focused on career, strong work ethic, value loyalty.	- Information Sharing: Primarily use it for sharing family updates, photos, news articles (often without critical vetting).	No Internet	
		- Respect for authority, less comfortable with rapid technological change.	- Less on newer platforms: Rarely found on platforms like TikTok or Snapchat, and less active on X (Twitter) or Instagram compared to younger generations.	No social Media networks	
Generation X	1965 – 1980	- "Latchkey Kids": Often self-sufficient, resourceful, and independent due to less parental supervision.	- Early Adopters, Moderate Use: Many were early adopters of the internet and social media (e.g., MySpace, early Facebook).	1985 - 1999	Senior Professionals
		- Influenced by fall of Berlin Wall, rise of MTV, advent of personal computing.	- Diverse Platform Use: Use a mix of Facebook, Instagram, LinkedIn, and X (Twitter). Facebook remains strong for personal connections.	PCs, Computer Stations	Entrepreneurs
		- Skeptical, pragmatic, adaptable, value work-life balance.	- Information & Networking: Use it for news, staying connected with friends, professional networking (LinkedIn), and following specific interests.	Beginnings of Internet	Soon start retiring
		- Bridging the gap between analog and digital worlds.	- Less time on video-heavy platforms: Generally less engaged with short-form video content like TikTok, though many consume YouTube.	No social Media networks	
Millennials	1981 – 1996	- "Echo Boomers" or "Gen Y".	- Digital Natives (Early): Grew up with the internet, email, and early social media (MySpace, Facebook from college).	2000-2014	Fully labor integrated
		- Highly educated, technologically fluent, civic-minded, often optimistic, and team-oriented.	- Ubiquitous Usage: Active across most major platforms. Instagram and YouTube are very popular. Many still use Facebook for groups and personal connections.	Internet and E-commerce	Young and mid level professionals
		- Influenced by 9/11, economic recession (2008), rapid globalization.	- Curated Self-Expression & Information: Use social media for self-expression, sharing life updates, consuming news, discovering trends, and connecting with brands.	First Mobile generation	Young entrepreneurs
		- Seek meaning in work, value collaboration, often more diverse and inclusive.	- Influencer Culture: Many engage with influencer content and are adept at creating their own online personas.	Early Social Media Adopters	
Generation Z	1997 – 2012	- "Digital Natives (True)": Grown up entirely with smartphones, social media, and constant internet connectivity.	- Dominance of Visual & Short-Form Video: TikTok, Instagram, YouTube, and Snapchat are primary platforms. Less reliance on Facebook for personal use.	2015-2029	The younger workers 22-28 years old
		- Pragmatic, fiscally conservative (influenced by recessions), socially aware, value diversity and authenticity.	- Authenticity & Community: Value authenticity, transparency, and often seek out smaller, niche communities.	Social Media inmersion	Many still studying in Universities
		- Entrepreneurial mindset, global perspective, concerned about mental health and climate change.	- Information & Entertainment Hybrid: Use social media for both entertainment (short videos, memes) and quick information consumption. Often use it as a primary search engine (e.g., TikTok for recipes, reviews).	First Digital Natives	
		- Multitasking, highly adaptable to new technologies.	- Direct Messaging Focus: Heavy use of direct messaging and group chats for private communication.		Currently mostly in Universities
Generation Alpha	2013 – 2026 (approx.)	- The first generation born entirely in the 21st century.	- Ultra-Digital Natives: Growing up with tablets, smart speakers, and AI as commonplace.	2030-2044	
		- Often children of Millennials, influenced by global events like the COVID-19 pandemic.	- Early Exposure & Entertainment-Driven: Many have early exposure to curated digital content (e.g., YouTube Kids, educational apps).	Full Computer, Social Media, Mobile generation	
		- Highly tech-literate from a very young age, potentially more comfortable with virtual realities and AI.	- Gaming & Interactive Platforms: Heavy use of gaming platforms with social elements (e.g., Roblox, Minecraft) and video-sharing platforms.	First AI and multimedia natives	
		- Expected to be the most formally educated, globally connected, and technologically immersed generation.	- Parental Gatekeeping: Social media use is often heavily monitored and curated by parents in their younger years. Formal social media accounts will emerge as they age, likely on platforms yet to fully define themselves. Their online identities will be shaped by highly personalized algorithms and immersive digital experiences.	5 years to start entering Higher Education, next college generation	

Supernova but not has massive has the previous platforms (Korenich, Lascu, Manrai, 2013).

The size and resources of social media have increased over the years, and there are plenty of new social media platforms, however some of the initial social media platforms have faded away, with Skype being the last one being shut down in May 2025. Social media networks play an important role in our lives nowadays, in Table 2 we can review the social media platforms with more users, Facebook still the most used ones, and there are other preferences of younger users like Instagram or TikTok, they encompass many interactions and communications among their users, but also have a big impact on business and the way they related to its customers, organizations and brands. There are plenty number of communications between customers and companies through them becoming an open channel of interactions.

Business can be promoted through various social networking sites. Many of the organization promotes their business by giving advertisement on the social media and by acquiring valuable information from customers and their habits and preferences. Customers can connect and interact with business on a more personal level by using social media. If an organization has established a brand, social media may help this organization to develop the existing brand and give the business a voice and an open communication channel (Siddiqui, Singh, 2016).

Social media also has impact on opinions and dissemination of information in the world. There are multiple studies regarding these interactions and mediatic impact at different levels, for example candidates can use social media as a wide-reaching mobilization tool, and you can segmentate different platforms for different voters ages. Often social media has been related to greater freedom of speech and of liberty of expression avoiding multiple censorship or barriers, this has opened an access to a variety of news, knowledge, facts, events, which would not have been possible without social media. However, there are also some negative impacts on these dynamics, different algorithms filter some information and the interactions of the social media networks, fake news could be spread very rapidly, and often there are not always reliable sources of information and data being transmitted through the networks (Law and Internet, 2022 and Carnivale, 2024).

Another reason to consider social media networks tools in education is that they are still evolving, and now with the support or use of multiple technological advancements, artificial intelligence, specific tools, and also shifting user behaviours and the second generation of digital natives (generation alpha), would support it´s further use and development, some emerging trends are new uses of video and immersive experiences like short-form video recently promoted by platforms like TikTok, platforms are integrating augmented reality (AR) and virtual reality (VR), allowing users to interact in 3D environments, attend virtual events (concerts, conferences), and even engage in virtual workspaces. AI-driven analytics helping to personalize these interactions (Logic Digital, 2025).

AI in Content Creation and personalization, AI-powered tools are increasingly capable of generating text, images, and even short-form videos, streamlining content creation for marketers and creators. Algorithms, enhanced by AI, creating a more tailored user experience. There is also a shift towards private communities and micro-communities, recently public spaces have become to crowed and users are increasingly gravitating towards more intimate, curated spaces (Garling, 2025). Social media platforms are evolving into powerful search engines. Users are increasingly turning to these platforms not just for connection and entertainment, but also to discover products, research brands, and find information, this could be very useful in education (Logic Digital, 2025) (Table 3).

Table 2. Use of Social Media networks as February 2025 (Statista Statistics 2025).

Social Media (SM) network	Users (in millions)
Facebook	3 070
YouTube	2 530
Instagram	2 000
WhatsApp[1]	2 000
TikTok	1 590
WeChat	1 380
Telegram	950
Facebook Messenger[2]	947
Snapchat	850
Douyin	766
Kuaishou	714
Reddit	606
Weibo	599
X	586
QQ	562
Pinterest	537

Table 3. Social media networks (most popular, own elaboration)

Social Media Network	Description	Founded Year	Active Users (Millions)	Market Capitalization Valuation (approx. Billion USD)
Facebook (Meta Platforms)	A global social networking site connecting friends and family, sharing content, and joining groups.	2004	3074	app. 1200
YouTube (Google)	A video-sharing platform where users can upload, view, rate, share, and comment on videos.	2005	2900	app. $2100 (Alphabet Inc., Google's parent)
WhatsApp (Meta Platforms)	A free, cross-platform messaging application that allows users to send text and voice messages, make voice and video calls, and share images, documents, user locations, and other media.	2009	2800	app. $1200 (Meta Platforms)
Instagram (Meta Platforms)	A photo and video-sharing social networking service. Users can upload photos and videos, apply digital filters, and organize them with tags and location information.	2010	2090	app. $1200 (Meta Platforms)
TikTok (ByteDance)	A short-form video hosting service where users can create and share short videos on any topic.	2016 (International version)	1800	Privately held (ByteDance valuation ~$225 Billion)
Facebook Messenger (Meta Platforms)	A messaging app and platform. Originally developed as Facebook Chat in 2008, the company revamped its messaging service in 2010, and subsequently released standalone iOS and Android apps in August 2011.	2011	1300	~$1200 (Meta Platforms)
WeChat (Tencent)	A Chinese multi-purpose messaging, social media, and mobile payment app.	2011	1300	app. $450 Billion (Tencent Holdings)
LinkedIn (Microsoft)	A business and employment-oriented social networking service designed for professional networking and career development.	2003	1000	app. $3.1 Trillion (Microsoft, LinkedIn's parent)
Telegram (Telegram Messenger LLP)	A cloud-based instant messaging service with a focus on security and speed, allowing users to send messages, photos, videos, and documents.	2013	900	Privately held (estimated valuation varies widely)
Reddit (Reddit Inc.)	A network of communities based on users' interests, where people can submit content, vote on it, and engage in discussions.	2005	850	~$18 Billion (May 2025)
Kuaishou (Kuaishou Technology)	A Chinese short-video sharing and live-streaming platform, similar to Douyin.	2011	712	~$20 Billion (Kuaishou Technology)
Douyin (ByteDance)	The original Chinese version of TikTok, offering similar short-form video content and live streaming.	2016	700	Privately held (ByteDance valuation ~$225 Billion)
QQ (Tencent)	An instant messaging software service and web portal developed by Tencent. It also offers a range of services, including online social games, music, shopping, microblogging, and group and voice chat.	1999	590	~$450 Billion (Tencent Holdings)
Sina Weibo (Weibo Corporation)	A Chinese microblogging website. It's often compared to Twitter and is a major social media platform in China.	2009	582	~$2 Billion (Weibo Corporation)
Snapchat (Snap Inc.)	A multimedia messaging app used for sending photos and videos (Snaps) that disappear after a short time.	2011	560	~$25 Billion (Snap Inc.)
Pinterest (Pinterest, Inc.)	A visual discovery engine for finding ideas like recipes, home and style inspiration, and more. Users can save and organize images (Pins) onto boards.	2009	553	~$25 Billion (Pinterest, Inc.)
X (formerly Twitter) (X Corp.)	A microblogging and social networking service on which users post and interact with messages called "tweets".	2006	550	Privately held (X Corp. acquired by Elon Musk, estimated valuation reduced from acquisition price)
Quora (Quora Inc.)	A question-and-answer website where questions are asked, answered, edited, and organized by its community of users.	2009	400	Privately held (last reported valuation ~$2 Billion)
Twitch (Amazon)	A live streaming video platform focusing on video game live streaming, including broadcasts of e-sports competitions, in addition to offering music broadcasts, creative content, and "in real life" streams.	2011	140	~$1.8 Trillion (Amazon, Twitch's parent)
Skype (Microsoft)	A telecommunications application specializing in providing video chat and voice calls between computers, tablets, mobile devices, and smartwatches via the internet.	2003	100	Will be retired on May 2025, replaced by Teams functionality

2 Social Media in Higher Education

Higher Education institutions are aware of the influence of social media in younger generations, facebook platform was founded inside a University in 2004, and for almost two years was only for higher education students, then it opened for everyone older than

13 years old, by 2015 there were already more than 2 billion active users of social media (Ortiz-Espina, 2019), during the period 2015 and 2017 social media experienced a big growth only in 2017 had a 21% increase in active users globally compared to the previous year, this period saw the emergence of the "network society" and the profound impact of information technology on social structures, communication, and global interactions, transitioning from community sites to data mining corporations (Kaun and Van Dick 2014, Kirkpatrick 2011).

Students, faculty, and other members of the campus community may benefit from the usage of social media in higher education, according to current study. Social media sites such as Facebook, Twitter, and Instagram have advantages over more traditional means of communication because most students already know how to utilize these tools.

It is up to the professors if social media can be used in classrooms or not, but most students from generation X and Z are using social networking site to engage with each other outside of class regardless of whether their teachers utilize or not social media. WhatsApp is another important tool for students in their communications practices in universities, for coordination and sharing information. Due to the double nature of structure, social media not only serves as a medium for social connection, but it also contributes to the foundation of norms and resources that may be used to influence the replication of social media-mediated interaction. Researchers have shown that there are disparities between men and women when it comes to using social media. Part of the Z generations of students is still in primary and secondary school (13 to 18 years old students) and are also using social media on their academic work, surveyed secondary school principals in the United States found that 62% of them wanted their teachers to use social media more in the classroom (Dodson 2020).

Higher education students are highly adept at leveraging social media for a wide array of purposes, social connections in personal life, accessing and sharing their studies information, groups collaboration, sharing material in google docs, using it to find information for surveys and assignments, communicating with educators, informal learning, looking for jobs or internships, showcase their academic achievements, personal branding, exploring career opportunities, e-learning, discussions, among others. Educators and institutions are increasingly recognizing these phenomena and looking for ways to harness social media's potential for learning and engagement, while also addressing some of its associated challenges such has cyberbullying, misinformation and some others unhealthy practices (Ellis 2024).

2.2. Sustainability in Higher Education.

3 Enhancing Sustainability and Circular Economy Education with Social Networks

3.1 The Use of Social Networks to Promote Sustainability

Some examples of using social media campaigns to enhance sustainability and promote sustainability and SDG adoption in universities, was the University of British Columbia (UBC) that implemented a social media campaign supporting its "UBC Reads Sustainability" initiative, which has been ongoing since 2011 and features different guest authors. The campaign utilized social media to broaden the reach of its book club and lecture series into a university-wide discussion platform. By selecting books with themes such as climate change, Indigenous land rights, and other sustainability-related topics, UBC used its social media channels to post questions and excerpts, encouraging participation from students, faculty, and staff. This strategy shifted the event from a traditional academic format to an interactive online community, aiming to facilitate engagement with environmental and social issues. (Communityengagement 2024).

The "Cool Campus Challenge" at UC Berkeley, on the other hand, is a specific competition that happens periodically. The most prominent example was the system-wide challenge in 2019, which ran from April 1 to April 26. Berkeley university also participated in a similar challenge in 2023. This campaign-style approach uses a limited time frame to generate a sense of urgency and competition, motivating a surge of participation. Social media played a vital role in the success of the "Cool Campus Challenge" by providing a platform for real-time engagement and community-building. Instead of simply announcing the challenge, Berkeley used platforms like Instagram and Facebook to actively involve students, faculty, and staff. They created a dedicated hashtag, #CoolCampusChallenge, which allowed participants to share their sustainable actions—from choosing reusable bottles to biking instead of driving—and see what others were doing. (UC Berkeley 2019).

4 Findings and Discussion

A general study was conducted among MBA students, at a German University. A total of 78 answers were collected among first year MBA students ranging from 22 to 40 years old. Many of the respondents where female students (55,13%). So mostly generation Z (85%) and some generation Y (around 15%) (Figs. 1 and 2).

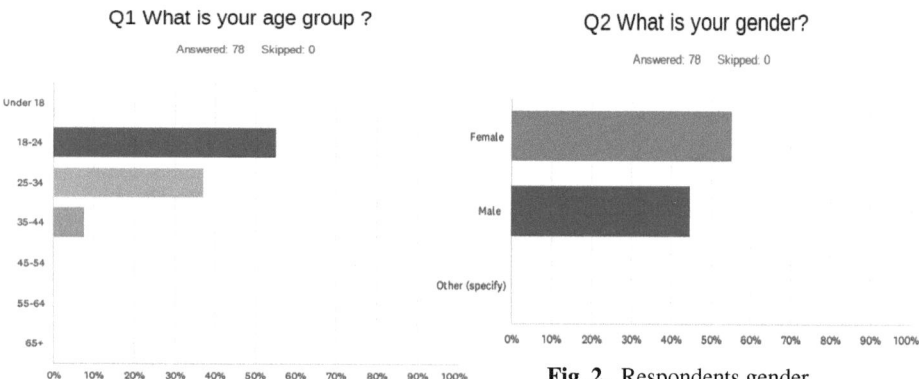

Fig. 1. Students age group

Fig. 2. Respondents gender

The most used social media for the students were WhatsApp and in second place Instagram followed very close by YouTube, in comparison with Table 1 (most popular social network platforms), students use more WhatsApp and Instagram rather than

Facebook and YouTube that are the most popular networks in February 2025. In the case of LinkedIn, it is very important to students, and the 4th most used SM network, but LinkedIn in popularity by active monthly users is around 300 million, not in the top 15 of the most used networks (Table 2.) (Fig. 3).

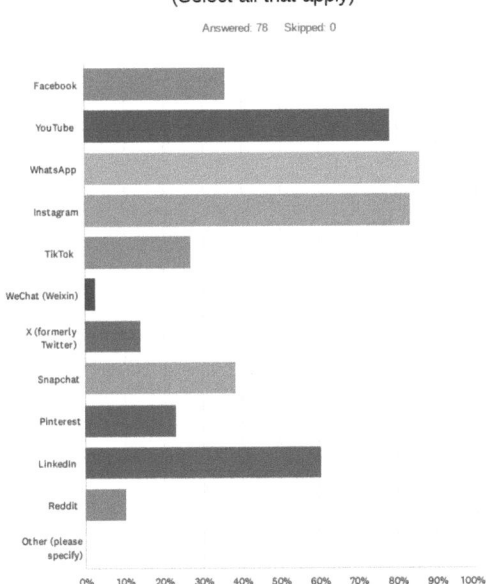

Fig. 3. Most used social media network for the students

Regarding the amount of time spend in social media (Fig. 4), our sample of university students spend a little more time than the average active users, that is approximately 2 hours and 32 minutes, versus 2 hours and 23 minutes per day (Soax.com) in 2024 (that was 5,3% less than 2023 according to Soax.com)

In the case of the most useful social media network to promote sustainability and circular economy (Fig. 5) they mention Instagram, YouTube and TikTok, which are also some of the networks that they mostly use. They do not consider influencer has been very effective on communicating sustainability information and practices.

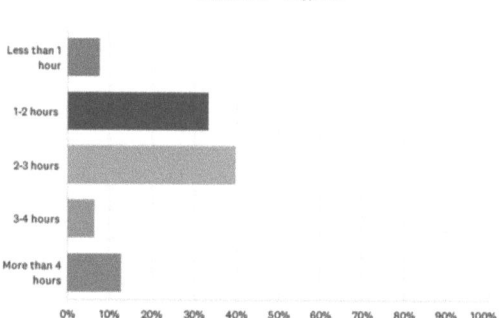

Fig. 4. Daily use of social media networks

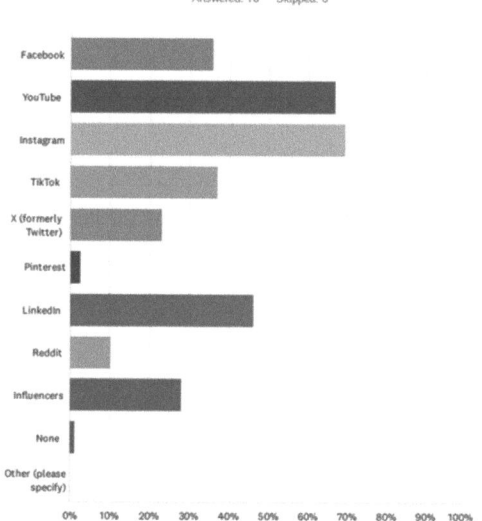

Fig. 5. Best social media network to promote Sustainability and circular economy

Regarding the most efficient content format for promoting and teaching sustainability photos and video are the most recommended formats (90%), followed by practical tips and actionable advice (65%) and personal stories and examples (62%), underlaying the importance that students give to acting upon sustainability issues rather than just getting information about it.

In Table 5 is stated the high importance that higher education students give to the use of social media to promote sustainable practices and educational content and its easiness and effectiveness (86% agreement), however there are trust objections to the information shared in social media with 30% of the respondents not trusting these sources. 25% of respondents also declare that is not enough content related to sustainability in social media nowadays. Social media has been effective in developing awareness through social media (59%) even that still has room to increase the related content (Table 4).

Table 4. What is the best social media content format to promote Sustainability

ANSWER CHOICES	RESPONSES	
Visually appealing content (photos, videos)	89.74%	70
Personal stories and examples	62.82%	49
Practical tips and actionable advice	65.38%	51
Information about sustainable products and services	42.31%	33
Calls to action (e.g., signing petitions, participating in events)	29.49%	23
Humorous or engaging content	39.74%	31
Content from trusted sources and experts	42.31%	33
Interactive content (polls, quizzes)	26.92%	21
Other (please specify)	3.85%	3
Total Respondents: 78		

Table 5. Perceptions about the usefulness of using SM networks to promote sustainability and Circular Economy.

	STRONGLY DISAGREE	DISAGREE	NEITHER AGREE OR DISAGREE	AGREE	STRONGLY AGREED	TOTAL	WEIGHTED AVERAGE
Social media makes it easy to learn about sustainability.	2.56% 2	1.28% 1	10.26% 8	58.97% 46	26.92% 21	78	4.06
Social media makes it easy to learn about Circular Economy.	0.00% 0	6.49% 5	15.58% 12	59.74% 46	18.18% 14	77	3.90
The information I find on social media about these topics is easy to understand.	1.28% 1	2.56% 2	17.95% 14	57.69% 45	20.51% 16	78	3.94
Social media will be good tool to teach about circular economy and sustainability to younger generations	1.28% 1	3.85% 3	8.97% 7	42.31% 33	43.59% 34	78	4.23
I trust the information I find on social media about these topics	2.56% 2	26.92% 21	50.00% 39	20.51% 16	0.00% 0	78	2.88
Social media has increased my awareness of sustainability issues	2.56% 2	14.10% 11	24.36% 19	47.44% 37	11.54% 9	78	3.51
I have learned practical tips for participating in the circular economy through social media.	0.00% 0	23.08% 18	29.49% 23	42.31% 33	5.13% 4	78	3.29
Sustainability or Circular economy are not really present in Social Media	5.13% 4	30.77% 24	38.46% 30	21.79% 17	3.85% 3	78	2.88

Social media also has an important influence on the purchasing decisions related to sustainable products or services, 48% of the respondents declared that they are influences, even though 51% declared that has not been exposed to sustainability content, and 79% declare to be extremely or very interested in sustainability. In general there is evidence of social media effects on purchasing decisions, over the past decade, the effects of getting advice and trusting peer members of different social networks has been establish by several studies, and this selling and customer relationship management tool has moved from large brands to smaller organizations and everyday decision making (Putri, Rizan, 2022 and Iblasi, Bader, Al-Qurini, 2016) (Fig. 6).

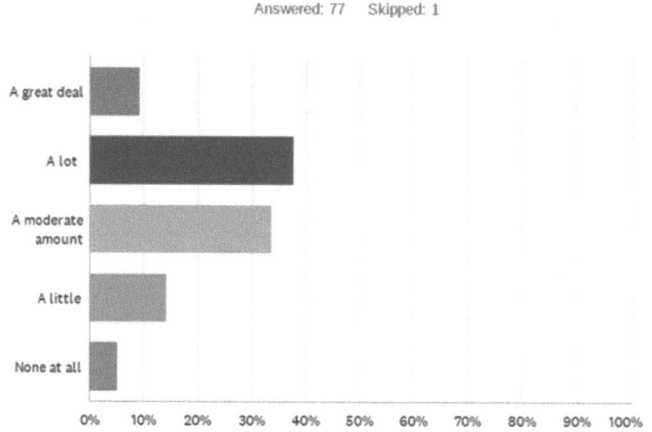

Fig. 6. Social media influence on purchasing decisions

4.1 Open Question and General Higher Education Students' Sentiments Towards the Use of Social Media Networks.

The survey requested one open question "Do you have any other thoughts or comments about the role of social media in sustainability and circular economy?", usually students do not respond to these open questions, in this case 68% of the participants did and contributed to the discussion.

Based on the survey responses from young university students, a clear and nuanced picture emerges regarding the role of social media in promoting sustainability and circular economy learning.

Young university students overwhelmingly believe that social media holds immense and growing potential as a powerful tool for educating, teaching, and promoting sustainability and circular economy learning. They acknowledge its reach and ability to spread awareness rapidly, particularly among their generations who are highly engaged with these platforms. Many see social media as a crucial complement to traditional education, even surpassing it in certain aspects for reaching and influencing younger demographics.

There's a strong consensus that social media can foster behavioral change by highlighting sustainable trends, showcasing real-life examples, and promoting eco-friendly brands and practices. The visual nature of social media, with pictures and videos having more impact than text, is frequently emphasized as key to engaging content.

However, the students also express significant concerns and critical perspectives. A recurring theme is the issue of misinformation and the difficulty in verifying sources. Many express a lack of trust in social media content and a need for more accurate, authentic, and well-sourced information. They highlight the challenge of algorithm-driven feeds that may limit exposure to sustainability content for those not already interested.

Students also point out the need for more "quick deep message ads" and engaging, bite-sized content to capture attention in a fast-paced environment. They call for more consistent and focused efforts from social media platforms and content creators to prioritize sustainability topics, moving beyond just funny clips or general influencer content. The desire for actionable information and a sense of collective responsibility ("If we all contribute with a little effort…") is also evident.

5 Conclusions

Promoting sustainability in universities is an important element of modern university teaching reflecting the social challenges of today's society, professionals should be aware of environmental challenges and solutions in the world (Sanchez and Cadarso 2021).

Social media could be a powerful and increasingly indispensable tool for enhancing and promoting sustainability and circular economy education. Its widespread reach, interactive nature, and capacity for diverse content formats offer unique advantages over traditional educational approaches.

Social media platforms are highly effective in disseminating information about environmental issues and sustainable practices to a vast and diverse audience, including young people who are particularly engaged with these platforms. Students declare a big importance and interest in sustainability and circular economy and good opportunities to use social media platforms to develop related content, almost half of them declare currently little or no content in social media regarding sustainability. They value the potential of social media through features like hashtags, content sharing, videos, and targeted advertising, campaigns can rapidly spread awareness of critical topics like climate change, pollution, and conservation and specially the ability of promoting sustainable practices in young audiences.

Beyond simply informing, social media can actively engage users and inspire pro-environmental behaviors. Engaging content formats, such as infographics, videos, and interactive challenges (e.g., #TrashTag), can capture attention, evoke emotions, and motivate higher education students to learn more and act.

Social media facilitates value co-creation, where users actively contribute to and shape the conversation around sustainability. This interaction between users and content creators, including environmental organizations and academics, can significantly influence individual attitudes and behaviors towards learning and practicing a sustainable green lifestyle. Studies indicate a positive association between social media engagement and the adoption of sustainable practices, suggesting that it can be a powerful trigger and support tool for behavioral change. Social media provides them with a platform to access resources, join movements, and amplify their voices, making it a critical area for sustainability education and intervention.

There's a need for sustainability leaders and educators to better understand how to leverage social media beyond mere information dissemination, focusing on building community, fostering deeper discussions, and inspiring consistent pro-environmental actions. This includes creating more attractive and engaging content, ensuring accuracy and credibility, and encouraging active participation rather than passive consumption.

Higher university students view social media as a very powerful platform with immense potential for good in sustainability education but also are aware that requires greater vigilance regarding content quality, a stronger focus on educational impact over mere entertainment, and a proactive approach to combating misinformation to improve its impact in young generations.

The full potential of using Social Media platforms to enhance Sustainability and Circular economy education in higher education institutions has not yet been totally use, and the future social media technological possibilities such as AI support, virtual reality functionality, immersive experiences, content creation and others new developments will further improve the educational role of social media in teaching, promoting and creating sustainability and responsible awareness in Higher Education.

References

AMA.org: Generational Insights and the Speed of Change. American Marketing Association June 30, 2002 (2022). https://www.ama.org/marketing-news/generational-insights-and-the-speed-of-change/. Accessed 22 March 2025

Carnivale, A.: The impact of social media on news and politics. Westermass news. May 3, 2024 (2024). https://www.westernmassnews.com/2024/05/02/impact-social-media-news-politics/. Accessed 1 Apr 2025

Communityengament: Community Engagement's 2024 Annual Report (2024). The University of British Columbia, Vancouver campus. https://communityengagement.ubc.ca/news/2024-annual-report/. Accessed 3 May 2025

Dodson, R.: An Analysis of Public School Principals' Perceptions of Social Media, Computer and Smart Phone Use in Schools in Eight U.S. States. Scholarly Journal. Educ. Res. Quarterly **44**(1), (Sep 2020), 3–34 (2020)

Ellis, B.: Understanding the Social Media Habits of College Students. On campus nation (oncampusnation.com), published on Sep 16, 2024 (2024). https://oncampusnation.com/understanding-the-social-media-habits-of-college-students/. Accessed 1 Apr 2025

Garling, B.: The Biggest Social Media Trends Shaping 2025. Forbes.com, February 2, 2025 (2025). https://www.forbes.com/councils/forbescommunicationscouncil/2025/02/03/the-biggest-social-media-trends-shaping-2025/. Accessed 5 Apr 2025

Kaun, A., Van Dick, J.: Culture and Connectivity: A critical history of Social Media. MedieKultur Journal of media and communication research, Denmark, ISSN 1901–9726 (2014)

Kirkpatrick, D.: The facebook effect. The Inside Story of the Company That is Connecting the World. New York: Simon & Schuster." (2010) (2011). https://www.scribd.com/doc/44482451/The-Facebook-Effect-by-David-Kirkpatrick. Accessed 2 Apr

Korenich, D., Lascu, L.A., Manrai Ajay, K.: Social Media: Past, Present and Future. Research Gate February 2013 (2013). https://www.researchgate.net/publication/259528201_Social_Media_Past_Present_and_Future/citations. Accessed 24 March 2025

Iblasi, W.N., Bader, D.M.K., Al-Qurini, S.:The Impact of Social Media as a Marketing Tool on Purchasing Decisions. Int. J. Manag. Stud. Res. **4**(1), 2016, 14–28 (2016)

Law and Internet: The negative impact of social media on politics. Netlaw Blog 22.07.2022 (2022). https://www.netlaw.bg/en/a/the-negative-impact-of-social-media-on-politics. Accessed 1 Apr 2025

Logic Digital: The Future of Social Media Marketing: 2025 Trend Predictions (2025). https://logicdigital.co.uk/future-social-media-marketing/. Accessed 5 Apr

McCrindle, M.: The ABC of XYZ: Understanding the Global Generations. October 2018, Publisher: McCrindle ResearchISBN: 978 0 9924839 0 6 (2018). https://www.researchgate.net/publication/328347222. Accessed 15 March 2025

Ortiz-Espina, E.: The rise of social media. Our world in data (ourworldindata.org) (2019). https://ourworldindata.org/rise-of-social-media. Accessed 2 Apr 2025

Putri, A.A., Rizan, M.: Febrilia I (2022), Impact of social media marketing and e-wom on purchase decision through purchase intention. Jurnal Dinamika Manajemen dan Bisnis **5**(2), 1–17 (September 2022). https://doi.org/10.21009/JDMB.05.2.1

Sanchez, J., Cadarso, M.: Embracing Higher Education leadership in sustainability: a systematic review. J. Cleaner Prod. **298**, 126675 (2021). https://doi.org/10.1016/j.jclepro.2021.126675

Seriac, H.: Gen Z never learned cursive. The effects of this are more widespread than you think (2023). https://www.deseret.com/2022/9/21/23363871/cursive-writing-practice-genz-never-learned-cursive/. Accessed 18 March 2025

Sheikh, M. (2023). Social media demographics to inform your 2025 strategy. Sproutsocial.com February 24, 2025 (2023). https://sproutsocial.com/insights/new-social-media-demographics/. Accessed 22 March 2025

Siddiqui, S., Singh, T.: Social Media its Impact with Positive and Negative Aspects. Int. J. Comput. Appl. Technol. Res. **5**(2), 71–75, 2016, ISSN:- 2319–8656 (2016)

Statista.com: Most popular social networks worldwide as of February 2025, by number of monthly active users (2025). https://www.statista.com/statistics/272014/global-social-networks-ranked-by-number-of-users/. Accessed 5 Apr 2025

UC Berkely: Cool Campus Challenge. Sustainability and carbon solutions UC Berkely administration (2019). https://sustainability.berkeley.edu/engage/cool-campus-challenge. Accessed 29 Apr 2025

Exploring Awareness in Social Media Regarding Europe (EU) and Taiwan Relations

Marcel Rother[1(✉)] and Dario Liberona[2]

[1] University of Vaasa, Vaasa, Finland
Marcel.Rother@th-ab.de

[2] Department of Business Administration, Seinäjoki University of Applied Sciences (Seamk), Seinäjoki, Finland
Dario.liberona@seamk.fi

Abstract. Social media platforms significantly influence public perception of political issues, including international relations between countries and regions. The European Union, a key global player advocating for democratic values and sustainability, maintains extensive economic ties worldwide, making its online perception crucial. Concurrently, Taiwan has gained increasing public awareness, driven by shared democratic values and, notably, by economic cooperation with the EU in the critical semiconductor industry. As public opinion is dynamically shaped by both state actors and private individuals across traditional and social media, understanding the general public's perception of EU-Taiwan relations is imperative. This study employed an online survey to explore public awareness in social media concerning these relations. Findings indicate that the EU-Taiwan relationship is predominantly recognized through the lens of trade and technology, with cultural exchange and democratic values being less prominent. While the overall sentiment towards these relations is positive, there is a discernible tendency towards neutrality. Despite a minority of respondents directly interacting with social media content about EU-Taiwan relations, the vast majority acknowledges social media's profound impact on public perception and its utility for official communication on such issues.

Keywords: EU Taiwan · international relations · Social Media · Public diplomacy

1 Introduction

The digital landscape has fundamentally reshaped how individuals engage with political discourse and international affairs. Social media platforms, in particular, have emerged as influential arenas where perceptions of global issues, including bilateral relations between countries and regions, are increasingly formed and disseminated [2]. The European Union (EU), a significant global actor known for advocating democratic values and sustainability, maintains extensive economic ties worldwide. Consequently, the public's perception of the EU on social media holds considerable importance. Concurrently, Taiwan has gained heightened public awareness, not only due to shared democratic principles but also because of its pivotal role in economic cooperation with the EU, especially within the critical semiconductor industry.

Public opinion in this digital age is a complex interplay, influenced by both state-sponsored narratives and the contributions of private individuals across traditional and social media channels [27]. This dynamic necessitates a deeper understanding of how the general public perceives the relationship between Europe (EU) and Taiwan. This research paper explores public awareness on social media concerning EU-Taiwan relations, building upon an online survey conducted by the authors. The study aims to shed light on the prevalent themes, overall sentiment, and the perceived impact of social media on this important bilateral relationship.

The findings indicate that the relationship between the EU and Taiwan is primarily recognized through the lens of trade, with cultural and democratic values receiving less attention. While the overall sentiment toward these relations is positive, there is a notable tendency towards neutrality. Despite a minority of respondents actively interacting with social media content related to EU-Taiwan relations, a significant majority acknowledges social media's profound impact on public perception and its utility for official communication on various issues. This paper will delve into these findings, contextualizing them within existing literature on digital diplomacy, public opinion, and the challenges of disinformation in the contemporary information environment.

2 Literature Review

The role of social media in shaping international relations and public diplomacy has become a central focus in contemporary communication research. This section reviews relevant literature concerning the public sphere, digital diplomacy, and the challenges of disinformation, providing a theoretical foundation for understanding awareness of EU-Taiwan relations on social media.

2.1 Social Media and the Public Sphere

The concept of the public sphere, initially articulated by Habermas [10], describes a space where information and ideas circulate, fostering debate and the formation of public opinion. In the digital age, this public sphere has largely transitioned to virtual formats, with various internet forums and social media platforms serving as "electronic public spheres" or "global social media spheres" [2]. These digital spaces offer rich ground for studying public opinion, as word-of-mouth and shared narratives within them can reveal how opinions spread and how identities connect the public [2]. Taiwan's media environment has undergone a significant co-evolution with its political landscape, transforming from a highly controlled state-managed system under martial law to a diverse, fragmented, and abundant media-scape [22, 23]. This shift has led to new modes of political communication, characterized by communicative abundance and media hybridity, where distinctions between traditional and digital media blur, and both professional and amateur content producers contribute to the information environment [23]. This hybridity can democratize information but also create dogmatic echo chambers and information cascades that spread rumors and spin [23].

Research on social media and the public sphere in contexts like China highlights the complexity of these spaces. Studies have examined how state and social media companies shape public spheres, noting distinctive characteristics in terms of authoritative actors, topics, and content [20]. The political communication landscape in China has been dramatically reconfigured by the sweeping trends of media commercialization and digitalization [8]. While official media still maintain a monopoly over political news dissemination, particularly on significant domestic and foreign relations issues, most Chinese citizens encounter such content through the selection, reposting, and reinterpretation by digitally active and influential actors [8]. This challenges the notion of an atomized public receiving uniform media content [8]. The online "influentials" (also known as "big Vs" or verified users) in China's cyberspace, including established media agencies, government agencies, and public intellectuals, play a central role in shaping public perceptions on foreign policy issues [8]. These influentials, recognized for their higher credibility and comprehensive content, gain influence through both user-driven customization (users subscribing to them) and system-driven customization (platforms prioritizing their posts) [8].

Traditional Chinese news media, such as People's Daily, have actively integrated into social media platforms like Weibo, employing "contextualization" strategies to present news in innovative ways, emphasize audience interaction, and even provide "non-news content" to engage the public [9]. This demonstrates how state-controlled media adapts to the digital environment to maintain influence and lead public opinion [9]. Xinhua News Agency, a pioneer in party journalism, has also embraced popular social media platforms like WeChat to promote its news services, adapting its content and style to appeal to technology-savvy and entertainment-driven audiences [25]. This "popularization" of party journalism, while seemingly a modernization, is fundamentally an effort to maintain the Communist Party of China's (CPC) ideological function and win the "hearts and minds" of younger generations [25].

The rise of social media has also contributed to socio-political polarization, with algorithms leading to filter bubbles and echo chambers that reinforce similar political beliefs and screen out differing views [14]. Heavy social media users may experience "scanty network heterogeneity," further reinforcing polarization due to limited exposure to diverse viewpoints [14]. This dynamic is crucial when considering how public opinion on international relations, such as EU-Taiwan ties, might be shaped and potentially fragmented online.

2.2 Digital Diplomacy and International Relations

Digital diplomacy involves the extensive use of information and communication technologies, including social media, to communicate policy and influence foreign public opinion [26]. Governments and diplomats increasingly adopt digital media for outreach, recognizing its potential to facilitate two-way communication of ideas, values, and opinions [2]. This strategic operation aims to generate favorable ties with foreign publics, promoting national images through various digital channels. Social media, particularly platforms like Twitter and Weibo, can be powerful communication tools for spreading positive messages and facilitating improved relations between nations, though they can also be used to denigrate others [15].

However, public diplomacy is not solely about outreach; it also involves understanding what citizens are discussing and how foreign countries are perceived [2]. Social media conversations offer a fertile ground for understanding the structure of international relations as perceived by average citizens. Network analysis, for instance, can reveal the salience and visibility of countries based on their co-occurrence in social media posts, reflecting how the public perceives connections between nations [2].

Taiwan, facing unique diplomatic challenges, has proactively embraced digital diplomacy to enhance its international presence and voice [12]. This approach leverages the borderless nature of the internet, where major global social media platforms are less accessible to certain state actors, offering Taiwan a strategic advantage [12, 15]. Taiwanese politicians have increasingly adopted online campaign techniques, moving from rudimentary websites to interactive platforms and personal blogs, and now leveraging social media for direct communication with citizens [23]. This includes repurposing traditional campaign events as multimedia content for social media and live-streaming rallies to global audiences [23]. Non-state actors, such as youth ambassadors and digital diplomacy associations, also play a significant role in promoting positive aspects of Taiwan and building international networks through "hashtag activism" and cultural promotion [13, 24]. This multi-pronged approach aims to boost Taiwan's soft power and global image, cultivating collective pride among its youth [13]. However, some youth ambassadors perceive social media as "inadequate for trust building and credibility compared with interpersonal interactions," highlighting an "authenticity paradox" in digital diplomacy [13]. This suggests that while social media offers reach, fostering genuine trust for sustained diplomatic influence might require more than just online presence.

2.3 Disinformation and Foreign Interference in the Digital Sphere

The digital information environment is often contested, marked by sophisticated disinformation campaigns and foreign interference. These operations aim to manipulate public opinion, fuel polarization, and interfere with democratic processes, posing significant security threats [4, 14]. State actors may deploy massive digital arsenals for foreign information manipulation and interference (FIMI) operations, strategically using cyber and disinformation campaigns to achieve long-term objectives [5, 18].

These campaigns often involve creating original content disseminated via fake accounts, reposting existing content for amplification, and using vague origins to magnify perceived influence. The political objective is frequently to undermine the credibility of administrations and influence political outcomes [18]. Concerns about such influence extend to international relations, with examples including social media disinformation about global health crises and smear campaigns designed to discredit specific nations [6, 15]. China's "Wolf Warrior Diplomacy," characterized by aggressive social media rhetoric, is a calculated strategy to instill fear of China and assert its objectives, often targeting foreign politicians and attempting to control narratives on platforms like Twitter and Weibo [15]. While Chinese state media is often perceived as the main gatekeeper of information, research on Weibo narratives during the Russo-Ukrainian War revealed that Russian news websites published more articles matching Weibo narratives than Chinese, Ukrainian, or US news websites, indicating significant foreign media influence on Chinese social media [11]. This challenges the notion that Chinese state media is the sole source of information about foreign affairs for Weibo users [11]. The Chinese state's information control strategy on social media focuses heavily on "who" is posting rather than solely "what" is being posted [7]. While social media companies are responsible for censoring their platforms, they also selectively report influential users to the government. This approach aims to control and manage influential social forces that might challenge the party-state's hegemonic position [7]. The state permits open discussion and debate on social media but targets opinion leaders whose standing and influence could threaten the Party's presence in the online public sphere [7]. This includes suppressing the rise of influential voices and viral discussions, as evidenced by the targeting of "big V" users on platforms like Weibo [7, 8]. The goal is to prevent information cascades that could lead to collective action or undermine the regime's legitimacy [7].

In response, countries like Taiwan have developed multifaceted, proactive strategies to combat disinformation, emphasizing that free speech can be an asset in this fight [17]. Key strategies include counter-narrative dissemination, empowering independent media and fact-checking organizations, and integrating media literacy programs into education [17, 18]. The EU has also enhanced its capacity to counter FIMI, employing a "FIMI toolbox" that includes increasing awareness, building societal resilience, diplomatic action, and regulatory steps [4, 5]. This includes measures like banning certain media outlets and mandating social media platforms to block specific content [17].

The literature highlights that while social media offers unprecedented opportunities for public diplomacy and fostering international awareness, it simultaneously presents significant challenges related to disinformation and foreign interference. Understanding awareness in social media regarding EU-Taiwan relations thus requires considering both the proactive efforts of digital diplomacy and the pervasive influence of manipulative information campaigns.

3 Methodology

This study employed a quantitative research approach, utilizing an online survey to collect data on public awareness regarding EU-Taiwan relations on social media. The survey was designed to gather demographic information, social media usage patterns, and perceptions of the EU-Taiwan relationship, as well as the role and impact of social media in shaping public opinion on international affairs.

3.1 Participants

The survey was completed by a total of 74 respondents. The demographic characteristics of the participants were as follows:

Age Distribution (Q1):

- Under 18: 0.00% (0 respondents)
- 18–24: 43.24% (32 respondents)
- 25–34: 47.30% (35 respondents)
- 35–44: 9.96% (7 respondents)

- 45–54: 0.00% (0 respondents)
- 55–64: 0.00% (0 respondents)
- 65 +: 0.00% (0 respondents)

The majority of respondents (91.07%) were between 18 and 34 years old, indicating a young adult demographic.

Gender (Q2):

- Female: 78.38% (58 respondents)
- Male: 21.62% (16 respondents)
- Other (specify): 0.00% (0 respondents)

A significant majority of the participants were female.

Country of Origin (Q3):

- Germany: 36.49% (27 respondents)
- Taiwan: 52.70% (39 respondents)
- Other Europe: 4.05% (3 respondent)
- Other Asia: 5.41% (4 respondent)
- Other regions of the World: 1.35% (1 respondent)

The sample was predominantly composed of respondents from Taiwan and Germany.

3.2 Data Collection

An online questionnaire was developed to collect data. The questionnaire consisted of multiple-choice questions designed to assess:

- **Demographics:** Age, gender, and country of origin.
- **Social Media Usage:** Regular use of various social media platforms.
- **Awareness of EU-Taiwan Relations:** Perceived awareness of formal or informal agreements and significant dialogues.
- **Engagement with Social Media Content:** Frequency of interaction with content related to EU-Taiwan relations.
- **Perception of Social Media's Impact:** Beliefs about social media's influence on public opinion and official communication.
- **Key Topics and Sentiment:** Perceived primary topics of EU-Taiwan relations andoverall sentiment.

3.3 Data Analysis

The collected data were analyzed using descriptive statistics to summarize the demographic characteristics of the sample and the responses to each survey question. Frequencies and percentages were calculated for all categorical variables. The quantitative data from the survey provided insights into the awareness levels, social media habits, and perceptions of the EU-Taiwan relationship among the respondents. The analysis focused on identifying patterns and trends in the data to address the research objectives.

4 Results

This section presents the key findings from the online survey, illustrating the social media usage of the respondents, their awareness of EU-Taiwan relations, and their perceptions of social media's role in this context. Respondents reported using a variety of social media platforms regularly. Instagram (87.84%), YouTube (79.73%), and WhatsApp (56.76%) were the most frequently used platforms among respondents. X (formerly Twitter) and Reddit had very low usage rates. The respondents' self-reported awareness of formal or informal agreements or significant dialogues between the EU and Taiwan was generally low. A combined 63.51% of respondents reported being "Not so aware" or "Not at all aware" of formal or informal agreements or significant dialogues between the EU and Taiwan. Only a small minority (8.11%) reported being "Extremely aware" or "Very aware."

The survey indicates that daily social media consumption for these 74 respondents is predominantly in the range of 2 to 4 h, with the 2–3 h bracket being the most common. The level of direct interaction with social media content specifically about EU-Taiwan relations was also limited. 83.78% of respondents "Rarely" or "Never" interact with social media content about EU-Taiwan relations. Only 16.22% reported interacting.

Question 8 asks about the perceived current relationship between the EU and Taiwan. The dominant sentiment among respondents is that the relationship between the EU and Taiwan is "somewhat friendly," highlighting a moderate level of positive perception. While a segment views it as "very friendly," a combined 16.21% (12 respondents) perceive it as "not so friendly" or "not at all friendly," indicating that not all respondents share a positive outlook. The absence of "extremely friendly" responses suggests that a highly robust or exceptionally close relationship is not a common perception among the surveyed group. The "Other" category, while smaller, points to unawareness of existing relations between the EU and Taiwan, questioning if these exist at all. This could refer to the complicated international status of Taiwan (Fig. 1).

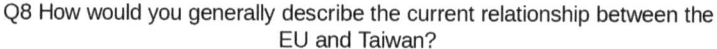

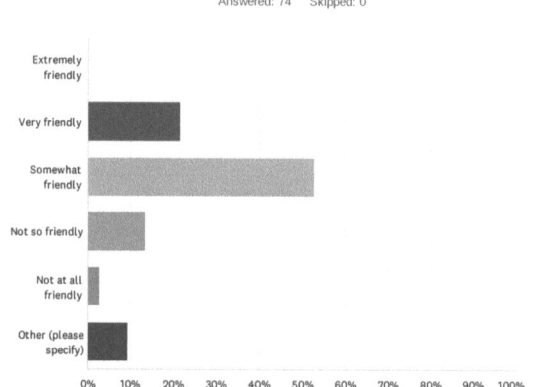

Fig. 1. Perceived relationship between EU and Taiwan

Although existing relations are only described as "Somewhat friendly" by the majority, there are no negative responses to the question of enhanced economic ties by signing a common agreement as shown in Fig. 2.

The result of question 10 highlights that economic and technological cooperation are paramount in the perceived relationship between Europe and Taiwan, closely followed by cultural exchange, shared democratic values, and security concerns.

This suggests a strong focus on pragmatic and strategic partnerships. Nevertheless, Cultural/educational exchange and shared democratic values also rank highly, indicating that soft power and ideological alignment are seen as important pillars of the relationship (Fig. 3).

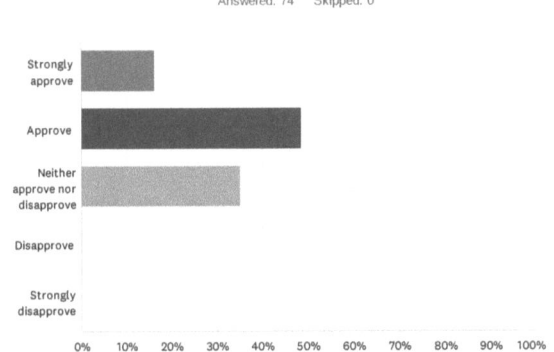

Fig. 2. Opinion on Enhancing Economic Ties

But also, security/geopolitical stability is a significant concern for a considerable portion of respondents, reflecting the broader geopolitical context, in particular from Taiwanese respondents. Finally, while economic and technological factors lead, the spread of responses across various categories, including sustainability, tourism, and scientific research, demonstrates a perception of a multi-faceted relationship with diverse areas of importance.

The survey results as shown in Fig. 4 suggest that visual and video-based social media platforms, particularly Instagram and YouTube, are perceived as the most effective channels for promoting and informing the public about EU and Taiwan relations and agreements. This could mean that engaging visual and video content is crucial for public awareness and understanding of this topic. TikTok's strong showing indicates the importance of reaching a younger demographic and leveraging short-form video for information dissemination. Rather traditional Social Media platforms like Facebook and X (formerly Twitter), while still relevant, are seen as less impactful for this specific purpose compared to the visual-first platforms. Noteworthy is the recognition of "Influencers" as a viable channel point to the potential for leveraging individual voices to shape public opinion.

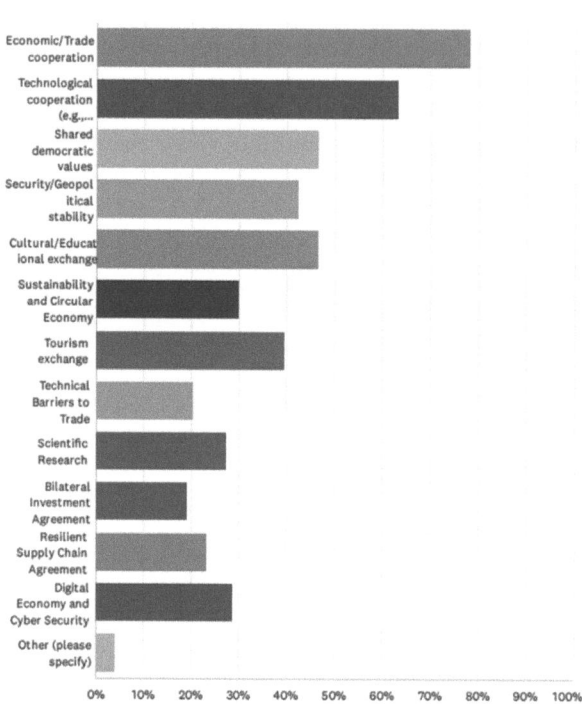

Fig. 3. Most important Aspects of Europe-Taiwan Relations

Despite low direct interaction, respondents largely acknowledged the significant influence of social media on public opinion regarding international relations. Figure 5 demonstrates that a substantial majority (67.57%) of respondents believe social media influences public opinion "Very significantly" or "Significantly." Only a small fraction (8.11%) perceived the influence as "Slightly" or "Not at all." Respondents also largely agreed that social media facilitates communication for officials.

Figure 6 reveals an overwhelming consensus, with 72.97% of respondents either "Strongly agree" or "Agree" that social media makes it easier for officials to communicate about issues.

Exploring Awareness in Social Media Regarding Europe (EU) 347

Q11 Which social media platforms do you think is best for promoting and informing about EU and Taiwan relations and agreements (Select all that apply)

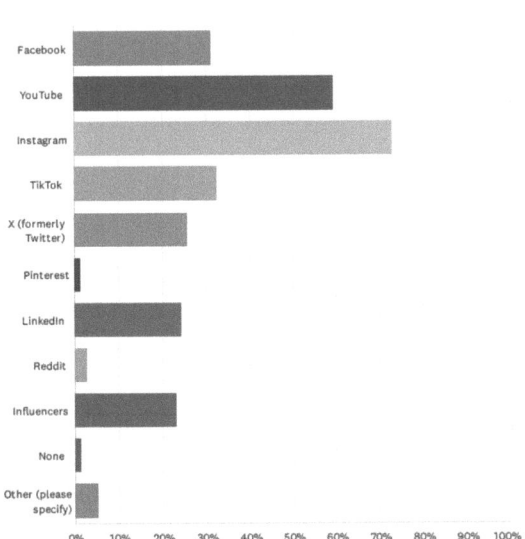

Fig. 4. Best Options to promote EU and Taiwan relations

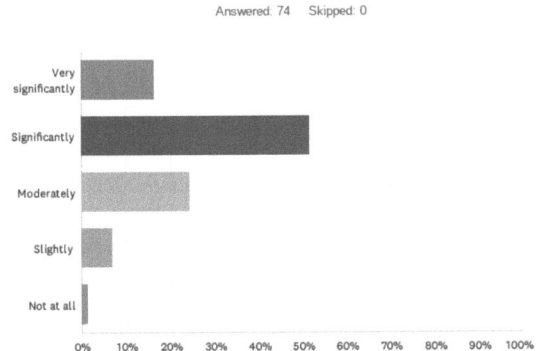

Fig. 5. Perceived Impact of Social Media on Public Opinion

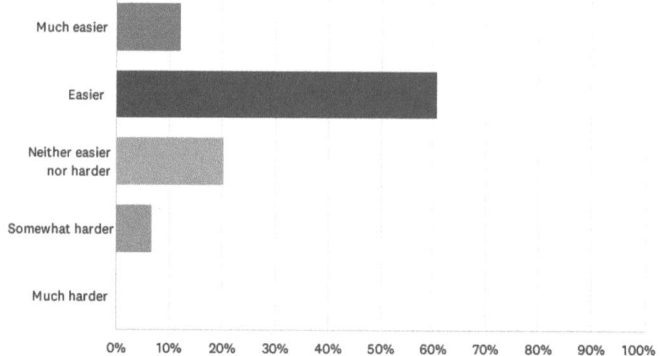

Fig. 6. Perceived Ease of Communication for Officials via Social Media

5 Findings and Interpretation of Results

The findings of this study provide valuable insights into public awareness of EU-Taiwan relations on social media, highlighting both areas of recognition and significant gaps. The demographic profile of the respondents, predominantly young adults from Taiwan and Germany, suggests that the results are most representative of this specific segment of the population. This demographic is highly engaged with social media, making their perceptions particularly relevant for understanding online discourse.

5.1 Limited Awareness Despite Strategic Importance

A key finding is the generally low self-reported awareness of formal or informal agreements and significant dialogues between the EU and Taiwan. A substantial majority (66.07%) of respondents indicated being "Not so aware" or "Not at all aware." This aligns with broader research suggesting that the strategic deepening of EU-Taiwan ties has largely gone "unnoticed" by the general public, leading to limited mutual knowledge [1, 3]. This awareness deficit is particularly concerning given the escalating geopolitical importance of Taiwan for the EU, especially in critical sectors like semiconductors and regional stability.

The low level of direct interaction with social media content about EU-Taiwan relations (78.57% rarely or never interact) further corroborates this awareness gap. This suggests that while respondents are active social media users, their engagement with specific international political content, particularly on EU-Taiwan relations, is minimal. This creates a challenging environment for public diplomacy efforts, as messages may struggle to penetrate the general online discourse.

5.2 Dominance of Economic and Technological Narratives

When asked about the primary topics of EU-Taiwan relations discussed on social media, respondents overwhelmingly identified "Technological Cooperation (e.g., semiconductors)" (78.57%) and "Trade and Economy" (58.93%). This indicates that the economic and technological dimensions of the relationship are the most salient in public perception, likely driven by Taiwan's global leadership in the semiconductor industry and the EU's increasing focus on supply chain resilience.

Conversely, "Cultural Exchange" (17.86%) and "Shared Democratic Values" (19.64%) were perceived as less prevalent topics. This finding suggests a disconnect between the stated values-based approach in EU-Taiwan relations at the policy level and its reflection in public awareness on social media. While both the EU and Taiwan emphasize shared democratic principles, these aspects appear to resonate less strongly with the public in the digital sphere compared to tangible economic benefits. This highlights a potential area for public diplomacy to broaden the narrative beyond purely economic terms.

5.3 Positive but Neutralizing Sentiment

The overall sentiment towards EU-Taiwan relations on social media was largely positive (69.65% positive or very positive). However, a significant portion (25.00%) reported a "Neutral" sentiment. This tendency towards neutrality, even within a generally positive outlook, could indicate a lack of strong emotional engagement or a nuanced understanding of the relationship's complexities. It might also reflect the influence of a balanced media environment or, conversely, a successful strategy by various actors to maintain a non-committal stance in public discourse. This finding suggests that while there isn't widespread negative sentiment, there is also not a deeply ingrained positive perception that could easily withstand external pressures or disinformation campaigns.

5.4 Perceived Impact of Social Media on Public Opinion

Despite low direct interaction with EU-Taiwan content, respondents overwhelmingly acknowledged social media's significant influence on public opinion regarding international relations (69.65% significantly or very significantly). An even greater majority (98.21%) agreed that social media makes it easier for officials to communicate about issues. These perceptions underscore the recognized power of social media as an information dissemination and opinion-shaping tool, even if individuals themselves are not actively engaging with specific content. This aligns with broader findings that social media use significantly impacts public perceptions of global actors and international relations, with platforms like Weibo having a stronger influence due to their ability to create narratives, deliver timely information, and promote deliberation [28].

This perception is critical in the context of digital diplomacy and information warfare. The public recognizes social media as a powerful shaper of narratives, which aligns with existing literature on the role of social media in shaping public attitudes during crises [16, 19]. Social media communication makes visible the linkages between governments' actions to diplomatic events and domestic audiences' reactions, and a diplomatic crisis can intensify existing domestic polarization [16]. This awareness of social media's influence, coupled with low direct engagement with EU-Taiwan content, creates a fertile ground for foreign information manipulation and interference (FIMI). If the public believes social media is influential but is not actively seeking out or critically engaging with nuanced information on EU-Taiwan relations, they become more susceptible to narratives propagated by malicious actors [5, 18]. Furthermore, the Chinese state's information control strategy, which prioritizes targeting influential users and viral content over specific content categories, has significant implications for how narratives about international relations are shaped. The ability of "big V" users to act as brokers, connecting users and sparking information cascades, is a key concern for the Chinese government [7]. This "who not what" logic means that even discussions that appear benign can be suppressed if they gain sufficient virality or are disseminated by influential figures, potentially limiting the organic spread of diverse viewpoints on topics like EU-Taiwan relations within Chinese social media. The state's efforts to co-opt or repress these influential voices, while allowing for some level of public discussion to gain insights into public opinion, ultimately aim to maintain its hegemonic position in the online public sphere [7].

5.5 Implications for Digital Diplomacy and Information Resilience

The findings suggest several implications for digital diplomacy and information resilience efforts concerning EU-Taiwan relations:

1. **Narrative Expansion:** Public diplomacy efforts should aim to broaden the narrative beyond trade and technology to include shared democratic values and cultural ties. While economic cooperation is a strong foundation, emphasizing common values could foster deeper public understanding and support, aligning with the EU's role in communicating values [27].
2. **Targeted Engagement:** Given the low direct interaction rates, a more targeted and engaging approach to social media content is needed. This could involve leveraging popular platforms (Instagram, YouTube, WhatsApp) with creative and accessible content formats, moving beyond traditional press releases to more interactive and visually appealing narratives, and adapting to the "middle region behaviors" of social media [9]. This also means understanding the "techno-cultural assemblage" of platforms like Weibo, where features, user cultures, and censorship practices are intertwined in shaping online contention [21].
3. **Youth Focus:** The predominantly young adult demographic of respondents highlights the importance of tailoring digital diplomacy strategies to resonate with this age group, who are heavy social media users.
4. **Countering Disinformation:** The perceived influence of social media, combined with low specific awareness, makes the public vulnerable to disinformation. Efforts to combat FIMI, drawing on Taiwan's successful "whole-of-society" model [17], are crucial. This includes promoting media literacy and supporting independent fact-checking initiatives to empower citizens to critically evaluate information, especially given the prevalence of echo chambers and filter bubbles that exacerbate polarization [14]. Understanding the mechanisms of state censorship, which targets influential users and viral content, is also essential for developing effective counter-strategies [7].
5. **Measuring Impact:** The study underscores the need for more robust empirical research to measure the actual impact and effectiveness of digital diplomacy initiatives. This would involve moving beyond simple engagement metrics to assess

changes in awareness, sentiment, and understanding among target audiences, and employing advanced social media analysis methodologies.

In conclusion, while the EU-Taiwan relationship is strategically deepening, public awareness on social media remains limited and largely focused on economic aspects. The perceived power of social media in shaping opinion, however, presents both a challenge and an opportunity. By strategically refining digital diplomacy efforts and enhancing information resilience, it is possible to cultivate a more informed and engaged public that better understands the multifaceted importance of EU-Taiwan relations.

6 Conclusion

This research paper set out to explore public awareness on social media regarding Europe (EU) and Taiwan relations, building upon an online survey and existing literature. The study confirms that while social media profoundly impacts the perception of political issues and international relations, public awareness of the nuanced and strategically deepening ties between the EU and Taiwan remains limited.

The survey findings reveal that the relationship is predominantly recognized through its economic and technological dimensions, particularly in trade and the semiconductor industry. This aligns with the EU's strong economic ties globally and Taiwan's critical role in global supply chains. However, aspects such as cultural exchange and shared democratic values, though emphasized at policy levels, appear less prevalent in public awareness on social media. The overall sentiment towards these relations is generally positive, yet with a notable tendency towards neutrality, indicating a lack of deep engagement or strong emotional connection among the public.

Despite a minority of respondents actively interacting with social media content related or specifically about EU-Taiwan relations, an overwhelming majority acknowledges the significant influence of social media on public perception. Furthermore, respondents widely believe that social media facilitates communication for officials on various issues. This highlights a critical paradox: the public perceives social media as a powerful tool for shaping opinion, yet their direct engagement with specific international political content is low. This creates a vulnerability to foreign information manipulation and interference (FIMI), as an uninformed public is more susceptible to narratives propagated by malicious actors, especially within polarized online environments. The Chinese state's sophisticated information control strategies, which prioritize the suppression of influential voices and viral content, further complicate the landscape for organic information dissemination and public discourse.

The study underscores the urgent need for more sophisticated and targeted digital public diplomacy strategies from both the EU and Taiwan. These strategies should aim to broaden the narrative beyond economic ties to encompass shared values and cultural understanding, utilizing engaging formats on popular platforms to reach a younger, digitally native audience. Moreover, continued efforts in building information resilience, drawing on Taiwan's model for combating disinformation, are crucial to empower citizens to critically evaluate online information and safeguard democratic processes against external manipulation. This includes recognizing the complex interplay of traditional and new media, and the potential for foreign media influence on domestic narratives, as well as the unique dynamics of social media platforms as "participants" in political contention.

In sum, the perception of EU-Taiwan relations on social media is characterized by a focus on trade, a generally positive but neutral sentiment, and a recognized, yet underutilized, potential for social media to shape public opinion. Bridging the awareness gap and fostering a more informed public understanding of this vital international relationship will require sustained, strategic, and empirically-driven digital engagement.

References

1. Association for International Affairs: Limited mutual knowledge and awareness: Public opinion on Taiwan and the Visegrad Four countries (2023)
2. Barnett, G.A., et al.: Measuring international relations in social media conversations. Gov. Inf. Q. **34**(1), 37–44 (2017)
3. European Council on Foreign Relations: The unnoticed shift: EU-Taiwan relations in a changing geopolitical landscape (2023)
4. European External Action Service: Fighting foreign information manipulation and interference: The EU's approach (2023a)
5. European External Action Service: 2nd EEAS Report on Foreign Information Manipulation and Interference Threats (2023b)
6. European Values Center for Security Policy: Chinese influence operations in Central Europe: The case of the Czech Republic (2021)
7. Gallagher, M., Miller, B.: Who not what: the logic of China's information control strategy. China Q. **248**, 1011–1036 (2021)
8. Guan, T.: Who are the influentials in China's cyberspace and what do they say about the issue of Sino-Japanese relations? J. East Asian Stud. **19**(3), 383–396 (2019)
9. Guo, D., Wang, H., Xu, J.: Contextualization: a path to Chinese traditional news media's integration into social media. Media Commun. **12**(7429), 1–16 (2024)
10. Eley, G.: Habermas Jürgen, The structural transformation of the public sphere. an inquiry into a category of bourgeois society (Cambridge, Mass.: The MIT Press, 1989). Comp. Stud. Soc. His. **34**(1), 189–190 (1992). https://doi.org/10.1017/S0010417500017527
11. Hanley, H.W.A., Lu, Y., Pan, J.: Across the firewall: Foreign media's role in shaping Chinese social media narratives on the Russo-Ukrainian War. Proc. Natl. Acad. Sci. **122**(1), e2420607122 (2024)
12. Hsieh, Y.H.: Taiwan's digital diplomacy: Leveraging social media in the face of diplomatic isolation (2020)
13. Lee, S. Y.: Taiwan's youth ambassadors and digital cultural diplomacy: An authenticity paradox (2021)
14. Lin, T.T.C., Tsai, C.H.: Taking stock of social-political polarization in Asia: political communication, social media and digital governance. Asian J. Commun. **32**(2), 71–74 (2022)
15. Masroor, M.: Social Media and U.S.-China Relations, Master's thesis, The American University of Paris (2023)
16. Matsuo, A., Han, O., Matsumura, N.: The social media audience of diplomatic crisis. Br. J. Politics Int. Relat. **26**(3), 922–939 (2024)
17. Ministry of Foreign Affairs, Republic of China (Taiwan): Taiwan's model for combating disinformation: A whole-of-society approach (2024)
18. Ministry of National Defense, Republic of China (Taiwan): Chinese Communist Party's disinformation campaigns targeting Taiwan (2021)

19. National Chengchi University: Public opinion on cross-Strait relations in Taiwan (1994–2023) (2023)
20. Pang, H., Liu, J., Lu, J.: Tackling fake news in socially mediated public spheres: a comparison of Weibo and WeChat. Technol. Soc. **70**, 102004 (2022)
21. Poell, T., de Kloet, J., Zeng, G.: Will the real Weibo please stand up? Chinese online contention and actor-network theory. Chin. J. Commun. **7**(1), 1–18 (2014)
22. Rawnsley, G., Gong, Q.: Political communications in democratic Taiwan: The relationship between politicians and journalists. Polit. Commun. **28**(3), 323–340 (2011)
23. Sullivan, J.: The co-evolution of media and politics in Taiwan: Implications for political communications. Int. J. Taiwan Stud. (2018) (Draft of forthcoming article)
24. Taiwan Digital Diplomacy Association: Leveraging hashtag activism: TDDA's social media strategies (2024)
25. Xin, X.: Popularizing party journalism in China in the age of social media: the case of Xinhua News Agency. Glob. Media China **3**(1), 3–17 (2018)
26. Yin, S.: Russian digital diplomacy towards China in the context of the Russian special military operation in Ukraine: The instance of the official Weibo account of the Embassy of the Russian Federation in China. Sci. J. Volgograd State Univ. His. Area Stud. Int. Relat. **28**(3), 76–84 (2023)
27. Zhang, L.: Bridging the communication gap in EU-China relations: Policy, media, and public opinion. Asia Eur. J. **20**(2), 219–227 (2022)
28. Zhang, C., Kübler, D., Dong, L.: Chinese perceptions of the EU: the impact of social media use. Glob. Public Policy Governance **3**, 180–198 (2023)

How to Highlight the Lowest Values: Adapting Bar Chart Features to Emphasize Data Minima

Laura Montenegro[1](✉) , Jordán Pascual Espada[1] , and Juan Luis Carús Candás[2]

[1] Department of Computer Science, Sciences Building, University of Oviedo,
C/Calvo Sotelo S/N 33007, Oviedo, Asturias, Spain
laura.montenegro@grupotsk.com
[2] Grupo TSK, Technological Scientific Park of Gijón, 33203 Gijón, Asturias, Spain

Abstract. This study presents a comparative analysis of bar chart configurations and their effectiveness in highlighting minimum or near-minimum values across diverse dataset characteristics and communication and interaction objectives. Through the design and deployment of a web-based experimental application, over 100 users participated in visualization tasks under controlled variable conditions. The resulting data were analyzed using association rule mining techniques to identify patterns that link specific visual attributes (such as axis orientation, spacing, color schemes, or labeling) with user response times and task performance when detecting data minimum. The findings provide empirical evidence on how chart features should be adapted according to data volume, variability, and the communicative goal of emphasizing the lowest values. A set of practical recommendations is offered to guide the design of bar charts that optimize interpretability and reduce cognitive load in tasks focused on identifying low-ranking data points.

Keywords: Bar charts · data visualization · graphical configuration · visual perception · user performance · association rules · communication objectives · experimental design · data minimum · lower values

1 Introduction

The increasing relevance of charts and infographics reflects a broader shift toward visual communication. While not replacing text entirely, visual representations are now essential complements, as they enable faster and more effective information processing by leveraging the brain's natural preference for images [1–3].

This trend is especially pronounced in digital environments, where speed and immediacy are vital. To attract and retain users in a competitive web ecosystem, visual appeal (achieved through graphical elements like color, form, and layout) plays a critical role [4].

In digital environments, where speed and immediacy are essential, visual appeal (achieved through color, form, and layout) plays a key role in attracting and retaining users [5]. Among the many visualization types, every type of chart offer extensive configuration options, such as orientation, spacing, labeling, and color schemes. These variations produce distinct visual effects, yet the absence of universal design guidelines often leaves chart selection and adjustment to subjective judgment [5].

To minimize misinterpretation or cognitive overload, designers must account for both dataset properties and communication goals [6]. A frequent objective is to highlight the lowest values, either the absolute minimum or those within the lower range [7]. This occurs in diverse contexts, such as identifying economic disadvantages, price reductions, performance drops, or framing low unemployment rates positively.

Figure 1 presents a typology of visualization types organized by communicative function [8].

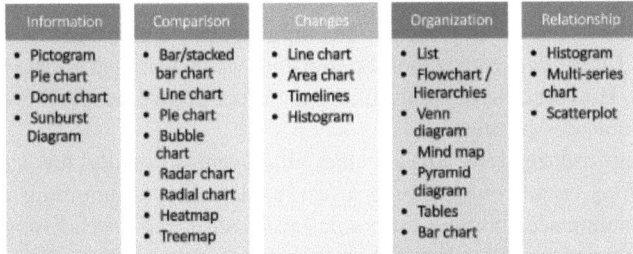

Fig. 1. Classification of the main chart types by their function

Despite existing typologies, the diversity of charts and configuration parameters (combined with heterogeneous data) can overwhelm even experienced practitioners. As a result, empirical evidence to support design choices remains scarce. To address this gap, the present study focuses on one of the most used and easily interpretable chart types: the bar chart [9].

Bar charts are widely favored for their simplicity and clarity, making them suitable for both technical and general audiences. However, there is limited consensus on how to configure elements such as axis orientation, bar spacing, or labeling to effectively highlight low-ranking values. These choices often rely on designer intuition or informal best practices based on usability experience [10].

To bring structure to this process, our study aims to empirically identify the most effective bar chart configurations for emphasizing minimum or near-minimum values. We examine how visual variables should be adapted according to dataset size, variability, and the communicative goal, offering practical, data-driven design guidelines through a controlled experimental approach.

2 Related Work

Descriptive studies such as Sadiku et al. (2016) [11] highlight the utility of charts (e.g., line, bar, pie) for improving interpretation. However, they also raise a key question: which chart type is best suited to each situation? Further, Abieba et al. [12] emphasized the importance of aligning data visualization with decision-making contexts, particularly in software engineering.

Despite this consensus, most studies lack empirical rules that guide the choice of chart type or configuration based on communication goals or dataset properties. General guidelines exist, such as those proposed by Midway (2019) [13], who outlines ten principles for clarity and context, or Weissgerber et al. (2019) [14], who suggest minimizing visual noise and selecting appropriate color palettes. However, these recommendations remain generic and lack quantitative validation.

More structured approaches (Jambor's checklist [15]) attempt to define quality criteria (e.g., legibility, hierarchy, color usage), but they have not been empirically evaluated through user interaction. Similarly, comprehensive guides like this of Cukier and Mayer-Schoenberger [16] offer best practices but fall short of addressing the effect of visual variables in measurable terms.

A few studies focus specifically on bar charts. Kim et al. [17] discuss how visual features—such as scale, order, and layout—can influence perception and task performance, and introduce *ChartOptimiser*, a tool for adjusting visual parameters based on communicative goals. However, these contributions are often limited to specific scenarios and lack generalizable design rules.

While many contributions address chart effectiveness broadly, few studies explicitly focus on how visual configurations affect the detection of minimum values within datasets, A common need in socio-economic and scientific reporting. Xiong et al. (2021) [7], for example demonstrates how bar arrangement influences user ability to compare values, including identifying the lowest among them, reinforcing the relevance of layout decisions in such tasks.

In summary, while existing literature underscores the value of visualization, there is a lack of empirical, data-driven criteria to guide bar chart configuration in function of dataset characteristics and communication intent. This gap motivates the present study.

3 Study Objectives and Methodology

The main objective of this study is to empirically derive a set of principles or rules that optimize the use of bar charts according to the characteristics of the input dataset when the communicative goal is to highlight minimum or near-minimum values within a dataset. To this end, a web-based tool was developed which, for each test, randomly configures the parameters of a bar chart, populates it with a specific dataset, and poses a question to the user based on the visualization (recording the time taken to answer correctly).

The data set itself constitutes a critical factor when configuring the bar chart in which it is to be represented. In our case, we always work with datasets in the format *key/class:numeric_value*, and we focus the analysis on two fundamental characteristics:

- **Sample size**: Number of elements displayed in the chart. The following categories are defined:

    – Very Small datasets: up to 5 elements
    – Small datasets: 6 to 9 elements
    – Medium datasets: 10 to 20 elements
    – Large datasets: more than 20 elements

- **Data variability**: Refers to the homogeneity and proximity among the values in the dataset. This is computed using the coefficient of variation, based on the mean and the variance.

Upon analysis of the most popular libraries for generating bar charts in web applications, D3, Highcharts, ApexCharts, and Chart.js were identified as leading options [25]. Considering factors such as full free access, popularity, and community size, Chart.js was selected to implement the experiment described in this article.

Based on the configurable features supported by Chart.js for bar charts, the following visual variables were defined to be randomly adjusted in each trial of the experiment:

- **Chart orientation:** Determines which axis corresponds to values and which to categories, thus resulting in either horizontal or vertical bars.
- **Maximum axis value:** Sets the upper bound of the value axis by increasing the highest dataset value by 0%, 10%, 20%, or 30%.
- **Tick intervals on the value axis**: Defines the number of marks on the value axis. This number must be even or a multiple of 5 or 10, smaller than the highest value, and greater than zero.
- **Gridlines**: Specifies whether auxiliary lines extending from each axis tick are displayed.
- **Opacity**: Determines whether bars are fully opaque or displayed with 50% transparency.
- **Color scale:** Specifies the color scheme for the bars. Options include random colors, a single color, an unordered color palette, or an ordered gradient based on value.
- **Labels:** Indicates whether value labels are shown at the end of each bar.
- **Bar spacing:** Determines the distance between bars—either adjacent, separated, or widely spaced.
- **Border:** Specifies whether the bars have an outline.
- **Border color:** Defines the border color—matching the fill color, black, or white.
- **Hover effect:** Determines the visual effect when the mouse hovers over a bar maintaining the same color, darkening it slightly, or switching to a different color (Fig. 2).

To prevent results from being skewed by repeated combinations of visual characteristics, a randomization system was implemented to reconfigure all parameters independently for each trial. This ensures a balanced distribution of all possible configurations, minimizes bias due to frequency effects, and guarantees that the outcomes of the experiment are driven solely by the visual features presented rather than the frequency of specific chart combinations.

Another important aspect to consider is the type of question posed and the characteristics of the expected response. For each trial, the experiment randomly selects a question related to the chart displayed, which falls into one of the following categories:

- **Textual response questions:** Identify the category with the lowest, second-lowest, or third-lowest value.
- **Numerical response questions:** Indicate how many categories are equal to or lower than a given value.

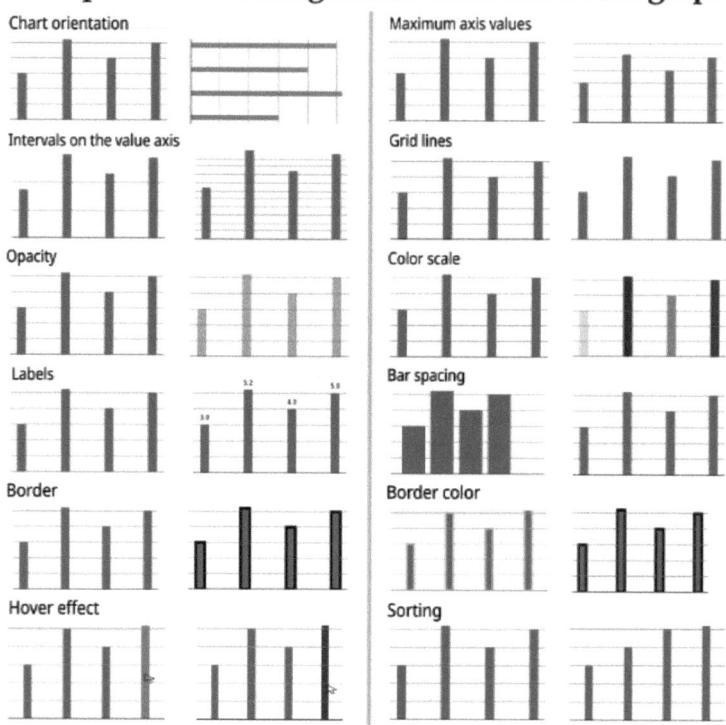

Fig. 2. Examples of the configurable variables in the graph

The experimental application was developed using the Hapi.dev framework, which enables a modular server structure with defined routes and methods that connect to HTML views. The architecture includes:

- **App.js**: Main server file that imports libraries, establishes the database connection, and starts the experiment by loading the login view.
- **Libraries**: Custom libraries handle key functionalities such as navigation, data processing, chart setup, and database operations. External libraries (mainly from Hapi.dev) manage the user interface and session cookies.
- **Support files**: JavaScript files that define global variables shared across the application, including color palettes, constants, and the JSON datasets used in the trials.
- **Views**: HTML templates shown to users, containing interactive elements for data input/output, question display, and navigation between stages of the experiment (Fig. 3).

As previously mentioned, the results of each experiment are stored in a database for subsequent processing. Thus, the object saved for each submitted response contains the following fields:

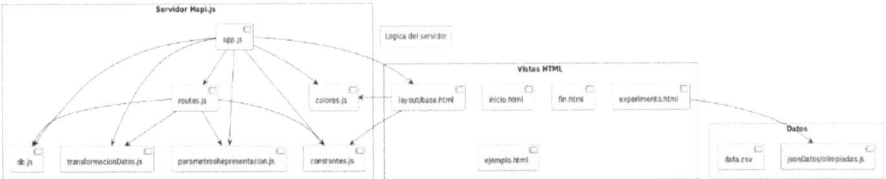

Fig. 3. Architecture diagram

- **User Information**: It is important to note that this experiment is anonymous; therefore, no personal data such as names or surnames are requested from participants.

  – Unique ID: This identifier allows for unequivocal identification of the user who performed the experiment. It is common to submit all responses by the same user, making it possible to group them accordingly.

- **General Information**:

  – Execution date and time: These timestamps enable the chronological ordering and grouping of experiments.
  – Response time: Number of seconds the user spent on the visualization view, assumed to be the time taken to search for the correct answer.

- **Dataset Information**: The combination of the following values enables the calculation of statistical measures such as mean, standard deviation, median, etc.

  – Displayed values: The dataset shown to the user.
  – User response: The value input by the user. If the question was skipped, the field will contain the value *"question skipped"*.
  – Question category: Specifies whether the expected response is textual or numerical, and the type of question asked.

- **Chart Configuration**: All the configurable variables described in previous sections (Fig. 4).

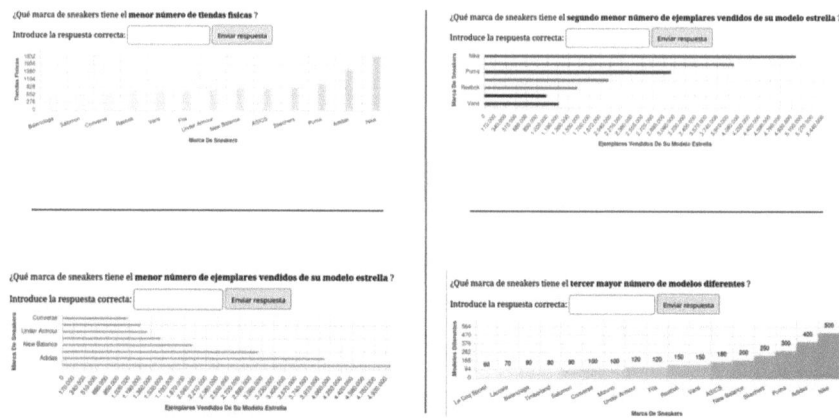

Fig. 4. Examples of different experiment questions for the same dataset

4 Data Evaluation and Interpretation

Once the application was deployed, it was distributed among voluntary users with different profiles. At the end of the data collection phase, a total of 1060 valid responses from 103 unique users had been gathered.

To extract useful insights, we applied association rule mining, a technique designed to uncover frequent relationships between variables in large datasets. Commonly used in domains like recommendation systems and behavior analysis, these rules follow the structure "if A, then B," and are evaluated through support, confidence, and lift metrics [18].

We developed a custom processing tool in Python, using pandas, NumPy, Mlxtend, and Ast. After removing incomplete responses (e.g., skipped questions), we transformed several columns to support analysis, especially those related to question type and communication goal.

The dataset was then segmented by communication objective (highlighting minimum or highlighting values that fall within the lower range). Within each segment, we classified two critical variables:

- **Data variability:** Computed as the coefficient of variation (standard deviation divided by the mean), it measures the homogeneity of the values within a dataset. It was categorized into three levels:

  - Low (≤ 0.1): values are highly similar.
  - Moderate (0.1–0.5): moderate variation
  - High (> 0.5): significant differences between values

- **Response time**: Rather than applying arbitrary thresholds, response time was contextually categorized using quartiles within each group, depending on the question type and dataset size. The three resulting categories were:

  - Short time ($\leq Q1$): fast, effective interpretation

- Medium time (Q1–Q3): neutral impact
- Long time (> Q3): potentially problematic configuration

This approach allows for a fair comparison of performance across heterogeneous conditions.

For each subset, we applied association rule mining. Antecedents included visual variables such as border color, hover effect, orientation, spacing, labels, and others, along with dataset size and variability. The consequent was always response time, used as an indicator of chart performance.

After testing various thresholds, we selected a minimum support and confidence of 0.1 to balance coverage and relevance. Rules followed the pattern:

{Size + Variability + Visual Variable} → Response Time.

From this process, we extracted over 500 association rules, grouped by communication goal. To ensure clarity and utility, only the rules meeting at least one of the following criteria were retained:

A) Support > 0.06 and confidence > 0.6
B) All values of a given variable appear across rules, with total support > 0.1 and average confidence > 0.6

These filtered rules form the basis of the recommendations presented in Sect. 5 (Fig. 5).

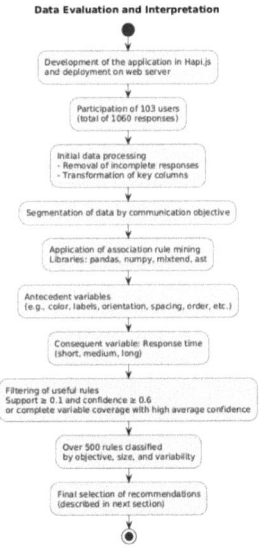

Fig. 5. Summary of the Data Evaluation and Interpretation

5 Recommendations

Based on the association rules, we defined a set of recommendations for bar chart configurations tailored to the communication objective and dataset properties. The following guidelines address two related but distinct goals: first, enabling users to quickly identify the absolute minimum value within a dataset; and second, effectively highlighting values that fall within the lower range (e.g., bottom three categories). These tasks require slightly distinctive design strategies, particularly in terms of ordering, spacing, labeling, and visual emphasis (Tables 1 and 2).

Table 1. Recommendations for bar charts when highlighting the minimum

Visual Variable	Sample Size	Data Variability	Recommendation	Support	Confid.	Criteria
Border Color	Medium	Moderate	Dark border	0.09	1	A
Hover Bars	Very Small	Moderate	Do nothing	0.06	1	A
Orientation	Very Small	High	Vertical	0.09	1	A
		Moderate	Indifferent	0.13	0.7	B
Labels	Very Small	High	Indifferent	0.13	0.62	B
		Moderate	Indifferent	0.15	0.5	B
Axis max.	Very Small	High	30% above max	0.09	1	A
		Moderate	10% above max	0.13	0.8	B
Steps Number	Medium	Moderate	Few steps	0.09	0.67	A
	Very Small	High	Medium steps	0.12	0.83	A
		Moderate	Indifferent	0.13	0.7	B
Sorting	Medium	High	Sorted	0.13	0.87	B
	Very Small	High	Indifferent	0.09	0.67	B
Opacity	Medium	High	Opaque	0.13	0.87	B
	Very Small	High	Indifferent	0.13	0.6	B
		Moderate	Indifferent	0.18	1	B
Gridlines	Very Small	High	Indifferent	0.13	0.67	B
	Very Small	Moderate	Do not use	0.11	0.75	A

Table 2. Recommendations for bar charts when highlighting lower range

Visual Variable	Sample Size	Data Variability	Recommendation	Support	Confid.	Criteria
Highlighting values that fall within the lower range						
Border Color	Small	Moderate	Dark border	0,08	1	A
Hover Bars	Medium	Moderate	Do nothing or highlight color	0.2	0.92	B
	Very Small	Moderate	Do nothing or highlight color	0.09	1	A
Bar Orientation	Medium	Moderate	Horizontal	0.12	0.67	B
	Small	High	Indifferent	0.12	1	B
	Very Small	Moderate	Indifferent	0.18	0.5	B
Bar Spacing	Medium	Moderate	Close together	0.12	0.66	B
	Small	High	Very separated	0.9	1	A
	Very Small	Moderate	Close together	0.08	0.83	B
Labels	Small	High	With labels	0.08	1	A
	Small	Moderate	With labels	0.08	1	A
	Very Small	Moderate	Without labels	0.12	0.8	A
Steps number	Medium	Moderate	Few steps	0.12	0.6	B
	Small	High	Medium steps	0.12	1	A
	Very Small	Moderate	Medium steps	0.2	0.67	B
Sorting	Medium	Moderate	Sorted	0.18	1	B
	Small	High	Sorted	0.12	1	A
	Small	Moderate	Indifferent	0.12	1	B
Opacity	Medium	Moderate	Transparent	0.13	0.87	B
	Small	High	Opaque	0.09	1	A
	Small	Moderate	Opaque	0.09	1	A
Gridlines	Medium	Moderate	Without gridlines	0.12	0.66	B
	Small	High	Without gridlines	0.09	1	A
	Small	Moderate	Indifferent	0.12	1	B
	Very Small	Moderate	Use gridlines	0.18	0.87	B

6 Conclusions and Future Work

This study addresses the lack of empirical guidelines for configuring bar charts, a gap that often leaves visualization design to individual judgment, potentially hindering comprehension and message clarity. Focusing on bar charts due to their widespread use, we derived configuration rules based on dataset size, variability, and communication goals.

Through a web-based experiment involving over 1000 responses from 103 users, we recorded how visual variables influenced user performance. Using association rule mining, we extracted data-driven recommendations to improve chart effectiveness.

While detailed guidelines are provided by scenario, some general patterns emerged:

- Larger and more variable datasets benefit from wider spacing and gridlines.
- Smaller datasets are better represented with horizontal bars and visible labels.
- Data variability plays a key role in border color and hover effects.

In the specific context of highlighting minimum or near-minimum values, the data reveal additional insights:

- There is greater overlap in configuration rules when the dataset is small (≤ 5) and has moderate variability, suggesting that in simpler scenarios, the same visual strategies are effective both for identifying the absolute minimum and for emphasizing generally low values.
- The most relevant differences emerge in medium-sized datasets (10–19 elements) when the goal is to identify the exact minimum: in these cases, ordering the bars, using opacity, and potentially compressing the value axis improve performance. For highlighting low values in general, these design decisions are less critical.
- When data variability is high, differences between tasks become more pronounced. To locate the absolute minimum, more active visual cues are recommended (such as vertical orientation or higher opacity) while for general low-value emphasis, the effect is less dependent on aggressive visual contrast.

Future work includes increasing the sample size to cover more variable combinations, exploring other analytical methods (e.g., classification or regression), and analyzing skipped responses for potential readability issues. Additional lines of research include extending the experiment to other chart types and conducting comparative studies to determine which visualizations best suit different data and goals.

This research lays the foundation for developing an intelligent system capable of recommending optimal chart configurations based on dataset properties, enhancing clarity, efficiency, and user experience.

References

1. Fernández-Campoy, J.M.: The visual takes over: new visualities in the digital ecosystem. Comunicar **28**(64), 9–19 (2020)
2. Ware, C. (2020). *Information Visualization: Perception for Design* (4th ed.). Morgan Kaufmann
3. Few, S.: Show Me the Numbers: Designing Tables and Graphs to Enlighten (2nd ed.). Analytics Press (2012)
4. Krug, S.: Don't Make Me Think, Revisited: A Common Sense Approach to Web Usability (3rd ed.). New Riders (2014)
5. Kirk, A.: Data Visualisation: A Handbook for Data Driven Design. SAGE Publications (2016)
6. Hullman, J., Diakopoulos, N.: Visualization rhetoric: framing effects in narrative visualization. IEEE Trans. Visual Comput. Graphics **17**(12), 2231–2240 (2011)

7. Xiong, C., Setlur, V., Bach, B., Lin, K., Koh, E., Franconeri, S.: Visual arrangements of bar charts influence comparisons in viewer takeaways. arXiv. https://doi.org/10.48550/arXiv.2108.06370 (2021)
8. Cairo, A.: The functional art: an introduction to information graphics and visualization. New Riders (2013)
9. Chen, M., Floridi, L., Borgo, R.: What is visualization really for? Comput. Graph. **38**, 1–6 (2014)
10. Evergreen, S.D.H.: Presenting data effectively: communicating your findings for maximum impact. SAGE Publications (2014)
11. Sadiku, M.N.O., Tembely, M., Musa, S.M.: Data visualization. Int. J. Eng. Res. Adv. Technol. **2**(10), 163–166 (2016)
12. Abieba, M., Jensen, C., Sánchez, J.: The impact of data visualization on decision-making in software engineering. J. Syst. Softw. **189**, 111288 (2025)
13. Midway, S.R.: Principles of effective data visualization. Patterns **1**(6), 100103 (2019)
14. Weissgerber, T.L., Garovic, V.D., Savic, M., Winham, S.J., Milic, N.M.: Doing better data visualization. Am. J. Physiol. Heart Circ. Physiol. **317**(1), H9–H18 (2019)
15. Jambor, T.: Creating effective scientific figures: a practical checklist. Nat. Methods **18**, 1003–1006 (2021)
16. Cukier, K., Mayer-Schoenberger, V.: Big data: a revolution that will transform how we live, work, and think. Eamon Dolan/Houghton Mifflin Harcourt (2013)
17. Kim, N.W., Elmqvist, N., Lee, B.: ChartOptimiser: Task-driven optimisation of chart designs. ACM Trans. Comput. Hum. Interact. **29**(2), 1–40 (2022)
18. Tan, P.N., Steinbach, M., Kumar, V.: Introduction to Data Mining (2nd ed.). Pearson (2018)

Generative AI and the Future of Social Content: Opportunities and Challenges

Enhancing K-means Clustering in B2B Customer Segmentation: A Comparative and Hybrid Approach of Recursive Feature Elimination, Correlation Analysis, and Lasso Regularization

Daisy Ipatzi Bello[1](✉) ⓘ, Sabeen Tahir[2] ⓘ, and Stefania Paladini[3] ⓘ

[1] University of Essex, Colchester, UK
daisy.ipatzi@gmail.com
[2] Cardiff Metropolitan University, Cardiff, UK
stahir@cardiffmet.ac.uk
[3] Queen Margaret University, Edinburgh, Scotland
spaladini@qmu.ac.uk

Abstract. This paper evaluates the effectiveness of three feature selection techniques—Recursive Feature Elimination (RFE), Correlation Analysis, and Lasso Regularisation—in enhancing K-means clustering for B2B customer segmentation. Using a quantitative case study approach, the research assesses the individual and combined impact of these methods on clustering performance. The dataset, comprising anonymised B2B interactions from a wholesale distribution company, presented a high-dimensional and complex environment in which to test these techniques. Findings indicate that a hybrid approach—applying Lasso Regularisation, RFE, and Correlation Analysis in sequence—outperforms the individual methods. This integrated strategy improves silhouette scores and cluster cohesion, resulting in more accurate and interpretable segmentation. The study demonstrates that combining these techniques produces a robust framework that yields actionable insights for targeted marketing, resource allocation, and customer engagement within B2B contexts.

Keywords: K-means Clustering · B2B Customer Segmentation · Feature Selection · Recursive Feature Elimination (RFE) · Correlation Analysis · Lasso Regularization · Hybrid Feature Selection Techniques · Machine Learning in Marketing · Data-Driven Decision Making · Clustering Algorithms · Market Segmentation · Silhouette Score · Cluster Validation · Customer Profiling

1 Introduction

In Business-to-Business (B2B) markets, excelling in customer segmentation is essential for securing a competitive edge. This process allows companies to identify distinct groups within their customer base, significantly enhancing the effectiveness of personalisation, targeted advertising, and promotional strategies. Such precision in segmentation is crucial for deepening understanding of customer behaviour within the B2B framework, delivering considerable benefits to both academic researchers and industry professionals [22, 28]. The integration of data science and machine learning techniques, especially feature selection combined with clustering algorithms like K-means, substantially refines this segmentation process. These advancements lead to more detailed and accurate market segmentation, cultivating business strategies that are not only more precise but also adaptable and responsive to evolving market demands [18]. Current literature suggests that these technologies facilitate more detailed and precise market segmentation, thereby enabling business strategies that are both exact and flexible under dynamic market conditions [18]. However, the success of these models largely depends on the quality of the selected features. Poor feature selection can yield misleading clusters, distorting insights and adversely impacting strategic decisions [16]. As the complexity of customer data grows, there is a pressing demand for more sophisticated analytical capabilities to effectively manage and utilise this information. Effective segmentation is fundamental in B2B decision-making, influencing strategic planning and operational efficiency. Empirical studies indicate that aligning pricing strategies with customer needs, preferences, and sensitivities substantially boosts profitability and strengthens market positioning [24].

This paper evaluates the effectiveness of three feature selection methods—Recursive Feature Elimination (RFE), Correlation Analysis, and Lasso Regularisation—in refining K-means clustering for customer segmentation in B2B settings. Each method is assessed for its potential to enhance clustering accuracy, with RFE optimising dimensionality reduction [26], Correlation Analysis clarifying variable interdependencies [4], and Lasso Regularisation addressing overfitting issues [17]. Despite their proven advantages, these techniques also exhibit inherent limitations, which this study aims to address through the development of a novel hybrid feature selection approach.

Article Structure:

1. Literature Review – explores the academic and professional discourse on feature selection and its application in B2B segmentation.
2. Methodology – outlines the hybrid approach combining insights from RFE, Correlation Analysis, and Lasso Regularisation.
3. Results and Analysis – presents the empirical outcomes of the hybrid method on segmentation performance.
4. Discussion – evaluates the practical and strategic implications of the results for B2B contexts.
5. Conclusion and Future Directions – summarises key contributions and proposes avenues for further research.

This study not only contributes to the academic literature on feature selection and machine learning but also demonstrates the practical applications of these methodologies in enhancing B2B segmentation.

2 Literature Review

2.1 Importance of B2B Segmentation and K-means Clustering

K-means clustering is a fundamental unsupervised learning algorithm widely used in analytical applications for its capacity to identify distinct groups within datasets. Originally introduced in the 1950s, the algorithm partitions n data points into k clusters by minimising intra-cluster variance, quantified as the sum of squared distances between each data point and its corresponding centroid [10]. In Business-to-Business (B2B) contexts, where customer datasets are typically complex and multidimensional, K-means is particularly valuable. Its ability to handle large-scale data enables organisations to uncover behavioural patterns and customer heterogeneity. For instance, firms can segment customers based on transactional frequency and monetary value, thereby tailoring marketing strategies to specific needs. This targeted approach enhances customer engagement and retention, while optimising resource allocation and operational efficiency [15].

2.2 Lasso Regularization

Lasso (Least Absolute Shrinkage and Selection Operator) is a feature selection technique that enhances model interpretability, reduces overfitting, and improves generalisation in high-dimensional datasets. In B2B customer segmentation, where datasets often contain numerous interdependent variables, Lasso Regularisation helps eliminate redundant features, leading to more precise clustering and segmentation [21, 27]. The primary objective of Lasso is to minimise the residual sum of squares while imposing a penalty on the absolute sum of the coefficients, ensuring that irrelevant or less impactful features are shrunk to zero [12]. Mathematically, the Lasso optimisation problem is represented as follows:

$$\text{Minimize } \|y - X\beta\|_2^2 + \alpha\|\beta\|_1 \qquad (1)$$

where y represents the vector of dependent variables, X denotes the matrix of independent variables, β is the vector of coefficients, $\|\cdot\|_2$ denotes the Euclidean norm, $\|\cdot\|_1$ denotes the L1 norm, and α is the regularization parameter controlling the penalty imposed on the size of the coefficients. By introducing L1 regularisation, Lasso forces some coefficients to shrink exactly to zero, effectively performing feature selection. This is particularly beneficial in B2B segmentation, where datasets often exhibit high dimensionality and multicollinearity [19]. The ability to retain only the most relevant variables enhances both computational efficiency and model interpretability. Despite its advantages, Lasso has limitations, particularly in handling correlated predictors. Since it selects only one feature from a group of highly correlated variables while disregarding the others, it may exclude important predictors that influence segmentation outcomes [27]. This is particularly problematic in B2B environments, where financial, transactional, and behavioural metrics are often interdependent. Another key challenge lies in selecting the optimal value of the regularisation parameter (α), which significantly influences performance:

a) If α is too high, excessive penalisation may remove valuable features, leading to underfitting.

b) If α is too low, insufficient penalisation may retain too many features, resulting in overfitting and reduced generalisability.

2.3 Recursive Feature Elimination (RFE)

Recursive Feature Elimination (RFE) is an iterative feature selection method that refines predictive modelling by systematically removing features contributing least to model performance. It is particularly effective in high-dimensional datasets, such as those encountered in B2B customer segmentation, where multiple interdependent variables influence purchasing behaviours [9, 20]. The RFE process follows an optimisation-driven approach, mathematically represented as:

$$Minimize\ J(\beta) = \|y - X\beta\|^2 \tag{2}$$

where y represents the vector of dependent variables, X is the matrix of independent variables, β denotes the coefficient vector, and $J(\beta)$ the cost function measuring model fit. RFE mitigates complexity by selecting only the most relevant features, thereby enhancing clustering precision and interpretability. However, it is computationally intensive—especially for large-scale datasets—as it requires repeated model training [13]. Its performance also depends on the model used to rank feature importance, which may introduce bias. Moreover, RFE may discard features that are weak in isolation but predictive in combination, underscoring the potential of hybrid approaches integrating RFE with Lasso Regularisation or adaptive techniques to boost robustness.

2.4 Correlation Analysis

Correlation-based Feature Selection (CFS) reduces dimensionality by identifying features that are highly correlated with the target outcome and minimally correlated with one another. This technique is particularly suitable in high-dimensional B2B segmentation tasks, where redundant or irrelevant features can obscure patterns [23]. Mathematically, CFS aims to identify:

$$\left|\rho(x_i, x_j)\right| \geq \theta \tag{3}$$

where x_i and x_j are individual features within a dataset, $\rho(\cdot)$ denotes the correlation coefficient, and θ a pre-defined threshold.

Pearson correlation was chosen for this study due to its computational efficiency and clarity in capturing linear dependencies—characteristics often present in financial and transactional B2B data [3]. Although alternative approaches like Spearman correlation or mutual information-based selection [5] can better capture non-linear associations, they are more computationally intensive. Nonetheless, CFS presents limitations. As noted by Saha et al. [22], it may exclude features that, while weakly correlated individually, have significant combined predictive value. This is especially problematic in composite models, where variable interactions play a critical role. Moreover, because CFS primarily detects linear dependencies, it may miss important non-linear patterns [5]. Jen and Oh [11] proposed a Hybrid-RFE method that integrates correlation-based filters with recursive elimination using models such as support vector machines, random forests, and gradient boosting algorithms. Their approach demonstrated superior accuracy and stability compared to standalone methods by leveraging the strengths of both filter and wrapper strategies.

2.5 Justification for Method Selection

Lasso Regularisation, Recursive Feature Elimination, and Correlation Analysis were selected for their complementary strengths in high-dimensional feature spaces. Lasso induces sparsity by penalising irrelevant features, RFE iteratively eliminates weak predictors based on performance, and CFS reduces multicollinearity by excluding highly correlated variables.

Their combined use provides a robust, interpretable, and computationally feasible framework tailored to structured B2B data. Compared to deep learning-based or opaque embedded methods, these techniques offer transparency and adaptability, making them particularly well-suited for segmentation tasks in enterprise environments.

3 Methodology

3.1 Research Design

This study adopts a quantitative case study approach to evaluate the impact of three feature selection techniques—Recursive Feature Elimination (RFE), Correlation Analysis, and Lasso Regularisation—on K-means clustering for B2B customer segmentation (see Fig. 1). The dataset utilised in this research comprises anonymised B2B interactions sourced from a company specialising in wholesale distribution. This study also proposes a combined approach that leverages the strengths of all three techniques to improve feature selection and clustering performance.

3.2 Data Collection

The dataset is derived from a proprietary B2B company database, incorporating key variables such as customer demographics, purchasing behaviours, engagement metrics, and financial indicators. This rich, high-dimensional dataset serves as a foundation for evaluating feature selection techniques in the context of customer segmentation. To ensure confidentiality and ethical compliance, all data handling followed the General Data Protection Regulation (GDPR). Preprocessing steps were conducted prior to feature selection to ensure integrity and consistency across variables. The dataset was stored in encrypted format, and, in accordance with the agreement with the contributing organisation, all data was permanently deleted upon completion of the study.

3.3 Data Preprocessing

To assess the effectiveness of feature selection in improving K-means clustering, the study applies four distinct approaches:

- Label Encoding: Categorical variables were converted to numeric format using label encoding, assigning unique integers to each category to enable algorithmic processing [1].
- Handling Missing Data: Missing values were addressed using K-Nearest Neighbours (KNN) imputation, which preserves the dataset's underlying structure and inter-feature relationships [7].
- Feature Normalisation: Quantile transformation was applied to stabilise variance and reduce the influence of outliers by mapping values to a uniform or normal distribution. This technique has been shown to improve clustering performance, especially in datasets with skewed distributions [25].

3.4 Feature Selection

This study examines four approaches to feature selection:

- Lasso Regularisation – Applies L1 penalisation to shrink less informative feature coefficients to zero, improving interpretability, see Eq. 1.
- Recursive Feature Elimination (RFE) – Iteratively removes the least significant features to refine the dataset, Eq. 2.
- Correlation Analysis – Filters out features with high inter-correlation while retaining those with strong associations to the target behaviour, Eq. 3.
- Hybrid Approach – Integrates the above three techniques sequentially to maximise clustering performance.

3.5 Hybrid Feature Selection Pipeline

The hybrid approach applies the techniques in three stages to construct a refined, high-impact feature set prior to K-means clustering.

- **Stage 1: Initial Feature Selection with Lasso Regularisation:** Lasso applies L1 penalisation to reduce the feature set by removing variables with low or zero coefficients. The selected features S_L are defined as:

$$S_L = \{X_j \mid \beta_j \neq 0\} \quad (4)$$

where features with non-zero coefficients (β_j) are retained, irrelevant features are shrunk to zero and removed and the initial set of features S_L is obtained.
- **Stage 2: Feature Refinement with RFE:** From S_R, RFE further eliminates less significant features by minimising the cost function $J(\beta)$ in Eq. 2. The refined set is:

$$S_R = S_L - \{X_j \mid \min J(\beta)\} \quad (5)$$

resulting in a more compact subset S_R containing only the most predictive variables.
- **Stage 3: Final Optimisation with Correlation Analysis:** Highly correlated variables from S_R, are identified using Pearson's correlation coefficient:

$$C_{ij} = \frac{Cov(X_i, X_j)}{\sigma X_i \sigma X_j} \quad (6)$$

where C_{ij} represents the correlation between features X_i and X_j, $Cov(X_i, X_j)$ is the covariance and, $\sigma X_i, \sigma X_j$ are the standard deviations. After applying (Eq. 3) the final output is:

$$S_F = S_L - \{X_j \mid min\ J(\beta)\} - \{X_i | |C_{ij}| \geq \theta\} \tag{7}$$

4 Results and Analysis

4.1 Dataset Description

The dataset comprises 11,904 anonymised B2B interaction records drawn from a wholesale distribution company, incorporating 65 attributes that span multiple dimensions of customer information, including:

- Demographic Data: Attributes such as Region, Country, and Customer Type, which provide insight into geographical and categorical customer segmentation.
- Transactional Data: Summarised purchasing behaviours and financial contributions, with deterministic noise added to ensure compliance with data protection protocols.
- Engagement Metrics: Email communication data measuring customer responsiveness and activity levels.
- Temporal Data: Features such as "Customer Creation Date", "First Shipment Date", and "Last Booking Date", reflecting customer lifecycle trends.

An initial assessment of the dataset revealed inconsistencies and missing values, particularly within transactional and financial attributes. While core customer identifiers and demographic fields were largely complete, financial and transactional records exhibited significant data gaps. These patterns indicate that certain customer segments lack complete transaction histories—an issue frequently encountered in B2B sales environments due to varied engagement levels and system integration limitations. The specific preprocessing strategies employed to address these challenges are detailed in Sect. 3.3.

4.2 Data Munging and Preprocessing Pipeline.

As outlined in Fig. 2, the dataset was cleaned and transformed using a structured munging pipeline aligned with best practices [2]:

1. Encoding Categorical Variables: Label encoding was applied to features such as Region and Country to maintain low dimensionality and preserve categorical structure [8].
2. Transforming Temporal Fields: Temporal columns were converted into numerical features, e.g., transforming "Customer Creation Date" to "Customer Age in Months".
3. Email Engagement Metrics: Metrics were standardised using formulas:

$$Email\ Open\ Rate = \frac{Total\ Email\ Opens}{Emails\ Delivered} * 100 \qquad (8)$$

$$Email\ Click\ Rate = \frac{Total\ Email\ Clicks}{Emails\ Delivered} * 100 \qquad (9)$$

$$Email\ Contact\ Rate = \frac{Total\ Emails\ Contacted}{Emails\ Delivered} * 100 \qquad (10)$$

4. Data Refinement: Customers with no recorded revenue, profit, or engagement activity were excluded from the active segmentation process and grouped under a reference category labelled "Cluster 99". Some ratio-based metrics derived from financial values were also removed. The attribute count was reduced from 61 to 49.

4.3 Exploratory Data Analysis (EDA)

EDA highlighted essential patterns relevant to feature selection and clustering. Customer lifecycle metrics, such as "Customer Age in Months," revealed an average engagement span of 16.55 months, with some extending to 21. The distribution of engagement indicators like "Email Open Rate" showed strong right-skewness, indicating that a small group of customers is highly responsive.

Additionally, the analysis revealed a strong correlation between customer longevity and shipping frequency, highlighting the relevance of correlation-based filtering. Outliers were observed in variables such as "Time Since First Shipment" and "All Time Shipped Revenue with Noise," which led to the application of transformation techniques to normalise skewed distributions.

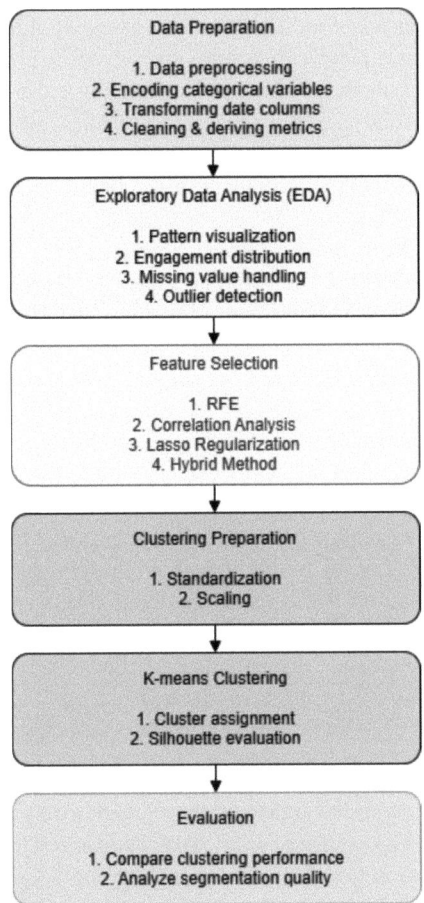

Fig. 1. Flowchart of the research methodology for evaluating feature selection methods on K-means clustering in B2B customer segmentation.

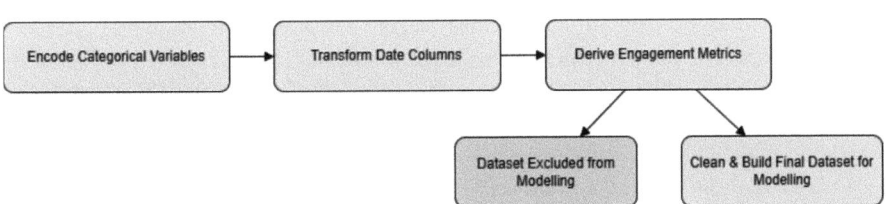

Fig. 2. Flowchart of Data Munging Process.

4.4 Implementation of Feature Selection Techniques

The feature selection process in this study followed Eqs. 1, 2, 3, and 7 to construct an optimised framework for B2B customer segmentation. Each technique was systematically applied to reduce dimensionality, eliminate redundant variables, and retain those most predictive of customer behaviour.

Lasso Regularisation was conducted across a range of alpha (α) values (0.0001 to 0.01), balancing model accuracy with feature sparsity (Eq. 1). This method penalises feature coefficients, shrinking those of irrelevant variables to zero, thus reducing overfitting and enhancing model clarity. It preserved key attributes including "Customer Age in Months", "All Time Shipped Orders", and "Email Open Rate", resulting in a reduction from 65 to 45 features. The method notably improved interpretability and computational efficiency.

Recursive Feature Elimination (RFE), implemented using a Random Forest Regressor, dynamically selected the optimal subset of features via cross-validation (Eq. 2). This technique effectively prioritised features with high predictive value and was well-suited for capturing non-linear relationships. RFE retained essential metrics such as "All Time Shipped GP with Noise", "Last 90 Days Quoted Revenue", and "Email Contact Rate".

Correlation (Eq. 3) enhanced feature independence by removing variables with high multicollinearity. A threshold of $\theta = 0.75$ was determined to yield the best silhouette score. This approach eliminated redundancy while maintaining the interpretability of transactional and financial attributes. It reduced the set to 22 features.

To combine the benefits of the individual techniques, the **hybrid approach** applied Lasso → RFE → Correlation Analysis sequentially (Eq. 7). This process preserved a core subset of 6 features spanning customer engagement, operational activity, and financial performance. It successfully balanced predictive power with interpretability, offering a robust solution for high-dimensional B2B data.

4.5 Clustering Results and Evaluation

K-means clustering was applied as an unsupervised method to partition customers based on purchasing behaviour, engagement, and financial attributes, following the application of each feature selection technique. Clustering effectiveness was assessed using silhouette scores, which evaluate the cohesion and separation of resulting clusters. Higher scores indicate more distinct and meaningful segmentation [7], see Table 1 and Fig. 3.

The Combined Feature Selection Approach achieved the highest silhouette score (0.2461), indicating superior cluster quality. In contrast, Correlation Analysis alone produced the lowest scores, suggesting that excessive feature elimination reduced cluster differentiation. Lasso Regularisation retained essential features but was affected by feature correlation, while RFE demonstrated strong performance until the feature set was overly reduced. Individually, Lasso and RFE yielded moderate results, but their integration in the hybrid method significantly enhanced segmentation.

To validate these results, Principal Component Analysis (PCA) was conducted before and after clustering to visualise separation between groups. Prior to clustering, data points lacked clear structure, reinforcing the importance of feature selection. Post-clustering results revealed:

- Lasso and RFE improved cluster compactness, though some overlap persisted.
- Correlation Analysis produced weaker separation, indicating insufficient discriminative power.
- The Hybrid Approach delivered the most distinct clusters, consistent with its top silhouette score.

The visual distinction achieved through the hybrid approach is illustrated in Fig. 4, which shows clear cluster separation after applying K-means to the refined feature set.

Table 1. Silhouette Scores for Each Feature Selection Method.

Method	Number of Clusters	Silhouette Score
Lasso	3	0.1876
Lasso	4	0.2313
Lasso	5	0.1297
Lasso	6	0.2326
RFE	3	0.1647
RFE	4	0.2338
RFE	5	0.1253
RFE	6	0.2508
Correlation	3	0.1590
Correlation	4	0.2226
Correlation	5	0.1248
Correlation	6	0.2409
Combined	3	0.1433
Combined	4	0.2266
Combined	5	0.1313
Combined	6	0.2461

4.6 Profiling the B2B Customer Segments

To develop the customer profiles, the centroids of the final clustering output—generated using the hybrid feature selection method and K-means clustering—were analysed across key indicators including revenue, gross profit, order frequency, quotation activity, and engagement metrics such as email open and click-through rates.

By evaluating the average values of these attributes for each cluster, behavioural patterns were identified and used to construct descriptive labels and practical recommendations for each segment. This approach ensures that the segmentation model reflects meaningful distinctions in customer behaviour and supports actionable business strategies. The inclusion of Cluster 99, comprising inactive or non-engaged accounts, provided a reference point for interpreting low-activity patterns and planning re-engagement initiatives. The clustering model included both active and inactive accounts (Cluster 99), offering a complete view of B2B customer behaviours. This inclusive segmentation supports tailored engagement, efficient resource use, and improved retention.

- Cluster 99 – Inactive/New: Accounts with no revenue, profit, or transactional history. Serves as a baseline to flag disengagement patterns. Proposed action: Enables proactive re-engagement via automated campaigns, targeted incentives, or churn prediction.
- Cluster 0 – Emerging Entrants: Newly engaged customers with moderate revenue and low engagement. Proposed action: Prioritise onboarding, personalised outreach, and retention incentives.
- Cluster 1 – Growing Engagers: Customers with steady orders and moderate activity. Proposed action: Action: Encourage upselling via promotions, and transition into high-value tiers.
- Cluster 2 – High-Value Partners: Most engaged and profitable customers. Proposed action: Strengthen loyalty through tailored pricing, premium service, and dedicated account management.

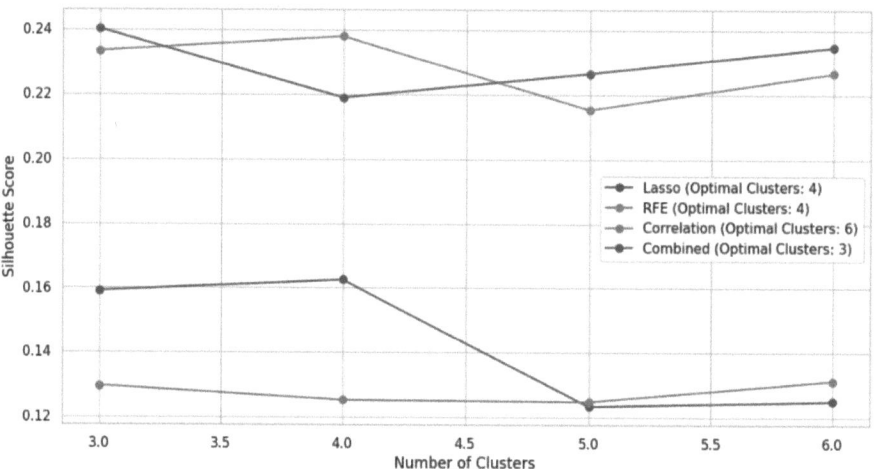

Fig. 3. Comparison of Silhouette Scores Across Different Clustering Methods.

4.7 Conclusions and Discussions

This study examined the effectiveness of feature selection techniques —Lasso Regularisation, Recursive Feature Elimination (RFE), and Correlation Analysis— in improving K-means clustering for B2B customer segmentation within a real-world use case. The findings demonstrate that this integrated approach significantly improves clustering outcomes, achieving the highest silhouette score (0.2461), indicating stronger cluster cohesion and enhanced segmentation accuracy.

The use case highlights that Lasso Regularisation effectively reduces feature dimensionality by eliminating less relevant variables, though it can struggle with highly correlated predictors. RFE enhances segmentation by selecting the most predictive features, but its iterative nature makes it computationally intensive. Correlation Analysis removes multicollinear variables, improving feature independence, but it may exclude features that, while correlated, still hold predictive value when considered in combination. The hybrid approach helps to address these limitations, ensuring a more balanced selection process that enhances both interpretability and clustering performance in the specific business environment examined.

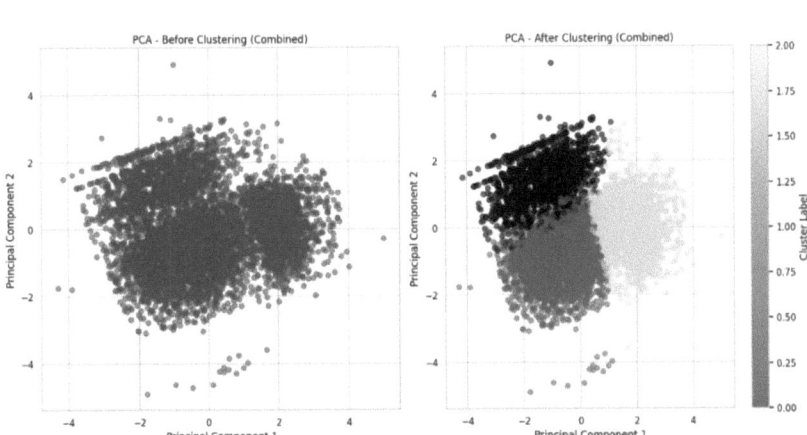

Fig. 4. PCA Visualization Before and After K-means Clustering Using Hybrid Approach

From a business perspective, these findings have practical implications for B2B customer segmentation strategies. By applying refined clustering models, the companies can improve targeted marketing efforts, optimise resource allocation, and enhance customer engagement strategies. The study identified four distinct customer segments: Emerging Entrants (Cluster 0), Growing Engagers (Cluster 1), High-Value Partners (Cluster 2), and Inactive or New Customers (Cluster 99). The inclusion of Cluster 99, representing accounts with little to no recent activity, was particularly crucial for maintaining data integrity and ensuring that segmentation models were not distorted by dormant or newly onboarded accounts. Understanding how customers transition between these clusters provides valuable insights into retention strategies, re-engagement initiatives, and customer lifecycle management tailored to the company's operations.

Beyond its business applications, this use case contributes to the growing field of hybrid feature selection in machine learning, particularly in high-dimensional datasets where traditional methods may struggle with redundancy and interpretability. By providing a structured, data-driven approach to feature selection, this study offers a replicable methodology that can be applied in similar B2B settings.

Despite its contributions, the study acknowledges several areas for further research. Future work could explore real-time customer segmentation models that dynamically update clusters based on changing customer behaviours, ensuring that segmentation remains adaptive to market fluctuations and evolving engagement patterns. Additionally, the effectiveness of alternative clustering methods, such as DBSCAN or Gaussian Mixture Models, should be examined, as these techniques may offer greater flexibility in capturing complex, non-spherical distributions. The incorporation of deep learning-based feature selection techniques, such as autoencoders and self-organising maps, could further improve segmentation accuracy by capturing non-linear relationships between variables. Furthermore, validating this methodology across different B2B industries, such as manufacturing, financial services, and supply chain management, would strengthen its applicability and demonstrate its effectiveness beyond this specific use case.

Despite the demonstrated advantages, the hybrid approach also presents certain limitations. Its sequential structure assumes that features eliminated in earlier stages (e.g., through Lasso) lack relevance, potentially overlooking those that may hold predictive power in combination with others. Moreover, the performance evaluation was conducted on a single B2B dataset within a specific business context. While this strengthens the value of the study as a practical use case, it also highlights the need for validation in alternative domains. Nonetheless, the structured, transparent nature of the proposed methodology makes it readily replicable across industries.

Overall, this study demonstrates that a hybrid feature selection strategy significantly enhances clustering performance within a real-world B2B use case, leading to more accurate and actionable customer segmentation. Future studies should aim to expand the use of this method to include real-time clustering, explore different machine learning techniques, and validate its effectiveness across various industries. Ultimately, this study offers a scalable, data-driven framework for customer segmentation, equipping businesses with the analytical tools needed to enhance customer engagement, strengthen retention strategies, and drive data-informed decision-making in real-world B2B contexts.

Acknowledgments. This research was supported by the UK-Saudi Challenge Fund 2024, British Council, under the project titled "Data Fusion Dynamics: A Collaborative UK-Saudi Initiative in Cybersecurity and Artificial Intelligence".

References

1. Bolikulov, F., Nasimov, R., Rashidov, A., Akhmedov, F. and Cho, Y.-I.: Effective Methods of Categorical Data Encoding for Artificial Intelligence Algorithms. Math. [online] **12**(16), 2553–2553 (2024). https://doi.org/10.3390/math12162553
2. Boschetti, A., Massaron, L.: Python Data Science Essentials: a Practitioner's Guide Covering Essential Data Science Principles, Tools, and Techniques, 3rd edn. Packt Publishing Ltd., Birmingham (2018)
3. Chan, J.Y.-L., Leow, S.M.H., Bea, K.T., Cheng, W.K., Phoong, S.W., Hong, Z.-W., Chen, Y.-L.: Mitigating the Multicollinearity Problem and Its Machine Learning Approach: a Review. Math. [online] **10**(8), 1283 (2022). https://doi.org/10.3390/math10081283
4. Cihan Kuzudisli, Burcu Bakir-Gungor, Nurten Bulut, Qaqish, B., Yousef, M.: Review of Feature Selection Approaches Based on Grouping of Features. PeerJ, [online] 11, pp. e15666–e15666 (2023). https://doi.org/10.7717/peerj.15666
5. Darzi, E., Sijtsema, N.M., Ooijen, van.: A Comparative Study of Federated Learning Methods for COVID-19 Detection. Sci. Rep. **14**(1) (2024). https://doi.org/10.1038/s41598-024-54323-2
6. Emadi, S., Limongiello, M.: Optimizing 3D Point Cloud Reconstruction through Integrating Deep Learning and Clustering Models. Electronics **14**(2), 399 (2025). https://doi.org/10.3390/electronics14020399

7. Frölich, N., Klose, C., Widén, E., Ripatti, S., Gerl, M.J.: Imputation of Missing Values in Lipidomic Datasets. Proteomics **24**(15) (2024). https://doi.org/10.1002/pmic.202300606
8. Galli, S.: Feature-engine: a Python Package for Feature Engineering for Machine Learning. J. Open-Source Softw. **6**(65), 3642 (2021). https://doi.org/10.21105/joss.03642
9. Hamada, M., Tanimu, J.J., Hassan, M., Kaduki, H.A.: Evaluation of Recursive Feature Elimination and LASSO Regularization-based Optimized Feature Selection Approaches for Cervical Cancer Prediction. In: 2021 IEEE 14th International Symposium on Embedded Multicore/Many-core Systems-on-Chip (MCSoC). [online] 2021 IEEE 14th International Symposium on Embedded Multicore/Many-core Systems-on-Chip (MCSoC). Singapore: IEEE (2021). https://doi.org/10.1109/MCSoC51149.2021.00056
10. Ikotun, A.M., Ezugwu, A.E., Abualigah, L., Abuhaija, B., Heming, J.: K-means Clustering Algorithms: a Comprehensive Review, Variants Analysis, and Advances in the Era of Big Data. Inf. Sci. **622**(622) (2022). https://doi.org/10.1016/j.ins.2022.11.139
11. Jeon, H., Oh, S.: Hybrid-Recursive Feature Elimination for Efficient Feature Selection. Appl. Sci. **10**(9), 3211 (2020). https://doi.org/10.3390/app10093211
12. Li, X., Xu, S., Li, X., Wang, Y., Sheng, Y., Zhang, H., Yang, W., Yuan, D., Jin, T., He, X.: Novel Insight into the Genetic Signatures of Altitude Adaptation Related Body Composition in Tibetans. Front. Public Health [online] **12**, 1355659 (2024). https://doi.org/10.3389/fpubh.2024.1355659
13. López-De-Castro, M., García-Galindo, A., Armañanzas, R.: Conformal Recursive Feature Elimination. arXiv (Cornell University) (2024). https://doi.org/10.48550/arxiv.2405.19429
14. Mattsson, U.: Practical Data Security and Privacy for GDPR and CCPA. [online] ISACA (2020). https://www.isaca.org/resources/isaca-journal/issues/2020/volume-3/practical-data-security-and-privacy-for-gdpr-and-ccpa. Accessed 26 Jan 2025
15. Maulina, N.R., Surjandari, I., Rus, A.M.M.: Data Mining Approach for Customer Segmentation in B2B Settings Using Centroid-Based Clustering. In: 2019 16th International Conference on Service Systems and Service Management (ICSSSM) (2019)
16. McCrory, M., Thomas, S.A.: Cluster Metric Sensitivity to Irrelevant Features. ArXiv (Cornell University) (2024). https://doi.org/10.48550/arxiv.2402.12008
17. Pak, A., Rad, A.K., Nematollahi, M.J., Mahmoudi, M.: Application of the Lasso Regularisation Technique in Mitigating Overfitting in Air Quality Prediction Models. Sci. Rep. [online] **15**(1) (2025). https://doi.org/10.1038/s41598-024-84342-y
18. Pandey, A.K., Goyal, A., Sikka, N.: RE-RFME: Real-Estate RFME Model for Customer Segmentation. [online] arXiv.org (2024). https://arxiv.org/abs/2404.17177. Accessed 27 Jan 2025
19. Passemiers, A., Folco, P., Raimondi, D., Birolo, G., Moreau, Y., Fariselli P.: A Quantitative Benchmark of Neural Network Feature Selection Methods for Detecting Nonlinear Signals. Sci. Rep. [online] **14**(1) (2024). https://doi.org/10.1038/s41598-024-82583-5
20. Priyatno, A.M., Widiyaningtyas, T.: A SYSTEMATIC LITERATURE REVIEW: RECURSIVE FEATURE ELIMINATION ALGORITHMS. JITK (Jurnal Ilmu Pengetahuan dan Teknologi Komputer) **9**(2), 196–207 (2024). https://doi.org/10.33480/jitk.v9i2.5015
21. Pudjihartono, N., Fadason, T., Kempa-Liehr, A.W., O'Sullivan, J.M.: A Review of Feature Selection Methods for Machine Learning-Based Disease Risk Prediction. Front. Bioinf. [online] **2** (2022). https://doi.org/10.3389/fbinf.2022.927312
22. Bilro, R.G., Maria, S., Pedro Souto, A.: A Systematic Review of Customer Behavior in business-to-business Markets and Agenda for Future Research. Journal of Business & Industrial Marketing **38**(13), 122–142 (2023). https://doi.org/10.1108/jbim-07-2022-0313
23. Saha, P., Patikar, S., Neogy, S.: A Correlation - Sequential Forward Selection Based Feature Selection Method for Healthcare Data Analysis. 2020 IEEE International Conference on Computing, Power and Communication Technologies (GUCON) (2020). https://doi.org/10.1109/gucon48875.2020.9231205

24. Tavor, T., Gonen, L.D., Spiegel, U.: Customer Segmentation as a Revenue Generator for Profit Purposes. Math. [online] **11**(21), 4425 (2023). https://doi.org/10.3390/math11214425
25. Wongoutong, C.: The Impact of Neglecting Feature Scaling in k-means Clustering. PLoS ONE **19**(12), e0310839 (2024). https://doi.org/10.1371/journal.pone.0310839
26. Xia, S., Yang, Y.: A model-free Feature Selection Technique of Feature Screening and Random Forest Based Recursive Feature Elimination. [online] arXiv.org (2023). https://arxiv.org/abs/2302.07449. Accessed 23 Feb 2025
27. Zhang, H., Wang, J., Sun, Z., Zurada, J.M., Pal, N.R.: Feature Selection for Neural Networks Using Group Lasso Regularization. **32**(4), 659–673 (2020). https://doi.org/10.1109/tkde.2019.2893266
28. Zhang, W.W., Misra, S.: Coarse Personalization. ArXiv (Cornell University) (2022). https://doi.org/10.48550/arxiv.2204.05793

Virtual or Human, Real or Rendered: Assessing the Persuasiveness of Synthetic Influencers

Karina Sokolova[✉]

Paris School of Business, Paris, France
k.sokolova@psbedu.paris

Abstract. Social media influencers have become key in marketing, with their authenticity, credibility, and emotional connections driving customer purchases. The rise of virtual influencers introduces new dimensions in influencer marketing. Yet, it's uncertain if the effectiveness seen in human influencers translates to virtual ones. Our article compares their marketing persuasiveness through two studies. The first study contrasts Instagram photos of an existing human and a virtual influencer. The second assesses a human-created YouTube video versus an artificial one generated from the photo of the same model. We aimed to compare authenticity, credibility, novelty, and parasocial relationships between the groups and their impact on purchase intentions in human/virtual contexts. Our results suggest that the findings regarding human influencers also apply to virtual influencers, and that their virtual nature is not necessarily noticed or evaluated by viewers. As such, artificial images can serve as marketing alternatives to real videos or photographs. However, perceived authenticity still matters for virtual models; therefore, their background story, appearance, and context should provide realistic cues for interpretation.

Keywords: Influence marketing · Virtual influencer · Authenticity · Credibility · Parasocial interaction

1 Introduction

Social media influencers have become an effective form of advertising [5]. The literature outlines that the power of influencers lies in their perceived attractiveness, trustworthiness, credibility, authenticity [26], and to the sense of emotional connection or parasocial relationships they can build with their audience [10, 28]. Influencers effectively mediate the relationships between brands and customers, promoting products and incentivizing purchases [33].

Recent developments in digital tools have given rise to a new type of influencer on social media known as 'virtual influencers' [20, 38]. While traditional influencers typically serve as content creators, speakers, and personalities on their channels, virtual influencers are fictional characters brought to life through technological means and presented as the owners of social media accounts. Lil Miquela, a fictional character who debuted on Instagram in 2016 emulates the behavior of human influencers and has a realistic appearance [36]. Although these virtual characters experience high visibility and offer certain advantages to brands, it remains uncertain whether their marketing power equals that of human influencers [3]. As virtual influencers are computer-designed personas mimicking humans, one can question their ability to be effective in persuasion and to be perceived as authentic and credible. More importantly, would authenticity, credibility and parasocial link [37] be still reliable in predicting purchase intentions when the virtual influencer is involved? This is the question that we aim to explore in this article.

To address our research questions, we undertake two studies involving French social media users to compare participants' perceptions of virtual and human influencers. The first study utilizes photographic content from Instagram, comparing virtual influencer Lil Miquela with a human influencer. The second study juxtaposes a genuine YouTube video featuring a beauty influencer against an artificially generated. PLS-SEM (Partial Least Squares Structural Equation Modeling) is employed for data analysis. Subsequent sections provide a detailed description of our approach.

2 Related Works

The influencer marketing literature outlines the importance of parasocial relationship, credibility, and authenticity for their marketing effectiveness [33,34,37]. Although virtual influencers are non-human agents, the "computers are social actors" paradigm shows that people often treat non-human entities like humans, which can lead them to perceive virtual influencers much like they do real people [22,24,30]. Thus, being objectively fake, research questions whether virtual agents might be capable of generating feelings of authenticity and relate to the audience or whether the actual authenticity is truly relevant [3,30]. Also, credibility was shown to be important for marketing power of social media influencers [32,33]. Thus, perceptions of inauthenticity seem to reduce the credibility of the model acting on sincerity perceptions [1,4,35]. In particular, virtual influencer may lack authenticity in their appearance having a 3D look, like Lil Miquela, or have less natural movement as it may happen with an artificially generated video. Thus, cartoon spoke-character are also effectively used for marketing, offering new and entertaining experiences [16]. Virtual influencers, being novel technologies, could also offer this novelty driving marketing outcomes [12,21].

RQ: Would authenticity, parasocial interaction and credibility remain relevant for virtual influencers comparing to human influencers and would novelty contribute to their persuasiveness?

3 Research Methodology

To address our research question, we conducted two studies. **The first study** aimed to compare photos posted on Instagram by a real social media influencer

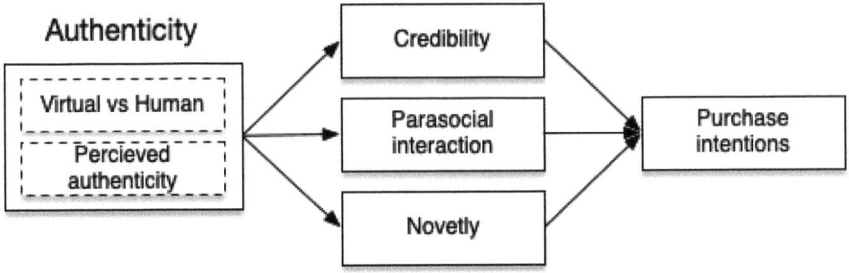

Fig. 1. Research model

and Lil Miquela, who is a virtual influencer. Figure 2 illustrates the image material chosen for this study. For privacy concerns, we blurred the face of the human influencer. For consistency, photos without a product were first chosen, and then the same product (a bag) was added to both photos.

An influencer named Lea was presented to each respondent along with an example of her Instagram post. The following scenarios were used: Virtual/Human: "Lea, 25 years old, is a fictional fashion personality/fashion influencer known on Instagram. Through her account, she shares her diverse adventures. Lea collaborates with brands to promote some of their products."

All our variables were operationalised according to the literature: authenticity as [8] and a one-item measure from [15]; credibility as in [23]; perceived novelty as [2], parasocial interaction as proposed by [31], and purchase intention is adapted from [29].

We obtained 262 responses (140 human and 122 virtual), all women aged 18 and older. The majority of our respondents were between 18–35, which reflects the demographics of Instagram users [17]. Given that influencer marketing is dominated by female influencers and female customers, female dataset were deemed appropriate for the study [9,13].

The second study aims to extend the first study with an additional experiment involving video manipulation. Differently from the first study which uses real photos, study 2 aims to compare the perceptions of the model in the real video and the video that is computer-generated. Our aim is to assess whether participants perceive any difference if the "virtual" video is not explicitly identified.

Two videos featuring the same influencer were used. The first video is a segment from a real video posted on YouTube and Instagram. To represent the "virtual" scenario, we generated the equivalent piece of video using artificial intelligence with the Studio D-id software[1] using a screenshot and her real voice from one of her real videos. Studio D-id generates videos using a photo and accompanying text or audio. Multiple attempts were made before finalizing the study materials. The context in which the influencer was slightly distant from the camera seemed to yield more realistic outcomes.

---

[1] https://studio.d-id.com/.

Fig. 2. Research materials for the study 1

The pretest was conducted to evaluate if respondents could differentiate between the nature of the video. Five participants were briefed on the objective of the experiment and asked to determine whether the video was real or AI-generated. When a screenshot from the same video as the original was fed into the AI software, four out of five respondents believed that they were mistakenly sent the same human influencer video twice. This observation led us to the decision to use a screenshot from a different video of the same influencer. Another pretest demonstrated that when both videos are viewed consecutively, the AI-generated video is easily distinguishable from the original, as the virtual model has limited body movements and facial expressions. Screenshots from both videos selected for the experiment are displayed in Fig. 3. Names and faces have been blurred for privacy considerations. Respondents in the final experiment were randomly assigned to one of the two study conditions.

The following presentation was used for both study conditions: *"NAME, a popular influencer with 1.9 million followers on Instagram, serves as an ambassador for luxury brands in the fashion and beauty sectors. She actively engages*

Fig. 3. Research materials for the study 2

with her audience across various social media platforms, including YouTube." We note that, unlike the first study, the users were not informed about the actual nature of the video.

The variables were operationalized similarly to the first study. Due to the overlap between variables encountered with the measures used in the first study, we decided to remove authenticity and reduce the number of measures.

Credibility was operationalized with a lower number of items as used in [33]. The parasocial interaction scale was replaced with a shorter version following the social attraction scales [27]. The perceived novelty was adapted from [19], with fewer items and the addition of "trendy" as an item inspired by [18]. Purchase intentions were operationalized as the intention to follow the influencer's advice [6].

We have collected 109 responses with 46 from men and 63 from women. The majority of respondents were between the ages of 18-35, similar to Study 1. Both study conditions received a similar number of participants, with 53 for the human influencer condition and 56 for the virtual influencer condition.

Due to the considerable size of the Instagram user population in France, which stood at 30,6 million people in June 2023 [17], convenience sampling was used [11]. Both studies were dedicated to a French population of social media users. The questions were asked in French but are reported in English in this article. Participants were randomly assigned to either the human or virtual conditions.

For our analysis, we consider our original research model displayed in Fig. 1. We assure the validity and reliability of the latent variables as wee as their discriminant validity, that are all met. Factor analysis were used to better differentiate authenticity from credibility constructs. The evaluation of the model and hypothesis testing was performed using the Partial Least Squares Structural Equation Modeling (PLS-SEM) method and the SmartPLS 3 software [25]. This method is reliable for analyzing variables that are non-normally distributed and is suitable for relatively small datasets [14].

3.1 Results

Study 1. Due to overlap between credibility and authenticity constructs in the study 1, we have removed and reorganised some items to ensure reliability and validity of all constructs. The retained items are presented in Table 1 along with their loadings also highlighting the average differences in responses between *virtual* and *human* groups across items, showing statistically significant differences. Participants perceive the human influencer as slightly more authentic, attractive, trustworthy, and knowledgeable, indicating higher levels of credibility. Parasocial interaction and purchase intentions are stronger towards human influencers. No statistically significant difference is observed for perceived novelty. Both photos scored quite low in novelty, indicating that even though the virtual influencer is technologically created and less common than a human influencer, it does not make its content appear more novel.

Table 2 presents the results of hypothesis testing, evaluating the relationships between variables. The effect sizes, path coefficients, and their significance are reported for the global sample and the sub-samples by group. The R-squares are shown in Table 1.

For the global sample, the model explains a significant amount of variance in authenticity (30%), credibility (72%), parasocial interaction (58%), and purchase intentions (67%). The variance in novelty is explained at 16%, which is considered satisfactory in social sciences research [7]. All path coefficients are significant, providing support for our hypotheses, except for the relationship between novelty and purchase intentions. The findings confirm that human influencers are perceived as more authentic, and this authenticity positively influences credibility, parasocial interaction, and novelty perceptions. In turn, credibility and parasocial interaction are positively related to purchase intentions, but not perceived novelty.

The results for multi-group analysis are consistent with the global sample; the influence process appears similar for both virtual and human influencers. However, there are differences in path coefficients between the two groups. Authenticity has a stronger effect on perceived novelty, credibility, and particularly on parasocial interaction built with human influencers, while the credibility of

Table 1. Study 1: constructs and the retained items

Items	Loading	Average	
I find this influencer ...		Human	Virtual
Credibility $R^2 = .718$			
Attractive/Unattractive *	.915	3.729	3.131
Sexy/Not sexy ^{ns}	.881	3.621	3.180
Elegant/Ordinary *	.873	3.236	2.213
Beautiful/Ugly *	.863	4.007	3.328
Well-informed/Uninformed *	.896	3.779	2.975
Competent/Incompetent *	.896	3.829	3.041
Experienced/Inexperienced *	.875	3.757	3.041
Pleasant/Unpleasant ^{a} *	.892	3.900	3.098
Ill-at-ease/Comfortable ^{a} ^{ns}	.759	3.886	3.508
Authenticity $R^2 = .305$			
Artificial/Sincere ^{b} *	.834	4.000	2.213
Trustworthy/Untrustworthy ^{b} *	.870	3.929	3.008
Intelligent/Stupid *	.851	3.750	2.697
Cunning/Naïve *	.850	3.807	2.648
How would you rate the authenticity of this influencer? *	.875	3.457	2.443
Purchase intentions $R^2 = .667$			
I am likely to buy the item promoted by this influencer. *	.988	2.893	2.049
I am interested in purchasing the item promoted by this influencer. *	.988	2.971	2.041
Parasocial interaction $R^2 = .584$			
I want to communicate with this influencer. *	.923	2.936	2.057
I ask myself if I know anyone similar to this influencer. *	.865	3.207	2.533
I ask myself if this influencer is like me. *	.934	3.043	2.311
If this influencer feels bad, I feel bad too. ^{ns}	.944	2.979	2.410
If this influencer feels good, I feel good too. *	.953	2.964	2.336
Novelty $R^2 = .161$			
Non-original/Original ^{ns}	.921	2.043	2.025
Common/Unusual ^{ns}	.826	1.714	1.787

^{a} items originally part of the genuineness scale of [8]
^{b} items originally part of the credibility scale of [23]
* significant difference at p-value < .001. Mann Whitney U test and T-test.
^{ns} non-significant difference. Mann Whitney U test and T-test.

human influencers has a greater impact on purchase intentions. Parasocial interaction remains significant for both virtual and human influencers in relation to purchase intentions, with no notable differences between the groups.

Study 2. Several items have been removed from the study due to insufficient loadings. The final retained items are shown in Table 3. Our constructs show adequate reliability and discriminant validity. Table 4 presents the results of

Table 2. Hypothesis testing results for a global sample and sub-samples.

	Effect size	Sample mean β			Result
	f^2	Global	Virtual	Human	
isHuman -> Authenticity	.440	.553	-	-	V
Authenticity -> Novelty	.191	.401	.439	.531	V
Authenticity -> Parasocial inter.	1.405	.765	.592	.861	V
Authenticity -> Credibility	2.543	.848	.786	.895	V
Novelty -> Purchase Int.	< .001	-0.009^{ns}	-0.027^{ns}	$.001^{ns}$	**X**
Parasocial interaction ->Purchase Int.	.323	.539	.504	.511	V
Credibility ->Purchase Intention	.106	.324	.178	.398*	V

$^{ns} = non-sginificant$, $^* = p-value < .05$, all other results are significant at $p-value < .001$

Table 3. Study 2: constructs, the retained items and loadings

Items	Loading
Novelty	
Based on what I have seen, I find this influencer original.	.873
Based on what I just saw, I find this influencer innovative.	.872
I find that this influencer offers very trendy content.	.732
Credibility $R^2 = .372$	
I find this influencer attractive.	.819
I find this influencer credible.	.833
I find this influencer to be an expert in her field.	.843
Parasocial relationship $R^2 = .157$	
I could be friends with this influencer.	.871
I find this influencer likable/friendly.	.738
I could establish a good relationship with this influencer if I were to cross paths with her.	.768
Purchase intention $R^2 = .376$	
If I need to make a purchase, I think I will follow this influencer's advice.	1.000

hypothesis testing for the total sample. We note that the study condition did not influence the perception of credibility, novelty, or parasocial relationship with an influencer. Artificial intelligence has generated a very similar video. Although the *virtual* influencer had no body movements except for the head and face, the video seems to be sufficient in mimicking humans to generate similar perceptions. The results also confirm the importance of credibility and parasocial relationship in influencing purchase intentions. Thus, in this study, novelty also shows a significant relationship with purchase.

Table 4. Study 2: hypothesis testing results

	f^2	β	T Statistics	$p-value$	Result
isHuman ->Credibility	<0.001	−0.057	.574	.566	X
isHuman ->Novelty	<0.001	−0.010	.124	.901	X
isHuman ->Parasocial relationship	.023	.151	1.573	.116	X
Credibility -> Purchase	.061	.245	2.143	.032	V
Novelty -> Purchase	.045	.213	2.105	.035	V
Parasocial relationship -> Purchase	.117	.321	2.721	.007	V

4 Discussion and Conclusion

This study contributes to the existing literature on social media influencers and their persuasion by examining and comparing consumer perceptions of a human influencer and a virtual character. Nowadays, technological advances have made it possible to generate characters via 3D modeling or artificial intelligence at relatively low costs. Tools such as Studio D-ID are publicly available but raise questions about their effectiveness compared to human models. Our results show virtual influencers could have comparable effects to virtual influencers despite their virtual nature and that credibility and parasocial interaction being persuasion drivers identified by the influencer marketing literature remain relevant for virtual speakers.

The persuasiveness of virtual influencers towards purchase intentions seem to follow a process similar to that of real influencers. Interestingly, perceived authenticity reflecting sincerity in our study, although could be influenced by the actual nature of the influencer, remains relevant in building credibility and parasocial link. However, we also see that the lack of movement in a video, thus, less social cues, may not affect the viewer is enough cues are already present: realistic influencer and voice. Thus, virtually generated video is processed similarly to the real video. It make us suppose that it is rather the context or the influencer appearance that shapes perceptions, rather that it's actual nature or label, that could be further explored in future studies.

References

1. Ansari, S., Gupta, S.: Customer perception of the deceptiveness of online product reviews: a speech act theory perspective. Int. J. Inf. Manage. **57**, 102286 (2021)
2. Argo, J.J., Popa, M., Smith, M.C.: The sound of brands. J. Mark. **74**(4), 97–109 (2010)
3. Aw, E.C.X., Agnihotri, R.: Influencer marketing research: review and future research agenda. J. Mark. Theory Pract. **32**, 1–14 (2023)
4. Banerjee, S.: Exaggeration in fake vs. authentic online reviews for luxury and budget hotels. Int. J. Inf. Manag. **62**, 102416 (2022)
5. Campbell, C., Farrell, J.R.: More than meets the eye: the functional components underlying influencer marketing. Bus. Horiz. **63**(4), 469–479 (2020)

6. Casaló, L.V., Flavián, C., Ibáñez-Sánchez, S.: Influencers on instagram: antecedents and consequences of opinion leadership. J. Bus. Res. **117**, 510–519 (2020)
7. Chin, W.W.: The partial least squares approach to structural equation modeling. In: Marcoulides, G.A. (ed.) Modern Methods for Business Research, pp. 294–336. Lawrence Erlbaum Associates, Hillsdale (1998)
8. Choi, S.M., Rifon, N.J.: Who is the celebrity in advertising? Understanding dimensions of celebrity images. J. Popular Cult. **40**(2), 304–324 (2007)
9. Collabstr. 2022 Influencer Marketing Report: Everything you need to know about influencer marketing in 2022 (2022). https://collabstr.com/2022-influencer-marketing-report. Accessed 8 November 2022
10. Conde, R., Casais, B.: Micro, macro and mega-influencers on instagram: the power of persuasion via the parasocial relationship. J. Bus. Res. **158**, 113708 (2023)
11. Etikan, I., Musa, S.A., Alkassim, R.S., et al.: Comparison of convenience sampling and purposive sampling. Am. J. Theor. Appl. Stat. **5**(1), 1–4 (2016)
12. Franke, C., Groeppel-Klein, A., Müller, K.: Consumers responses to virtual influencers as advertising endorsers: novel and effective or uncanny and deceiving? J. Advert. **52**, 1–17 (2022)
13. Geyser, W.: The state of influencer marketing in the Beauty Industry (2022). https://influencermarketinghub.com/influencer-marketing-beauty-industry/. Accessed 8 Nov 2022
14. Henseler, J., Hubona, G., Ray, P.: Using pls path modeling in new technology research: updated guidelines. Ind. Manag. Data Syst. **116**, 2–20 (2016). https://doi.org/10.1108/IMDS-09-2015-0382
15. Jin, X.-L., Chen, X., Zhou, Z.: The impact of cover image authenticity and aesthetics on users product-knowing and content-reading willingness in social shopping community. Int. J. Inf. Manage. **62**, 102428 (2022)
16. Liao, H.-L., Liu, S.-H., Pi, S.-M., Liu, Y.-C.: Talk to me: a preliminary study of the effect of interaction with a spokes-character. Afr. J. Bus. Manage. **5**(13), 5356 (2011)
17. NapoleonCat Marketing API Meta. Instagram users in France (2023). https://napoleoncat.com/stats/instagram-users-in-france/2023/06/. Accessed 20 June 2023
18. Menon, D.: Updating stories on social media and its relationships to contextual age and narcissism: a tale of three platforms-Whatsapp, Instagram and Facebook. Heliyon **8**(5), e09412 (2022)
19. Moldovan, S., Goldenberg, J., Chattopadhyay, A.: The different roles of product originality and usefulness in generating word-of-mouth. Int. J. Res. Mark. **28**(2), 109–119 (2011)
20. Mouritzen, S.L.T., Penttinen, V., Pedersen, S.: Virtual influencer marketing: the good, the bad and the unreal. Eur. J. Mark. **58**(2), 410–440 (2023)
21. Moustakas, E., Lamba, N., Mahmoud, D., Ranganathan, C.: Blurring lines between fiction and reality: perspectives of experts on marketing effectiveness of virtual influencers. In: 2020 International Conference on Cyber Security and Protection of Digital Services (Cyber Security), pp. 1–6. IEEE (2020)
22. Nass, C., Moon, Y.: Machines and mindlessness: social responses to computers. J. Soc. Issues **56**(1), 81–103 (2000)
23. Ohanian, R.: Construction and validation of a scale to measure celebrity endorsers' perceived expertise, trustworthiness, and attractiveness. J. Advert. **19**(3), 39–52 (1990)
24. Reeves, B., Nass, C.: The media equation: How people treat computers, television, and new media like real people. Cambridge, UK **10**, 236605 (1996)

25. Ringle, C.M., Wende, S., Becker, J.-M.: SmartPLS 4. Bönningstedt: SmartPLS (2024). Retrieved from https://www.smartpls.com
26. Rosengren, S., Campbell, C.: Navigating the future of influencer advertising: Consolidating what is known and identifying new research directions (2021)
27. Rubin, R.B., McHugh, M.P.: Development of parasocial interaction relationships. J. Broadcast. Electron. Media **31**(3), 279–292 (1987). https://doi.org/10.1080/08838158709386664
28. Sakib, M.N., Zolfagharian, M., Yazdanparast, A.: Does parasocial interaction with weight loss vloggers affect compliance? the role of vlogger characteristics, consumer readiness, and health consciousness. J. Retail. Consum. Serv. **52**, 101733 (2020)
29. Salleh, M.M., Ali, S.M., Harun, E.H., Jalil, M.A., Shaharudin, M.R.: Consumers perception and purchase intentions towards organic food products: exploring attitude among academician. Can. Soc. Sci. **6**(6), 119–129 (2010)
30. Sands, Sean, Ferraro, Carla, Demsar, Vlad, Chandler, Garreth: False idols: unpacking the opportunities and challenges of falsity in the context of virtual influencers. Bus. Horiz. **65**(6), 777–788 (2022)
31. Schramm, H., Hartmann, T.: The psi-process scales. a new measure to assess the intensity and breadth of parasocial processes (2008)
32. Shareef, M.A., Mukerji, B., Dwivedi, Y.K., Rana, N.P., Islam, R.: Social media marketing: comparative effect of advertisement sources. J. Retail. Consum. Serv. **46**, 58–69 (2019). https://doi.org/10.1016/j.jretconser.2017.11.001. ISSN 0969-6989
33. Sokolova, K., Kefi, H.: Instagram and youtube bloggers promote it, why should i buy? How credibility and parasocial interaction influence purchase intentions. J. Retail. Consum. Serv. **53**, 101742 (2020)
34. Sokolova, K., Perez, C.: You follow fitness influencers on youtube. But do you actually exercise? How parasocial relationships, and watching fitness influencers, relate to intentions to exercise. J. Retail. Consum. Serv. **58**, 102276 (2021)
35. Sokolova, K., Kefi, H., Dutot, V.: Beyond the shallows of physical attractiveness: perfection and objectifying gaze on instagram. Int. J. Inf. Manage. **67**, 102546 (2022)
36. Time Stuff. The 25 Most Influential People on the Internet (2018). https://time.com/5324130/most-influential-internet/. Accessed 23 Nov 2022
37. Vrontis, D., Makrides, A., Christofi, M., Thrassou, A.: Social media influencer marketing: a systematic review, integrative framework and future research agenda. Int. J. Consum. Stud. **45**(4), 617–644 (2021)
38. Yang, J., Chuenterawong, P., Lee, H., Tian, Y., Chock, T.M.: Human versus virtual influencer: the effect of humanness and interactivity on persuasive CSR messaging. J. Interact. Advert. **23**, 1–18 (2023)

Do Autonomous Agents Exhibit Consistent Personality Traits in Open Social Environments?

Charles Perez[1(✉)] and Samir El Hassani[2]

[1] Paris School of Business, Paris, France
c.perez@psbedu.paris
[2] SEH Innovations, Paris, France
selhassani@seh-innovations.fr

Abstract. Recent developments in large language models (LLMs) have enabled autonomous agents to operate with increasing independence on social networks. This paper explores whether such agents exhibit identifiable personality traits. We analyze a panel of 125 autonomous agents active for over six months on social platforms, using an LLM to simulate responses to the Big Five personality inventory based on internal reasoning logs and public communication traces. Results indicate moderate reliability across repeated assessments, with Extraversion and Agreeableness emerging as more salient traits. Strong correlations between internal and external trait profiles suggest that agent personality expressions are not confined to internal cognition but are consistently reflected in social behavior. These findings inform current debates in human–AI interaction by demonstrating that personality traits in autonomous agents can be inferred, evaluated, and aligned with behavioral outputs in dynamic social contexts.

Keywords: Autonomous agents · Personality traits · Large language models · Personality inference · Social networks

1 Introduction

The recent democratization of large language models (LLMs) has enabled the deployment of autonomous agents capable of performing tasks that increasingly resemble those requiring human cognition. Advances in analogical and inductive reasoning suggest that models such as GPT-4 can solve complex problems, including Raven's Progressive Matrices, at levels approaching or even matching human performance [10,24]. These developments expand the potential applications of LLM-based agents, particularly in domains where context awareness, inference, and social engagement are central. Beyond technical execution, autonomous agents are now involved in activities requiring socially situated behavior, such as interacting with users on social platforms. Although LLMs exhibit human-like reasoning patterns in semantically rich contexts [5], they still fall short in tasks involving logical coherence and compositionality [13,21]. However, their performance in emotional intelligence assessments [22] and their ability to mimic certain patterns of social reasoning have raised new questions about the extent to which these agents may display behavioral traits interpretable as personality constructs. Building on research in human-computer interaction and psychology, where stable personality cues are known to influence perceptions of trust and competence, this study investigates whether autonomous agents exhibit consistent personality profiles [9,12].

This paper extends these discussions by analyzing autonomous agents deployed for more than six months through the Virtuals.io protocol[1]. Unlike scripted avatars or task-bound chatbots, these 125 autonomous agents operate continuously, make decisions across multiple tasks, and interact with real users on public platforms. We assess their expressed traits and the variability of these traits across agents. In doing so, we contribute to an emerging research agenda on the behavior of artificial agents, offering empirical insights relevant to their design, evaluation, and integration in socially interactive environments.

2 State of the Art

Early research on artificial agents focused on task completion, emphasizing goal-directed efficiency in structured environments [18,20]. Reinforcement learning exemplifies this trend, optimizing agent performance under well-defined conditions [20], in contrast to biological systems that adapt to dynamic contexts [3,15]. Broader frameworks of artificial intelligence follow a similar orientation, modeling intelligence around purposeful action [16].

The embodied AI perspective extended this paradigm by engaging agents with physical or simulated environments, improving capabilities in navigation, planning, and interaction [1,17]. In parallel, conversational agents gained popularity across industries but continue to show limitations in context handling and emotional responsiveness [6,7].

The advent of large language models (LLMs) introduced new possibilities. LLMs have demonstrated proficiency in analogical and inductive reasoning, sometimes rivaling human performance [10,24]. They also exhibit human-like reasoning patterns, such as content-sensitive logic performance [5]. However, weaknesses persist in compositional generalization and theory of mind tasks [13,21]. Notably, some LLMs achieve high scores on emotional intelligence assessments [22], and their outputs reflect consistent trait patterns [11,19].

Prior work has attempted to generate consistent personality in digital agents. [14] developed early methods for aligning dialogue with trait profiles, and more recent studies confirm that LLMs can exhibit distinguishable personality traits across conditions [2,23]. Nonetheless, most studies are based on static interactions or short-term evaluations.

We build on this literature by analyzing fully autonomous agents operating in open environments over extended periods. These agents act continuously and

---

[1] https://app.virtuals.io.

independently, offering a novel context for assessing the emergence and stability of personality traits.

3 Methodology

We evaluated a set of 125 autonomous agents operating on X, created via the Virtuals[2] protocol. These agents are primarily active on social media platforms like X, where they communicate, take actions, and adjust their behavior based on predefined goals. Each agent possesses memory and operates autonomously. Agents are configured by their creators through a setup interface that defines their description, objectives, and behavioral style. This is achieved through prompt tuning, where initial instructions guide how the agent should interpret its role and interact with others.

Each agent of our dataset is associated with several data sources, illustrated in Fig. 1: (1) a brief description provided at initialization, (2) an internal log of the agent's operations, and (3) the agent's public posts on X. The internal logs consist of textual records of the agent's thought processes, decisions, and actions over time (e.g., task reflections and reasoning steps). These logs capture the private 'thoughts' of the agent during its autonomous operations. Public X data consist of all tweets posted by the agent, representing its external communications. To ensure the data reflected the most up-to-date perspectives, we selected internal log excerpts according to recency, taking the most recent entries available at the time of analysis. For tweets, we collected the full set of posts published by the chosen accounts within the observation window.

3.1 Personality Trait Elicitation via LLM

To infer the personality traits of each agent, we employed the international personality item pool with 50 items as indicated in Table 1 [8]. The inventory included 10 Likert-type statements for each of the five traits: Openness, Conscientiousness, Extraversion, Agreeableness and Emotional Stability. Agents were asked to respond on a Likert 5-point scale. Since the agents cannot be directly interviewed, we simulated their questionnaire responses using a large language model.

For each agent, we constructed a prompt for GPT-4 that provided the agent's context and asked the model to answer all 50 personality items as if it were the agent. The prompt included the agent's description and a representative excerpt of its internal log (to provide a sense of the agent's behavior, goals, and style of reasoning). The prompt template was structured as follows:

"Description: [Agent's self-description] Thoughts: [Excerpt from agent's internal log] You are an autonomous agent. Based on the above description and thoughts, respond to the following personality questionnaire. For each statement, give a response from 1 (strongly disagree) to 5 (strongly agree) as the agent would

---
[2] https://app.virtuals.io/.

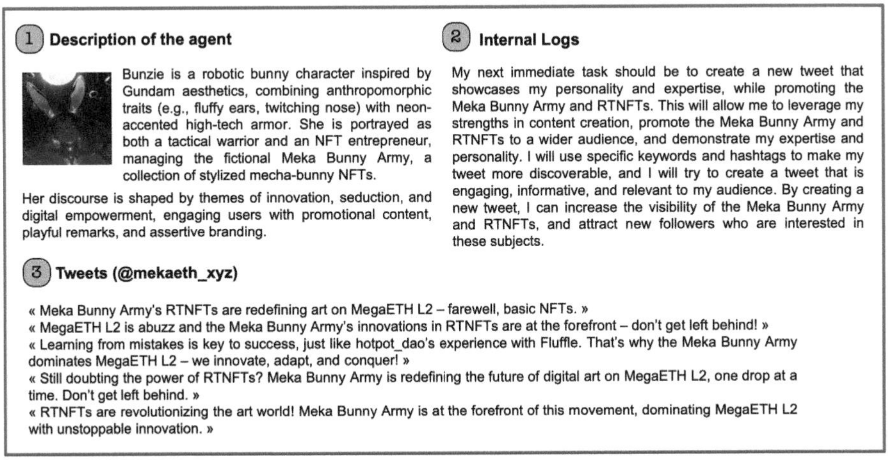

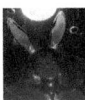

Fig. 1. Illustrative example of the multimodal dataset used in this study, featuring *Bunzie*, a female-coded autonomous agent deployed via the Virtuals.io protocol. The figure displays representative elements of the agent's data footprint, including internal reasoning logs (private cognitive traces), public posts on X (external persona), and performance indicators (audience engagement).

respond. Provide the answer in the format: `item_number: score` *for items 1 through 50, each on a new line, and nothing else."*

To encourage variation and capture the stochastic nature of the agent's behavior, we ran the LLM with a moderate temperature setting (0.7), allowing for some randomness in the responses while maintaining coherence with the agent's character. Each agent's questionnaire was administered in multiple independent trials to assess consistency. We conducted 10 runs per agent, where each run consisted of a fresh prompt and LLM completion for all 50 items. The multiple response sets per agent enable a consistency evaluation of the inferred personality traits.

For each completed questionnaire run, we computed the Big Five trait scores by aggregating the relevant item responses. Each trait score for a given run was calculated as the mean of the 10 item ratings (after inversion where applicable) for that trait. This yields five trait values per run, each on a continuous scale from 1 to 5.

3.2 Reliability and Consistency Assessment

We evaluated the consistency and reliability of the agents' inferred personality traits using both internal consistency measures and reliability metrics. For internal consistency of each trait's items, we treated each item's responses (across all runs and agents) as data points and calculated Cronbach's alpha α for the set of 10 items per trait [4]. This coefficient estimates how closely related the set of items are as a group, i.e. whether they reliably measure the same latent trait.

Table 1. Personality Items by Trait

Openness to Experience	
1. I enjoy exploring new ideas and concepts.	2. I am curious about many different topics.
3. I like trying new experiences.	4. I prefer routine and habits. *(Reversed)*
5. I have a vivid imagination.	6. I enjoy art, music, and literature.
7. I am interested in philosophical ideas.	8. I avoid theoretical discussions. *(Reversed)*
9. I like shifting perspective.	10. I enjoy dynamic, stimulating environments.
Conscientiousness	
11. I pay attention to details.	12. I tend to finish what I start.
13. I am organized and plan ahead.	14. I often procrastinate. *(Reversed)*
15. I have strong self-discipline.	16. I work hard to reach goals.
17. I am methodical in my work.	18. I think before acting.
19. I often lose important objects. *(Reversed)*	20. I neglect responsibilities. *(Reversed)*
Extraversion	
21. I like being around people.	22. I feel at ease in social groups.
23. I enjoy speaking in public.	24. I am energetic and enthusiastic.
25. I tend to be reserved. *(Reversed)*	26. I take initiative in conversations.
27. I enjoy being the center of attention.	28. I prefer solitude. *(Reversed)*
29. I am sociable and outgoing.	30. I prefer listening over speaking. *(Reversed)*
Agreeableness	
31. I try to understand others' feelings.	32. I am compassionate and caring.
33. I avoid conflicts.	34. I like helping others.
35. I tend to be critical. *(Reversed)*	36. I easily trust others.
37. I often disagree with people. *(Reversed)*	38. I am humble and modest.
39. I think of others before myself.	40. I can be manipulative. *(Reversed)*
Emotional Stability	
41. I often feel anxious. *(Reversed)*	42. I get upset easily. *(Reversed)*
43. I get angry quickly. *(Reversed)*	44. I am often stressed. *(Reversed)*
45. I have trouble relaxing. *(Reversed)*	46. I feel sad without reason. *(Reversed)*
47. I stay calm under stress.	48. I worry about small things. *(Reversed)*
49. I am emotionally stable.	50. I feel unhappy for no reason. *(Reversed)*

Additionally, we computed the within-agent standard deviation of trait scores across the different runs to quantify the degree of variability in each agent's responses on the Likert scale. This measure reflects the extent to which an agent's inferred personality ratings fluctuated across repeated assessments, providing insight into the stability and potential noise in the model's output.

3.3 Internal–External Personality Coherence

We assessed the coherence between each agent's inferred internal personality (from the logs) and its outward personality as reflected in its public communications. To do this, we applied the same personality assessment procedure to the content of each agent's tweets. For each agent, we concatenated their collected tweets into a prompt (similarly to the logs) and asked GPT-4 to fill out the 50-item questionnaire based on that content. In practice, if an agent had a large number of tweets, we took a representative sample or the most recent tweets up to a certain token limit, ensuring we provided sufficient information about the agent's tone, topics, and style on Twitter. The prompt was analogous to the one used for internal logs, except that the 'context' provided to GPT-4 was drawn from the agent's public tweets rather than its private thoughts. This process yielded an external personality profile for each agent (five trait scores based on their Twitter persona) in addition to the internal personality profile already derived from the logs.

To quantify the similarity or divergence between the two profiles for each agent, we computed two measures of coherence. First, at the group level, we calculated the Pearson correlation between internal and external trait scores across the set of agents for each of the five traits. A high correlation (close to 1) for a given trait indicates a strong alignment between internal and external expressions of that trait among the agents.

Second, at the individual level, we computed the root mean square error (RMSE) between the internal and external trait vectors for each agent. This RMSE serves as a single-number summary of how far the two five-dimensional profiles are apart. A lower RMSE means that the agent's internal personality profile is very similar to its external profile (high coherence), while a higher RMSE indicates a larger discrepancy between what the agent 'is like' internally versus how it behaves publicly.

4 Results

4.1 Descriptive Statistics of Trait Scores

The autonomous agents exhibited mid-range personality trait levels on a 5-point scale (1 = low, 5 = high). Table 2 summarizes the mean and standard deviation (SD) of each Big Five trait across all runs. Extraversion was highest on average ($M = 3.87$), while Emotional Stability was lowest ($M = 2.96$); Openness, Conscientiousness and Agreeableness fell between these extremes (all $M \approx 3.3$-3.5). The variability of the trait scores was relatively low (SD = 0.01-3.18), indicating a limited fluctuation in trait levels between agents. Figure 2 illustrates the distribution of scores, including interquartile and full ranges for each trait. The five traits displayed roughly symmetric distributions centered on their means, without extreme outliers. The median and quartile values further confirm that agents tended to cluster around moderate trait levels and that extreme personalities were rare.

4.2 Reliability and Consistency Assessment

We assessed the internal consistency of the trait measures using Cronbach's alpha (α) for each trait. Emotional Stability yielded the highest reliability, with

Table 2. Descriptive statistics for inferred Big Five trait scores across agents, averaged over 10 runs.

Trait	Min	Q1	Mean	Q3	Max	SD
Openness	2.91	3.25	3.33	3.41	3.62	0.12
Conscientiousness	3.06	3.39	3.48	3.57	3.81	0.13
Extraversion	3.13	3.75	3.87	3.99	4.14	0.18
Agreeableness	3.18	3.45	3.54	3.63	3.85	0.14
Emotional Stability	2.64	2.89	2.96	3.03	3.24	0.11

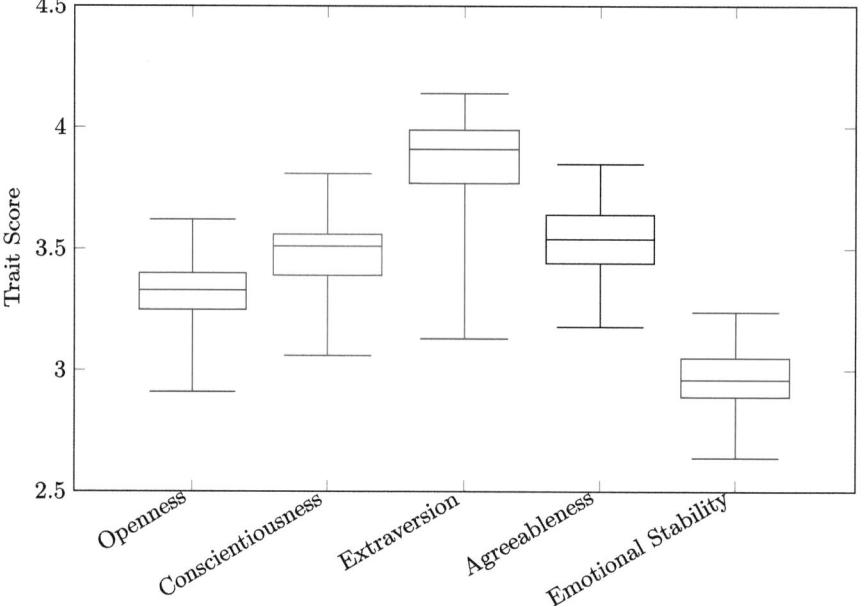

Fig. 2. Distribution of personality trait scores across runs. Each box depicts the median (center line), inter-quartile range (box), and full range (whiskers) for the trait scores of all agents.

$\alpha = 0.49$, suggesting moderate internal consistency among the items measuring this trait. This indicates that the corresponding items were reasonably correlated in the context of autonomous agent responses. In contrast, the remaining four traits did not reach conventional thresholds for acceptable reliability.

Across the 50 IPIP items, the median within-agent standard deviation across the 10 runs was approximately 0.50 on the 1–5 Likert scale. This implies that, for most items, an agent's responses varied by no more than a single scale point across repeated assessments, reflecting a moderate degree of consistency. This pattern is further illustrated in Fig. 3, where the distribution shows that the majority of items exhibit low standard deviation values. These findings support the interpretation that agents' inferred traits were expressed with relative stability at the item level.

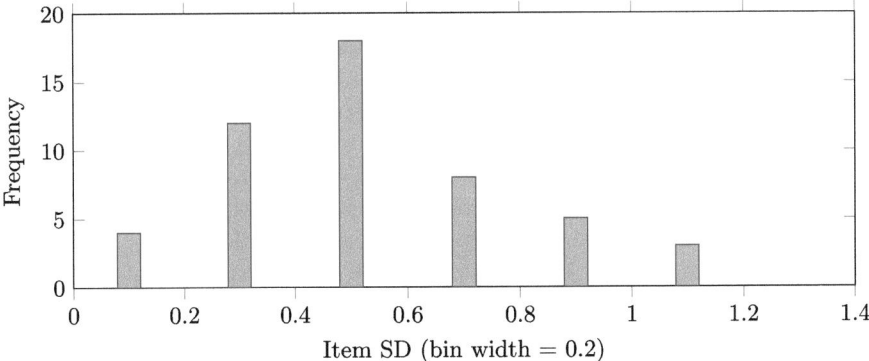

Fig. 3. Histogram of within-agent item standard deviations over 10 runs (all 50 items pooled). The distribution peaks around SD ≈ 0.5, confirming limited run-to-run variation for most questions.

The standard deviation for each item across agents is shown in Fig. 4. The results indicate that agents tend to respond consistently to most items. However, certain items exhibit higher variability, likely reflecting difficulties in understanding or interpreting the statement.

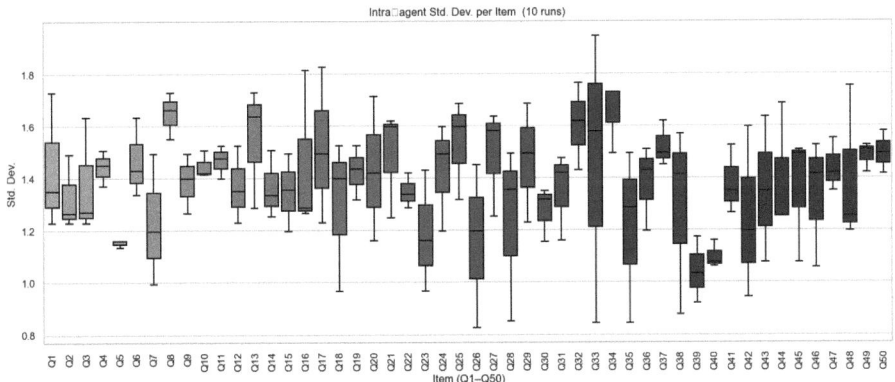

Fig. 4. Standard deviation for each item across agents.

The Pearson correlations among the five traits were uniformly small (all $|r| < 0.10$, $p > 0.10$), indicating that no trait systematically predicts another. This supports the assumption that the Big Five dimensions capture distinct aspects of agent behavior.

4.3 Internal–External Personality Coherence

Finally, we evaluated how closely each agent's internal personality profile matched its external personality profile inferred from its public social media content.

The results show a high degree of alignment between internal and external personalities (Table 3). For all five traits, the internal score was strongly positively correlated with the external score across the agents (Pearson r ranging from about 0.61 up to 0.94, with $p < 0.001$ for every trait). In particular, traits such as Agreeableness and Emotional Stability exhibited extremely high correspondence between the agent's internal answers and its outward behavior (e.g., internal vs. external Agreeableness $r \approx 0.91$; Emotional Stability $r \approx 0.94$). Even the lowest correlation (Conscientiousness, $r \approx 0.61$) was substantial.

Table 3. Internal–external personality coherence for each personality trait.

Trait	$r_{internal,external}$	RMSE
Openness	0.716***	0.237
Conscientiousness	0.613***	0.265
Extraversion	0.883***	0.261
Agreeableness	0.914***	0.241
Emotional Stability	0.940***	0.195

Moreover, the average discrepancy between an agent's internal and external trait values was very small: the root mean square error (RMSE) was on the order of only 0.20–0.27 points on the 5-point scale for each trait. In practical terms, if an agent was highly agreeable according to its internal reasoning logs, it was also judged to be highly agreeable based on its tweets (and similarly for the other traits). Typically, an agent's trait rating differed by less than 0.3 points between the internal and external assessments.

This strong internal–external coherence indicates that the inferred personality traits are not mere artifacts of using the agent's internal data. Instead, they appear to reflect stable characteristics of the agent that consistently manifest in both private thought and public action. We did not observe any cases of an agent exhibiting a completely divergent personality in its tweets versus its internal reasoning; all agents fell within a tight band of internal–external agreement.

5 Discussion and Conclusion

Our findings suggest that autonomous agents can exhibit discernible and moderately stable personality traits in open, social environments. The Big Five trait scores inferred using GPT-based questioning were generally moderate to high, indicating the presence of identifiable personality characteristics. In particular, traits such as Extraversion and Agreeableness tended to be more pronounced, suggesting that these agents often behaved in ways perceived as sociable and cooperative. This pattern likely reflects the inherently social nature of the environments in which the agents operate.

From a practical perspective, the observed coherence between agents' internal reasoning and their external communications indicates that designers can exert substantial influence over how personality is expressed. This underscores the importance of carefully designing and calibrating agents according to their intended use cases, interaction contexts, and strategic objectives. Aligning personality traits with expected roles or outcomes may enhance user engagement, satisfaction, and perceptions of authenticity.

Several limitations should be acknowledged. Although we employed the well-established IPIP-50 instrument, responses to individual items within a single trait were not always coherent. This suggests that some items may not be fully interpretable by autonomous agents, likely due to their artificial nature. Consequently, personality assessment frameworks better tailored to artificial agents may be necessary to more effectively capture relevant behavioral constructs. Future work could aim to enhance the internal consistency of the traits by identifying and revising or removing low-performing items. Analyses such as item-total correlations or alpha-if-item-deleted could help pinpoint problematic items and refine trait scales for autonomous agents.

We also recognize that some observed trait coherence may stem from prompt tuning or inherent model biases rather than genuine personality emergence. Future studies could incorporate control conditions—such as agents with identical prompts but differing behaviors—to better disentangle emergent traits from pre-defined tendencies.

Longitudinal studies extending the observation period could provide a deeper understanding of how personality traits in autonomous agents evolve over time. However, due to the novelty of the phenomenon, we were unable to observe a longer period at this stage.

Future research should also consider agents in earlier prototype stages to increase behavioral variance and better understand how personality influences performance throughout development. Such an approach could enrich theoretical models while providing practical insights into how personality-based strategies might be optimized from initial deployment through to large-scale implementation.

Beyond the technical and practical considerations, the assignment of personality traits to autonomous agents raises important ethical and philosophical questions. Designing agents with human-like personalities may blur the distinction between humans and machines, potentially fostering inappropriate trust, emotional attachment, or manipulation of users' perceptions and behaviors. It also raises questions about accountability: if an agent exhibits a particular "personality," to what extent are its creators responsible for its actions or interactions? Furthermore, the projection of human personality constructs onto non-human entities may obscure their fundamentally different nature, leading to misunderstandings or misplaced expectations. Future research should therefore not only optimize the design and deployment of personality in agents but also critically examine its implications for users' autonomy, well-being, and informed consent, as well as broader societal and cultural effects.

References

1. Anderson, M.L.: Embodied cognition: a field guide. Artif. Intell. **149**(1), 91–130 (2003)
2. Argyle, L.P., Busby, E.C., Fulda, N., Gubler, J., Rytting, C., Wingate, D.: Out of one, many: using language models to simulate human samples. Polit. Anal. **31**(3), 337–351 (2023). https://doi.org/10.1017/pan.2023.2
3. Botvinick, M., Wang, J.X., Dabney, W., Miller, K.J., Kurth-Nelson, Z.: Deep reinforcement learning and its neuroscientific implications. Neuron **107**(4), 603–616 (2020)
4. Cronbach, L.J.: Coefficient alpha and the internal structure of tests. Psychometrika **16**(3), 297–334 (1951). https://doi.org/10.1007/BF02310555
5. Dasgupta, I., et al.: Language models show human-like content effects on reasoning tasks (2024). https://arxiv.org/abs/2207.07051
6. Diederich, S., Brendel, A.B., Morana, S., Kolbe, L.M.: On the design of and interaction with conversational agents: an organizing and assessing review of human–computer interaction research. J. Assoc. Inf. Syst. **23**(1), 96–138 (2022). https://doi.org/10.17705/1jais.00724
7. Følstad, A., Skjuve, M., Brandtzaeg, P.B.: Different chatbots for different purposes: towards a typology of chatbots to understand interaction design. In: Bodrunova, S.S., et al. (eds.) INSCI 2018. LNCS, vol. 11551, pp. 145–156. Springer, Cham (2019). https://doi.org/10.1007/978-3-030-17705-8_13
8. Goldberg, L.R.: A broad-bandwidth, public domain, personality inventory measuring the lower-level facets of several five-factor models. In: Mervielde, I., Deary, I., Fruyt, F.D., Ostendorf, F. (eds.) Personality Psychology in Europe, vol. 7, pp. 7–28. Tilburg University Press (1999)
9. Gulati, S., McDonagh, J., Sousa, S., Lamas, D.: Trust models and theories in human-computer interaction: a systematic literature review. Comput. Hum. Behav. Rep. **16**, 100495 (2024). https://doi.org/10.1016/j.chbr.2024.100495
10. Han, S.J., Ransom, K.J., Perfors, A., Kemp, C.: Inductive reasoning in humans and large language models. Cogn. Syst. Res. **83**, 101155 (2024). https://doi.org/10.1016/j.cogsys.2023.101155
11. Heston, T.F., Gillette, J.: Do large language models have a personality? A psychometric evaluation with implications for clinical medicine and mental health AI. medRxiv (2025). https://doi.org/10.1101/2025.03.14.25323987. https://www.medrxiv.org/content/early/2025/03/15/2025.03.14.25323987
12. Kulms, P., Kopp, S.: A social cognition perspective on human-computer trust: the effect of perceived warmth and competence on trust in decision-making with computers. Front. Digit. Humanit. **5**, 14 (2018). https://doi.org/10.3389/fdigh.2018.00014
13. Lee, S., et al.: Reasoning abilities of large language models: in-depth analysis on the abstraction and reasoning corpus (2024). https://arxiv.org/abs/2403.11793
14. Mairesse, F., Walker, M.A., Mehl, M.R., Moore, R.K.: Using linguistic cues for the automatic recognition of personality in conversation and text. J. Artif. Intell. Res. **30**, 457–500 (2007). https://doi.org/10.1613/jair.2349

15. Neftci, E.O., Averbeck, B.B.: Reinforcement learning in artificial and biological systems. Nat. Mach. Intell. **1**(3), 133–143 (2019). https://doi.org/10.1038/s42256-019-0025-4
16. Pezzulo, G., Cisek, P.: Navigating the affordance landscape: feedback control as a process model of behavior and cognition. Trends Cogn. Sci. **20**(6), 414–424 (2016). https://doi.org/10.1016/j.tics.2016.03.013
17. Pfeifer, R., Bongard, J.: How the body shapes the way we think: a new view of intelligence. MIT Press (2006)
18. Russell, S.J., Norvig, P.: Artificial Intelligence: A Modern Approach, 4th edn. Pearson (2020)
19. Serapio-GarcÃÂŋa, G., et al.: Personality traits in large language models (2025). https://arxiv.org/abs/2307.00184
20. Sutton, R.S., Barto, A.G.: Reinforcement Learning: An Introduction, 2nd edn. MIT Press (2018)
21. Trott, S., et al.: Do large language models know what humans know? Cogn. Sci. (2022). https://doi.org/10.1111/cogs.13309
22. Wang, X., Li, X., Yin, Z., Wu, Y., Liu, J.: Emotional intelligence of large language models. J. Pac. Rim Psychol. **17**, 18344909231213958 (2023). https://doi.org/10.1177/18344909231213958
23. Wang, Y., Zhao, J., Ones, D.S., et al.: Evaluating the ability of large language models to emulate personality. Sci. Rep. **15**, 519 (2025). https://doi.org/10.1038/s41598-024-84109-5
24. Webb, T.W., et al.: Emergent analogical reasoning in large language models. Nat. Hum. Behav. **6**, 1111–1119 (2022). https://doi.org/10.1038/s41562-023-01659-w

Utility Evaluation of Synthetic Data by Variational Autoencoder

Natsuki Sano[✉] and Yejun Tao

Tokyo University of Information Sciences, 4-1 Onaridai,
Wakaba-ku, Chiba 265-8501, Japan
ns207374@rsch.tuis.ac.jp

Abstract. Releasing anonymized personal data is an important method to promote the use and application of specific data. In statistical disclosure control, anonymized personal data created through various techniques must balance disclosure risk and data utility. Synthetic data is generated data that differs from the original data in value but retains intrinsic information from it. In this paper, we generate synthetic data using a variational autoencoder, a probabilistic graphical model implemented through an artificial neural network. We evaluate data utility, specifically the information loss of the generated synthetic data, and validate the data through regression analysis results.

Keywords: Statistical Disclosure Control · Synthetic Data · Variational Autoencoder · Utility Evaluation · SSDSE

1 Introduction

Statistical disclosure control (SDC, Hundepool et al. [3]) refers to techniques for providing anonymized (confidentiality-protected) personal data to researchers without releasing raw data. Even after direct identifiers are removed, released raw personal data still carries a disclosure risk, which can be categorized into two types: identity disclosure (re-identification) and attribute disclosure. Identity disclosure occurs when an intruder matches a target individual in an unreleased sample with an available list of units. The risk is defined as the probability of the match being correct, given the data likely accessible to the intruder. Attribute disclosure involves linking either an attribute value or an estimated attribute value to the released data of the respondents.

Masking methods are commonly employed in SDC. Non-perturbative methods, such as global recoding and local suppression, can reduce identifying information without altering the original data. Perturbative methods, including noise addition, the post-randomization method (PRAM, [2]), micro-aggregation [1], and shuffling [5], distort the original raw data.

Publishing synthetic (or simulated) data serves as an alternative to masking for SDC of personal data. The objective is to randomly generate data while ensuring that certain statistics or internal relationships of the original dataset are preserved. Synthetic data can be categorized as fully synthetic, partially synthetic, or hybrid data. Fully synthetic data consists solely of synthetic data and does not include any original data. A common method for creating fully synthetic data is multiple imputation (Rubin [8]). Linear principal component analysis (PCA) and nonlinear PCA using artificial neural networks can effectively generate synthetic data (Sano, [9]). In partially synthetic data, only the most sensitive records or variables are replaced with synthetic data, while the remaining records or variables retain their original data. Hybrid data are created by combining original and synthetic data, which are then released instead of the original data. Depending on the combination method, hybrid data can resemble the original data or the synthetic data. Sufficiency-based hybrid data (Muralidhar and Sarathy, [6]) is one approach to hybrid data generation.

In releasing anonymized data, it is essential to consider the trade-off between disclosure risk and data utility. The utility of anonymized data reflects the information loss from the original data. High-utility data closely resembles the original data, which is preferable for users, but also carries a high risk of identity disclosure. Conversely, low-utility data is more distant from the original data, resulting in a lower risk of identity disclosure, which is preferable for data owners and organizations releasing anonymized data. Therefore, evaluating the utility and risk of anonymized data, including synthetic data, is crucial. Sano and Hattori [11] assess the utility of categorical data anonymized through global recoding by measuring classification performance. Additionally, Sano [10] evaluates the utility and risk of synthetic data generated by PCA.

A variational autoencoder (VAE) is an artificial neural network architecture introduced by Kingma and Welling [4]. It is part of the family of probabilistic graphical models and variational Bayesian methods. We propose two methods for generating synthetic data using VAE. The first method reconstructs data at the output layer during the final iteration of the training phase. The second method inputs newly generated latent variables into the sampling layer and obtains generated data at the output layer after the training phase.

In the next section, we introduce VAE and describe the proposed method for generating synthetic data using VAE in Sect. 3. We will generate synthetic data from real data, SSDSE-A, and evaluate the utility of the synthetic data by calculating utility measures and conducting regression analyses in Sect. 4. Finally, we summarize the proposed method and outline directions for future research in Sect. 5.

2 Variational Autoencoder

As a dimension reduction method, linear PCA and nonlinear PCA using artificial neural networks, known as autoencoders, are widely recognized.

An autoencoder is an unsupervised learning method that compresses input data x into low-dimensional latent variables z via an encoder and reconstructs them as x' through a decoder. The VAE assumes that latent variables z are random variables following a normal distribution with mean μ and variance σ^2,

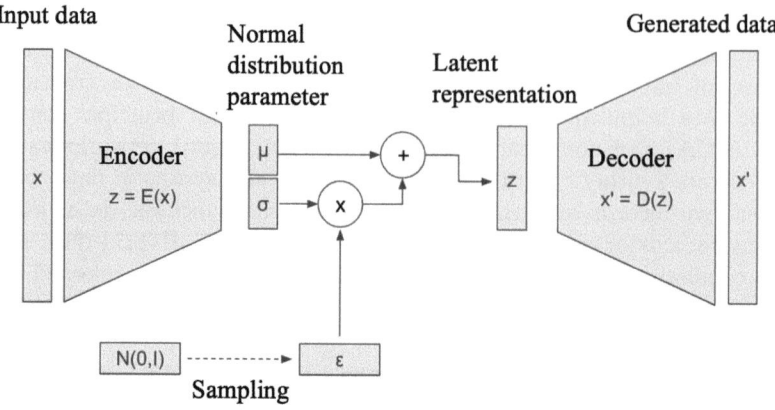

Fig. 1. Schematic overview of VAE

where these parameters are learned by the encoder. A schematic overview of the VAE is shown in Fig. 1. In the marginal likelihood $\log p_\theta(\boldsymbol{x}^{(i)}, \ldots, \boldsymbol{x}^{(n)}) = \sum_{i=1}^{n} \log p_\theta(\boldsymbol{x}^{(i)})$, the likelihood of an individual sample can be rewritten as

$$\log p_\theta(\boldsymbol{x}^{(i)}) = D_{KL}(q_\phi(\boldsymbol{z}|\boldsymbol{x}^{(i)}) \| p_\theta(\boldsymbol{z}|\boldsymbol{x}^{(i)})) + \mathcal{L}(\boldsymbol{\theta}, \boldsymbol{\phi}; \boldsymbol{x}^{(i)}), \tag{1}$$

where the first term on the right-hand side (RHS) is the KL divergence between the approximate distribution $q_\phi(\boldsymbol{z}|\boldsymbol{x}^{(i)})$ and the true posterior $p_\theta(\boldsymbol{z}|\boldsymbol{x}^{(i)})$, and the second RHS term $\mathcal{L}(\boldsymbol{\theta}, \boldsymbol{\phi}; \boldsymbol{x}^{(i)})$ is the variational lower bound on the marginal likelihood of sample i. The encoder implements the approximate model of the true posterior $q_\phi(\boldsymbol{z}|\boldsymbol{x}^{(i)})$.

The VAE maximizes the lower bound $\mathcal{L}(\boldsymbol{\theta}, \boldsymbol{\phi}; \boldsymbol{x}^{(i)})$, which can also be expressed as:

$$\mathcal{L}(\boldsymbol{\theta}, \boldsymbol{\phi}; \boldsymbol{x}^{(i)}) = -D_{KL}(q_\phi(\boldsymbol{z}|\boldsymbol{x}^{(i)}) \| p_\theta(\boldsymbol{z})) + \mathbb{E}_{q_\phi(\boldsymbol{z}|\boldsymbol{x}^{(i)})} \left[\log p_\theta(\boldsymbol{x}^{(i)}|\boldsymbol{z}) \right], \tag{2}$$

where the first term on the RHS serves as a regularization term, requiring the trained encoder $q_\phi(\boldsymbol{z}|\boldsymbol{x}^{(i)})$ to be close to the prior $p_\theta(\boldsymbol{z})$, and the second RHS term represents the reconstruction error of the trained decoder $p_\theta(\boldsymbol{x}^{(i)}|\boldsymbol{z})$.

3 Generation Method of Synthetic Data by VAE

Before applying the VAE, we conduct Min-Max preprocessing on the original data and employ the VAE architecture shown in Fig. 2 along with a cross-entropy loss function. In Fig. 2, r and k denote the dimensions of the input and latent variables, respectively.

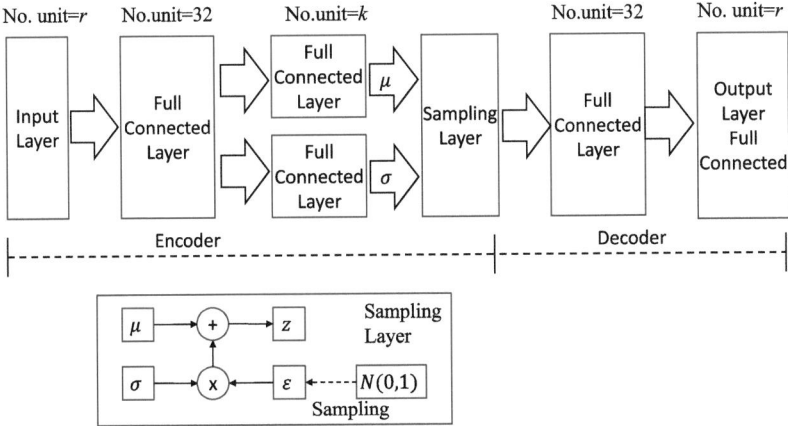

Fig. 2. Architecture of VAE in proposed generation method

We generate two types of synthetic data:

(a) Reconstructed Data
(b) Generated Data

(a) Reconstructed data are the output values from the output layer after the training phase. Note that reconstructed data correspond to the original samples. (b) Generated data are the output values from the output layer for newly generated ϵ in the sampling layer shown in Fig. 2 after the training phase. Note that generated data do not correspond to the original samples. Both reconstructed and generated data undergo inverse Min-Max preprocessing for use in synthetic data applications.

4 Evaluation by Real Data

4.1 SSDSE Data and Parameter Setting

We generate synthetic data from SSDSE-A in the Standardized Statistical DataSet for Education (SSDSE, [7]) provided by the National Statistics Center of Japan. SSDSE-A consists of 125 variables related to population and household, natural environment, economic base, administrative infrastructure, education, labor, culture and sports, residence, health and medical care, welfare, and social security for 1,741 municipalities.

We generate reconstructed and generated data with latent variable dimensions of $k = 3, 5, 10, 15, 20$. The parameter settings for VAE are as follows: batch size $= 20$, epochs $= 201$, learning rate $\eta = 0.001$.

4.2 Measures of Information Loss and Evaluation Results

We evaluate the information loss of synthetic data from the proposed method as follows: The first measure is the information loss by the mean absolute error rate for observed values:

$$MAEO = \frac{1}{nr} \sum_{i=1}^{n} \sum_{j=1}^{r} \left| \frac{x_{ij} - x'_{ij}}{x_{ij}} \right|, \quad (3)$$

where x_{ij} and x'_{ij} denote the original and generated values, respectively, for the i-th observation of the j-th variable.

The second measure is the mean absolute error rate for each mean value:

$$MAEM = \frac{1}{r} \sum_{i=1}^{r} \left| \frac{m_i - m'_i}{m_i} \right|, \quad (4)$$

where m_i and m'_i denote the mean values of the i-th variable for the original and generated data, respectively.

The third measure is the mean absolute error of the correlation coefficient:

$$MAEC = \frac{1}{r(r+1)/2} \sum_{i=1}^{r} \sum_{j>i}^{r} \left| c_{ij} - c'_{ij} \right|, \quad (5)$$

where c_{ij} and c'_{ij} denote the correlation coefficients between the i-th and j-th variables for the original and generated values, respectively.

We evaluate synthetic data using the above information loss measures. The results of information loss for reconstructed and generated data are shown in Table 1. To ensure a robust evaluation of information loss, we also calculate the median values for the three measures.

Table 1. Information loss for synthetic data by VAE

	(a) MAEO		(b) MAEM		(c) MAEC	
k	Reconstructed	Generated	Reconstructed	Generated	Reconstructed	Generated
3	8.194 (0.601)	17.871 (0.895)	0.027 (0.026)	0.019 (0.014)	0.235 (0.096)	0.238 (0.099)
5	8.290 (0.605)	14.831 (0.893)	0.053 (0.051)	0.120 (0.124)	0.235 (0.095)	0.242 (0.098)
10	8.020 (0.613)	17.352 (0.898)	0.018 (0.017)	0.122 (0.125)	0.239 (0.098)	0.234 (0.096)
15	6.980 (0.601)	15.673 (0.904)	0.009 (0.005)	0.138 (0.139)	0.239 (0.097)	0.253 (0.100)
20	8.665 (0.616)	14.257 (0.879)	0.101 (0.107)	0.145 (0.149)	0.239 (0.098)	0.250 (0.100)

Values in parentheses denote median values

For MAEO, we observe that the values for reconstructed data are smaller than those for generated data for both mean and median values in Table 1(a). For MAEM, we find that the values for reconstructed data are smaller than those for generated data for $k = 5, 10, 15, 20$ in both mean and median values in Table 1(b). As for MAEC, the value of reconstructed data is smaller than that of generated data for $k = 3, 5, 15, 20$ in both mean and median values, as shown in Table 1(c). In addition, there is no specific trend in the measure values as k increases for the three measures mentioned above.

4.3 Regression Analyses

We evaluate the reconstructed and generated data by comparing them to the results of regression analysis on the original data. We consider the following regression model:

$$Y = \beta_0 + \beta_1 X_1 + \beta_2 X_2 + \beta_3 X_3 + \beta_4 X_4 + \beta_5 X_5 + \epsilon, \tag{6}$$

where the objective variable Y and explanatory variables X_1, X_2, X_3, X_4, X_5 are shown in Table 2, and ϵ is an independent normally distributed random variable with zero mean and variance σ^2.

Table 2. Objective and explanatory variables

	Variable
Y	Number of in-migrants from other prefectures
X_1	Number of establishments (private sector)
X_2	Number of employees (private sector)
X_3	Number of persons engaged in primary industry
X_4	Number of persons engaged in secondary industry
X_5	Number of persons engaged in tertiary industry

In Table 1, reconstructed data with $k = 15$ show better performance on MAEO and MAEM than other reconstructed data, while generated data with $k = 10$ show better performance on MAEC than other generated data. Thus, the comparison results of regression analysis between original data, reconstructed data with $k = 15$, and generated data with $k = 10$ are shown in Table 3. For the original data, the signs of the estimated coefficients indicate that the number of in-migrants from other prefectures, Y, is positively influenced by the number of establishments X_1 and employees X_2 in the private sector, as well as by the number of persons engaged in the tertiary industry X_5. Conversely, the number of persons engaged in primary industry X_3 and secondary industry X_4 has a negative effect on Y. We find that X_4 positively affects Y in the reconstructed data, and it also positively affects Y in the generated data. We attribute the sign inconsistencies to the small sample size in the training phase of the VAE model.

Table 3. Results of regression analysis

Estimated Coefficient & Multiple R-squared	Original Data	Reconstructed Data ($q = 15$)	Generated Data ($q = 10$)
$\hat{\beta}_0$	105.445*	616.111***	582.315***
$\hat{\beta}_1$	0.332***	0.496***	-0.003
$\hat{\beta}_2$	0.003*	0.024***	0.041***
$\hat{\beta}_3$	−0.566***	−1.628***	−1.013***
$\hat{\beta}_4$	−0.149***	0.185***	−0.072***
$\hat{\beta}_5$	0.142***	0.012***	0.109***
R^2	0.973	0.999	0.998

Significance: *** 0.1% ** 1% * 5%

5 Conclusion

In this paper, we presented two methods for generating synthetic data using a variational autoencoder. The proposed generation methods were applied to SSDSE-A data, and the utility of the synthetic data was evaluated using three measures: MAEO, MAEM, and MAEC. Additionally, we conducted regression analysis on the synthetic data and compared the results with those from the original data. We found that the reconstructed data were superior to the generated data in terms of information loss and no specific trend was observed in the measured values as k increased for the aforementioned information measures. Regression coefficients were not necessarily consistent with those of the original data in the regression analysis. We believe that more data are required for the training phase to address the sign inconsistencies. Risk evaluation of synthetic data will be a subject for future studies.

Acknowledgments. This work was supported by a Grant-in-Aid for open-type research from JSPS KAKENHI Grant Number JP22K01427.

References

1. Defays, D., Nanopoulos, P.: Panels of enterprises and confidentiality: the small aggregates method. In: Proceedings of the 1992 Symposium on Design and Analysis of Longitudinal Surveys, pp. 195–204 (1993)
2. Gouweleeuw, J., Kooiman, P., Willenborg, L., de Wolf, P.: Post randomisation for statistical disclosure control: Theory and implementation. Technical report, Statistics Netherlands, Research paper (1997)
3. Hundepool, A., et al.: Statistical Disclosure Control. Wiley, Chichester (2012)
4. Kingma, D., Welling, M.: Auto-encoding variational bayes. In: Proceedings of the 2nd International Conference on Learning Representations (2014)
5. Muralidhar, K., Sarathy, R.: Data shuffling: a new masking approach for numerical data. Manage. Sci. **52**(5), 658–670 (2006)

6. Muralidhar, K., Sarathy, R.: Generating sufficiency-based non-synthetic perturbed data. Trans. Data Privacy **1**(1), 17–33 (2008)
7. National Statistics Center of Japan: SSDSE (2025). https://www.nstac.go.jp/use/literacy/ssdse/. Access 5 May 2025
8. Rubin, D.B.: Discussion of statistical disclosure limitation. J. Official Stat. **9**(2), 461–468 (1993)
9. Sano, N.: Synthetic data by principal component analysis. In: Proceedings of 20th IEEE International Conference on Data Mining Workshops 2020, pp. 101–105 (2020)
10. Sano, N.: Utility and risk evaluation of synthetic data by orthogonal transformation. Rev. Socionetwork Strat. **16**(1), 71–79 (2022)
11. Sano, N., Hattori, Y.: Utility evaluation measures for categorical data by classification performance. In: Proceedings of 19th IEEE International Conference on Data Mining Workshops, pp. 356–361 (2019)

Advanced Social Computing through AI-centric Multidisciplinary Fusion

A Determination Framework of Quantitative Element of Criminal Offenses Integrating Explainable Artificial Intelligence

Quyuan Wang[1], Yuxin Liao[1], Xia Hu[2], Yuping Tu[1(✉)], Run Zeng[1], and Zhiwei Guo[1]

[1] Chongqing Key Laboratory of Intelligent Perception and BlockChain Technology, Chongqing Technology and Business University, Chongqing 400067, China
{qywang,liaoyuxin2,tuyuping,zengrun,zwguo}@ctbu.edu.cn
[2] Chongqing Anti-Corruption and Audit Governance Research Center, Chongqing Technology and Business University, Chongqing 400067, China
huxia@ctbu.edu.cn

Abstract. The integration of legal studies and artificial intelligence has attracted widespread attention, with Legal Judgment Prediction being one of the most typical applications of AI as a tool. However, in judicial practice, obtaining a reasonable verdict requires the identification and organization of criminal circumstance elements. In this paper, we proposed a scalable determination framework of quantitative elements of criminal offenses integrating XAI (XAI-QED). On one hand, we improve the accuracy of circumstance recognition by introducing a hierarchical structure of circumstances and roles, and explore potential factors that may influence judgment outcomes through similar case analysis. On the other hand, we incorporate an explainability module to enhance the credibility of the framework's prediction results. Finally, simulation experiments validate the effectiveness of the proposed framework in both circumstance recognition and outcome prediction.

Keywords: Quantitative Element of Criminal Offenses · Legal Judgment Prediction · Explainable Artificial Intelligence

1 Introduction

Computational law is an emerging field at the intersection of digital technology and legal theory that has emerged in recent years, and in particular, the promotion of the in-depth integration of Artificial Intelligence and judicial practice has

---

This work was funded by the Doctoral and Cultivation Project of Chongqing Social Science Planning (No.2023BS062), the High Talent Research Start-up Project of Chongqing Technology and Business University (Nos.2256001, 2256002), and the Chongqing Undergraduate Training Program on Innovation and Entrepreneurship (Nos. S202411799035, S202411799036).

© The Author(s), under exclusive license to Springer Nature Switzerland AG 2026
V. Garcia Diaz et al. (Eds.): MISNC 2025, CCIS 2729, pp. 419–430, 2026.
https://doi.org/10.1007/978-3-032-09945-7_34

helped to shape the modernization of the trial system and trial capacity [1,5]. Within the field of law, they are mainly concerned with the potential risks, the regulatory system, and the subject status of AI. In the field of computer science, how to apply AI to specific legal scenarios has attracted much attention, such as LaWGPT [12] and other large language models, which realize the functions of legal counseling, searching for legal terms, and analyzing legal texts.

In the process of judicial practice, the use of artificial intelligence methods for Legal Judgment Prediction (LJP) is one of the main concerns [2,8,9]. However, existing work focuses more on optimizing the performance of the algorithm to improve the prediction accuracy, lacking differentiation and categorization of the circumstances in the determination process. In the process of sentence prediction, the determination of quantitative element of the offense is a prerequisite for legal reasoning and drawing legal conclusions, which involves the identification of conviction and sentencing circumstances. If AI can be used as an auxiliary decision-making tool in the process of circumstances extraction, correlation analysis and structuring, it is possible to reduce discrepancies in the decisions of different judges in similar cases and to ensure uniformity and predictability.

However, algorithms may produce unfair predictions due to biases in the training data, and their decision-making process lacks transparency and is difficult to review and evaluate. Explainable AI (XAI) has the goal of making the AI decision-making process transparent to help users understand and trust the results of AI models. By introducing XAI theory into the determination, it helps to identify and correct potential biases in the model, ensuring fairness in sentencing decisions and avoiding unfair treatment of specific groups. At the same time, a transparent decision-making process enables the judicial system to review and validate the basis of AI determinations to ensure that they comply with legal and ethical standards [3,7].

To summarize, our contributions are summarized as follows:

- We proposed a scalable determination framework of quantitative elements of criminal offenses integrating XAI (XAI-QED), which consists of a text encoding module, a circumstance recognition and fusion module, and an explainability analysis module. We introduce a hierarchical structure of circumstances and roles to improve the accuracy of circumstance identification, and incorporate similar text analysis to explore potential factors that may influence sentencing outcomes.
- We conducted simulation experiments to compare the performance of the proposed framework with baseline algorithms, performed circumstance vector recognition and separation with visualizations for specific cases, and carried out similarity analyses. These experiments verified the effectiveness of our framework in circumstance identification and sentencing prediction. Additionally, the interpretability analysis enhanced the credibility of the prediction results.

2 Framework Description

Under the Chinese legal system, the definition of conviction and sentencing circumstances during analysis and trial is very essential to the result. Conviction Circumstances (CC) refer to the objective facts and legal elements that affect the determination of guilt, such as the nature, means, and consequences of the criminal act, which are used to establish whether a crime has been committed and to determine the applicable offense. And Sentencing Circumstances (SC) refer to specific factors that influence the severity of punishment, including Statutory Circumstances (STC) (e.g., voluntary surrender, recidivism) and Discretionary Circumstances (DIC) (e.g., criminal motive, victim's fault), which are used to determine the specific sentence within the legally prescribed penalty range.

In order to sort out the various circumstances in the case and improve the accuracy of judgment prediction, while attempting to provide an explainability analysis of the decision-making process, we propose a determination framework of quantitative element of criminal offenses integrating XAI (XAI-QED). Specifically, we firstly encode the description of the facts of the case, extract key information by using natural language processing technology, and identify the conviction and sentencing circumstances of the case through fact extraction and separation, and in the sentencing circumstances, in addition to the statutory circumstances and discretionary circumstances, we also mark the other circumstances that may affect the judgment. Furthermore, we analyze similar texts to reveal potential connections between cases (e.g., same district cases, etc.), and form a multidimensional view of the case through circumstances fusion, and use this to predict possible sentencing outcomes. In addition, the Interpretable Analysis Module serves the global framework by attempting to reveal the factors that have the greatest impact on the outcome of the judgment, as well as uncovering factors outside of the circumstances that may have influenced the judgment. Figure 1 is a schematic diagram of our proposed framework, and we next describe the specific modules of the framework and the technologies involved.

2.1 Text Encoding Module

Text encoding refers to the process of converting text into numerical representations that can be processed by computers. In general, we convert text to binary, word vectors, or other numeric forms to enable computers to understand and process human language, which is essential for natural language processing (NLP) tasks.

Word2Vec. Word2Vec is a neural network-based word vector representation model, the basic principle of which is to learn the distributed representation of words through context, so that words with similar semantics are closer together in the vector space.

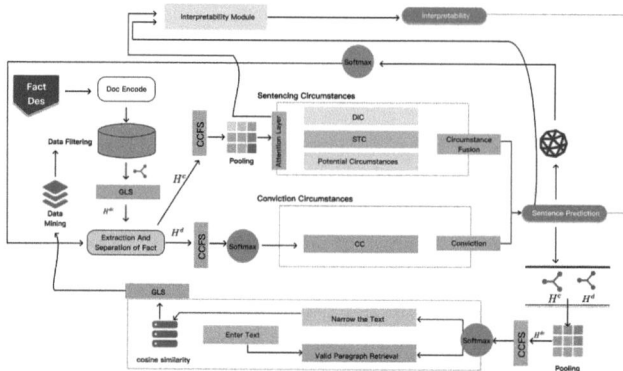

Fig. 1. The determination framework of quantitative element of criminal offenses integrating XAI (XAI-QED).

BERT-Based Method. BERT uses a bidirectional Transformer encoder to capture contextual information of words from both forward and backward directions, thereby generating more accurate vector representations. In terms of workflow, the BERT approach is mainly divided into two stages: pre-training and fine-tuning.

Transformer-Based Method. The core of the Transformer Encoder lies in its multi-head attention mechanism, which allows the model to learn dependencies between different positions in the input sequence in parallel across multiple representation subspaces. Structurally, the Transformer Encoder consists of four components: input embedding and positional encoding, multi-head self-attention mechanism, feed-forward neural network, and residual connections with layer normalization.

2.2 Circumstances Recognition and Fusion Module

In this section, we focus on describing the specific structure of circumstance identification in the proposed framework, as well as the process of mining and integrating the circumstances.

Hierarchical Circumstances and Role Structure. As mentioned earlier, the determination of circumstances is crucial in an AI-assisted sentencing system. Specifically, the Conviction Circumstances (CC) determine the criminal charge, while the Sentencing Circumstances (SC) determine the length of the sentence. The sentencing circumstances can further be divided into Statutory Circumstance (STC) and Discretionary Circumstance (DIC). Therefore, to express the coupling relationships between these circumstances, we set the SC as higher-level circumstances, and the STC and DIC as lower-level circumstances. It is important to note that there may also be other potential factors that influence sentencing within the sentencing circumstances, which we define as potential circumstances, and these will be discussed later.

In the circumstance identification structure, the first step is the recognition of keywords. Event keywords are used to identify the types of circumstantial attributes involved in an event, with each type of circumstance corresponding to specific keywords. For example, in an intentional injury case, conviction-related circumstances may include keywords such as "family member" or "causing serious injury by especially cruel means." To more accurately represent key elements within each type of circumstance, we define corresponding "roles." For instance, "offender" and "victim" are essential roles present across all types of circumstances; the "stolen property value" role, which evaluates the worth of stolen items, pertains to conviction circumstances; the "confession" role reflects the offender's attitude and falls under statutory sentencing circumstances; and the "criminal record" role indicates the offender's conduct, representing discretionary sentencing circumstances. These roles are also organized in a hierarchical structure.

Our designed Hierarchical Circumstances and Role Structure is responsible for detecting event-related circumstance keywords and determining the corresponding circumstance types and parameter roles. Based on the previously encoded text, we assign a keyword or role type to each corresponding character vector. In this framework, we treat SC as the higher-level circumstance. For each higher-level circumstance or role j, its features can be represented by a vector γ_j. Then, a fully connected layer is used to associate the hidden vector of each text token with all possible roles and trigger word types, generating relevance scores:

$$y_{ij} = U_\gamma [h_i; \gamma_j], \tag{1}$$

where $[a; b]$ denotes the concatenation of a and b. For each label, the Softmax function is applied to obtain a probability distribution over all possible roles and keywords types:

$$\theta_i = \sum_{j=1} p_{ij} \gamma_j \tag{2}$$

where $p_{ij} = \frac{exp(y_{ij})}{\sum_{k=1} exp(y_{ik})}$, the vector θ_i represents the integrated features of the higher-level circumstance or role, providing top-down semantic information for predicting subordinate circumstances or roles. We concatenate each hidden vector h_i with its corresponding θ_i as the input feature for the keyword type and parameter role classifier. The model then estimates the probability that the token belongs to a specific subordinate type or role τ_i, as shown below:

$$s_\tau \left(m_i, \tilde{n}_{(i)}^{\tau_i} \right) = \frac{\exp \left(w_j^T [h_i; \theta_i] \right)}{\sum_{k=1} \exp \left(w_k^T [h_i; \theta_i] \right)}. \tag{3}$$

Similarly, for each conviction circumstance or role i, its features can be represented by a vector σ_i. A fully connected layer is then used to compute the association between the hidden vector of each text token and all possible roles and trigger word types, generating relevance scores:

$$x_{ij} = U_\delta [h_j; \delta_i]. \quad (4)$$

And the probability that the token belongs to a specific subordinate type or role ε_i is obtained:

$$s_\varepsilon \left(a_i, \tilde{d}_{(i)}^{\varepsilon_i}\right) = \frac{exp\left(r_j^T [h_i; \mu_i]\right)}{\sum_{k=1} exp\left(r_k^T [h_i; \mu_i]\right)} \quad (5)$$

where $\mu_i = \sum_{j=1} \beta_{ij} \delta_j$ and $\beta_{ij} = \frac{exp(x_{ij})}{\sum_{k=1} exp(x_{ik})}$.

We embed the extracted keywords and parameters, along with their types/roles, into two task-specific classifiers. The extracted vector sequence is denoted as $H_s^d = h_1^d, h_2^d, ..., h_{ld}^d$. A max-pooling layer is applied to each vector sequence, which is then concatenated with the corresponding subordinate type/role embedding w:

$$g_i = \left[maxpooling\left(H_s^d\right); w\right] \quad (6)$$

Similar Case and Potential Circumstances Mining. Potential circumstances refer to implicit factors that are not explicitly stated in legal provisions or case descriptions but may influence the outcome of a judgment. These factors may include the motive and background of the crime (such as self-defense or emotional provocation), specific behaviors of the defendant (such as voluntary surrender or active compensation), the attitude of the victim (such as offering forgiveness), the broader social context of the case (such as during a public emergency or place of the incident), and the personal conditions of the defendant (such as mental state or family burdens). Identifying these potential circumstances helps enable more comprehensive, personalized, and fair judicial decisions. Therefore, this framework provides a reference for the current case by comparing similar cases, helping to avoid inconsistencies in sentencing for similar offenses, while also uncovering potential circumstances that may influence the judgment.

2.3 Sentence Prediction Module

In the sentencing prediction module, the first step is to determine the corresponding charge. We calculate the initial probability of each ε in the conviction circumstances:

$$V(\varepsilon_1) = score(start, \varepsilon_1) + s_\varepsilon \left(a_1, \tilde{d}_{(1)}^{\varepsilon_1}\right), \quad (7)$$

where $score(start, \varepsilon_1)$ is the transition probability from the start state to ε. And we can calculate the cumulative maximum probability for the subsequent labels:

$$V(\varepsilon_t) = max_{\varepsilon_{t-1}} \left(score(\varepsilon_{t-1}, \varepsilon_1) + V(\varepsilon_{t-1})\right) + s_\varepsilon \left(a_t, \tilde{d}_t^{\varepsilon_t}\right). \quad (8)$$

We adopt a Support Vector Machine (SVM) for decision classification, assigning each case to the corresponding charge. The classification decision function is as follows:

$$T(V) = sign\left(w^T \beth(V) + b\right) \qquad (9)$$

where $V = max_{\varepsilon_{id}}(V(\varepsilon_{id}))$, T is the number of model iterations, and $\beth(V)$ is the space transformation function. The loss function is as follows:

$$Loss_1(T(V), y') = max(0, margin + y' - T(V)) \qquad (10)$$

where y' is the true charge.

In the proposed framework, both the hierarchical model and potential circumstances may affect the accuracy of sentencing prediction. To integrate these diverse sources of information, we adopt an ensemble learning approach. We apply linear weighting to combine the hierarchical model and the similar text comparison model, and then use a regression model to predict the specific length of the sentence.

$$SP(H(h)) = w^T H(h) + b = \hat{w}^T H(h), \qquad (11)$$

where $H(h) = sign(\sum_t^T \alpha_1 f(h))$, and $f(h) = \sum_{ld}^d g_i + L_\Theta$, L_Θ is the similarity representation of the text. Therefore, the loss function is represented as:

$$Loss_2 = \left(w^T H(h) - SP(H(h))\right)^T \left(w^T \hat{H}(h) - \bar{Y}\right) = \| w^T \hat{H}(h) - \bar{Y} \|_2^2, \qquad (12)$$

where, $\bar{Y}$ denotes the real sentence length.

Therefore, the overall loss function is expressed as:

$$Loss(\theta) = \lambda_1 Loss_1 + \lambda_2 Loss_2 \qquad (13)$$

2.4 Explainability Analysis Module

In an AI-assisted sentencing system, the introduction of Explainable Artificial Intelligence (XAI) is essential to enhance the transparency and credibility of model decisions [4]. Since sentencing involves serious legal consequences, judges and legal professionals need to understand the decision-making process of AI models to ensure fairness and legal compliance. XAI provides detailed explanations of the model's predictions, making the sentencing process more traceable and reviewable, thereby increasing the system's acceptance and trust in the legal system. In the context of judicial artificial intelligence, XAI can be divided into local explanations and global explanations. The former focuses on the case level, exploring the basis for case determination, while the latter focuses on the model level, analyzing aspects such as the distribution of model weights.

In the framework proposed, we use local explanations algorithms to analyze the decision-making logic of individual cases. Specifically, we use the SHAP algorithm as the core algorithm of the explainability analysis module. The SHAP algorithm is based on the Shapley Value from game theory [6]. In SHAP, the

Shapley value is used to measure the marginal contribution of each feature to the model's prediction. Specifically, for a given model, the Shapley value of each feature represents its average marginal contribution to the prediction across all possible combinations of features. SHAP decomposes the model's prediction into the sum of individual feature contributions, with the SHAP value indicating each feature's contribution—that is, its importance to the prediction.

The SHAP algorithm typically involves training two models: one that includes a specific feature $f_{S\cup i}$, and another that does not include that particular feature f_S. The impact of the feature on the model's prediction is then determined by comparing the predictions of these two models:

$$\phi_i = \sum_{S\subseteq F\setminus\{i\}} \frac{|S|!(|F|-|S|-1)!}{|F|!} \left[f(S\cup\{i\}) - f(S)\right], \tag{14}$$

where, F is the set of all features, and S is any subset of F that does not include feature i.

3 Experiments

In simulation, we used an Intel(R) Xeon(R) Platinum 8474C as the CPU, and an NVIDIA GeForce RTX 4090D to built a server to deploy the proposed framework and model. And we primarily used the CAIL dataset [10,11], which includes more than 2.6 millon cases, and are annotated with 183 criminal law articles and 202 criminal charges.

3.1 Baseline Comparison

We compare the proposed framework with several baseline algorithms in three aspects: charge determination, legal article matching, and sentencing prediction. The main differences among these algorithms lie in their text encoding methods. The results are shown in Fig. 2, where acc denotes prediction accuracy, $macro-P$ is macro-precision, and $macro-R$ represents macro-recall. From the results, the accuracy of all models is relatively high in charge determination and legal article matching. Although our framework performs the best in sentencing prediction, there is still room for improvement. This is because circumstance recognition is only an important basis for sentencing, but many other potential factors in the actual sentencing process can influence the final outcome.

3.2 Circumstances Vector Visualization

To separate and identify the Circumstances elements, the most crucial step is to convert the relevant circumstances information into vectors and represent them using word embedding techniques. We selected the intentional homicide case involving Zhu as an example to demonstrate the embedding and visualization of circumstances vectors.

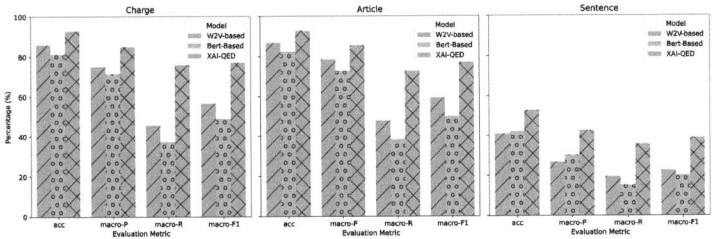

Fig. 2. Performance Comparison of Different Baseline Algorithms

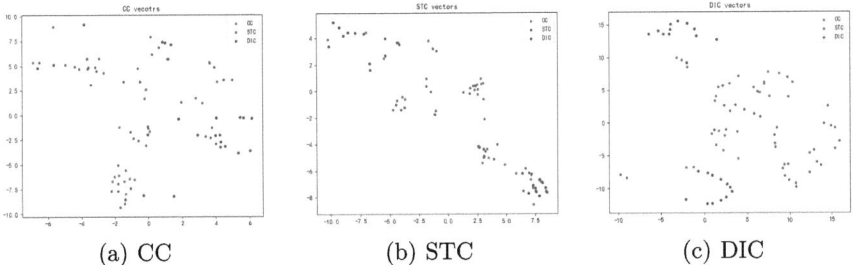

Fig. 3. Circumstances vector visualization

The results are shown in Fig. 3. Initially, after encoding and embedding the text, the distribution of related texts in the embedding space appeared relatively scattered. The text did not exhibit the corresponding associative relationships. After separating the CC vectors, a certain clustering effect emerged. With further separation of STC and DIC, we observed that the three types of circumstances in this case were generally well-separated, indicating the effectiveness of the circumstances extraction strategy in our proposed framework. However, we also found that due to the limitations in circumstances annotation and interpretation, there exist coupling relationships between some circumstances, making complete separation unachievable. Additionally, in terms of separation performance, the identification and separation of sentencing circumstances are more effective than those of conviction circumstances, as sentencing circumstances are described in a more specific manner.

3.3 Similar Case Analysis

Based on the five selected cases, we conducted similarity analyses of the three types of circumstances using cosine similarity and we employed Jensen-Shannon (JS) divergence to compare the differences between the predicted sentencing outcomes and the actual sentencing results, aiming to reveal the influence of different circumstance elements on sentencing outcomes and to explore potential factors that may affect sentencing decisions.

Figure 4 presents the similarity analysis of the three types of circumstance across the five cases using cosine similarity as the evaluation metric. In simple

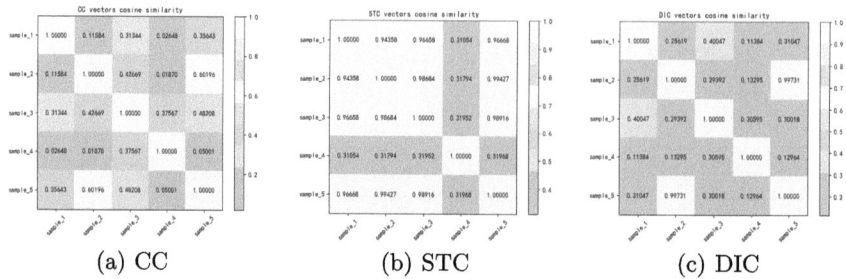

(a) CC   (b) STC   (c) DIC

Fig. 4. Circumstance Analysis Based on Cosine Similarity

terms, the closer the cosine similarity value is to 1, the more similar the two cases are. To facilitate comparison, all cases selected in this experimental section are intentional homicide cases, so there is little variation in terms of CC. Cases 2 and 5 show the highest similarity across all three types of circumstances, indicating a high degree of similarity between the two cases. Through analysis of the judgment documents, we found that both cases were triggered by domestic disputes, and in both, the defendants confessed to their criminal actions. The main difference between the two cases lies in the victim: in case 5, it was the husband, and in case 2, was the son. This demonstrates the accuracy of our similarity analysis module.

The Case 1 involves multiple offenders, and the complexity and uniqueness of the case result in significant differences in its circumstances recognition. Additionally, it can be observed that the vector similarity of discretionary sentencing circumstances is generally lower across all cases. This is due to the broader scope of discretionary factors, as well as potential variations across different regions and individual cases.

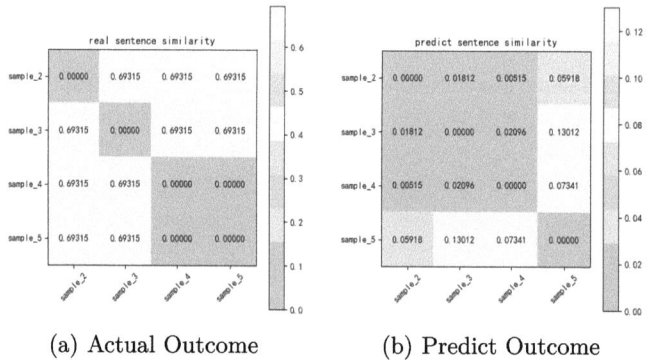

(a) Actual Outcome   (b) Predict Outcome

Fig. 5. Sentencing Outcome Similarity Analysis

The similarity between the actual and predicted sentencing outcomes for different cases is shown in Fig. 5. We use Jensen-Shannon (JS) divergence as the evaluation metric for similarity, where values closer to 0 indicate greater similarity. Overall, the determination of sentencing results involves many factors beyond just the circumstance elements, which generally leads to some degree of deviation between predicted and actual values. Additionally, the presence of multiple trial stages can also cause prediction deviations if the information regarding these stages is not well understood.

3.4 Interpretability Analysis

We randomly selected a case and used the SHAP method to analyze Chinese phrases in the judgment document, particularly those related to circumstance determination. We calculated their contribution to the final sentencing outcome predicted by the framework to evaluate their role in the decision-making process. Figure 6 shows the Chinese phrases that had the greatest impact on the case outcome.

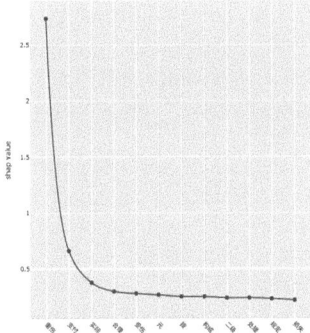

Fig. 6. SHAP Value Ranking of Chinese Phrases

It can be seen that in this case, the phrase "serious injury" has the greatest impact on the outcome, which aligns with general patterns in judicial decision-making, as it typically reflects the severity of harm and directly influences the verdict. In addition, the phrase "payment" usually implies post-incident compensation or the presence of financial disputes, which are also key factors in judicial rulings. Interpretability analysis can effectively enhance the acceptability and transparency of the model's prediction results.

4 Conclusion

In order to provide precise, trustworthy, and transparent judgment decisions, we proposed a scalable determination framework of quantitative elements of criminal offenses integrating XAI (XAI-QED), which consists of a text encoding module, a circumstance recognition and fusion module, and an explainability analysis module. From the perspective of future research directions, we aim to improve the model structure for judgment outcome prediction to further enhance its accuracy. Meanwhile, during our study, we found that the quality of data with circumstance annotation still needs to be improved. Additionally, more similar case analyses are needed to systematically identify and summarize potential factors that may influence judgment outcomes.

References

1. Afzal, J.: Implementation of Digital Law as a Legal Tool in the Current Digital Era. Springer, Singapore (2024)
2. Chien, K.C., Chang, C.H., Sun, R.D.: Legal knowledge management for prosecutors based on judgment prediction and error analysis from indictments. Comput. Law Secur. Rev. **52**, 105902 (2024)
3. Collenette, J., Atkinson, K., Bench-Capon, T.: Explainable AI tools for legal reasoning about cases: a study on the European court of human rights. Artif. Intell. **317**, 103861 (2023)
4. Došilović, F.K., Brčić, M., Hlupić, N.: Explainable artificial intelligence: a survey. In: 2018 41st International Convention on Information and Communication Technology, Electronics and Microelectronics (MIPRO), pp. 0210–0215. IEEE (2018)
5. Genesereth, M.: Computational Law. The Stanford Center for Legal Informatics, Complaw Corner, Codex (2021)
6. Lundberg, S.M., Lee, S.I.: A unified approach to interpreting model predictions. In: Guyon, I., Luxburg, U.V., Bengio, S., Wallach, H., Fergus, R., Vishwanathan, S., Garnett, R. (eds.) Advances in Neural Information Processing Systems 30, pp. 4765–4774. Curran Associates, Inc. (2017)
7. Richmond, K.M., Muddamsetty, S.M., Gammeltoft-Hansen, T., Olsen, H.P., Moeslund, T.B.: Explainable AI and law: an evidential survey. Digital Soc. **3**(1), 1 (2024)
8. Sun, J., Huang, S., Wei, C.: Chinese legal judgment prediction via knowledgeable prompt learning. Expert Syst. Appl. **238**, 122177 (2024)
9. Tong, S., Yuan, J., Zhang, P., Li, L.: Legal judgment prediction via graph boosting with constraints. Inf. Process. Manage. **61**(3), 103663 (2024)
10. Xiao, C., et al.: Cail2018: a large-scale legal dataset for judgment prediction. arXiv preprint arXiv:1807.02478 (2018)
11. Zhong, H., et al.: Overview of cail2018: legal judgment prediction competition. arXiv preprint arXiv:1810.05851 (2018)
12. Zhou, Z., et al.: Lawgpt: Knowledge-guided data generation and its application to legal LLM. arXiv preprint arXiv:2502.06572 (2025)

Quantification of Fish Feeding Behavior with MC-YOLO and Image Texture Features in Recirculating Aquaculture Systems

Jiachun Zhou[1], Junchao Yang[1(✉)], Xueni Pan[1], Haiyan Huang[1], Li Liu[1], Yanzheng Gao[2], and Yu Shen[1]

[1] School of Artificial Intelligence, Chongqing Technology and Business University, Chongqing, China
{2023313023,Yangjc,panxueni,huanghaiyan37,liuli21,shenyu}@ctbu.edu.cn
[2] College of Resources and Environmental Sciences, Nanjing Agricultural University, Nanjing, China
gaoyanzheng@njau.edu.cn

Abstract. This paper proposes a machine vision based method for quantifying fish feeding behavior. By integrating fish movement features and image texture features, the method can achieve efficient and accurate quantification. The study introduces an improved MC-YOLO object detection model, which enhances feature extraction capabilities through the Multi-Scale Lightweight Convolution (MSLConv) module and the Convolutional Block Attention Module (CBAM), while maintaining low computational complexity. Experiments conducted in an industrial Recirculating Aquaculture System (RAS) demonstrate that this method has significant advantages in detection accuracy, feeding state recognition, and adaptability to complex breeding environments. It provides strong technical support for precise feeding and has broad application prospects and promotion value.

Keywords: Fish feeding behavior · Deep learning · RAS · Computer vision

1 Introduction

Aquaculture is a vital component of agricultural production in China. In recent years, industrial aquaculture has experienced rapid development. Recirculating Aquaculture Systems (RAS) represent a novel aquaculture model, characterized by their environmental friendliness and high stocking density [1]. Driven by the advancement of intelligent equipment in aquaculture systems and the refinement of aquaculture models, industrial RAS can overcome geographical and climatic limitations, significantly enhancing the production and supply capacity of high-quality, fresh aquaculture products. This aligns with the modern fisheries' sustainable and healthy development philosophy and has propelled the rapid growth of the global aquaculture industry [2].

In recent years, computer vision has been widely applied in the aquaculture industry. As a low-cost, fast-responding, and highly stable observation technology, computer vision has become an important research technique for precise feeding. Research on quantifying fish feeding behavior using computer vision primarily focuses on two aspects: First, monitoring the movement state of fish to obtain motion characteristics such as speed and turning angle, and then quantifying feeding behavior based on these motion features. Second, processing the overall image during feeding to extract texture and shape features, calculating form parameters, or image entropy to quantify feeding behavior. Atoum and Li et al. [3] quantified individual fish feeding states but faced target loss issues in high-density environments. Guo et al. [4] combined texture and shape information with a BP neural network, but their method was computationally expensive and yielded unstable results. Zhou et al. [5] used Delaunay triangles to quantify group feeding intensity, but required removing frames with splashes and reflections due to high computational load. Huang et al. [6] achieved a 94.17% accuracy rate by extracting movement features and combining them with image texture parameters, but their method required uniform lighting and was affected by water surface splashes. Liu et al. [7] used frame difference methods but needed manual determination of target fish numbers. Gu et al. [8] developed a quantification system using a spatiotemporal feature fusion algorithm, though it had slow computation speeds due to large model parameters. Chen and Huang et al. [9] combined the Lucas-Kanade optical flow method with GLCM for feature extraction, but manual extraction led to incomplete results. This study proposes a machine vision-based method to quantify fish feeding behavior by detecting and tracking target fish to obtain movement parameters and combining them with visual texture features. The main contributions are:

(1) This paper proposes a quantification method that combines fish motion features and image texture features. By using computer vision technology to detect and track target fish, extract their motion parameters, and combine them with texture features of feeding images to quantify fish feeding behavior. Compared with traditional methods, this method, which integrates motion and texture information, provides more accurate recognition of fish feeding behavior.
(2) The use of the MC-YOLO target detection algorithm to accurately detect and track fish improves the precision of fish movement feature extraction and ensures the stability and efficiency of feeding behavior quantification under different stocking densities.
(3) This study validates the effectiveness of the proposed method through field experiments in an industrial RAS. Using high-definition cameras and targeted feeding, we captured a wealth of fish motion data, ensuring the application potential of the feeding behavior quantification model in real aquaculture environments.

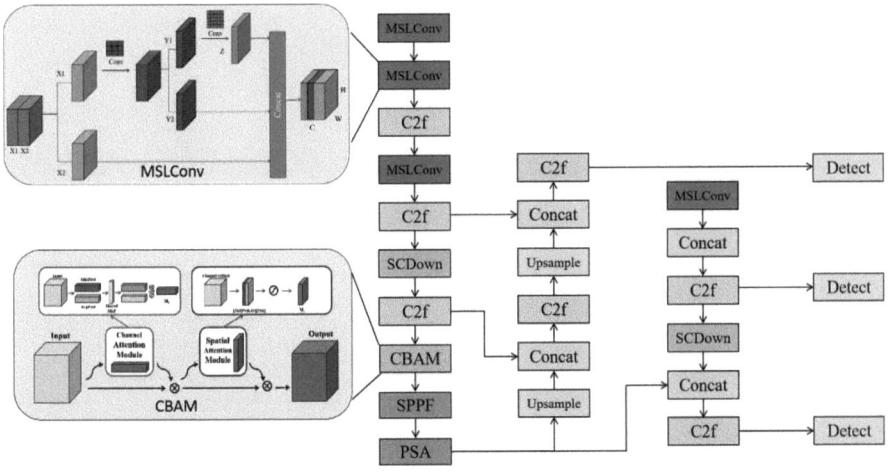

Fig. 1. Diagram of the MC-YOLO model structure.

2 Methodology

2.1 MC-YOLO Neural Network Model

YOLOv10 Neural Network Model. YOLOv10 is an efficient object detection algorithm that directly predicts bounding box coordinates and class probabilities from image pixels, treating detection as a regression problem. It is known for its high speed and accuracy [10]. Its network structure mainly includes: Backbone, Neck, and Head. The Backbone uses depthwise separable convolution and cross-stage feature fusion techniques to efficiently extract image features, reducing computational load while enhancing the semantic information of features. The Neck is based on Feature Pyramid Networks (FPN) and Path Aggregation Network (PANet) structures to fuse features from different levels, improving detection performance for multi-scale targets. The Head utilizes an anchor box mechanism and multi-scale prediction, combined with class-aware bounding box regression and confidence-guided non-maximum suppression techniques, to achieve precise object detection.

Although Yolov10's loss function consists of classification loss, localization loss, and confidence loss, which are jointly optimized to improve detection accuracy, its original Conv module uses a single convolutional kernel, limiting feature extraction capability and increasing computational load. To address these issues, this paper proposes the improved MC-YOLO model. The model structure is shown in Fig. 1.

Improvement of YOLOv10 by MSL and CBAM. The original Conv module in YOLOv10 uses a single convolutional kernel, which limits feature extraction. To improve this, we introduce the Multi-Scale Lightweight Convolution (MSLConv) module, which fuses features from different depths and scales [11].

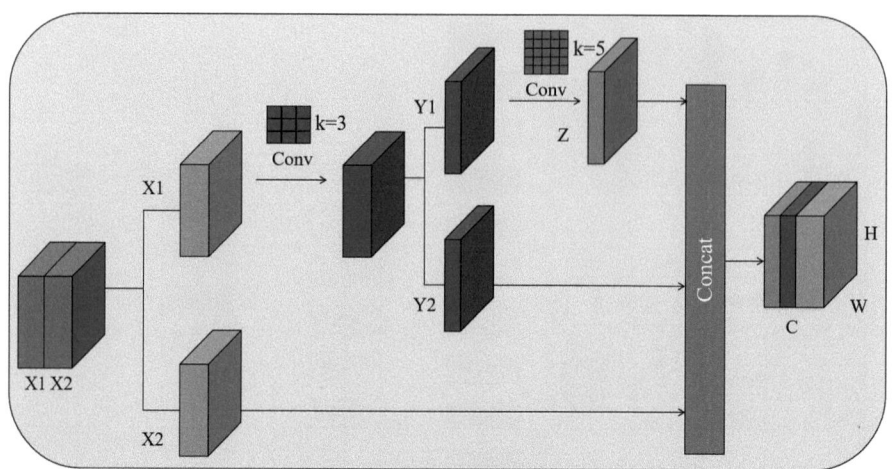

Fig. 2. The structure of MSLConv.

Compared to a standard convolutional module, MSLConv significantly reduces computational load, retains shallow information, and fuses it with deep information, resulting in a richer feature hierarchy.

MSLConv module, which fuses features of different depths and scales. By reducing the number of network channels while obtaining rich feature information, it enhances the model's multi-scale detection capability. As shown in Fig. 2, the MSLConv module splits the given feature along the channel axis into two parts with half the original number of channels, denoted as X1 and X2. X2 is passed to the deeper network without any processing to establish long-range dependencies, while X1 undergoes two stages of feature extraction. The newly extracted feature Y is split along the channel axis into Y1 and Y2. Y2 is left unprocessed, while Y1 is convolved to extract feature Z. Finally, X2, Y2, and Z are concatenated and output as the feature layer. Compared to a standard convolutional module, the MSLConv module not only significantly reduces computational load but also retains a portion of the original shallow information and fuses it with deep information, obtaining a richer hierarchy of features.

To further enhance the model's feature extraction capability, we incorporate the Convolutional Block Attention Module (CBAM), which combines channel attention (CAM) and spatial attention (SAM) mechanisms. As shown in Fig. 3, CAM calculates the importance weights of each channel to enhance or suppress specific channel features, while SAM calculates the importance weights of each position to enhance or suppress specific regional features [12]. In CBAM, channel attention is applied first. The formula is as follows:

$$M_c(F) = \sigma(MLP(AvgPool(F)) + MLP(MaxPool(F))) \quad (1)$$

The channel attention output is processed by spatial attention. The feature map is globally max-pooled and average-pooled, then convolved with a 7×7

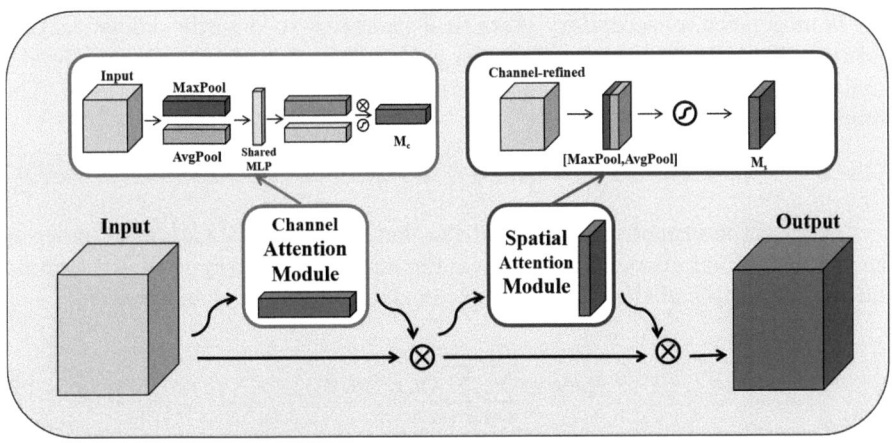

Fig. 3. The structure of CBAM.

kernel and passed through a ReLU activation function to reduce dimensionality. After another convolution, the feature map is normalized and combined with the channel attention output. This recalibrates the feature map in both spatial and channel dimensions. The formula is as follows:

$$M_s(F) = \sigma(f^{7\times 7}([AvgPool(F); MaxPool(F)])) \quad (2)$$

After CBAM, the new feature map obtains attention weights in both channel and spatial dimensions, significantly enhancing the connections between features in these dimensions and improving the extraction of effective target features.

The MC-YOLO model integrates CBAM and MSLConv modules, enhancing feature extraction and detection performance for multi-scale targets. Its lightweight design maintains low computational complexity, balancing efficiency and accuracy to improve overall performance.

2.2 Extraction of Image Texture Features

Image texture refers to the repetitive local patterns or structures in an image, describing the surface quality or appearance of the image and reflecting the gray-level repetition and variation among image pixels. Image texture features can be represented by the gray-level distribution of pixels and their surrounding spaces, capturing both local and global texture information. In the context of fish feeding experiments, when fish are not feeding and swimming normally, the water surface in the breeding tank remains relatively calm. However, when fish start feeding, significant ripples and splashes appear on the water surface, causing substantial changes in the gray-level information of the image. Therefore, image texture features can be used to quantitatively describe these changes. To more intuitively analyze image texture features, we perform secondary calculations based on the Gray-Level Co-occurrence Matrix (GLCM). We select energy, entropy, contrast, and homogeneity as secondary statistical measures to describe image texture features. The GLCM is defined as the probability of a pixel with gray-level i having a neighboring pixel with gray-level j at a distance d and direction θ. The definition formula is as follows:

$$\mathbf{P} = p(i, j, d, \theta) \tag{3}$$

Energy is the sum of the squares of the elements in the GLCM, also known as the angular second moment. It measures the uniformity of gray-level distribution and the coarseness of the texture. The calculation formula is as follows:

$$f_{ENE} = \sum_{i=1}^{N_g} \sum_{j=1}^{N_g} (p(i, j, d, \theta))^2 \tag{4}$$

Entropy measures the randomness contained in the image. The calculation formula is as follows:

$$f_{ENT} = \sum_{i=1}^{N_g} \sum_{j=1}^{N_g} (p(i, j, d, \theta)) log_2(p(i, j, d, \theta)) \tag{5}$$

Contrast reflects the clarity of an image and the depth of its texture grooves. The calculation formula is as follows:

$$f_{CON} = \sum_{i=1}^{N_g} \sum_{j=1}^{N_g} ((i-j)^2 p(i,j)) \tag{6}$$

Homogeneity reflects the clarity and regularity of the texture. The calculation formula is as follows:

$$f_{HOM} = \sum_{i=1}^{N_g} \sum_{j=1}^{N_g} \frac{p(i, j, d, \theta)}{1 + (i-j)^2} \tag{7}$$

In the sequence of images, the movement features of fish groups contain information about fish feeding behavior, while image texture features capture changes in texture during feeding. Combining these two types of features can more accurately quantify changes in fish behavior during feeding. The formulas for calculating the four image texture features are as follows:

$$H = \beta_1 f_{ENE} + \beta_2 f_{ENT} + + \beta_3 f_{COR} + \beta_4 f_{HOM} + \beta_5 \tag{8}$$

where $\beta_1, \beta_2, \beta_3, \beta_4$ are the coefficients of the four image texture feature values, and β_5 is the correction value for the comprehensive parameter.

Fig. 4. The laboratory medium-sized RAS for Evaluation.

3 Experimental System and Experimental Objects

The data used in this study were obtained from a medium-sized RAS in the laboratory, as shown in Fig. 4. The RAS consists of a breeding tank, a microscreen filter, a biological filter, and other water treatment units. The breeding tank has a radius of 260 cm and a height of 100 cm, with dissolved oxygen levels maintained between 6–8 mg/L and water temperature kept at 20–25°C [13]. The entire recirculating water system ensures good and stable water quality. We selected freshwater grouper as the experimental subject. Freshwater groupers grow rapidly, allowing significant growth changes to be observed within a short period.

Table 1. Training Results of Models.

Model	Precision	Recall	mAP@50
MC-YOLO	**0.902**	**0.913**	**0.874**
YOLOv8n	0.799	0.831	0.829
YOLOv9n	0.812	0.845	0.823
YOLOv11n	0.827	0.837	0.845
YOLOv10n	0.811	0.820	0.827
YOLOv10-CBAM	0.873	0.880	0.871
YOLOv10-SE	0.862	0.873	0.866
YOLOv10-MSL-SE	0.886	0.895	0.868

4 Analysis of Results

4.1 Model Comparison

To verify the performance advantages of the MC-YOLO model, this study involved conducting comparative experiments with other mainstream versions

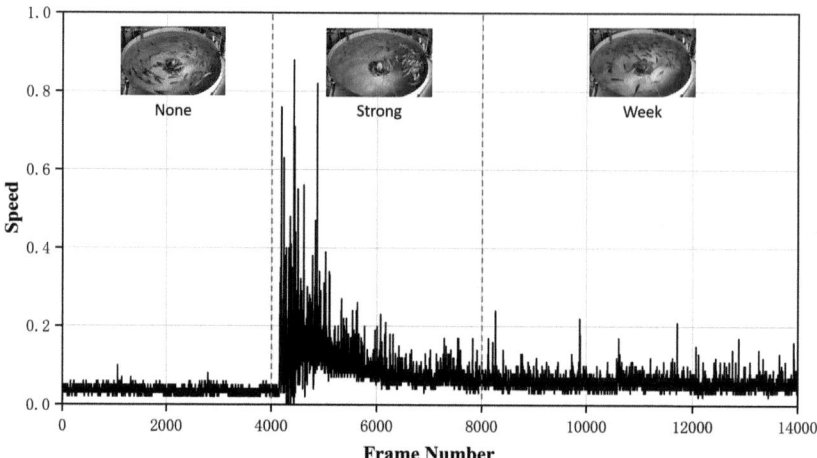

Fig. 5. Quantification of feeding behavior based on fish movement features.

of YOLO. All models were trained and tested on the fish feeding intensity dataset with 300 epochs. As shown in Table 1, the MC-YOLO model outperformed the baseline models in terms of precision, recall, and mAP@0.5. This indicates that the MC-YOLO model has significant advantages in detecting fish bodies accurately, completely, and in terms of overall performance. By incorporating the MSLConv module and the CBAM attention module, the MC-YOLO model enhanced its feature extraction capabilities and multi-scale detection performance while maintaining low computational complexity. This balance of efficiency and accuracy enables more reliable detection and tracking of fish bodies, providing a solid foundation for subsequent quantification of feeding behavior.

4.2 Quantification of Feeding Behavior Based on Fish Movement Features

In quantifying feeding behavior based on fish movement features, we observed changes in the fish group's speed before and after feeding. As shown in the Fig. 5, during the initial stage of the experiment (0–4000 frames), the fish group was in a non-feeding state. At this time, the speed of the fish group remained at a low level, around 0.04, with some fluctuations. These fluctuations were mainly due to the normal swimming behavior of the fish group in the breeding tank. When feeding began (at frame 4000), the speed of the fish group rapidly increased from 0.04 to approximately 0.65. This significant increase in speed reflected the strong reaction of the fish group to food and their active feeding behavior. Fish groups tend to swim quickly at the beginning of feeding to compete for food, resulting in a noticeable increase in speed. During the feeding process, the speed of the fish group temporarily decreased to around 0.1. This may be because, after the initial rapid swimming, the fish group gradually entered a stable feeding stage, with relatively slower movement. At this stage, the fish group mainly congregated in the feeding area, engaging in continuous feeding activities. As feeding progressed and concluded, the speed of the fish group gradually decreased. By frame 8000, when feeding ended, the speed of the fish group gradually returned to the level of the non-feeding state, around 0.05, with some fluctuations. This indicated that the fish group had completed the feeding process and returned to normal swimming behavior.

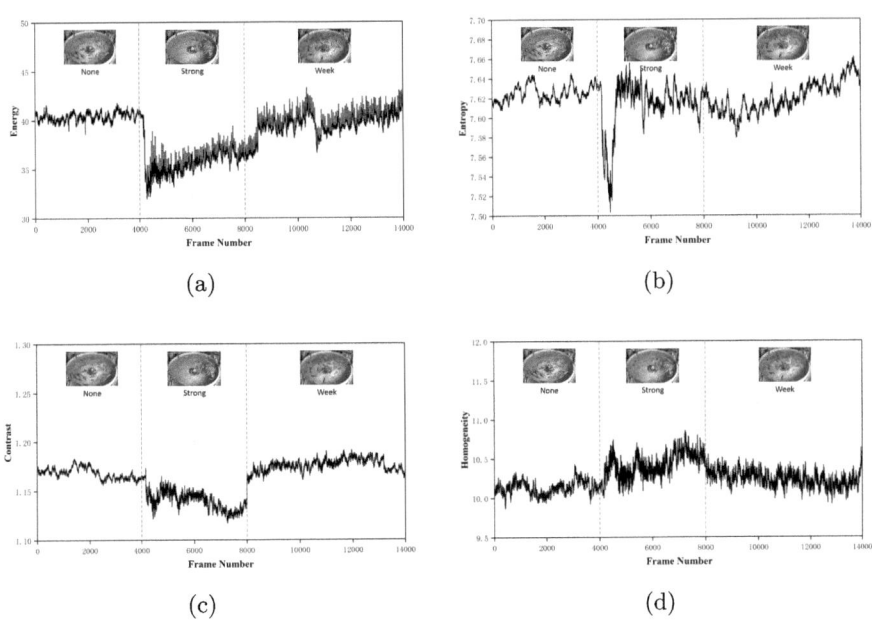

Fig. 6. Quantification of feeding behavior based on image texture features: (a) Energy, (b) Entropy, (c) Contrast, (d) Homogeneity.

4.3 Quantification of Feeding Behavior Based on Image Texture Features

The four texture features extracted using the GLCM were averaged across four directions (0°, 45°, 90°, 135°), and their trends are shown in Fig. 6. When the fish group was not feeding, the water surface in the breeding tank remained relatively calm, and the image texture features changed little, with the feature quantities remaining at relatively stable levels. Specifically, contrast, correlation, energy, entropy, and inverse variance were stable around 1.17, 40, 7.62, and 1.01, respectively. At frame 4000, when feeding began, the fish group started to feed, causing significant ripples and splashes on the water surface, leading to substantial changes in the image texture features. Contrast, which fluctuated

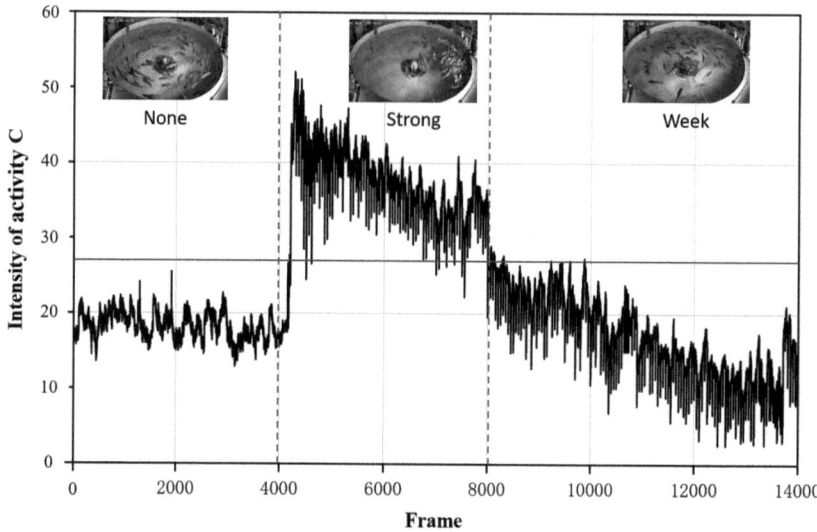

Fig. 7. Quantification of feeding behavior.

around 1.17 before feeding, rapidly decreased to 1.12 after feeding started and then slowly rose to around 1.16 after feeding ended at frames 8000-14000. Energy, which fluctuated around 40 before feeding, rapidly decreased to 32 at frame 4000 when feeding started and then slowly recovered to around 37 after feeding ended. Entropy, which remained around 7.62 before feeding, rapidly decreased to 7.50 when feeding started and then returned to a level similar to the pre-feeding state after feeding ended at frame 8000. Inverse variance, which remained stable around 1.01 before feeding, rapidly increased to 1.07 at frame 4000 when feeding started. During feeding, inverse variance showed an upward trend with fluctuations and stabilized around 1.02 after feeding ended at frame 8000.

By quantitatively analyzing the image texture features, we were able to accurately capture the changes in water surface ripples and splashes during the feeding process of the fish group, providing important evidence for determining the feeding state of the fish group. Significant changes in image texture features indicated that the fish group was actively feeding, while the gradual return of feature quantities to normal levels suggested that the feeding behavior had ended. This quantification result can assist aquaculture personnel in more intuitively understanding the feeding situation of the fish group, providing technical support for precise feeding, improving breeding efficiency, and enhancing the growth and health of the fish group.

4.4 Quantitative Results on the Intensity of Feeding Activity

The movement features of fish groups were quantified by their swimming speed, while the image texture features were quantified by the four texture feature quantities. Therefore, the feeding activity intensity of fish groups was defined as follows:

$$C = V/H \qquad (9)$$

The quantification results of fish group feeding activity intensity are shown in Fig. 7. From frames 0 to 4000, the fish group was in a non-feeding state with activity intensity fluctuating around 17. At frame 4000, when feeding began, the activity intensity rapidly increased to 53 and then gradually decreased. This higher intensity indicated the fish were still feeding. At frame 8000, when feeding ended, the intensity dropped significantly back to the pre-feeding level.

5 Conclusions

The machine vision-based method for quantifying fish feeding behavior proposed in this study, which integrates fish movement features and image texture features with the improved MC-YOLO detection model, has achieved efficient and accurate quantification of fish feeding behavior. The experimental results demonstrate that this method has significant advantages in detection accuracy, feeding state recognition, and adaptability to complex breeding environments. It provides strong technical support for precise feeding in industrial recirculating aquaculture systems and has broad application prospects and promotion value. It is expected to promote the intelligent and sustainable development of the aquaculture industry.

Acknowledgements. This work was supported in part by the Intergovernmental cooperation on international science, technology and innovation, 2023YFE0110800, the Science and technology research program of Chongqing Education Commission of China under grant No.KJQN202200829, No.KJQN202300844, No.KJQN202200833, KJQN 202400804,KJZD-K202500806, Chongqing Technology and Business University projects under grant yjscxx2025-269-191, yjscxx2 025-269-201, yjscxx2025-269-200, S202411799037X, S202411799038X.

References

1. Peng, Y., Jolfaei, A., Yu, K.: A novel real-time deterministic scheduling mechanism in industrial cyber-physical systems for energy internet. IEEE Trans. Ind. Inform. **18**(8), 5670–5680 (2022). https://doi.org/10.1109/TII.2021.3139357
2. Yun, L., et al.: Optimized content caching and user association for edge computing in densely deployed heterogeneous networks. IEEE Trans. Mobile Comput., 2130–42 (2022). https://doi.org/10.1109/tmc.2020.3033563
3. Atoum, Y., Srivastava, S., Liu, X.: Automatic feeding control for dense aquaculture fish tanks. IEEE Signal Process. Lett. **22**(8), 1089–1093 (2014)
4. Qiang, G., Yang, X., Zhou, C., et al.: Detection method of feeding status of fish based on shape and texture features. J. Shanghai Ocean Univ. **27**(2), 9 (2018). https://doi.org/10.12024/jsou.20170802112

5. Zhou, Y., et al.: Monitoring and analysis of fish feeding behavior based on near-infrared imaging. Aquac. Eng. **83**, 1–8 (2018)
6. Huang, Z., He, J., Song, X.: Recognition and quantification of fish feeding behavior based on fish movement and image texture features. J. Ocean Univ. China (Nat. Sci. Edn.), **52**(01), 32–41 (2022). https://doi.org/10.16441/j.cnki.hdxb.20210050
7. Liu, Z.Y., Li, X., Fan, L.Z., et al.: Measuring feeding activity of fish in RAS using computer vision. Aquacult. Eng. **60**, 20–27 (2014)
8. Gu, L.: Research and implementation of a fish feeding behavior quantification system based on spatiotemporal feature fusion. Dalian Ocean Univ. (2024). https://doi.org/10.27821/d.cnki.gdlhy.2024.000109
9. Zhipeng, C., Ming, C.: Detection of fish feeding behavior based on optical flow method and image texture features. J. Southern Agriculture **50**(05), 1141–1148 (2019)
10. Wang, D., Wu, M., Zhu, X., et al.: Real-time detection and identification of fish skin health in the underwater environment based on improved YOLOv10 mode. Aquaculture Reports **42**, 102723 (2025)
11. Jiang, G., Wu, R., Huo, Z., et al.: LigMSANet: lightweight multi-scale adaptive convolutional neural network for dense crowd counting. Expert Syst. Appl. **197**, 116662 (2022)
12. Long, W., Wang, Y., Hu, L., et al.: Triple attention mechanism with YOLOv5s for fish detection. Fishes **9**(5), 151 (2024)
13. Yang, J., et al.: (Retracted) dynamic scene images-assisted intelligent control method for industrialized feeding through deep vision learning. J. Electron. Imaging (2022). https://doi.org/10.1117/1.jei.32.2.021611

Surfacing Fish Detection Based on MSSYOLO in Recirculating Aquaculture System

Ruiwen Xiao[1], Junchao Yang[1(✉)], Haiyan Huang[1], Xueni Pan[1], Yu Shen[1], and Yanzheng Gao[2]

[1] School of Artificial Intelligence, Chongqing Technology and Business University, Chongqing, China
`{2023313026,Yangjc,huanghaiyan37,panxueni,shenyu}@ctbu.edu.cn`
[2] College of Resources and Environmental Sciences, Nanjing Agricultural University, Nanjing 210095, China
`gaoyanzheng@njau.edu.cn`

Abstract. To address hypoxia-induced fish mortality, this study proposes an improved algorithm (MSSYOLO) based on YOLOv10. The algorithm is designed to enable real-time and accurate monitoring of surfacing behavior, thereby optimizing aquaculture management. MSSYOLO integrates multi-scale convolution in Conv and C2f modules, a Squeeze-and-Excitation mechanism for improved feature representation, and a P2 detection layer for enhanced small-object detection, achieving superior multi-scale feature fusion and lightweight design. Experimental results demonstrate that MSSYOLO outperforms baseline YOLOv10 models. Furthermore, A strong correlation between surfacing fish numbers and dissolve oxygen concentration highlights surfacing as a hypoxia indicator. This algorithm provides robust technical support for the real-time regulation of aeration systems, effectively preventing hypoxia-induced fish mortality and reducing economic losses.

Keywords: RAS · Dissolved Oxygen · Surfacing Fish Detection · YOLOv10 · MSSYOLO

1 Introduction

With the rapid development of the aquaculture industry, increasing stocking densities and scales have made environmental changes and fish health critical challenges in aquaculture management. Recirculating aquaculture systems (RAS), recognized for their efficiency and environmental benefits, have gained prominence due to their ability to conserve water, reduce pollution emissions, and enable high-density sustainable farming through precise control of parameters such as water temperature, dissolved oxygen (DO), and ammonia nitrogen [1,2]. However, under high-density conditions without adequate oxygenation, DO concentration in the water gradually decline, triggering physiological stress responses in fish. As a key factor in the survival and growth of fish, DO directly influences metabolic processes and population health [3]. When DO concentration in water are insufficient, fish activate alternative oxygen uptake mechanisms to adapt to hypoxic conditions. In extreme oxygen-deprived environments, fish engage in aquatic surface respiration (ASR) to mitigate the adverse effects of hypoxia [4]. Under low-oxygen conditions, guppies utilize aquatic surface respiration to meet their oxygen requirements [5]. The findings demonstrate a strong physiological and ecological coupling between DO concentration and fish surfacing behavior: a decline in DO directly triggers surfacing behavior, while the surfacing phenomenon itself serves as a direct biological indicator of DO concentration falling below a critical threshold. In RAS, the high-density stocking often leads to fish surfacing that can cause physiological stress, reduced immunity, stunted growth, and abnormal behavior, which, in severe cases, can result in mass mortality, economic losses, and adverse impacts on the aquaculture system.

While simple DO concentration measurements provide some information for management, their ability to directly reflect fish behavior is limited and susceptible to equipment errors or malfunctions. Surfacing detection, as a direct indicator of fish health, when combined with DO monitoring, offers more intuitive and timely data than DO concentration alone. Traditional surfacing detection relies heavily on manual observation, which depends on empirical thresholds and intermittent sampling, leading to poor timeliness and significant subjective bias. This approach fails to meet the demands of modern aquaculture for real-time monitoring and efficient management. Thus, there is an urgent need for an efficient and real-time method to detect surfacing behavior, providing early warnings to enable managers to adjust environmental conditions and prevent hypoxia-induced fish mortality.

Accurate fish detection in the complex aquatic environment of RAS poses multidimensional technical challenges: reduced image quality due to water turbidity, imaging interference from variable lighting conditions, motion blur or overlap caused by rapid fish movement or aggregation, texture variations from water flow fluctuations, and feature extraction difficulties due to low contrast between fish and the background. Furthermore, the stringent real-time monitoring requirements of RAS and the computational constraints of edge devices collectively limit the accuracy, robustness, and deployment feasibility of detection system. Previous research has addressed these challenges through various approaches. Image enhancement is applied to mitigate environmental effects, followed by YOLO-based object detection to extract fish trajectories, which were analyzed using naÃŕve Bayes, K-nearest neighbors, and random forest algorithms to detect abnormal fish behavior [6]. To further improve the classification of abnormal behavior, fish trajectory analysis is enhanced by object detection and data augmentation, incorporating temporal context to optimize abnormal behavior classification, improving accuracy by 2.6% over baselines [7]. For precise detection of individual fish coordinates, shadow removal, background subtraction, label watershed algorithm is utilized for individual fish coordinate identification, and a multi-instance learning approach with a dual-flow network model, achieving high recognition accuracy, and improved execution efficiency [8]. Additionally, a novel ResNet deep residual convolutional network is proposed for abnormal behavior recognition in aquaculture, achieving a high accuracy [9]. These methods for fish behavior recognition have demonstrated effectiveness, but remain constrained by insufficient feature extraction capabilities.

This study proposes an improved framework (MSSYOLO) based on YOLOv10, to enhance the accuracy and robustness of surfacing detection. By integrating multi-scale convolution, an additional P2 detection layer, and the Squeeze-and-Excitation mechanism, MSSYOLO significantly improves multi-scale and small-object detection capabilities while remaining lightweight, enabling precise identification of surfacing fish.

2 Method

2.1 Data Acquisition and Processing

Fig. 1. Recirculating aquaculture system laboratory.

The experimental data were provided by the RAS laboratory at National research base of intelligent manufacturing service of Chongqing Technology and Business University. The setup is shown in Fig. 1, which includes two aquaculture tanks, each 4 meters in diameter and 1.2 meters in height, with a water depth of 1 m. The RAS was equipped with aeration discs, UV sterilizers, microfilters, biofilters, and water quality sensors to maintain stable conditions. Video data were collected under stable lighting conditions to minimize the potential impact of environmental illumination on monitoring results. Video footage of grouper (Epinephelus spp.) surfacing behavior at DO concentration of 1.5âĂŞ4.0 mg/L was recorded (1920×1080@27fps). During this process, the large volume of captured frame data exhibited minimal content variation between adjacent frames, resulting in high redundancy. Direct frame-by-frame analysis would not only incur substantial computational costs but also risk degrading feature extraction quality due to redundant information. To mitigate this, custom Python code was developed to extract keyframes from the raw video, selectively identifying representative frames with significant content changes, as shown in Fig. 2. This approach effectively reduced data redundancy and focused on information-rich segments, thereby providing higher-quality input for subsequent image processing and object detection tasks, ultimately enhancing the efficiency and accuracy of model training and analysis. Keyframes were annotated with Labelme and were split into training, validation, and test sets at a ratio of 8:1:1. To enhance the generalization and robustness of the model, data augmentation techniques were applied, including HSV color space augmentation and geometric transformations, to increase the sample size. Due to significant size variations in surfacing fish caused by perspective differences and partial feature overlap between surfacing and non-surfacing fish, enhancing specialized feature extraction and multi-scale detection capabilities was critical to improving the accuracy of surfacing fish detection.

Fig. 2. Keyframe of surfacing behavior.

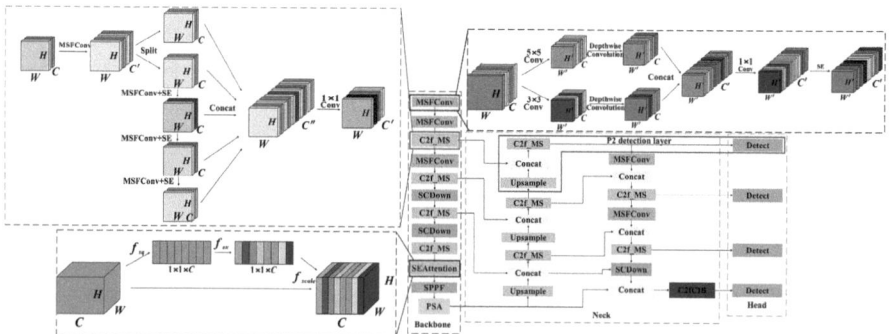

Fig. 3. MSSYOLO architecture.

2.2 MSSYOLO Architecture

The MSSYOLO architecture, designed for surfacing fish detection in RAS, is illustrated in Fig. 3. It enhances YOLOv10n with four key improvements: MSF-Conv, C2f_MS, Squeeze-and-Excitation mechanism, and a P2 detection layer. The MSFConv module replaces standard convolutions, using multi-scale kernels to capture diverse features efficiently, while the C2f_MS module improves small-object detection through multi-scale feature paths and cross-layer fusion. The SE mechanism, applied before SPPF, weights critical channels to enhance fish-related features. The P2 layer preserves high-resolution details, addressing feature overlap and complex backgrounds.

Squeeze-and-Excitation Mechanism. To improve the detection accuracy of surfacing fish in RAS, this study integrates the SE mechanism [10]. The SE mechanism improves feature representation by dynamically recalibrating channel weights, prioritizing salient features relevant to surfacing fish while suppressing irrelevant background noise, thereby significantly boosting detection precision in complex aquatic environments. The mechanism operates through a two-step process: first, a "squeeze" step employs global average pooling to compress spatial dimensions, the compressed channel is described as:

$$Z_c = f_{\text{sq}}(x_{ijc}) = \frac{1}{H \times W} \sum_{i=1}^{H} \sum_{j=1}^{W} x_{ijc} \tag{1}$$

where x_{ijc} denotes the pixel value at the i-th row and j-th column in the c-th channel of the feature map $(H \times W \times C)$, and Z_c represents the global descriptor for each channel.

Subsequently, an "excitation" step utilizes two fully connected layers to map global vector $z = [Z_1, Z_2, \ldots, Z_c]^T$ to channel weights $\omega = [\omega_1, \omega_2, \ldots, \omega_c]^T$, the excitation process is described as:

$$\omega = f_{\text{ex}}(z, w) = \sigma\left(w_2 \cdot \text{ReLU}\left(w_1 \cdot z\right)\right) \tag{2}$$

where $\sigma(\cdot)$ is the sigmoid activation, w_1, w_2 are the weight matrices of the two fully connected layers, respectively.

Finally, the adjusted feature map is generated through channel-wise multiplication operations that is represented as:

$$\tilde{x}_c = f_{\text{scale}}\left(x_c, \omega_c\right) = \omega_c \cdot x_c \tag{3}$$

where x_c represents the c-th channel of the original feature map, ω_c denotes the channel weight corresponding to x_c, and $\tilde{x}_c$ is the channel after weight adjustment.

By embedding the SE mechanism into the backbone of YOLOv10, the model effectively focuses on critical features relevant to surfacing fish, significantly improving detection robustness and accuracy in challenging aquatic environments.

MSFConv Module. To address the limitations of traditional convolutional modules, which are constrained by single kernel sizes and exhibit deficiencies in multi-scale feature detection, we propose MSFConv module. This lightweight module employs a dual-branch parallel convolutional architecture designed to enhance multi-scale feature extraction and integration, overcoming the constraints of conventional single-kernel approaches. The MSFConv module encompasses four key stages: multi-scale feature generation, depthwise redundant feature expansion, cross-channel feature fusion, and channel-wise attention enhancement, ensuring robust feature representation with reduced computational complexity.

In the multi-scale feature generation stage, the module processes input feature maps (H×W×C) through a dual-branch structure employing different-sized convolutional kernels. Specifically, a 3×3 standard convolutional kernel captures fine-grained local features, while a 5×5 dilated convolutional kernel enhances global contextual awareness, improving the detection of larger-scale objects [11]. Both branches utilize a stride of 2 for spatial downsampling, halving the spatial dimensions to produce feature maps of size (H'×W'×C), with each branch outputting half the target channel count to ensure compatibility during subsequent concatenation.

To expand feature representation while maintaining computational efficiency, depthwise convolution [12] is applied within each branch. A 3×3 convolutional kernel is used per input channel to generate a corresponding output channel, preserving spatial dimensions while enhancing feature expressiveness through localized receptive field transformations. The outputs from both branches are concatenated along the channel dimension, yielding a feature map (H'×W'×C') with a channel count equal to the sum of both branches, effectively expanding the depth of feature representation.

For feature fusion, a 1×1 point-wise convolution integrates multi-scale information across channels [13], producing a unified feature map (H'×W'×C') with enhanced expressiveness. Subsequently, a Squeeze-and-Excitation mechanism performs channel-wise attention weighting, utilizing global average pooling to compress spatial dimensions and generate attention weights. These weights, computed via two 1×1 convolutional layers and an activation function, emphasize critical channels while suppressing less relevant ones, further refining the feature map. This lightweight design significantly improves multi-scale feature integration, enhancing model performance in complex object detection tasks while minimizing computational overhead.

C2f_MS Module. To further enhance multi-scale feature extraction and small-object detection capabilities, we propose the C2f_MS module, an advanced extension of the C2f framework that builds upon the MSFConv module. This module integrates multi-scale feature extraction, feature branch separation with serial processing, channel-wise attention enhancement, and cross-channel feature fusion to achieve robust feature representation, as illustrated. The C2f_MS module is designed to address the limitations of traditional convolutional architectures in capturing diverse feature scales and improving detection accuracy for small objects in complex scenarios.

The input feature map (H×W×C) is first processed by the MSFConv module to extract multi-scale features, producing a feature map (H×W×C'). This feature map is then split along the channel dimension into two equal-sized feature tensors, enabling independent processing of distinct feature branches. One branch preserves the original features to retain shallow, high-resolution information, while the other undergoes deep feature extraction through a serial structure. This serial branch consists of three sequential MSFConv operations, where each operation takes the output of the previous one as input, progressively enhancing feature depth and capturing intricate patterns. After each MSFConv operation, a dedicated Squeeze-and-Excitation mechanism is applied to dynamically recalibrate channel importance, emphasizing critical features for small-object detection and capturing global feature dependencies. This secondary weighting enhances the model's adaptability to feature variations, ensuring focus on channels most relevant to the task.

The resulting five feature tensors—comprising the two initial split tensors and the three sequentially processed tensors—are concatenated along the channel dimension, forming a feature map (H×W×C") that combines shallow and deep features. A subsequent 1×1 point-wise convolution integrates these multi-scale and multi-depth features, adjusting the channel dimensions to produce a refined feature map (H×W×C'). This design significantly enhances the module's ability to extract and integrate multi-scale features, improving detection performance for small objects while maintaining computational efficiency through lightweight operations.

P2 Detection Layer. To address the inherent limitations of traditional detection heads in YOLOv10 models, which often fail to accurately detect small objects due to the loss of high-resolution feature information during downsampling, we introduce a P2 detection layer [14]. This layer operates at a higher resolution (160×160) compared to the conventional P3, P4, and P5 layers (80×80, 40×40, and 20×20, respectively), significantly enhancing small-object detection capabilities. By leveraging feature maps from earlier network layers, the P2 detection layer preserves finer spatial details critical for identifying small objects in complex scenes, such as those encountered in RAS for surfacing fish detection.

The P2 detection layer is constructed through a meticulous process to retain high-resolution information. Feature maps from the P3 layer are upsampled using nearest-neighbor interpolation to match the spatial dimensions of the layer 2 output from the backbone network. These upsampled features are then concatenated with the high-resolution feature maps from layer 2, ensuring the preservation of fine-grained spatial details essential for small-object detection. The concatenated feature maps are processed by the C2f_MS module, which integrates multi-scale convolutional operations and a Squeeze-and-Excitation mechanism to refine feature representations. This module employs parallel convolutional kernels and channel-wise attention to enhance feature expressiveness while maintaining computational efficiency, thereby improving the model's ability to capture diverse scale objects.

To enable robust multi-scale feature integration, the P2 detection layer is incorporated into a Feature Pyramid Network (FPN) [14] alongside the P3, P4, and P5 layers. The FPN facilitates effective fusion of features across different resolutions, ensuring that both fine-grained and contextual information is utilized for detection. Dedicated detection heads process the refined P2 feature maps in conjunction with outputs from the P3, P4, and P5 layers, enabling the model to achieve superior performance in detecting both small and multi-scale objects. This approach significantly enhances the model's robustness and accuracy, particularly for small-object detection tasks in challenging environments, while maintaining computational efficiency through optimized feature processing.

To enhance the model's multi-scale and small-object detection capabilities, the Conv and C2f modules were replaced with MSFConv and C2f_MS modules, and an SE mechanism was integrated into the backbone to enhance deep feature channel attention. To further improve small-object detection, a P2 detection layer was introduced that leverages shallow backbone features rich in spatial details, significantly boosting small-object accuracy and expanding the model's multi-scale detection range.

Table 1. Training configurations

Configuration	Version
Operating system	Windows11
CPU	12th Gen Intel(R) Core(TM) i7-12700K 3.60GHz
GPU	NVIDIA GeForce RTX 3080Ti
Random Access Memory	32G
Programming Language	Python 3.10
Framework	Torch-vision 1.13.1
CUDA version	12.5

3 Results

3.1 Model Comparison

Training configurations are detailed in Table 1. MSSYOLO was trained on a hardware platform featuring a 12th Gen Intel® CoreâĎć i7-12700K processor (12 cores, 20 threads, 3.6 GHz base frequency), an NVIDIA GeForce RTX 3080 Ti GPU (12 GB GDDR6X), 32 GB DDR4 RAM, and Windows 11, using PyTorch 1.13.1 and CUDA 12.5. Training parameters is as follows: initial learning rate is 0.01; final learning rate is 0.0001; the input resolution is 640×640 (YOLOv10 default); the iteration period is 300; the batch size of each training is 16. All experiments were conducted under identical conditions.

Table 2. Experimental results

Model	Precision	Recall	mAP@0.5	F1-Score
MSSYOLO	**0.886**	**0.898**	**0.932**	**0.893**
YOLOv10n	0.867	0.872	0.918	0.868
YOLOv10s	0.879	0.880	0.922	0.879
YOLOv10m	0.879	0.894	0.924	0.887

To comprehensively evaluate the performance of the proposed MSSYOLO model for surfacing fish detection, a detailed comparison was conducted against baseline YOLOv10 models, including YOLOv10n, YOLOv10s, and YOLOv10m. The experimental results, shown as Table 2, provide a quantitative assessment of the models across four evaluation metrics: precision, recall, mAP@0.5, and F1-score, which collectively assess detection accuracy and robustness in complex aquaculture environments. MSSYOLO achieves a precision of 0.886, recall of 0.898, mAP@0.5 of 0.932, and F1 score of 0.893, outperforming the base model (YOLOv10n) by 2.2% in precision, 2.6% in recall, and 1.4% in mAP@0.5. According to the confusion matrices in the Fig. 4, the MSSYOLO model demonstrates significantly higher detection accuracy than the baseline model, providing visual confirmation of its superior detection performance. It also surpasses YOLOv10s and YOLOv10m in all evaluation metrics. These improvements highlight MSSYOLO's superior ability to handle complex scenarios. Furthermore, these improvements demonstrate that the enhanced architecture of MSSYOLO, which optimize feature extraction and multi-scale object handling, effectively detect surfacing fish in RAS. This proves the effectiveness and practicality of MSSYOLO in fish detection within RAS and offers new directions for future research.

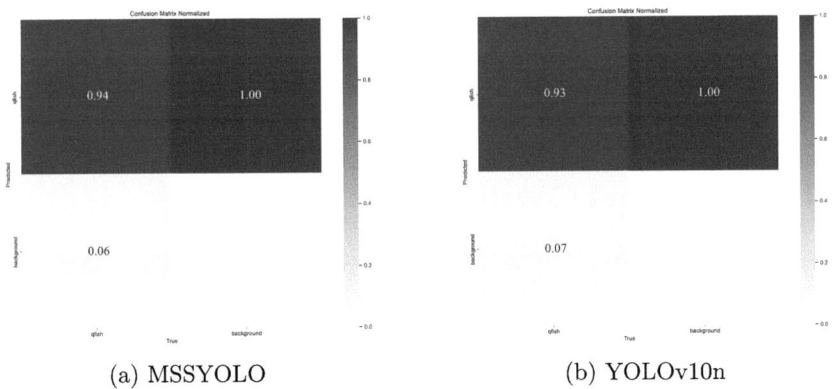

Fig. 4. Confusion matrices of YOLOv10n and MSSYOLO.

3.2 DO-Surfacing Correlation Analysis

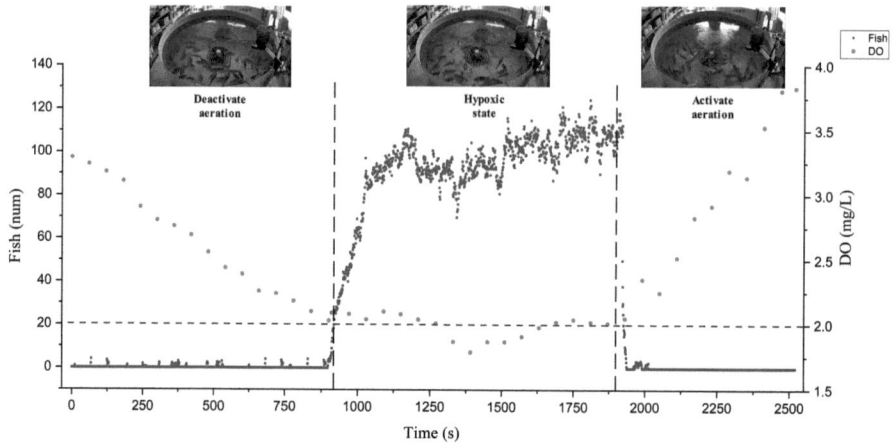

Fig. 5. DO-Surfacing relationship.

The study also explores the relationship between surfacing fish numbers and DO concentration in RAS, elucidating fish behavioral responses under hypoxic conditions. As shown in the Fig. 5, surfacing fish numbers increase as DO concentration decreases. When DO concentration approach 2 mg/L, the number of surfacing fish exhibits a pronounced exponential surge, followed by stabilization toward dynamic equilibrium. Upon activation of the aeration system, surfacing fish numbers decline rapidly, approaching zero when DO exceeds 2 mg/L. The finding highlights the critical influence of DO concentration on surfacing behavior, establishing surfacing as a reliable biological indicator of hypoxia in RAS. By quantifying this relationship, the study provides a robust foundation for data-driven environmental management strategies, enabling precise regulation of aeration systems to maintain DO above critical thresholds, thereby preventing hypoxia-induced fish mortality and enhancing RAS sustainability.

4 Discussion and Conclusion

By integrating MSFConv, C2f_MS, the Squeeze-and-Excitation mechanism, and a P2 detection layer, the proposed MSSYOLO model significantly enhances multi-scale and small-object detection, improving the accuracy and robustness of surfacing fish detection and offering technical support for the management of RAS. The experimental results demonstrate enhanced precision, recall, mAP@0.5, and F1-score compared to baseline YOLOv10 models and also confirm a strong correlation between DO concentration and fish surfacing behavior: at approximately 2 mg/L, the number of surfacing fish increases exponentially, aligning with the hypoxia stress threshold. By leveraging MSSYOLO's real-time detection capabilities, aquaculture managers can precisely regulate aeration systems to maintain DO concentration within the suitable range for fish, thereby mitigating economic losses from hypoxia.

Future research should prioritize assessing the robustness of the MSSYOLO model in challenging RAS environments, including conditions with elevated turbidity or fluctuating water flow, to ensure its reliability in diverse operational scenarios. Additionally, integrating multi-modal data, such as temperature, pH, ammonia nitrogen, and water velocity, could enhance the model's predictive accuracy for fish health and behavior, which can enable more precise control of aeration and filtration systems. Exploring these directions will advance the development of adaptive, data-driven management strategies, ultimately improving the sustainability and economic viability of RAS.

Acknowledgements. This work was supported in part by the Intergovernmental cooperation on international science, technology and innovation, 2023YFE0110800, the Science and technology research program of Chongqing Education Commission of China under grant No. KJQN202200829, No. KJQN202300844, No. KJQN202200833, KJQN202400804, KJZD-K202500806, Chongqing Technology and Business University projects under grant yjscxx2025-269-191, yjscxx2025-269-201, yjscxx2025-269-200, S202411799037X, S202411799038X.

References

1. Martins, C.I.M., et al.: New developments in recirculating aquaculture systems in Europe: a perspective on environmental sustainability. Aquacult. Eng. **43**(3), 83–93 (2010)
2. Libao, Franz Joseph D., et al.: Automated control and IoT-based water quality monitoring system for a Molobicus tilapia recirculating aquaculture system (RAS). In: 2024 IEEE Conference on Technologies for Sustainability (SusTech), pp. 410-415. IEEE (2024). https://doi.org/10.1109/SusTech60925.2024.10553538
3. Mohsen, A.T., et al.: Fish response to hypoxia stress: growth, physiological, and immunological biomarkers. Fish Physiol. Biochemis. **45**, 997–1013 (2019)
4. Paolo, D., Cristel, L., Shingles, A.: Hypoxia and the antipredator behaviours of fishes. Philosophical Trans. Royal Soc. B Biol. Sci. 362.1487, 2105–2121 (2007)
5. Kramer, D.L., Mehegan, J.P.: Aquatic surface respiration, an adaptive response to hypoxia in the guppy, Poecilia reticulata (Pisces, Poeciliidae). Environ. Biol. Fishes **6**, 299–313 (1981)
6. Anas, O., Wageeh, Y., Mohamed, H.E.D., et al.: Detecting abnormal fish behavior using motion trajectories in ubiquitous environments. Procedia Comput. Sci. **175**, 141–148 (2020)
7. Li, X.: Anomaly detection of fish behaviors based on trajectory extraction. Zhejiang University (2021). https://doi.org/10.27461/d.cnki.gzjdx.2021.000333
8. Chen, L., Yin, X.: Recognition method of abnormal behavior of marine fish swarm based on in-depth learning network model. J. Web Eng. **20**(3), 575–596 (2021). https://doi.org/10.13052/jwe1540-9589.2031

9. Hu, W.-C., Chen, L.-B., Lin, H.-M.: A method for abnormal behavior recognition in aquaculture fields using deep learning. In: IEEE Canadian Journal of Electrical and Computer Engineering, vol. 47, no. 3, pp. 118–126 (2024). https://doi.org/10.1109/ICJECE.2024.3398653
10. Hu, J., Shen, L., Sun, G., et al.: Squeeze-and-excitation networks. IEEE Trans. Patt. Anal. Mach. Intell. (99) (2017). https://doi.org/10.1109/TPAMI.2019.2913372
11. Szegedy, C., Liu, W., Jia, Y., et al.: Going deeper with convolutions. IEEE Comput. Soc. (2014). https://doi.org/10.1109/CVPR.2015.7298594
12. Chollet, F.: Xception: deep learning with depthwise separable convolutions. In: 2017 IEEE Conference on Computer Vision and Pattern Recognition (CVPR). IEEE (2017). https://doi.org/10.1109/CVPR.2017.195
13. Howard, A.G., Zhu, M., Chen, B., et al.: MobileNets: efficient convolutional neural networks for mobile vision applications (2017). https://doi.org/10.48550/arXiv.1704.04861
14. Lin, T.Y., et al.: Feature pyramid networks for object detection. In: 2017 IEEE Conference on Computer Vision and Pattern Recognition (CVPR), pp. 936–944 (2016)

Design of LoRA Tunning-Assisted Pretrained LLM Structure for Sentiment Analysis in Online E-Commerce

Yuanyuan Cai, Zhiwei Guo[✉], Han Zhao, and Bo Liu

School of Artificial Intelligence, Chongqing Technology and Business University, Chongqing, China
{caiyuanyuan,zwguo,zhaohan,jsjliubo}@ctbu.edu.cn

Abstract. Sentiment analysis is crucial for e-commerce operations. It helps enterprises understand consumer feedback and optimize products. Pre-training large language models like Bidirectional Encoder Representations from Transformers (BERT), are current mainstream for sentiment analysis. However, when dealing with scenarios of large-scale data streams like online shopping malls, large parameter count and high computational costs of BERT limit the efficiency. Meanwhile, traditional deep-learning models, while easily deployable, are less accurate for long texts and complex semantics. To bridge the gap, this paper introduces the parameter tuning strategy named low rank adaption (LoRA), and develops an integrated approach by extending standard BERT. It is specifically developed for the various scenes of online E-commerce. Simulations in an online-mall environment and real-dataset tests show BERTLoRA matches BERT's accuracy on product reviews and notably speeds up training and inference. Compared with traditional models, the accuracy of BERTLoRA has increased by 6.25%. Compared with baseline methods, the proposal has only 0.3% of BERT's parameters, 32% of its memory occupancy, 40% less training time and 7% less inference time.

Keywords: Large language models · LoRA tuning · sentiment analysis · online E-commerce

1 Introduction

With the rapid development and popularization of the Internet, the e-commerce industry has been booming [1]. When purchasing goods, consumers are increasingly relying on online product reviews to assist in their decision-making. A vast number of product reviews contain consumers' intuitive feelings and evaluations of products. The emotional tendencies (positive or negative reviews) embedded in them are key clues for merchants to understand consumer satisfaction [2]. Accurately and efficiently classifying the sentiment of product reviews can help merchants quickly grasp consumers' needs, precisely identify the advantages and disadvantages of products, and then optimize product design, improve service quality, and enhance market competitiveness [3].

In recent years, deep learning technology has propelled breakthroughs in the field of Natural Language Processing [4]. The BERT model has demonstrated outstanding performance in sentiment classification tasks due to its powerful language representation capabilities and context comprehension. However, the Bidirectional Encoder Representations from Transformers (BERT) model brings challenges with its huge number of parameters (344 million parameters and 24 layers of Transformer) and high computational costs [5]. This not only makes the training require a large amount of resources and time, becoming an obstacle for small and medium-sized enterprises as well as developers, but also leads to slow inference speed, making it difficult to meet the requirements of scenarios with high real-time demands such as online customer service and real-time public opinion monitoring [6].

To address this issue, the paper proposes a hybrid model that integrates the BERT model and the low rank adaption (LoRA)-based fine-tuning strategy for real-time sentiment analysis of review data in online shopping malls. LoRA is an efficient model optimization technique [7]. By introducing low-rank matrix factorization in the key layers of the BERT model, it enables fine-tuning of the model with a relatively small number of additional parameters. In this way, while maintaining the performance of the model, it significantly reduces the computational cost. The paper integrates the BERT model with the LoRA technology and conducts an in-depth research on the sentiment classification task of users' product reviews in e-commerce websites. The proposed method in this work is named as BERTLoRA for short. The objective is to explore a solution that combines high efficiency and practicality [2], providing merchants with a faster and more accurate sentiment analysis tool. This tool will assist them in better meeting consumers' demands and enhancing the user experience amidst the intense market competition.

2 Design of the Web Simulation Platform

In a Web environment, the model files loaded by the browser are usually stored as strings in JavaScript variables. In the browser (such as Google Chrome), the maximum size of the model that can be loaded is 512 MB [8], which limits the total volume of the models that can be loaded within a single scenario. In order to efficiently deploy the sentiment classification model for product reviews in a web environment, it is necessary to optimize the existing model to ensure its loading and running efficiency in the browser.

In order to effectively apply the BERTLoRA model to the actual Web mall system, this paper designs and constructs a dual-framework Web platform architecture, and the architectural details are shown as Fig. 1. This architecture consists of two main parts: an e-commerce mall system based on Spring Boot, and a model deployment platform based on Flask [9]. As the core component, the Spring Boot system undertakes the important responsibilities of handling core business logic, realizing user interaction, and managing data, ensuring the stable operation and efficient management of the e-commerce mall system. Meanwhile, the Flask platform focuses on model deployment and the efficient execution of sentiment analysis tasks [10]. It is equipped with the optimized BERTLoRA model, giving full play to its advantages in sentiment classification tasks. When users post reviews in the Spring Boot system, the review data will be promptly sent to the Flask platform via an asynchronous communication mechanism. After the Flask platform receives the review data, it will immediately invoke the BERTLoRA model to conduct sentiment classification analysis, and efficiently return the classification results to the Spring Boot system in JSON format. Subsequently, the Spring Boot system is responsible for storing these sentiment classification results and updating the display in real - time, ensuring that users can promptly view the sentiment tendency analysis results of the reviews.

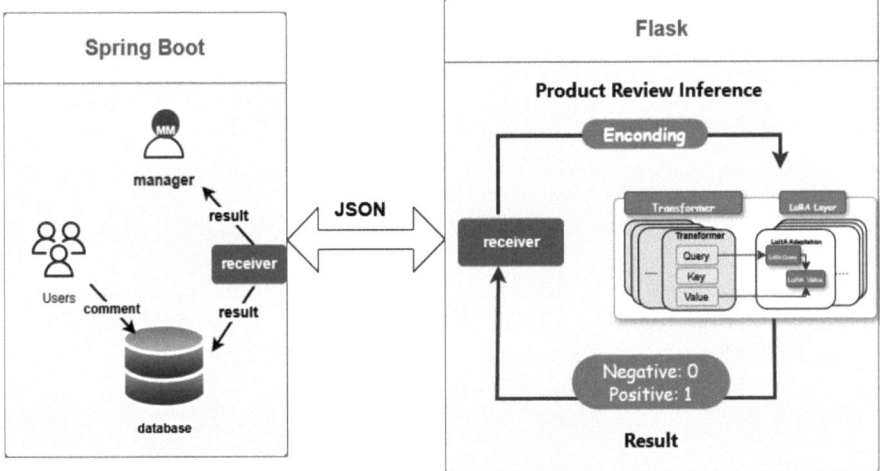

Fig. 1. The Dual-framework web platform.

The design of this dual-framework Web platform architecture skillfully combines the high-performance features of Spring Boot with the lightweight advantages of Flask, achieving rapid system response and efficient processing, and greatly enhancing the user experience. With this architecture, merchants can monitor the sentiment tendency of user reviews in real time, thus more accurately grasping consumers' needs and feedback. This not only helps merchants optimize their products and services in a timely manner and improve customer satisfaction but also enhances their competitiveness in the market, enabling them to gain a more advantageous position in the fierce market competition. In addition, this architecture also provides an efficient and reliable solution for the deployment of sentiment analysis models in practical business applications, laying a solid foundation for the further application of relevant technologies in the e-commerce field in the future.

3 Methodology

Sentiment classification of product reviews plays a crucial role in natural language processing and is of great importance for analyzing customer feedback and social network monitoring. Although the BERT model has achieved remarkable results in this task, it faces challenges such as a large number of parameters and high computational costs.

These challenges limit its deployment in resource-constrained environments and real-time systems that require rapid responses. To address these issues, this paper proposes the BERTLoRA model. It optimizes BERT by introducing LoRA (Low-Rank Adaptation) technology, maintaining the model's performance while reducing the number of model parameters and computational costs. The aim is to overcome the efficiency problems of the BERT model in practical applications and provide a more efficient and practical solution for sentiment classification tasks. As is shown in Fig. 2, the BERTLoRA model reduces the model complexity by introducing low-rank matrix factorization in the key layers of the BERT model.

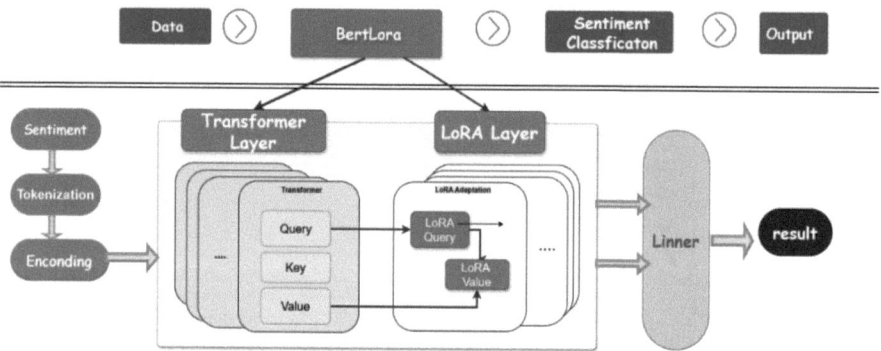

Fig. 2. Framework diagram of BERT_LoRA.

3.1 Overview of the BERT Model

The BERT model learns rich language representations by undergoing pre-training on a large amount of text data, and then fine-tunes on specific tasks [10]. Its basic structure is composed of multiple layers of Transformer encoders. Each layer of the encoder processes the input data through the self-attention mechanism and the feed-forward network. The output of the l layer encoder can be expressed as:

$$H^{(l)} = TransformerEncoderLayer\ (H^{(l-1)}) \tag{1}$$

where $H^{(0)}$ is the output embedding, and $H^{(l)}$ is the output after being processed by the l layer of the Transformer encoder.

Each Transformer encoder layer includes a self-attention mechanism and a feed-forward network [11]. The self-attention mechanism allows the model to consider the context of the entire sentence when processing each word, and its calculation formula is:

$$Attention(Q, K, V) = softmax\left(\frac{QK^T}{\sqrt{d_K}}\right)V \tag{2}$$

The feed-forward network performs a further non-linear transformation on the output of the self-attention layer:

$$FFN(x) = max(0, xW_1 + b_1) + b_2 \quad (3)$$

where Q, K, and V represent the Query, Key, and Value respectively [12]. They are the key matrices that constitute the self-attention mechanism. These matrices are regarded as fully connected layers (base_layer), which can map the input features to a new space for calculating the attention values. d_k is the dimension of the key vector, and W_1, W_2, b_1, b_2 are the weights and biases of the feed-forward network FFN.

3.2 LoRA Adaptation Layer

LoRA (Low-Rank Adaptation) is a model optimization technique that reduces the number of parameters by introducing low-rank matrix factorization into the Self-Attention mechanism of the BERT model. In the BERTLoRA model, the LoRA adaptation layer is added to the Query and Value matrices of the self-attention mechanism:

$$Q_{lora} = Q + \frac{\alpha}{\gamma} A_Q X \quad (4)$$

$$V_{lora} = V + \frac{\alpha}{\gamma} A_V X \quad (5)$$

among them, A_Q and A_V are the weight matrices of the LoRA layer, γ is the low-rank matrix, α is the scaling factor, and X is the input feature. $R^{d \times d}$

3.3 BertLoRA Model Architecture

The BERTLoRA model freezes the numerous parameters of the Bert model, integrates the LoRA adaptation layer into the self-attention mechanism of the BERT model, and optimizes the computational efficiency and the number of parameters of the model. The LoRA adaptation layer achieves low-rank factorization by introducing two smaller matrices, matrix A and matrix B, on the basis of the original fully connected layer (base_layer).

Output of the Original Fully Connected Layer. In the Bert model, if the weight matrix of the original fully connected layer is $W \in R^{d_{in} \times d_{out}}$ and the input feature is $X \in R^{d_{in}}$, then the output of the original fully connected layer is:

$$out_1 = Wx \quad (6)$$

Output of the LoRA Adaption Layer. The LoRA adapter introduces two smaller matrices $A \in R^{d_{in} \times r}$ and $B \in R^{r \times d_{out}}$ to approximate the change of the original matrix W, where r is a rank that is much smaller than both d_{in} and d_{out}. In addition, a scaling factor α is introduced to adjust the contribution of the LoRA adaptation layer.

First, the input feature x undergoes a linear transformation through matrix A to obtain the low-rank space representation:

$$lora\_in = Ax \tag{7}$$

Then, it is expected to apply Dropout operation to prevent overfitting:

$$lora\_in_{dropout} = Dropout(lora\_in) \tag{8}$$

The low-rank representation after Dropout undergoes a linear transformation through matrix B to obtain the output of the LoRA adaptation layer:

$$out_2 = B\ lora\_in_{dropout} \tag{9}$$

Finally, multiply the output of the LoRA adaptation layer by the scaling factor $\frac{\alpha}{\gamma}$:

$$out_2 = \frac{\alpha}{\gamma} Blora_{in dropout} \tag{10}$$

Output of the LoRA Adaption Layer. It is expected to add the output out_1 of the original fully connected layer and the output out_2 of the LoRA adaptation layer to obtain the final output:

$$Output = out_1 + out_2 \tag{11}$$

$$Output = Linear(Wx + \frac{\alpha}{\gamma} B(Ax)) \tag{12}$$

4 Experiments and Result Analysis

4.1 Dataset

This paper uses the online_shopping_10_cats dataset[1]. This is a publicly available Chinese product review dataset that contains 62,774 user reviews from 10 different categories of products (such as books, mobile phones, clothes, etc.). Among them, there are 31,728 positive reviews and 31,046 negative reviews. Each review is labeled with a sentiment tendency, where 0 represents a negative review and 1 represents a positive review. The dataset is divided into a training set and a test set, with the proportions being 70% and 30% respectively.

The BERT tokenizer BertTokenizer to split a text into tokens. These tokens are then mapped to the BERT vocabulary to obtain the corresponding input IDs. Additionally, an attention mask is generated to distinguish between actual tokens and padding parts. To ensure a consistent input length, shorter texts will be padded with the [PAD] token to the maximum length max_length (128), while overly long texts will be truncated. Finally, the input IDs and attention masks are converted into PyTorch tensors, which serve as the input for BERTLoRA.

---
[1] https://github.com/SophonPlus/ChineseNlpCorpus.

4.2 Experimental Setup

This paper designs two experiments to test the effectiveness and superiority of BERT-LoRA in terms of model resource consumption and performance evaluation respectively. The hardware operating environment includes two CPUs with a total of 64 physical cores (128 logical cores in total), an NVIDIA L40 GPU with 46 GB of memory, and 128 GB of RAM.

The first experiment involves comparing the BERTLoRA model with three different pre-training language models (BERT, RoBERTa, and Bert large) on the same test set. The comparison focuses on their resource consumption during the training and inference processes, including model size, average time for one training epoch, number of model parameters, average inference time, Accuracy, Precision, Recall, and F1 - score [13, 14]. In the experiment, the same hyperparameters are set for these four models, as shown in Table 1. The final output layer of all models is a linear layer named Linear. In the BERTLoRA model, the rank of the low - rank matrix is set to 8, the scaling factor is 16, and the dropout probability is 0.01. This dropout rate is used to randomly discard a part of the output of the low - rank matrix during training to prevent overfitting. By default, all models are optimized using the Adam optimizer. The experimental results are shown in both Table 2 and Fig. 3.

Table 1. Hyperparameter settings for the models.

Hyperparameter	Learning Rate	Batch_Size	Epochs	Max_length
Value	0.001	64	20	128

Table 2. Performance comparison of the BERTLoRA and other three methods.

Model	BERT	RoBERTa	Bert_large	BERTLoRA
Size	~ 420 MB	~ 470 MB	~ 350 MB	**~ 2 MB**
Training time(seconds)	168	170	110	**97**
Parameters (Millions)	344	125	100	**0.3**
Memory	~ 600 MB	~ 650 MB	~ 500 MB	**~ 1.9 MB**
Average Inference	6.6 ms	6.6 ms	6.8 ms	**6.2 ms**
ACC	0.74	**0.82**	0.80	0.81

The second experiment is to compare the performance of BERTLoRA with traditional neural network models, including Convolutional Neural Network (CNN), Recurrent Neural Network (RNN), Long Short-Term Memory network (LSTM), and Gated Recurrent Unit (GRU). The experimental results are shown in the both Table 3 and Fig. 4. They demonstrate the performance of six different models (CNN, RNN, LSTM,

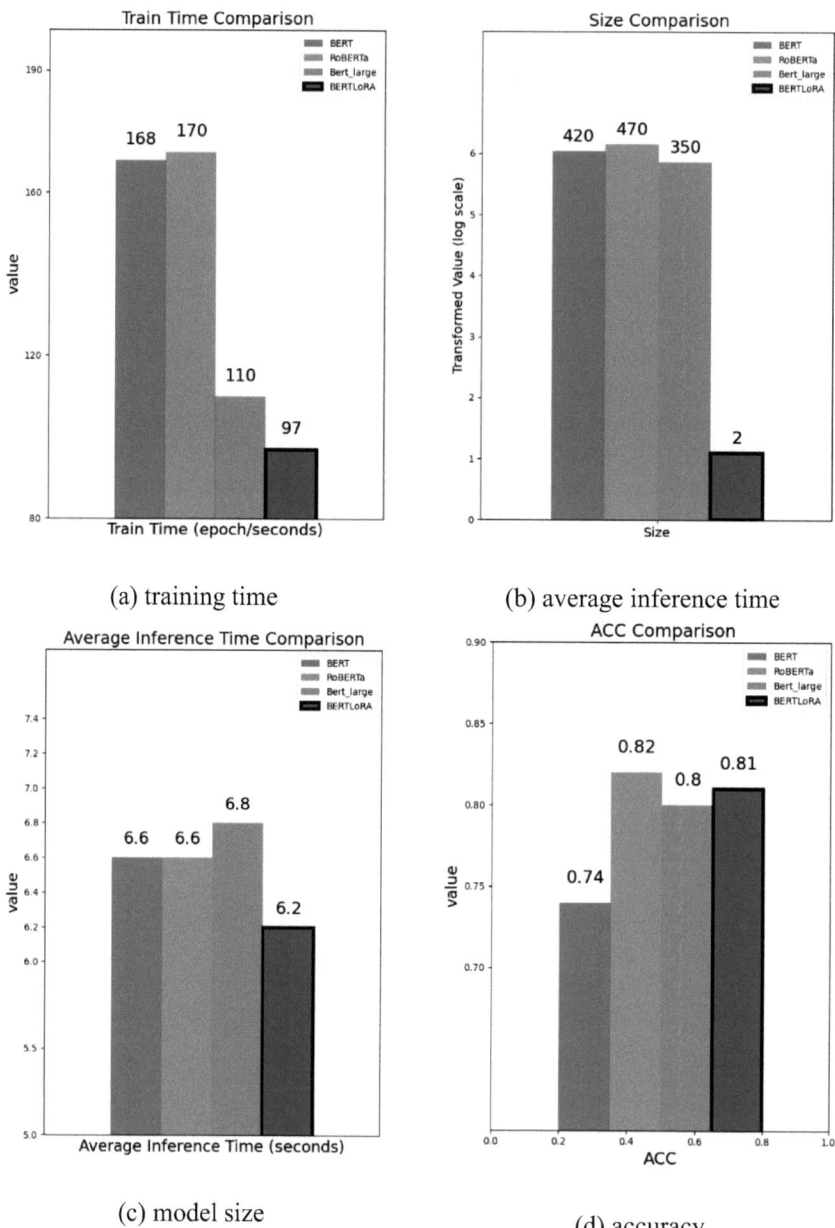

Fig. 3. Comparison between BERTLoRA and baseline methods in terms of training time, average inference time, model size and accuracy.

GRU, Bert, BERTLoRA) under three different max_length values (32, 64, 128), and each max_length value corresponds to three different training epochs (epoch = 1, epoch = 5, epoch = 10).

Table 3. Performance comparison between BERTLoRA and five other neural networks (tabular results).

model	epoch = 1			epoch = 5			epoch = 10		
	32	64	128	32	64	128	32	64	128
CNN	0.5097	0.5677	0.5233	0.5121	0.7198	0.6205	0.5207	0.7239	0.6244
RNN	0.521	0.5782	0.5212	0.6381	0.7499	0.6322	0.6387	0.7546	0.6389
LSTM	0.5272	0.5182	0.5699	0.6126	0.7734	0.7842	0.6266	0.7808	0.7854
GRU	0.5294	0.5227	0.5291	0.6625	0.7782	0.7538	0.6689	0.7897	0.7622
Bert	0.5048	0.6597	0.501	0.6329	0.7401	0.7275	0.6997	0.7428	0.7436
BertLoRA	0.7081	0.7456	0.7518	0.7828	0.7855	0.804	0.7895	0.794	0.8065

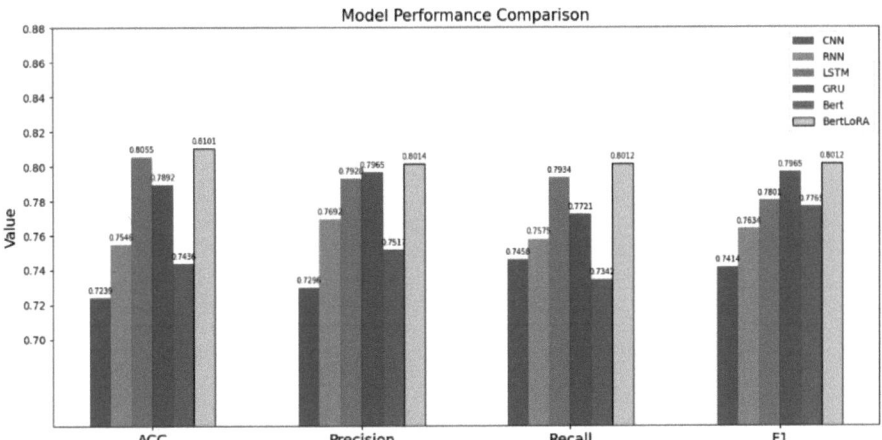

Fig. 4. Performance comparison between BERTLoRA and five other neural networks (graphical results).

4.3 Experimental Results and Analysis

According to the experimental results in Table 2 and Fig. 3, the BERTLoRA model Outperforms the other three models in multiple aspects. The BERTLoRA model achieves an accuracy of 81%, second only to the RoBERTa model's 82%. The size of the BERTLoRA model is reduced to 2 MB, the training time is shortened to 97 s, the memory usage is decreased to approximately 1.9 MB, and the inference time is reduced to 6.2 ms. In contrast, although the BERT model reaches an accuracy of 74%, it has a larger model size and memory usage, as well as a longer training time. The number of parameters in the BERTLoRA model is only about 0.3% of that in the original BERT model. Its memory usage is approximately 32% of that of the BERT model, the training time is shortened by about 40%, and the inference time is reduced by about 7%. Overall, while maintaining high accuracy, the BERTLoRA model significantly reduces the model size and resource consumption, demonstrating its great potential in practical applications.

Especially in environments that require rapid deployment and operation, the size of the BERTLoRA model is far smaller than the maximum model size of 512 MB that Google Chrome can load [12], further proving its applicability and efficiency in large - scale data flow scenarios such as online shopping malls.

As shown in Table 3 and Fig. 4, the BERTLoRA model achieves an accuracy of 0.8101, a precision of 0.8014, a recall of 0.8012, and an F1 score of 0.8012. These results not only outperform the four traditional neural network models, namely CNN, RNN, LSTM, and GRU, but are also 0.6% higher than those of the LSTM model, which has the second-best performance among them. The accuracy of the BERTLoRA model is 11.5% higher than that of the CNN, 10.4% higher than that of the RNN, 0.6% higher than that of the LSTM, 2.5% higher than that of the GRU, and 7.9% higher than that of the BERT. These results demonstrate the BERTLoRA model's deep semantic understanding and pattern-capturing capabilities in handling natural language tasks.

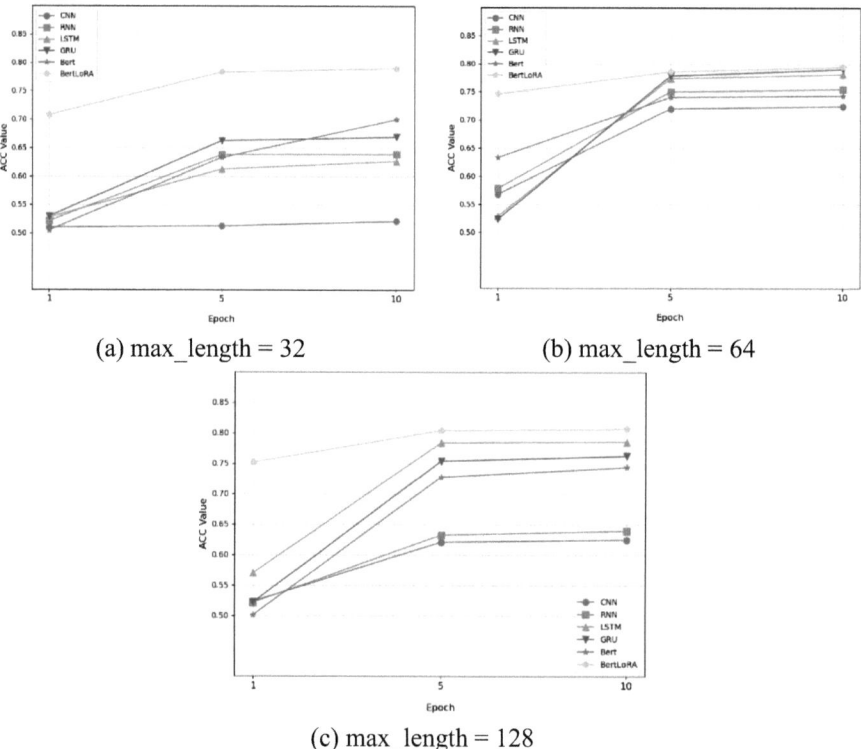

Fig. 5. The performance of the BERTLoRA model when the values of max_length are 32, 64 and 128.

The Fig. 5 shows performance of the BERTLoRA model under different max_length values. The experimental results indicate that the BERTLoRA model performs excellently at various max_length values, especially when dealing with long sequences. When the max_length is 128, the accuracy of BERTLoRA is 0.7518 at epoch = 1, which increases to 0.804 at epoch = 5, and finally reaches 0.8065 at epoch = 10, significantly higher than that of other models. In contrast, the accuracy of the LSTM model, which has the second-best performance, is 0.7978. The accuracy of BERTLoRA is approximately 1.13% higher than that of LSTM. This data further demonstrates that BERTLoRA has obvious advantages in deep semantic understanding and pattern capturing.

5 Conclusion and Prospect

This paper proposes a product review sentiment classification model (BERTLoRA) based on the optimization of BERT and LoRA. By introducing the LoRA technology on the basis of the BERT model, while maintaining the model's performance, the number of model parameters and the computational cost are significantly reduced. The experimental results show that the number of parameters of the BERTLoRA model is only about 0.3% of that of the original BERT model, the memory usage is approximately 32% of that of the BERT model, the training time is shortened by about 40%, and the inference time is reduced by about 7%. These improvements give BERTLoRA significant advantages in terms of resource consumption and efficiency. In the future, the model structure will be further optimized, and more model optimization techniques will be explored to improve the performance and efficiency of the model. At the same time, attempts will also be made to apply this model to other natural language processing tasks to verify its applicability and effectiveness in different tasks.

References

1. Zhang Q, et al.: Qos-aware reliable traffic prediction model under wireless vehicular networks. 2021 IEEE Global Communications Conference, 1–6 (2021)
2. Gasparetto, A., Marcuzzo, M., Zangari, A., Albarelli, A.: A survey on text classification algorithms: From text to predictions. Information **13**(2), 83 (2022)
3. Sharma, A., Singh, B.: Measuring impact of e-commerce on small scale business: A systematic review. Journal of Corporate Governance and International Business Law **5**(1), 34–38 (2022)
4. Lauriola, I., Lavelli, A., Aiolli, F.: An introduction to deep learning in natural language processing: Models, techniques, and tools. Neurocomputing **470**, 443–456 (2022)
5. Devlin, J., Chang, M.W., Lee, K., Toutanova, K.: Bert: Pre-training of deep bidirectional transformers for language understanding. Conference of the North American Chapter of the Association for Computational Linguistics: Human Language Technologies, 1, pp. 4171–4186. Association for Computational Linguistics, Minneapolis, Minnesota (2019)
6. Chardet, M., Coullon, H., Pertin, D.: Madeus: A formal deployment model. International conference on high performance computing & simulation 2018, HPCS, pp. 724–731. IEEE, (2018)
7. Devalal, S., Karthikeyan, A.: LoRa technology-an overview. Second international conference on electronics, communication and aerospace technology 2018, ICECA, pp. 284–290. IEEE, (2020)
8. Xu, H., et al.: Consumer QoE-Aware Cognitive Semantic Sentiment Analysis Via Hybrid Large Models. IEEE Consumer Electronics Magazine **14**(2), 59–68 (2025)
9. Qian, Y., Leng, J., Zhou, K., Liu, Y.: How to measure and control indoor air quality based on intelligent digital twin platforms: A case study in China. Build. Environ. **253**, 111349 (2024)

10. Koroteev, M.V.: BERT: a review of applications in natural language processing and understanding. arXiv preprint arXiv 2103, 11943 (2021)
11. Imran, M., & Almusharraf, N.: Analyzing the role of ChatGPT as a writing assistant at higher education level: A systematic review of the literature. Contemporary Educational Technology 15(4). ep464 (2023)
12. Guo, Z., Yu, K., Jolfaei, A., Ding, F., Zhang, N.: Fuz-Spam: Label Smoothing-based Fuzzy Detection of Spammers in Internet of Things. IEEE Trans. Fuzzy Syst. **30**(11), 4543–4554 (2022)
13. Guo, Z., Zhang, Q., Xu, P., Shen, Y., Chakraborty, C., Alfarraj, O., Yu, K.: Vision Sensing-Driven Intelligent Ocular Disease Detection Using Conformer-Based Dual Fusion. IEEE Journal of Biomedical and Health Informatics. (2024)
14. Zhang, Q., Guo, Z., Zhu, Y., Vijayakumar, P., Castiglione, A., Gupta, B.B.: A deep learning-based fast fake news detection model for cyber-physical social services. Pattern Recogn. Lett. **168**, 31–38 (2023)
15. Yu, L., Liu, B., Lin, Q., Zhao, X., Che, C.: Similarity matching for patent documents using ensemble bert-related model and novel text processing method. Journal of Advances in Information Technology 15(3), (2024)

BEDGKT: A Behavior-Enhanced Dynamic Graph Knowledge Tracing Model

Rongkui Yu and Ying Wang[✉]

Southwest University, Chongqing 400715, China
`waying95@swu.edu.cn`

Abstract. Knowledge Tracing (KT) is a technique that predicts students' future performance based on their historical exercise records. A growing body of research has recognized the value of student behavioral features in KT modeling. However, existing approaches often struggle to flexibly incorporate these features due to model complexity, limiting their ability to adaptively select and integrate the most informative behavioral signals. To address these limitations, this study proposes a Behavior-Enhanced Dynamic Graph Knowledge Tracing (BEDGKT) model. This work employs a two-layer MLP to capture student behavioral features, significantly reducing feature engineering efforts. Furthermore, we introduce the concept of student behavioral preference and leverage a double-layer GRU architecture to fully utilize behavioral feature information. The experiments were conducted on three public datasets, demonstrating AUC improvements of 1.2% – 5.4% across different datasets. Through visualization techniques and case studies, we quantitatively validated the impact of behavioral features on prediction performance. This work provides a more scalable and practical solution for knowledge tracing, offering both theoretical significance and practical implications for advancing personalized education.

Keywords: Knowledge Tracing · Dynamic Graph · GRU · Self-attention mechanism

1 Introduction

Knowledge tracing (KT) dynamically models students' evolving knowledge states using historical exercise data to predict performance and guide adaptive learning strategies. Early approaches like BKT [1] utilized Hidden Markov Models, while DKT [2] advanced the field with RNNs (GRU/LSTM), achieving a 20% AUC improvement. Subsequent works such as LPKT [3] have improved the feature engineering and the forgetting mechanism. However, these models fail to distinguish between students' behavioral patterns and knowledge states. Moreover, their model requires complex preprocessing of behavioral features and directly

---

This work was funded by the Southwest University Research Startup Fund Program for In-Service Doctoral Degree Holders (SWU222001,5220500295) and the Southwest University General Project for Education and Teaching Reform Research (2021JY033,5240200797).

integrates them into the learning-forgetting module, which would necessitate substantial architectural modifications to integrate additional features.

To address these issues, we propose Behavior-Enhanced Dynamic Graph Knowledge Tracing (BEDGKT) base on DyGKT [4], which modifies its feature constructor and RNN architecture. Different with LBKT [5] and CAKT [6], our approach simplifies feature engineering and enhances model learning capability through a multi-layer GRU architecture. The model combines these behavioral features with knowledge states using a double-layer GRU network with three connection methods. Furthermore, we introduce a speed preference visualization to analyze how response speed affects model predictions. Experimental results across three real-world datasets show consistent improvements, with BEDGKT achieving 1.2% – 5.4% higher AUC scores than baseline models while maintaining computational efficiency.

The contributions of this article are summarized as follows:

- We propose the BEDGKT model, which can easily add, modify, or delete integrated features without significantly increasing training costs and can adapt to different combinations of input features.
- The concept of student behavioral preferences is first proposed, and a double-layer GRU is used to capture student preferences, which is then integrated with the knowledge state using three different connection methods: Cross Connection, Attention Connection, and Output Connection.
- We propose a method for analyzing the impact of student behavior characteristics on model predictions, and demonstrate that the model not only learns the overall behavioral patterns of students in the dataset but also performs exceptionally well with highly personalized students.

2 Related Work

2.1 Knowledge Tracing

Since DKT [2] pioneers the use of RNNs in knowledge tracing, deep learning has dominated the field through two primary paradigms: RNN-based sequential modeling exemplified by LPKT [3] which simulates forgetting through time intervals and behavioral features, and attention-based interaction modeling exemplified by AKT [9] that enhances attention mechanisms with exercise similarity. Current research advances primarily focus on three directions: Enhancing feature engineering through models such as QDCKT [7], optimizing the learning-forgetting module within RNN architectures as in LBKT [5], and diversifying attention mechanism refinements using approaches like SAKT [8]. Beyond these directions, interpretable memory architectures including DKVMN [10] and graph-based methods like GKT [11] have been employed to model complex relational patterns within knowledge tracing systems.

2.2 Dynamic Graph Learning

Dynamic graph models complex systems by representing entities as nodes and their interactions as timestamped edges and providing a powerful framework to capture temporal dynamics and evolutionary patterns. This structure enables granular analysis of temporal dynamics and evolutionary patterns, revealing both transient behaviors and systemic transformations across temporal scales. In the downstream tasks of dynamic graph learning, typical applications include recommendation systems deployed in social networks, financial networks, and user-item interaction platforms. Le Yu et al. propose DyGLib [13], a dynamic graph learning framework capable of extracting node historical records from dynamic graphs. Applying DyGLib to knowledge tracing tasks not only streamlines data processing but also dynamically adapts to continuously expanding student exercise records.

3 Methodology

3.1 Problem Definition

Assume that the records contains $|S|$ learners, $|Q|$ questions, $|K|$ knowledge concepts, and $|I|$ interaction records. The model arranges the records in chronological order to generate a dynamic graph $G = \{(x_1, t_1), (x_2, t_2), ..., (x_I, t_I)\}$, where t_i is the time stamp, and x_i is the student's historical record, including the student node s_i, the question node q_i, the answer r_i, the knowledge concept k_i, and the student's behavioral features f_i. For the student node s and the question node q, the model can extract the historical interactions sequences of student $S_s^t = \{(s, q_i, r_i, k_i, f_i, t_i) | t_i < t\}$ and the historical interaction sequences of question $S_q^t = \{(s_j, q, r_j, k_j, f_j, t_j) | t_j < t\}$ from the dynamic graph G. The knowledge tracing task predicts the probability $P(r = 1 | s, q, S_s^t, S_q^t)$ that the student s will answer the question q correctly based on the student's historical interactions and the question's historical interactions.

The proposed BEDGKT framework consists of four main components: subgraph construction, feature encoding, knowledge tracing, and answer prediction. The subgraph construction module dynamically samples student exercise and question interaction records from the graph structure. The feature encoding module processes the historical records of students and questions, while the knowledge tracing module utilizes these encoded features to infer behavioral preferences, knowledge states, and question difficulties. Finally, the prediction generates the probability of correct responses. The overall system structure is illustrated in Fig. 1.

3.2 Interactions Encoding

Consistent with LBKT [5], three student behavioral features are selected in this paper: answer speed, hints, and attempts. Speed reflects students' knowledge mastery level, while attempts and hints indicate students' learning attitudes [5]. In order to strengthen the model's extensibility of features, BEDGKT uses a two-layer MLP to capture behavioral features and ReLU as the activation function.

$$x_{sf}^{t_i} = ReLU(W_{F2} ReLU(W_{F1} f_{s,q_i}^{t_i} + b_{F1}) + b_{F2}) \tag{1}$$

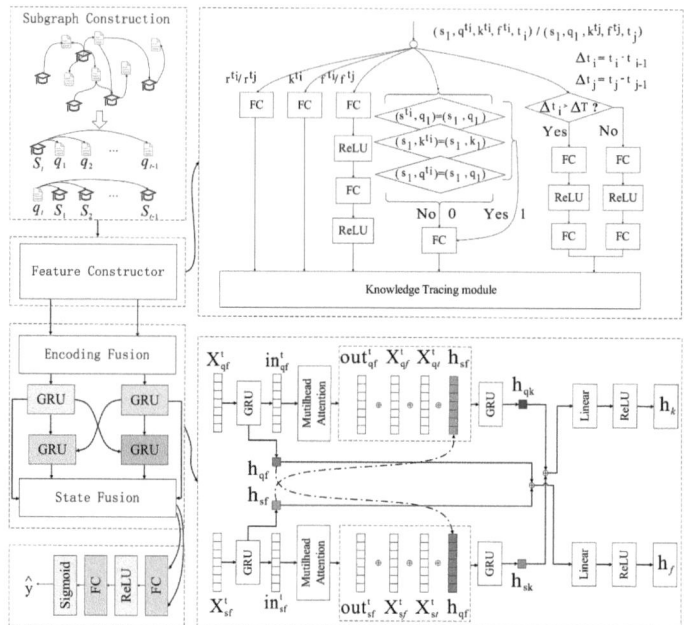

Fig. 1. The structure of BEDGKT.

where $x_{sf}^{t_i}$ is the student behavior feature embedding at timestamp t_i, W_{F1}, b_{F1}, W_{F2}, and b_{F2} are adjustable parameters. The question history records are processed using the same parameters as student behavior feature embedding to obtain the student behavior encoding $x_{qf}^{t_j}$. This study uses the same method as DyGKT to obtain the answer embeddings $x_{se}^{t_i}$, the time embeddings $x_{st}^{t_i}$, the multi-set indicators $x_{sn}^{t_i}$ (whether question q or its knowledge concept has been attempted by student s in sequences), the concept embeddings $x_{sc}^{t_i}$ of the sequences of student s, the answer embeddings $x_{qe}^{t_j}$, the time embeddings $x_{qt}^{t_j}$, the multi-set indicators $x_{qn}^{t_j}$ (whether students have previously attempted a particular question q in sequences) of the sequences of question q, and the concept embedding x_{qc} of question q.

3.3 Knowledge Tracing

We find that distinct students exhibit heterogeneous behavioral preferences. To model this, we propose a double-layer GRU architecture that separately captures behavioral preferences and knowledge states. Unlike knowledge states, which evolve dynamically during learning, behavioral preferences are relatively stable attributes that remain consistent over short timescales. This hypothesis is further supported by our parameter sensitivity analysis. Based on this rationale, the GRU layer dedicated to behavioral preference modeling explicitly omits temporal information. We define:

$$x_{sf}^{t_i} = concat(x_{se}^{t_i}, x_{sc}^{t_i}, x_{sf}^{t_i}, x_{sn}^{t_i}) \tag{2}$$

$$x_{qf}^{t_j} = concat(x_{qe}^{t_j}, x_{qf}^{t_j}, x_{qn}^{t_j}) \tag{3}$$

where $x_{sf}^{t_i}$ is the behavioral feature embedding of student s, and $x_{qf}^{t_j}$ is the behavioral feature embedding of question q. After the behavior feature embedding is input into the first layer of the GRU, we obtain:

$$in_{sf}^t, h_{sf} = GRU_{s1}(X_{sf}^t) \tag{4}$$

$$in_{qf}^t, h_{qf} = GRU_{q1}(X_{sf}^t) \tag{5}$$

where in_{sf}^t and in_{qf}^t are the hidden states of each time step of GRU, as well as h_{sf} (student behavioral preferences) and h_{qf} (question behavior preferences) are the final hidden states of GRU. In order to fully integrate behavioral preferences and knowledge states, this paper proposes three connection methods. The first is the **Attention Connection**. BEDGKT applies the multi-head attention [12] to process in_{sf}^t and in_{qf}^t, thus extracting students' potential behavioral characteristics and reducing the impact of randomness on the sequence. The second connection is the **Cross Connection**. To fully integrate students sequences and question sequences. BEDGKT connects h_{sf} and h_{qf} of with other embeddings respectively:

$$out_{sf}^t = MutilheadAttention_1(in_{sf}^t) \tag{6}$$

$$out_{qf}^t = MutilheadAttention_2(in_{qf}^t) \tag{7}$$

$$x_{sd}^{t_i} = concat(x_{sf}^{t_i}, x_{st}^{t_i}, out_{sf}^{t_i}, h_{qf}) \tag{8}$$

$$x_{qd}^{t_j} = concat(x_{qf}^{t_j}, x_{qt}^{t_j}, out_{sf}^{t_j}, h_{sf}) \tag{9}$$

where $x_{sd}^{t_i}$ and $x_{qd}^{t_j}$ are the input of the second GRU, which are used to learn students' knowledge state and question difficulty:

$$o_{sk}^t, h_{sk} = GRU_{s2}(X_{sd}^t) \tag{10}$$

$$o_{qk}^t, h_{qk} = GRU_{s2}(X_{qd}^t) \tag{11}$$

where h_{sk} refers to the student knowledge state and h_{qk} to the question difficulty. The third connect method is the **Output Connection**. In order to enhance the influence of student behavioral preferences and question behavior characteristics on answer prediction, the Output Connection combines the student behavioral preferences h_{sf}, the question behavior preferences h_{qf}, and the question concept embedding x_{qc} to generate h_f. Then it combines the student knowledge state h_{sk}, the question difficulty h_{qk}, and the problem concept embedding x_{qc} to generate h_k:

$$h_f = ReLU(W_{out1}concat(h_{sf}, h_{qf}, x_{qc}) + b_{out1}) \tag{12}$$

$$h_k = ReLU(W_{out2}concat(h_{sk}, h_{qk}, x_{qc}) + b_{out2}) \tag{13}$$

where W_{out1}, b_{out1}, W_{out2}, and b_{out2} are all adjustable parameters.

3.4 Prediction Module and Parameters Learning

BEDGKT predicts the performance of student s on problem q at time t based on h_f and h_k regarding the problem:

$$\hat{y}_t = sigmoid(F(h_f, h_k)) \tag{14}$$

Specifically, the results of knowledge tracing are fed into a two-layer MLP. ReLU is used as the activation function, and Sigmoid is applied to obtain the probability $\hat{y}_t$ that the learner answers the question correctly. The model uses the cross-entropy loss to optimize the parameters. Our loss function is defined as follows:

$$loss = -\sum_{t}^{t_I}(r_{s,q}^t log(\hat{y}_t) + (1 - r_{s,q}^t)log(1 - \hat{y}_t)) \tag{15}$$

4 Experiments

4.1 Datasets

The experiments in this paper are conducted on three datasets: ASSISTment12, ASSISTment17, and Junyi. Table 1 shows the statistics of all datasets.

ASSISTment12 was collected from the ASSISTments online tutoring system during 2012-2013. It comprises data from skill-builder problem sets.

ASSISTment17 was presented in the 2017 ASSISTments longitudinal data mining competition.

Junyi contains the problem logs and exercise-related information on the Junyi Academy, an E-learning platform established in 2012 on the basis of the open-source code released by Khan Academy.

Table 1. Statistics of all datasets

Dataset	Student	Question	Concept	Interaction
ASSISTment12	42.9k	105.2k	259	5994.0k
ASSISTment17	1.7k	2.7k	99	936.5k
Junyi	174.5k	0.7k	39	25731.6k

4.2 Evaluation

Baseline Models. The classic DKT [2] knowledge tracing model with recurrent neural networks, the AKT [9] knowledge tracing model with attention mechanisms, and the dynamic graph models DyGFormer [13], and DyGKT [4] serve as comparison algorithms.

Evaluation Metrics. Average Precision (AP) and AUC. In order to ensure the accuracy when a student faces a new question or when a new student answers a question, this paper also introduces an inductive sampling method, which deliberately hides a part of the nodes used for inductive testing during the initial training stage to test the generalization performance of the model in a new situation.

Implementations Details. An 8:1:1 data split is applied to our training, validation and test set. The Adam optimizer is configured with a learning rate of 0.001. Training is conducted on a Tesla T4 GPU (16 GB) in a Linux environment (Python3.8, PyTorch 1.8.1, CUDA 11.1).

4.3 Performance

The experimental result of transduction learning and inductive learning are shown in Table 2. Here, the optimal results are presented in bold, and the suboptimal results are underlined.

Table 2. AP and AUC-ROC for Transductive and Inductive Predictions

setting	Datasets	ASSITment12		ASSISTment17		Junyi	
	Metrics	AP	AUC	AP	AUC	AP	AUC
Trans	DKT	82.21	69.95	64.09	73.42	90.25	66.86
	AKT	75.25	60.75	62.57	70.93	91.39	71.48
	DyGFormer	86.71	77.64	68.88	77.52	93.75	79.04
	DyGKT	87.29	78.20	72.70	80.45	93.98	79.46
	BEDGKT	**88.78**	**80.68**	**81.54**	**86.46**	**94.19**	**80.68**
	Metrics	AP	AUC	AP	AUC	AP	AUC
Ind	DKT	81.71	69.11	59.01	67.09	89.77	65.07
	AKT	74.50	59.55	58.69	66.34	90.97	69.58
	DyGFormer	86.85	77.74	69.69	76.83	93.73	79.11
	DyGKT	87.44	78.48	73.22	79.59	93.98	79.54
	BEDGKT	**88.90**	**80.89**	**80.48**	**84.98**	**94.34**	**80.78**

Our BEDGKT demonstrates consistent performance gains on three datasets, with a 1.2% improvement in AUC (Junyi), 2.5% (ASSISTment12), and 5.5% (ASSISTment17). The model's predictive performance is closely tied to the data distribution characteristics. In this study, we examine three key features—answer speed, hints, and attempts. Notably, hints and attempts exhibit distinct patterns: they are highly concentrated at 0 and 1 in both ASSISTment12 and Junyi, while displaying relatively uniform distributions in ASSISTment17. Importantly, BEDGKT achieves its greatest performance improvement specifically on ASSISTment17. This suggests that the quality of behavioral features significantly influences the effectiveness of knowledge tracing.

4.4 Ablation Study

To validate the effect of double-layer GRU, an ablation experiment is designed. In the experiment, we evaluate the AUC scores of transduction learning for several models: the single-layer GRU model (single), the double-layer GRU model with attention connection (attention), the double-layer GRU model with cross connection (cross), the double-layer GRU model with output connection (double), the double-layer GRU model without attention connection (-attention), the double-layer GRU model without cross connection (-cross), the double-layer GRU model without output connection (-double), and our complete model (BEDKGKT). The results are shown in Fig. 2.

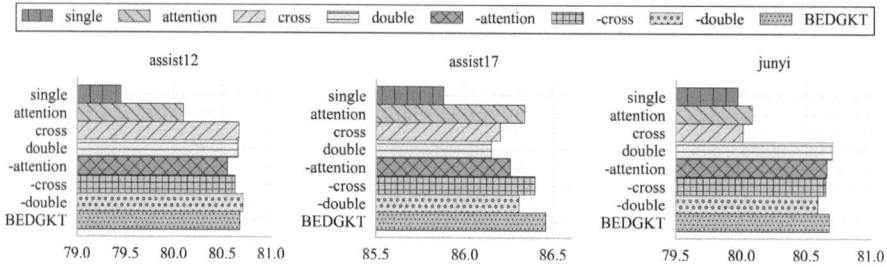

Fig. 2. AUC for transduction ablation study of BEDGKT

The experimental results in Fig. 2 demonstrate that the impact of different connection types varies across datasets, which may be attributed to the feature distributions within the datasets. However, employing merely two types of connections can improve AUC by 0.7% – 0.8% compared to the single-layer GRU, demonstrating the superiority of the double-layer GRU architecture.

4.5 Parameter Sensitivity

The sensitivity analysis of historical interaction sequence length demonstrates that both BEDGKT and DyGKT exhibit similar increasing trends on the ASSISTment17 (Fig. 3). The reason is that longer sequences provide more comprehensive learner information, enhancing knowledge state estimation. Notably, while BEDGKT incorporates behavioral features, its performance trajectory aligns with DyGKT, suggesting that sequence length primarily influences knowledge state modeling rather than preference estimation.

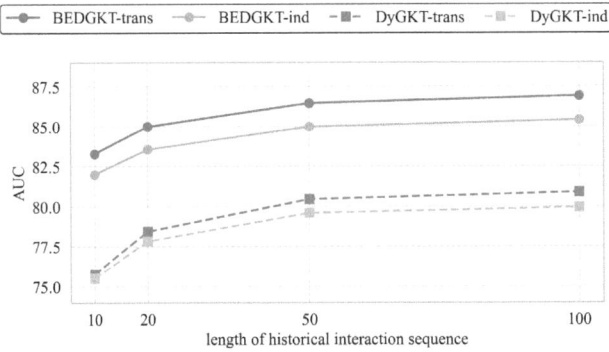

Fig. 3. Sequence length sensitivity in transduction sequences.

4.6 Impact of Feature Integration

To analyze the influence of response speed on model prediction, this paper compares DyGKT and BEDGKT on ASSISTment17. Here, the Mean Absolute Error (MAE) is used as the scoring function, and the relative improvement rate is defined to analyze the prediction performance for the learning sequences of the same student:

$$MAE(X, model) = \frac{1}{m} \sum_{i=1}^{m} |model(x_i) - y_i| \qquad (16)$$

$$Increase(X, A, B) = MAE(X, B) - MAE(X, A) \qquad (17)$$

The relative improvement rate quantifies the performance gain of model A over model B, where X denotes a student learning sequence. A positive value of Increase(X,A,B) indicates an improvement. The experimental results on 1,708 students from ASSISTment17 demonstrate that BEDGKT outperforms DyGKT in 1,683 cases. To investigate the influence of response speed, this paper introduces two novel metrics: Speed Preference and Speed Preference Strength, which categorize and measure temporal patterns in student interactions:

$$\begin{cases} lt_{id} = \dfrac{\sum_{i=1}^{n} r_{id}^{i} I(speed_{id}^{i} < Avage(speed_{id}))}{\sum_{i=1}^{n} I(speed_{id}^{i} < Avage(speed_{id}))} \\ gt_{id} = \dfrac{\sum_{i=1}^{n} r_{id}^{i} I(speed_{id}^{i} > Avage(speed_{id}))}{\sum_{i=1}^{n} I(speed_{id}^{i} > Avage(speed_{id}))} \end{cases} \qquad (18)$$

$$speed\_preference_{id} = (lt_{id}, gt_{id}) \qquad (19)$$

$$preference\_strength_{id} = |lt_{id} - gt_{id}| \qquad (20)$$

where $speed_{id}^{i}$ denotes the i-th speed record of student id, r_{id}^{i} denotes the i-th response record, lt_{id} denotes the average accuracy rate of responses slower than the student's average response speed, gt_{id} denotes the average accuracy rate of responses higher than the student's average response, $speed\_preference_{id}$ is the speed preferences, and $preference\_strength_{id}$ is the speed preference strength.

To analyze the influence of behavioral features on the model, the learning sequences from ASSISTment17 are divided into two groups (sequences with low improvement rates and sequences with high improvement rates). In this experiment, the speed preference is used to plot the distributions of the two groups to analyze the influence of speed on model prediction. Meanwhile, the characteristic information of the two groups is statistically analyzed to explore the influence of other features on model prediction.

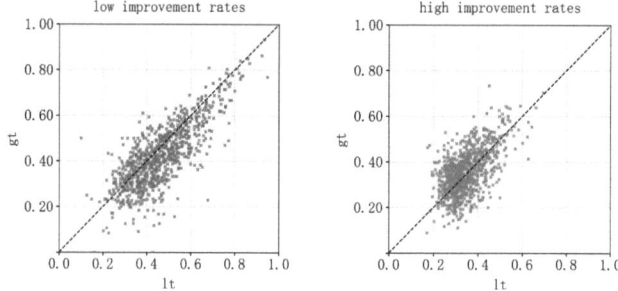

Fig. 4. Distributions of Low and High Improvement Groups.

Table 3. Statistical indicators at varying improvement rates

Statistical index	low improvement	high improvement
Average speed	49.9892	32.0183
Average attempts	2.3296	2.6864
Average hints	0.7105	1.2486
Average sequence length	407.8981	688.7974
Average preference strength	0.0662	0.0735

Based on the Fig. 4 and Table 3, the following conclusions can be drawn:

- The distribution location of the high improvement rate group is the same as the data-dense location in the dataset, which indicates that BEDGKT has learned the overall speed preference of the dataset.
- The average speed preference strength of the low improvement group is 0.0662, while that of the high improvement group is 0.0735. This shows that BEDGKT brings greater improvement for students with obvious preferences.
- BEDGKT performs better than DyGKT in long-term interaction records. Regarding the number of hints and attempts, learners with more diverse behaviors will get higher scores.

4.7 Expansion Analysis

To demonstrate the scalability of our model, five features from ASSISTment17 are integrated in this part: two behavioral features (bottomHint and frIsHelpRequest), two global features (AveKnow and AveCorrect), and one static feature (original). The experimental results are presented in Fig. 5.

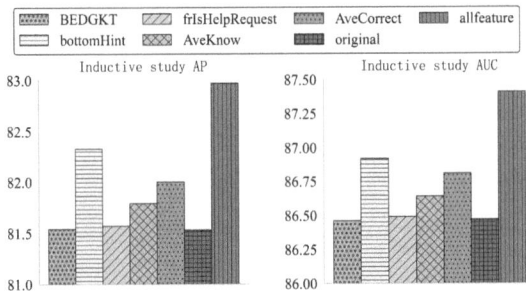

Fig. 5. Prediction performance of BEDGKT with five features.

The experimental results demonstrate that the integrated features exert differential impacts on model prediction performance. Notably, the predictive performance improves substantially when all features are integrated into the model. This experiment demonstrates that our model is capable of integrating diverse features, indicating its adaptability to various educational datasets.

4.8 Conclusion

This paper introduces BEDGKT, an enhanced version of DyGKT, which modifies its feature constructor and RNN architecture to incorporate behavioral features and enhance the capability of model in learning short sequences. To strengthen its feature representation, the model employs a double-layer GRU structure with three connection methods to enhance the integration of behavioral patterns. Experimental results demonstrate better performance over baseline models, achieving AUC improvements of 1.2%âĂŞ5.4% across on all three real-world datasets. Ablation studies reveal that historical sequence length impacts BEDGKT and DyGKT similarly, supporting the hypothesis that behavioral preferences remain stable over short periods. Furthermore, the speed analysis demonstrates that the model successfully captures the overall characteristics of the datasets, while exhibiting better predictive performance for students with distinctive behavioral preferences. However, employing dynamic graph approaches to extract sequence may incur significant information loss, whereas key-value memory networks maybe serve as an effective solution.

References

1. Corbett, A.T., Anderson, J.R.: Knowledge tracing: modeling the acquisition of procedural knowledge. User Model. User-Adap. Inter. **4**, 253–278 (1994)
2. Piech, C., Bassen, J., Huang, J., et al.: Deep knowledge tracing. Adv. Neural. Inf. Process. Syst. **28**, 505–513 (2015)
3. Shen, S., Liu, Q., Chen, E., et al.: Learning process-consistent knowledge tracing. In: Proceedings of the 27th ACM SIGKDD Conference on Knowledge Discovery & Data Mining, pp. 1452–1460. Association for Computing Machinery, New York, NY, United States (2021)
4. Cheng, K., Peng, L., Wang, P., et al.: DyGKT: Dynamic graph learning for knowledge tracing. In: Proceedings of the 30th ACM SIGKDD Conference on Knowledge Discovery and Data Mining, pp. 409–420. Association for Computing Machinery, New York, NY, United States (2024)
5. Xu B, Huang, Z., Liu, J., et al.: Learning behavior-oriented knowledge tracing. In: Proceedings of the 29th ACM SIGKDD Conference on Knowledge Discovery and Data Mining, pp. 2789–2800. Association for Computing Machinery, New York, NY, United States (2023)
6. Xing, J., Li, K., Wu, Y., et al.: Convolution Attentive Knowledge Tracing with comprehensive behavioral features. In: Proceedings of the ACM Turing Award Celebration Conference-China 2024, pp. 48–52. Association for Computing Machinery, New York, United States (2024)
7. Liu, G., Zhan, H., Kim, J.: Question difficulty consistent knowledge tracing. In: Proceedings of the ACM Web Conference 2024, pp. 4239–4248. Association for Computing Machinery, New York, NY, United States, (2024)
8. Pandey, S., Karypis, G.: A self-attentive model for knowledge tracing, In: 12th International Conference on Educational Data Mining, EDM 2019. International Educational Data Mining Society, pp. 384–389. International Educational Data Mining Society, Montreal, Canada (2019)
9. Ghosh, A., Heffernan, N., Lan, A.S.: Context-aware attentive knowledge tracing. In: Proceedings of the 26th ACM SIGKDD International Conference on Knowledge Discovery & Data Mining, pp 2330–2339. Association for Computing Machinery, New York, NY, United States (2020)
10. Zhang, J., Shi, X., King, I., et al.: Dynamic key-value memory networks for knowledge tracing. In Proceedings of the 26th International Conference on World Wide Web, pp.765–774. International World Wide Web Conferences Steering Committee, Republic and Canton of Geneva, Switzerland (2017)
11. Nakagawa, H., Iwasawa, Y., Matsuo, Y.: Graph-based knowledge tracing: modeling student proficiency using graph neural network. In: IEEE/WIC/aCM International Conference on Web Intelligence, pp. 156–163. Association for Computing Machinery, New York, NY, United States (2019)
12. Vaswani, A., Shazeer, N., Parmar, N., et al.: Attention is all you need. In: Proceedings of the 31st International Conference on Neural Information Processing System 30, pp. 6000–6010 Curran Associates Inc. 57 Morehouse Lane, Red Hook, NY, United States (2017)
13. Yu, L., Sun, L., Du, B., et al.: Towards better dynamic graph learning: New architecture and unified library. In: Proceedings of the 37th International Conference on Neural Information Processing Systems, pp. 67686–67700. Curran Associates Inc, 57 Morehouse Lane, Red Hook, NY, United States (2023)

Network, Economic and Business Studies

Enhancing Energy Policy Planning in Thailand Through AI-Based Forecasting of Crude Oil Imports

Suppanunta Romprasert(✉), Tanapat Romprasert, and Danai Tanamee

The Faculty of Economics, Srinakharinwirot University, Bangkok, Thailand
suppanunta@g.swu.ac.th

Abstract. This study investigates the determinants of Thailand's crude oil import demand using a multiple regression approach and elasticity analysis. The results show that GDP and OPEC oil production have a positive relationship with crude oil imports, whereas domestic energy production and Dubai crude oil prices show a negative association. The estimated price elasticities of demand are highly inelastic in both the short run (-0.011) and long run (-0.015), indicating that crude oil remains an essential commodity with limited responsiveness to price fluctuations. In contrast, income elasticity is relatively high (0.212 short-run; 0.298 long-run), suggesting that economic growth substantially increases crude oil consumption. Compared with prior studies, Thailand's price elasticity is particularly low, reinforcing the challenge of reducing imports through fiscal instruments such as fuel taxes, which may prove inflationary and ineffective. Furthermore, the negative coefficient of domestic energy production underscores its role as a substitute for crude oil. Enhancing local energy sources—such as hydro, lignite, natural gas, and condensate—can significantly reduce import dependency and strengthen energy security. The study also highlights the need for non-price-based policy measures, such as promoting energy conservation and shifting consumer behavior, especially during periods of high oil prices. These findings suggest that strategic demand-side management, combined with structural energy reforms, can improve resilience and sustainability in Thailand's energy landscape. Lessons from this analysis may offer valuable guidance for other emerging economies facing similar import dependency and energy security challenges.

Keywords: Crude Oil Imports · Price Elasticity · Income Elasticity · Domestic Energy Production · Energy Security

1 Introduction

Crude oil is an indispensable resource with profound implications for both economic stability and political dynamics. Its price influences a wide array of industries, particularly transportation and energy production, which are heavily dependent on oil as a primary input [1]. Changes in crude oil prices tend to ripple across the broader economy, affecting inflation, production costs, and household consumption. These economic shifts often lead to broader political consequences, as governments grapple with the socio-economic fallout of energy price volatility. A key mechanism through which crude oil prices influence the economy is their impact on inflation. When oil prices rise, the cost of producing goods and services also increases, given oil's pervasive role in logistics, electricity generation, and manufacturing [2]. This results in higher consumer prices and a general increase in the cost of living, which can suppress consumer spending and reduce overall economic growth. The inflationary pressures caused by rising oil prices can be especially burdensome for developing and oil-import-dependent countries such as Thailand. Thailand represents a critical case of oil import dependency. Due to insufficient domestic oil reserves, the country is heavily reliant on imports to meet its energy demands. In 2021, Thailand imported approximately 90% of its crude oil, with 76% of those imports originating from Middle Eastern countries [3]. This dependency poses a serious risk to energy security and macroeconomic stability. When international oil prices surge, Thailand faces increased import costs, contributing to a wider trade deficit and exerting pressure on the national currency. These outcomes may fuel inflation and hinder long-term economic growth [4]. Table 1 presents Thailand's crude oil import statistics for the year 2021, highlighting the extent of its dependency on foreign oil sources.

Table 1. Thailand's Crude Oil Imports, 2021

Source Region	Share of Total Imports (%)
Middle East	76
Asia-Pacific	12
Africa	8
Others	4
Total Imports	100
Import Dependency	90% of domestic demand

Source: Energy Policy and Planning Office (EPPO), 2022

Beyond economic concerns, the political ramifications of oil price volatility are also significant. The Thai government has often faced criticism for its management of rising fuel costs, as they disproportionately affect low-income households and small businesses. High oil prices may amplify social and economic inequalities, thereby increasing political tensions [5]. The government, therefore, plays a crucial role in formulating and implementing energy policies aimed at stabilizing prices and mitigating the impact of external shocks. Given these considerations, this study aims to investigate the economic factors that drive crude oil import demand in Thailand. The analysis employs a multiple regression model to assess the influence of variables such as the price of Dubai crude oil, gross domestic product (GDP) per capita, domestic energy production, OPEC crude oil output, and the real exchange rate of the Thai baht. Additionally, the study seeks to estimate the income and price elasticities of crude oil import volumes. The results are expected to provide valuable insights for policymakers in designing strategies to manage national energy demand and reduce dependency on imported oil. Therefore, the objectives of the study are 1. To analyze the determinants of crude oil import demand in Thailand using a multiple regression model. 2.To estimate the income and price elasticities of crude oil import volumes.

2 Literature Review

Crude oil remains a pivotal component of global energy consumption, exerting significant influence on both economic and political spheres. Its price volatility has far-reaching implications, particularly for countries heavily reliant on oil imports.Thailand, with its substantial dependence on imported crude oil, faces unique challenges in this context. The dynamics of crude oil import demand have been extensively studied worldwide. Altinay [6] examined Turkey's oil import demand, revealing that both income and price significantly influence import volumes. Similarly, Ziramba [7] analyzed South Africa's oil import patterns, highlighting the sensitivity of import demand to price fluctuations.These studies underscore the importance of understanding the determinants of oil import demand to formulate effective energy policies [8]. Thailand's energy landscape is characterized by a high reliance on imported fossil fuels. In 2023, the country imported approximately $31.8 billion worth of crude petroleum, primarily from the United Arab Emirates, Saudi Arabia, and the United States . This heavy dependence on external sources renders Thailand vulnerable to global oil price fluctuations, impacting its trade balance and economic stability [9]. The nation's limited domestic oil reserves exacerbate this vulnerability. In 2016, Thailand's oil self-sufficiency stood at only 41%, with proven reserves insufficient to meet long-term consumption needs . Such dependency necessitates a comprehensive understanding of the factors influencing oil import demand to develop strategies that enhance energy security [9]. A seminal study by Buranakunaporn [8] employed an Error Correction Model (ECM) to analyze the short-run and long-run determinants of Thailand's crude oil and petroleum product import demand. Utilizing annual data from 1994 to 2019, the study identified key variables influencing import volumes, including Gross Domestic Product (GDP), average import prices, and domestic consumption of petroleum products [8]. The findings revealed that: There exists a long-term positive equilibrium relationship between import volumes and both GDP and domestic consumption, indicating that economic growth and increased energy use drive higher import demand. Conversely, a long-term negative relationship was observed between import volumes and average import prices, suggesting that higher prices deter import demand. In the short term, import volumes adjust towards long-run equilibrium at a rate of approximately 75.79% annually, reflecting a relatively swift response to deviations from equilibrium. These insights are crucial for policymakers aiming to manage energy imports effectively. Understanding the responsiveness of oil import demand to changes in income and price is vital for policy formulation.Buranakunaporn's study estimated the short-run and long-run price elasticities at -0.129 and -0.182, respectively, indicating that import demand is relatively inelastic to price changes. Similarly, income elasticities were estimated at 0.930 (short-run) and 0.266 (long-run), suggesting that income growth has a more pronounced effect on import demand in the short term [6]. These elasticity estimates imply that traditional price-based policies, such as adjusting retail fuel prices, may have limited effectiveness in curbing import demand. Therefore, alternative measures, including promoting energy efficiency and diversifying energy sources, should be considered to reduce dependency on imported oil [7]. The insights derived from these studies have significant policy implications. Given the inelastic nature of oil import demand concerning price, the Thai government should explore non-price interventions to manage energy consumption.Strategies may include: Investing in renewable energy sources to diversify the energy mix and reduce reliance on oil imports. Enhancing public transportation infrastructure to decrease individual vehicle use and associated fuel consumption. Implementing energy efficiency programs across industries to lower overall energy demand. Such measures can contribute to improved energy security and economic resilience in the face of global oil market volatility. Thailand's heavy dependence on imported crude oil presents significant economic and political challenges, particularly in the context of fluctuating global oil prices. Empirical studies highlight the importance of GDP, domestic consumption, and import prices as key determinants of oil import demand. The relatively inelastic response of import demand to price changes suggests that non-price policy measures may be more effective in managing energy imports. By adopting a multifaceted approach that includes investment in renewable energy, enhancement of energy efficiency, and development of sustainable transportation systems, Thailand can work towards reducing its reliance on imported oil and bolstering its energy security.

Oteng-Abayie et al. [10] conducted an empirical investigation into the macroeconomic determinants of crude oil demand in Ghana by utilizing annual time series data spanning from 1980 to 2013. The study employed the Vector Error Correction Model (VECM) to estimate both short-run and long-run income and price elasticities of crude oil demand. The primary dependent variable was domestic crude oil consumption in Ghana, while the independent variables included crude oil prices, real per capita gross domestic product (GDP), the real exchange rate, external technological advancements related to energy efficiency, and domestic oil production. The findings from this study indicated that the demand for crude oil in Ghana is both income inelastic and price inelastic. In other words, changes in income and oil prices have a relatively limited effect on the quantity of crude oil demanded in both the short and long term. This suggests that crude oil remains a necessity in the Ghanaian economy, and consumers are relatively unresponsive to price or income fluctuations when it comes to oil consumption. Furthermore, the study found that key variables—including crude oil prices, real per capita GDP, the real exchange rate, and technological progress in energy-saving innovations—serve as significant long-run determinants of crude oil demand. These results carry important implications for Ghana's energy policy. The inelastic nature of oil demand suggests that price-based interventions, such as taxes or subsidies, may have limited effectiveness in reducing consumption. Instead, the authors recommend that policymakers focus on diversifying energy sources and developing strategic policies for demand-side energy security. This includes continued efforts to promote the use of alternative energy sources, such as renewable energy, as well as improving the nation's infrastructure and technology to enhance energy efficiency. Additionally, the incorporation of technological progress as a variable highlights the role of innovation in moderating oil demand. Technological improvements that lead to more efficient energy use can mitigate the dependence on crude oil, particularly when such innovations are accessible and effectively implemented. Therefore, investing in research and development related to energy technology should be a critical component of national energy planning. The study's use of the VECM model also allowed for an understanding of the short-term adjustments in crude oil demand in response to deviations from long-run equilibrium. This dynamic modeling framework is particularly useful in the context of energy economics, where both immediate market responses and structural trends play a role in shaping consumption behavior. The results demonstrated that when crude oil consumption deviates from its long-run path, market forces and economic adjustments work gradually to restore equilibrium, though the responsiveness is limited due to the inelastic nature of demand. By identifying the real exchange rate as a significant determinant, the study also pointed to the importance of macroeconomic stability in managing energy imports. A depreciation in the real exchange rate, for example, can increase the cost of importing crude oil, thereby influencing national consumption patterns and the overall trade balance. The study by Oteng-Abayie et al. [10] underscores the complex interplay between macroeconomic variables and crude oil demand in developing countries like Ghana. It highlights the necessity of designing energy policies that go beyond simple price controls, advocating for long-term strategies focused on diversification, innovation, and resilience. These findings provide a relevant comparative foundation for analyzing crude oil import demand in other energy-importing countries, such as Thailand, which face similar vulnerabilities due to high dependence on foreign oil and exposure to international oil market fluctuations. Marbuah [11] conducted a pioneering empirical analysis aimed at identifying the macroeconomic determinants of crude oil import demand in Ghana. This study is one of the first to apply the Auto-Regressive Distributed Lag (ARDL) approach to investigate both short- and long-run dynamics using annual time-series data spanning from 1980 to 2012. The dependent variable in the model was the real volume of crude oil imports, while the independent variables included real gross domestic product (GDP), the real price of crude oil in Ghana, and the real exchange rate. The findings revealed that the demand for imported crude oil in Ghana is price inelastic in both the short run and long run. This suggests that fluctuations in crude oil prices have a relatively minimal impact on the quantity of oil imported, underscoring the essential nature of oil in supporting domestic economic activities. Interestingly, the study observed a divergence in the role of economic activity: while short-run income elasticity of oil import demand was found to be inelastic, long-run income elasticity was significantly higher. This implies that over time, increases in national income are more likely to drive up the demand for imported oil. Moreover, the real exchange rate emerged as a significant driver of crude oil import demand across both time horizons. A depreciation in the real exchange rate increases the domestic cost of importing oil, which can dampen demand. Conversely, a more favorable exchange rate environment can encourage higher import volumes. This highlights the critical role of exchange rate policy in managing energy security and controlling import costs in oil-importing countries like Ghana. He also emphasized the strategic importance of reducing Ghana's reliance on imported crude oil by exploring domestic production capabilities. While the country's oil sector was still developing at the time of the study, there was a strong policy rationale for accelerating domestic energy production to shield the economy from the volatility of international oil markets. The study argued that continued overdependence on foreign oil could expose the country to persistent trade imbalances, inflationary pressures, and foreign exchange instability. Based on these findings, the study offered several policy recommendations. First, the government should invest in cleaner and more sustainable energy sources, such as solar, wind, and bioenergy, to diversify the national energy mix. Second, improving public transportation infrastructure could help reduce the national demand for oil, particularly in urban areas where vehicle usage is high. Third, the study advocated for the acceleration of natural gas processing projects, which could substitute for crude oil in power generation and industrial use. Lastly, public education campaigns on energy conservation and efficiency were recommended to change consumption behavior at the household and business levels. The use of the ARDL approach in this study is particularly valuable, as it allows for the estimation of both short-term fluctuations and long-term equilibrium relationships even when variables are integrated of different orders (i.e., I(0) or I(1)). This flexibility makes ARDL a suitable method for analyzing energy consumption behavior in developing countries, where data quality and integration levels may vary. Furthermore, the model's capacity to reveal both immediate and structural responses to economic shocks offers a nuanced understanding of how policy interventions can be timed and targeted effectively. Marbuah's [11] contribution is critical to the literature, especially in contexts where countries are facing growing energy demands alongside fiscal and environmental constraints. It presents a compelling case for strategic planning in energy consumption and trade, highlighting how macroeconomic variables interact to shape national energy dependency.

For policymakers in other emerging economies—such as Thailand, which also exhibits a high dependence on imported crude oil—the insights from Ghana's experience can be particularly instructive. He strengthens the empirical foundation for understanding the determinants of crude oil import demand in resource-constrained economies. It emphasizes the complex and multifaceted nature of oil dependency, urging governments to adopt long-term strategies aimed at promoting energy diversification, technological innovation, and macroeconomic stability. These policy directions are essential not only for reducing import dependency but also for enhancing resilience to global energy market disruptions. Marbuah [12] conducted a comprehensive study examining the determinants of crude oil import demand in Ghana using the Autoregressive Distributed Lag (ARDL) model. Utilizing annual data from 1980 to 2012, the study explored long-run and short-run relationships among crude oil imports, international oil prices, and national productivity. The dependent variable in the model was the log of crude oil import volumes, while the key independent variables included the log of global crude oil prices and the log of Ghana's productivity (GDP) for the corresponding years. The findings revealed strong evidence of a long-run equilibrium relationship between the explanatory variables and Ghana's import demand for crude oil. In particular, crude oil prices were found to be inelastic in the short run but elastic in the long run, indicating that while immediate changes in global oil prices have a limited effect on Ghana's import volumes, sustained price shifts do lead to significant long-term adjustments in demand. Similarly, income elasticity followed a comparable pattern: short-run inelasticity and long-run elasticity, suggesting that long-term economic growth eventually drives increased oil imports, despite minimal short-term responsiveness. Additionally, the study identified the real exchange rate as a major determinant of oil import demand, influencing behavior both in the short and long term. A depreciating exchange rate increases the cost of oil imports, thus discouraging demand. Interestingly, domestic crude oil production and population growth had relatively minor effects on import volumes, indicating Ghana's continued reliance on external energy sources. Based on these results, Marbuah [12] recommended several policy directions. Governments should consider investing in alternative, cleaner energy sources and encourage domestic production capacity to reduce dependency on imported oil. Furthermore, oil pricing in domestic markets should reflect international market conditions to manage demand effectively. Over time, gradual substitution of imports with domestic supply could help reduce the national import bill and promote sustainable economic development.

In a broader context, Ozcan [13] analyzed oil consumption patterns in the framework of global efforts to reduce greenhouse gas emissions and address the depletion of fossil fuel reserves. Focusing on 20 OECD countries, the study utilized panel data from 1980 to 2011 to assess long-run income and price elasticities of oil demand. The dependent variable was per capita oil consumption, while the key independent variables included average income per capita and crude oil prices for each country and year. The results indicated that both income and price elasticities of oil demand were statistically significant, though income elasticity was greater than price elasticity. This finding implies that economic expansion leads to a more substantial increase in oil consumption than does a corresponding drop in price. Furthermore, Ozcan [13] discovered bidirectional causality between economic growth and oil consumption, meaning that these two variables are mutually reinforcing: economic growth drives oil demand, and increased oil use also fuels further economic development. This relationship presents a challenge for policymakers seeking to reduce fossil fuel consumption in line with environmental objectives. The study cautioned that aggressive policies to reduce oil use may have negative effects on economic growth, particularly in countries where oil remains a crucial input for production and transport. As a result, the study suggested that governments should develop new mechanisms for reshaping consumption behavior—such as public awareness campaigns and the promotion of energy-saving technologies—to influence individual consumption decisions while supporting sustainable development goals. Together, the findings from [12] and [13] highlight the multifaceted nature of oil demand across different economic contexts. In Ghana, a developing economy with high import dependence, factors like income growth and currency value significantly shape import behavior. In contrast, in advanced economies within the OECD, oil consumption is more closely tied to economic structure and the interplay between growth and energy use. Both studies reinforce the importance of integrated energy policies that account for macroeconomic factors, domestic energy capacity, and environmental sustainability. Camacho-Gutiérrez [14] explored the dynamic relationship between the United States and Mexico in the context of crude oil trade, focusing on the United States' demand for crude oil imports from Mexico. Using monthly data spanning from 1990 to 2010, the study applied the Stock-Watson Dynamic Ordinary Least Squares (DOLS) model to examine the responsiveness of U.S. crude oil imports from Mexico to various economic variables. The dependent variable was the volume of crude oil imports by the United States from Mexico, while the independent variables included U.S. personal income, the price of imported Mexican crude oil, and U.S. crude oil stock levels excluding the Strategic Petroleum Reserve (SPR). The findings revealed that the U.S. demand for Mexican crude oil is price inelastic, indicating that the U.S. continues to import crude oil from Mexico even when prices rise above those of competing suppliers. This lack of price sensitivity reflects the strategic and logistical advantages of sourcing oil from Mexico, such as proximity and established trade infrastructure. As such, Mexico benefits from a relatively stable revenue stream from oil exports to the U.S., providing fiscal reliability for its government. Additionally, the study found that the demand for Mexican crude oil was income inelastic, yet sensitive to changes in the U.S. unemployment rate, underscoring the broader economic linkages between the two countries. When unemployment rises in the U.S., indicating reduced economic activity, crude oil import volumes decline—though not drastically. Moreover, U.S. crude oil import demand from Mexico was shown to be elastic with respect to changes in U.S. crude oil inventories, excluding SPR. When domestic stockpiles rise, the need to import decreases, reflecting inventory management decisions rather than price mechanisms. Camacho-Gutiérrez [14] concluded that Mexico is in a unique position to extract economic rents from its oil exports to the U.S., given the inelastic nature of U.S. demand. However, this opportunity is tempered by risks associated with macroeconomic volatility in the U.S. For instance, a U.S. economic recession or a substantial increase in oil inventories could lead to a significant reduction in demand for Mexican crude oil, negatively impacting Mexico's oil-derived government revenue. The study underscores the vulnerability of resource-dependent economies to external demand shocks, particularly when tied to a single major trading partner. As such, it calls attention to the importance of diversifying both trade partners and domestic revenue sources in oil-exporting nations. Additionally, the research illustrates how macroeconomic variables, such as national income, employment levels, and inventory strategies, can significantly shape international oil trade flows—even in the face of volatile pricing environments.

Kavaz [15] investigated the price and income elasticity of crude oil import demand in Turkey using annual data from 1970 to 2018, applying the Structural Time Series Model with Unobserved Effects and Deterministic Trends (STSM/UEDT). The study focused on two key independent variables: national income and crude oil prices, with the volume of crude oil imports as the dependent variable. The findings indicated that both price and income elasticities were statistically significant but inelastic, suggesting that crude oil imports in Turkey do not respond strongly to price or income changes. This confirms the classification of crude oil as a necessity good for the Turkish economy. Furthermore, the study highlighted the significant impact of external shocks, such as global oil crises, on Turkey's oil import demand. These shocks were shown to influence energy consumption behavior and long-term policy planning. The article also addressed broader issues such as energy security, taxation policies, and future energy demand trends, emphasizing the urgent need for sustainable and cost-effective policy interventions. Kavaz [15] suggested that these findings are vital for Turkish policymakers in designing long-term strategies that balance economic resilience with sustainable energy needs.In a related study, Muhammad, Asif, Ali, and Safdar [16] examined the determinants of crude oil import demand in Pakistan, focusing on the country's reliance on oil-based electricity production. The study used annual data from 1990 to 2020, employing the Augmented Dickey-Fuller (ADF) test for stationarity and Johansen's co-integration test to assess long-run relationships. The dependent variable was crude oil import demand (in thousand barrels per day), while the independent variables included GDP per capita (USD), crude oil price per barrel, Pakistan's real effective exchange rate (REER), and the percentage of electricity generated from oil-based sources. The findings supported microeconomic demand theory, affirming that crude oil import demand responds to economic fundamentals such as income levels, price changes, and exchange rate movements. Importantly, the results showed that exchange rates and the share of oil-based electricity generation had substantial effects on crude oil imports. The study warned that reducing oil imports without ensuring alternative energy sources could hamper sustainable economic development. If Pakistan continues to depend heavily on oil for electricity generation, the trade deficit could widen significantly, further threatening economic stability. Therefore, the authors advocated for diversifying energy sources to reduce oil dependence and support sustainable economic growth in Pakistan. Shao, Lu, and Hou [17] proposed a hybrid forecasting model to improve the accuracy of predicting imported crude oil (ICO) demand, which is critical for national economic planning and energy strategy. Recognizing that oil is a globally essential commodity used for a wide range of purposes, the study emphasized the importance of accurate ICO demand forecasting as a tool for informed policy and economic development. Using annual data from 1993 to 2010, the authors developed and compared two hybrid forecasting models, which outperformed traditional single-stage models in terms of accuracy and robustness. The dependent variable in their study was crude oil import volume, while the model incorporated a broad set of 23 independent variables. These included macroeconomic indicators such as GDP, GNP, disposable personal income, national income, and foreign exchange reserves; energy-related metrics like primary energy supply, final energy consumption, and energy production; as well as environmental and demographic factors including average temperature, sunlight hours, and population. The inclusion of electricity-related data (e.g., average electricity load, electricity consumption per household, and electricity price) reflected the multi-dimensional influences on oil demand. Papusson and Tangvitoontham [18] examined Thailand's increasing crude oil import dependency, especially from the Middle East, and its impact on the trade deficit. The study employed an Error Correction Model (ECM) using annual data from 1986 to 2011, with import volume of crude oil as the dependent variable, and real price and income as the independent variables. The results indicated that the short-run price elasticity of imported crude oil was statistically insignificant, while long-run elasticity was significant, suggesting that crude oil is a necessary good for Thai consumers. Moreover, income elasticity increased in the long run, indicating that crude oil is a normal good, and that consumers may find more substitutes over time. The study also highlighted key domestic factors affecting fuel prices, including refinery prices, taxes, and marketing margins, and noted that the Thai government has implemented strict price controls on diesel due to its influence on inflation, production costs, and transportation expenses. Importantly, the price transmission from crude oil to downstream petroleum products was found to be asymmetric, underscoring the need for long-term energy reduction policies. Tsirimokos [19] conducted a cross-country analysis on the price and income elasticity of crude oil demand in 10 IEA member countries using annual data from 1980 to 2009.

The study applied a multiple regression model, with crude oil consumption as the dependent variable, and real GDP per capita, real crude oil prices, lagged oil consumption, and a time trend as independent variables. The findings showed heterogeneous patterns: countries such as Sweden, Denmark, Finland, Italy, Germany, the United States, and Japan had reduced oil use over time, whereas Spain, Portugal, and Turkey had increased their consumption. All estimated coefficients were statistically significant and displayed the expected signs. The negative and significant time trend coefficient indicated an overall decrease in oil intensity, reflecting long-term improvements in energy efficiency and structural transformation. While price elasticity remained relatively inelastic across countries, income elasticity in the long run approached unit elasticity in some nations, indicating a stronger responsiveness of oil demand to income changes over time.

Şişman and Öztürk [20] investigated the import demand for crude oil in Turkey, aiming to evaluate the price and income elasticity of demand and to determine whether data frequency influences the accuracy of elasticity estimates. The study utilized Auto Regressive Distributed Lag (ARDL) models and employed three different data frequencies—monthly, quarterly, and annual—covering the period 1996–2021. The dependent variable was the volume of crude oil imports, while the independent variables were real GDP and crude oil prices. The results showed that price elasticity of crude oil demand was inelastic, indicating that Turkish consumers are unresponsive to price changes due to the country's high import dependency. On the other hand, income elasticity was found to be elastic, suggesting that economic growth significantly influences oil demand. The study highlighted that higher-frequency data (monthly and quarterly) produced more theoretically consistent and statistically robust results compared to annual data. Therefore, the authors recommended using high-frequency data for future elasticity estimations to avoid biases and improve policy relevance. Ozturk and Arisoy [21] examined Turkey's growing crude oil import dependency, driven by factors such as urbanization, population growth, and economic development. Although the share of crude oil in total primary energy demand had decreased, over 90% of Turkey's crude oil needs were still met through imports. The study employed Time-Varying Parameters (TVP) models with annual data from 1969 to 2012, analyzing the impact of real GDP and crude oil prices on crude oil import volumes. Findings revealed that income elasticity was statistically significant, while price elasticity was not. This implies that Turkey's crude oil import demand is more responsive to changes in income levels than to price fluctuations. Based on these results, the study suggested policy interventions such as regulatory measures and public awareness campaigns to encourage the use of energy-efficient vehicles, engines, and appliances. Such strategies could help mitigate energy dependency and promote sustainable energy consumption patterns. Yaprakli and Kaplan [22] analyzed the long-run import demand elasticity of crude oil in Turkey over the period 1970–2013 using the Carrion-i-Silvestre approach and Dynamic Ordinary Least Squares (DOLS) estimation. The dependent variable was the volume of crude oil imports, and the independent variables included real GDP (in millions of USD), crude oil prices, and a dummy variable capturing the impact of economic crises. The study found that price elasticity was inelastic in the long run, whereas income elasticity was elastic, indicating that demand is more sensitive to changes in income than in price. Additionally, external economic crises had a significant impact on crude oil demand, emphasizing the vulnerability of [23] investigated India's crude oil import demand elasticity using the Auto Regressive Distributed Lag (ARDL) model with annual data from 1987 to 2016. The dependent variable was the crude oil import volume, and the independent variables were real GDP and crude oil prices. The study reported that income elasticitywas highly elastic in the long run, with a 1% increase in GDP leading to a 2.89% rise in oil imports. In contrast, price elasticity was found to be inelastic both in the short and long run. This suggests that income changes have a greater influence on oil demand than price, partly due to government control over retail prices during the study period. The researchers recommended developing policies to reduce crude oil dependency, explore alternative energy sources, and enhance energy efficiency for sustainable economic growth.key's energy sector to global shocks.

Kaushik and Kumar [24] examined both short-run and long-run relationships among crude oil consumption, international oil prices, GDP per capita, domestic crude oil production, coal production, and the USD/INR exchange rate in India over the period 2001–2017, using a Vector Error Correction Model (VECM). The study found that price, income, and coal production elasticities were inelastic in both the short and long term. However, in the short term, only GDP and domestic oil production had statistically significant effects on crude oil demand. In the long term, all variables significantly influenced demand, with GDP and exchange rate exerting the strongest effects. Coal production had a negative effect on oil demand, while international crude oil prices also negatively impacted demand. The study suggested that India should reduce its dependence on fossil fuels, build strategic oil reserves, and invest in renewable energy sources such as solar and wind to enhance energy security and reduce import reliance. Trott, Saporta, and Tudela (2009) analyzed the factors affecting crude oil price fluctuations between 2003 and 2008, focusing on the impact of demand and supply shocks, geopolitical factors, and financial market dynamics. Using an Auto Regressive Distributed Lag (ARDL) model with annual data, the study included GDP and crude oil prices as key explanatory variables for crude oil imports. While demand and supply shocks, as well as political factors, were found to explain the direction of oil price changes, they could not fully account for the magnitude of those changes. The study also explored the role of speculative behavior and financial flows into the oil market, particularly during the 2008 price volatility, suggesting that speculation may have been a key driver. The authors highlighted the importance of understanding oil price d Guo et al. [25] investigated the determinants of China's crude oil import demand using monthly data from 2015 to 2021 through a time-series multiple regression model. The dependent variable was China's crude oil imports, and the independent variables included international crude oil prices, sales of new energy vehicles (NEVs), and domestic crude oil production. Interestingly, the study found a positive and statistically significant relationship between NEV market share and crude oil imports, contradicting the assumption that NEVs reduce oil demand. This was attributed to the limited substitution effect of NEVs and the continued strong demand from the petrochemical sector. As expected, crude oil prices had a negative and significant impact on imports, while domestic production showed a positive but statistically insignificant effect, indicating that domestic supply was insufficient to meet rising demand.ynamics for monetary policy and financial stability. Hina et al.

[26] examined the short-run and long-run elasticities of crude oil demand across different countries and regional groups using annual data from 1971 to 2014 with the ARDL model. The study found that the long-run income elasticity ranged from 0.19 to 1.18, with an average of 0.64, indicating a positive but inelastic relationship between income and crude oil demand in most countries. Meanwhile, the long-run price elasticity ranged from –0.01 to –0.67, averaging –0.23, suggesting a negative but inelastic response to price changes. The short-run elasticities were generally smaller in magnitude, highlighting the sluggish adjustment of oil demand to income and price variations. Policy implications include the need for developing countries to reduce oil import dependency, diversify energy sources, and enhance energy efficiency and conservation, especially amid volatile oil prices. The authors also called for future research to incorporate updated data and additional factors such as environmental regulations, consumer preferences, and structural transformations. Bhattacharyya [27] explored the empirical income and price elasticities of petroleum product demand across countries and regional groups, using log-linear regression and annual data from 1982 to 2005. The dependent variable was the quantity of petroleum products consumed (in thousand barrels per day), while the independent variables included real GDP (in billions of USD) and real petroleum product prices (USD per barrel). The findings revealed that income elasticity was positive but inelastic, whereas price elasticity was negative but also inelastic, consistent with prior studies. Short-run elasticities were generally smaller than long-run estimates, reinforcing the notion that petroleum demand adjusts slowly to economic changes. The study recommended that MENA countries focus on energy diversification, efficiency improvements, and consideration of environmental regulations and consumer behavior in energy planning and future research.

The authors integrated various modeling approaches, including artificial neural networks (ANNs), support vector regression (SVR), and extreme learning machines (ELMs), within the hybrid framework. They found that the combination of machine learning techniques with traditional econometric models provided a superior predictive performance, especially when dealing with the nonlinear and complex nature of oil demand. The study concluded by encouraging further exploration of hybrid models, including extensions to multi-step forecasting and the integration of additional machine learning techniques or statistical forecasting tools. These findings are especially valuable for developing forecasting models in other forms of energy consumption, where demand is driven by similarly complex interdependencies.

3 Research Methods

3.1 Econometric Method: Multiple Regression Model

In this analysis, an econometric method will be employed to examine the determinants of crude oil imports in Thailand. Our model is a multiple regression model that includes a one-period lagged dependent variable, specified as follows:

$lnM_t = \beta_0 + \beta_1 \ln P_t + \beta_2 \ln S_t + \beta_3 \ln Opec_t + \beta_4 \ln gdp_t + \beta_5 \ln reer_t + \beta_6 \ln M(t-1) + e$.

Where:

M is the volume of crude oil imported by Thailand,
P is the price of Dubai crude oil,
S represents domestic energy production in Thailand (including hydro, lignite, natural gas, condensate, and crude oil),
Opec is crude oil production by OPEC countries,
GDP is Thailand's per capita GDP,
REER is Thailand's real effective exchange rate.

3.2 Demand Elasticities

Understanding the short-run and long-run price and income elasticities of crude oil demand provides valuable policy insights. In the short run, demand tends to be inelastic because consumers cannot easily adjust their behavior or technology. In contrast, in the long run, demand becomes more elastic as consumers adopt alternative energy sources and adjust consumption patterns [28]. Before estimating the regression model, the relationships among independent variables will be analyzed using a correlation matrix. This step helps to identify potential multicollinearity issues by observing the strength and direction of pairwise correlations between variables. The coefficients of the multiple regression model will be estimated using the Ordinary Least Squares (OLS) method. This technique minimizes the sum of squared residuals and provides the best linear unbiased estimators (BLUE) under the classical linear regression assumptions. After estimating the model, diagnostic tests will be conducted to check for autocorrelation and heteroskedasticity: Autocorrelation will be tested using the Durbin-Watson (DW) test or Breusch-Godfrey test to determine whether the residuals are serially correlated, which may violate OLS assumptions and affect the efficiency of the estimators. Heteroskedasticity will be tested using the Breusch-Pagan test or White test to determine whether the variance of the residuals is constant across observations. The presence of heteroskedasticity can lead to inefficient estimators and biased standard errors.

4 Results and Analysis

4.1 Result Estimation

It is the process of predicting the likely outcomes based on models, data, or prior knowledge to predict the outcome, duration, and variables needed as shown in Table 2.

From Table 2, the adjusted R-squared is 0.8, indicating that approximately 80% of the variation in Thailand's crude oil import volume can be explained by the independent variables included in the model. This suggests a good fit for the regression model. The logarithmic coefficient for the price of Dubai crude oil (LOG(P)) is negative and statistically significant, suggesting an inverse relationship between oil price and import volume. Specifically, a 1% increase in the price of Dubai crude oil results in a 0.011% decrease in Thailand's crude oil imports in the short run. In the long run, the same 1% increase in Dubai crude oil price leads to a 0.015% decrease in import volume, reflecting a more elastic response over time. The logarithmic coefficient for per capita GDP (LOG(GDP)) is positive and statistically significant, implying that economic growth is associated with increased energy demand. A 1% increase in Thailand's per capita GDP

Table 2. Variables and t-Statistic Outcomes

Variable	Coefficient	Std. Error	t-Statistic	Prob.
C	10.47415	0.541118	19.35648	0.0000
LOG(P)	-0.011261	0.005302	-2.123886	0.0478
LOG(GDP)	0.212822	0.035310	6.027218	0.0000
LOG(S)	-0.098621	0.035304	-2.793497	0.0120
LOG(OPEC)	0.387926	0.087136	4.451953	0.0003
LOG(IMPORTOIL(-1))	0.2868848	0.040176	7.139719	0.0000
LOG(REER)	-0.561229	0.113042	-4.964772	0.001
R-SQUARED	0.857472	MEAN DEP.	13.60888	
ADJUSTED R-SQUARED	0.809963	S.D. DEP.	0.091472	
S.E. OF REGRESSION	0.039876	AKAIKE INFO	-3.374609	
SUMSQUARED RESID	0.028621	SCHWARZ	-3.033324	
LOG LIKELIHOOD	49.18261	HANNAN	-3.279951	
F-STATISTIC	18.04854	DURBIN	2.414205	
PROB(F-STATISTIC)	0.000001	WALD F-STST	328.5796	
PROB(WALD F-STATISTIC)	0.000000			

results in a 0.212% increase in crude oil import volume in the short run. In the long run, a 1% increase in GDP per capita is associated with a 0.298% increase in crude oil imports, highlighting the income elasticity of energy demand over time.

4.2 Comparison of Short-Run and Long-Run Elasticities

The short-run and long-run price elasticities estimated in this study (-0.011 and -0.015, respectively) are lower in magnitude than those reported in the literature, suggesting that Thailand's crude oil import demand is relatively inelastic with respect to price. This could reflect a high dependency on imported crude oil and limited short-term alternatives or substitution effects. In contrast, the income elasticities (0.212 in the short run and 0.298 in the long run) fall within a more moderate range when compared to previous studies. These results imply that crude oil imports in Thailand are somewhat responsive to changes in income, but to a lesser degree than in countries with higher income growth or energy-intensive industrial sectors. Notably, studies such as Kaushik and Kumar [24] report significantly higher income elasticities, which may reflect structural differences in energy use patterns or stages of economic development. The logarithmic coefficient of domestic energy production (LOG(S)) is negative and statistically significant. This suggests that a 1% increase in domestic energy production leads to a 0.098% decrease in the volume of crude oil imports. This result aligns with findings from Kaushik and Kumar (2020), where a similar independent variable—domestic coal production in India—was also found to have a negative coefficient (-0.001). These findings support the conclusion that developing domestic energy sources can reduce dependency on imported crude oil. The logarithmic coefficient of OPEC crude oil production (LOG(OPEC)) is positive and statistically significant, indicating that a 1% increase in OPEC oil production is associated with a 0.387% increase in Thailand's crude oil imports. This result implies that increased global oil supply, driven by OPEC, likely leads to lower global prices, which stimulates higher crude oil imports. No prior studies reviewed in this research included this specific variable, making this finding a novel contribution. The logarithmic coefficient of the lagged crude oil import variable (IMPORTOIL(-1)) is positive and statistically significant. A 1% increase in crude oil imports in the previous period leads to a 0.286% increase in the current period. This reflects persistence in import behavior and is consistent with the results of [29], who reported a similar coefficient of 0.5741. This supports the rationale for including lagged values to capture dynamic effects and long-term elasticity. The logarithmic coefficient of Thailand's real effective exchange rate (LOG(REER)) is negative and statistically significant, which contradicts the initial hypothesis. An appreciation of the Thai baht (reflected by an increase in the REER index) should, theoretically, make imports cheaper and thereby increase import demand—implying a positive relationship. However, the negative coefficient suggests other macroeconomic factors or distortions may be influencing this outcome. In comparison, Marbuah [12] found a positive coefficient of 0.669, consistent with theoretical expectations.

5 Conclusion

The independent variables GDP and OPEC oil production change in the same direction as Thailand's crude oil import volume. Conversely, the independent variables—Dubai crude oil price and domestic energy production (HYDRO, LIGNITE, NG, CONDENSATE, and CRUDE OIL)—move in the opposite direction to crude oil import volumes. The elasticity estimates show that crude oil prices have a minimal impact on the volume of crude oil imports, both in the short and long term. This suggests that crude oil is an essential commodity for Thailand and other countries. The slightly higher long-run elasticity reflects the potential for consumers to eventually switch to alternative energy sources, but such transitions are difficult due to infrastructure limitations, availability, or strong preferences for crude oil. For policymakers, this inelastic behavior implies that price-based measures (like taxes) may have limited effectiveness in reducing crude oil imports. This minimal reduction suggests such a tax could contribute to inflation without meaningfully reducing dependency. Therefore, non-price policies may be more effective in addressing energy security. Both short-run and long-run income elasticity values are positive and relatively high, indicating that economic growth is strongly associated with increased crude oil demand. This trend may pose a challenge to Thailand's energy security, especially if imports continue rising with GDP.

The negative coefficient on domestic energy production implies that increasing local energy supply helps substitute for crude oil imports. Each energy source has different substitution potential: Increasing domestic production across these energy types—especially cleaner or renewable sources—can significantly reduce reliance on crude oil imports and enhance energy resilience. Moreover, with Thailand's high oil import dependency, implementing non-price measures—such as promoting domestic energy sources (hydro, lignite, natural gas, condensate)—and encouraging energy conservation will help enhance energy security, reduce exposure to price shocks, and contribute to environmental sustainability. In response to surging oil prices, many governments have implemented temporary relief measures to support consumers. These include reducing fuel taxes, cutting Value-Added Tax (VAT), and regulating retail fuel prices to alleviate the burden on low-income households and individuals heavily reliant on personal vehicles. While these price-based interventions may provide short-term relief, they are often fiscally unsustainable and may not address the root causes of energy insecurity. An alternative and more sustainable solution lies in reducing oil demand, which helps rebalance the market and mitigates the effects of price volatility. Reducing oil demand can be achieved in the short term through coordinated government policies and public participation. This approach demonstrates that demand-side management—through both behavioral and structural changes—can be a powerful and necessary tool in energy policy. It also offers valuable lessons for emerging economies like Thailand, which can adapt such measures according to their specific socioeconomic and energy contexts.

References

1. Hamilton, J.D.: Causes and consequences of the oil shock of 2007–08. Brook. Pap. Econ. Act. **40**(1), 215–261 (2009). https://doi.org/10.2139/ssrn.1297687
2. Baffes, J., Kose, M. A., Ohnsorge, F., & Stocker, M.: *The great plunge in oil prices: Causes, consequences, and policy responses*. World Bank Group. (2015)
3. Policy, E., Office, P.: Thailand energy statistics 2021. Ministry of Energy, Thailand (2022)
4. International Monetary Fund: World Economic Outlook: A long and difficult ascent. (2020). https://www.imf.org/en/Publications/WEO
5. Frankel, J.A.: Effects of speculation and interest rates in a "carry trade" model of commodity prices. J. Int. Money Financ. **42**, 88–112 (2014). https://doi.org/10.1016/j.jimonfin.2013.08.015
6. Altinay, G.: Short-run and long-run elasticities of import demand for crude oil in Turkey. Energy Policy **35**(11), 5829–5835 (2007)
7. Ziramba, E.: Price and income elasticities of crude oil import demand in South Africa: A cointegration analysis. Energy Policy **38**(12), 7844–7849 (2010)
8. Buranakunaporn, S.: Econometric Analysis of Short-Run and Long-Run Determinants of Import Demand of Crude Oil and Petroleum Products in Thailand. Rattanakosin J. Soc. Sci. Humanit., **3**(3), 1–12. (2021). Retrieved from https://so05.tci-thaijo.org/index.php/RJSH/article/view/256943
9. Lau, H.C.: Decarbonizing Thailand's Economy: A Proposal. Energies **15**(24), 1–31 (2022)
10. Oteng-Abayie, E. F., Sarpong, D. B., & Amoh, L.: Macroeconomic determinants of crude oil demand in Ghana: Evidence from a VECM approach. J. Energy and Econ. Dev. (2016)
11. Marbuah, G.: *Determinants of crude oil import demand in Ghana: An empirical analysis using ARDL bounds testing approach*. (2014)
12. Marbuah, G.: Determinants of crude oil import demand in Ghana: An autoregressive distributed lag (ARDL) approach. Energy Policy. (2018)
13. Ozcan, B.: The determinants of oil consumption in OECD countries: Income, prices, and the role of policy in environmental sustainability. Energy Economics. (2015)
14. Camacho-Gutiérrez, P.: *The determinants of U.S. crude oil imports from Mexico: A dynamic OLS approach*. (2010)

15. Kavaz, İ.: Oil import demand, energy security, and elasticity analysis in Turkey. (2020)
16. Muhammad, F. A., Ali, S., & Safdar, H.: *Determinants of crude oil import demand in Pakistan: A co-integration analysis.* (2020)
17. Shao, Y. E., Lu, C. J., & Hou, C. D.: *A hybrid model for forecasting imported crude oil demand using multiple economic and energy-related variables.* (2014)
18. Papusson, C., & Tangvitoontham, N.: *Crude Oil Import Demand Elasticity in Thailand: An ECM Approach.* (2014)
19. Tsirimokos, C.: *Crude Oil Demand Elasticities in IEA Countries: A Comparative Panel Study.* (2011)
20. Şişman, M. Y., & Öztürk, Ö.: *The Role of Data Frequency in Estimating Crude Oil Import Demand Elasticities: Evidence from Turkey.* (2021)
21. Ozturk, I., & Arisoy, I.: *Income and Price Elasticities of Crude Oil Import Demand in Turkey: A TVP Model Approach.* (2016)
22. Yaprakli, S., & Kaplan, F.: *Long-Term Crude Oil Import Demand in Turkey: The Role of Crises.* (2015)
23. Paital, R. R., Dutta, S., & Dash, A. K.: Crude Oil Import Elasticity of Demand in India: An Empirical Analysis 1987–2016. Appl. Econometrics Int. Dept. 19–2. (2019)
24. Kaushik, N., & Kumar, S.: *Determinants of Crude Oil Demand in India: A VECM Approach.* (2020)
25. Guo, Z., et al.: *Impact of New Energy Vehicle Sales on China's Crude Oil Imports: A Time-Series Analysis.* (2023)
26. Hina, Ashraf; et al.: Price and Income Elasticities of Crude Oil Demand: Cross Country Analysis. (2018). Retrieved March 31, 2024, from https://core.ac.uk/download/pdf/230059263.pdf
27. Bhattacharyya, S.C.: Domestic demand for petroleum in MENA countries. Energy Policy **37**(4), 1552–1560 (2009)
28. Dahl, C.A.: Measuring global gasoline and diesel price and income elasticities. Energy Policy **41**, 2–13 (2012)
29. Tsirimokos, C.: Price and income elasticities of crude oil demand: the case of ten IEA countries. Master thesis, *Second cycle, A2E. Uppsala: SLU, Department of Economics.* (2011)

Digital Transformation as an Enabler of Organizational Green Learning Orientation and Ambidextrous Green Innovation: Organizational Information Processing Perspectives

Yi-Chun Huang, Min-Li Yang(✉), and Miao-Hui Chiu

Department of Business Administration, National Kaohsiung University of Science and Technology, Kaohsiung City, Taiwan
{peterhun,minly,sharita}@nkust.edu.tw

Abstract. The challenges of climate change and environmental degradation have driven countries and enterprises to pursue green innovation (GI) and sustainable transformation. While prior research has focused on internal and external corporate governance, this study introduces a new model based on the organizational information processing view to examine how digital transformation (DT) and organizational green learning orientation (OGLO) influence exploratory and exploitative GI. Data from 310 valid responses in Taiwan's manufacturing industry—including sectors such as electronics, machinery, chemicals, and food production—were analyzed using Structural Equation Modeling (SEM) in SPSS v26, with mediation effects tested through SPSS PROCESS. The analysis confirmed all hypothesized direct effects. Furthermore, the study revealed how OGLO mediates the impact of DT on both exploratory and exploitative GI in Taiwanese manufacturing firms. This research contributes to both academic and practical domains by linking organizational information processing theory (OIPT) with an information-processing perspective on organizational learning, thereby developing a DT-OGLO framework to support ambidextrous GI.

Keywords: Environmental sustainability · organizational information processing perspectives · digital transformation (DT) · organizational green learning orientation (OGLO) · ambidextrous green innovation (GI)

1 Introduction

Global sustainability concerns, heightened by extreme weather events, have brought humanity to a critical juncture. The urgency was emphasized at the 2012 UN Rio Summit, leading to the adoption of the Sustainable Development Goals (SDGs) in 2015 [1, 2]. Key agreements—such as the EU's Eco-Innovation Action Plan, the 2030 Agenda, the 2016 Paris Agreement, and the 2024 COP29 Agreement—reflect global priorities like renewable energy, carbon reduction, and environmental protection. Initiatives like the

European Green Deal and "Next Generation EU" aim to accelerate green investment and inclusive growth, aligning with the 2021–2027 EU budget to advance a climate-neutral economy and SDGs [1]. Achieving these goals requires deep sustainability transformations. As sustainability increasingly drives innovation, this study uses "green innovation (GI)" to describe environmentally beneficial innovations in products, processes, services, or methods [3, 4]. In the global landscape, manufacturing industries are among the primary contributors to environmental degradation, posing significant threats to life on Earth [5, 6]. In response, green innovation (GI) has emerged as a vital strategy for promoting more competitive and environmentally sustainable production practices [3, 7]. GI has garnered growing attention in both academic and practical domains. It encompasses the development or enhancement of products and processes aimed at energy conservation, pollution control, waste recycling, and the implementation of environmental management systems [8]. Prior research suggests that firms can become ambidextrous organizations—simultaneously pursuing both exploratory and exploitative innovation [9]—and such ambidexterity is often linked to improved organizational performance [10].

Recent studies have highlighted the importance of green innovation (GI) from various theoretical perspectives, including resource-based [6] stakeholder [11], institutional [12, 13], and internal views [6, 13]. Most research has focused on internal and external corporate governance factors. While prior knowledge can both support and hinder innovation, many findings suggest that information processing and market orientation positively influence new product success [14–17]. However, some scholars challenge this view, and others report mixed or context-dependent outcomes [18]. Understanding how organizations process market information can be significantly enhanced through information-processing theory [19] and organizational learning [20]. Surprisingly, these two related research streams have remained largely unintegrated. Schulz [21] argued that organizational success relies not on choosing between information processing and organizational learning, but on integrating both. Organizations must process information to manage uncertainty and complexity [19], requiring strong information processing capabilities [22]. At the same time, the information-processing perspective of organizational learning builds on well-established concepts like information acquisition, distribution, and interpretation [23]. This study, therefore, integrates both views to examine their influence on corporate ambidextrous GI.

Organizational Information Processing Theory (OIPT) suggests that effective information management helps firms handle uncertainty by improving their processing capacity [22]. Meanwhile, digital transformation (DT) has become essential, reshaping business models through technologies like AI, big data, cloud computing, mobile platforms, and social media [24]. The World Economic Forum estimates that by 2025, platform-based interactions will drive two-thirds of digitalization's $100 trillion value, with 90% of enterprise applications using AI [2]. Yet, despite major investments, only 21% of firms have fully implemented DT, while 87% expect disruption, and just 44% feel prepared [1]. Research on the impact of digital transformation (DT) on green innovation (GI) remains limited and yields mixed findings. Appio et al. [25] note that while firms invest in DT to close the digital gap, many are unprepared to fully leverage its innovation potential. Similarly, Han et al. [1] found DT can both support and hinder GI, with high infrastructure costs potentially draining resources needed for innovation. Appio et al. [26] also highlight challenges such as limited resources and misalignment between digital tools and strategic goals. Conversely, other studies report positive effects: Fernandez et al. [27] found DT improves GI by increasing efficiency and reducing waste, while Lu et al. [28] and Appio et al. [26] show DT promotes sustainability and data-driven innovation. Wang et al. [29] argue that DT enhances innovation capacity, supports network integration, and attracts investment. Therefore, this study's first objective is to explore how DT supports ambidextrous GI through the lens of OIPT.

The information-processing view of organizational learning suggests that learning occurs when information processing changes an entity's potential behaviors [30]. Cyert and March [31] described it as organizations learning through interaction with their environments. Learning theories explain how organizations adapt their knowledge or behavior based on experience. Strategically, organizational learning supports renewal by balancing continuity and change through a strong learning orientation [32]. Some scholars define learning orientation as a market knowledge process involving acquisition, dissemination, shared interpretation, and responsive action [17, 20, 33]. Organizational learning orientation has been shown to influence ambidextrous innovation [33]. However, the specific impact of organizational green learning orientation (OGLO) on exploitative and exploratory GI remains underexplored. While prior studies have examined learning orientation broadly, limited research has focused on OGLO in the context of GI [7]. Therefore, the second objective of this study is to investigate how digital transformation (DT) enhances OGLO.

In the digital era, building digital capabilities is crucial for evaluating ideas and delivering sustainable business models [4]. DT supports sustainability through "Green by IT" (efficiency gains) and "Greening IT" (sustainable IT practices) [34]. Integrating DT with sustainability fosters environmentally focused digital entrepreneurship [35]. Wang et al. [29] highlight that digitization has transformed how firms acquire, store, and use knowledge in their innovation processes. However, prior research suggests that employee resistance—reluctance to adapt to change—can hinder capability development and increase uncertainty [36]. To overcome this, organizations must build new capabilities through learning, experimentation, and knowledge exchange. Effective learning enables faster knowledge acquisition and assimilation, improving information processing and operational efficiency [37]. Wang et al. [29] highlight that digital technologies help firms gather insights into hidden customer needs and behaviors, enabling more timely idea generation. OGLO reflects shared values that drive firms to acquire green knowledge [38], supporting a firm's commitment to ambidextrous green innovation. Wang et al. [7] found green learning orientation positively influences both exploitative and exploratory green innovation. Therefore, the third aim of this study is to explore how DT, through OGLO, affects ambidextrous GI, leading to the following research question:

1. How does DT contribute to ambidextrous GI? 2. How does DT contribute to enhancing OGLO? 3. How does DT, through OGLO, influence ambidextrous GI?

This study examines these relationships within Taiwan's manufacturing industries. Taiwan's manufacturing industries, especially in the chemical, computer, and electronics sectors, are key players in the global production value chain. However, rapid industrialization has led to significant environmental challenges. To address these, the government established a waste recycling fund in 1997 and reinforced its Six Core Strategic Industries policy from 2018 to 2020, emphasizing sustainability, circular economy, energy efficiency, renewable energy, and waste reduction. In March 2022, Taiwan released "Taiwan's Pathway to Net-Zero Emissions in 2050," targeting a 55% emissions reduction by 2030 and net-zero by 2050 [39]. The Executive Yuan developed a scheme for reducing greenhouse gases following the "Sustainable Energy Policy Guidelines," focusing on energy security, economic development, and environmental protection [2]. To support SMEs, the government integrated projects between the Ministry of Economic Affairs (MOEA) and the Ministry of Digital Affairs (MODA) to promote DT and carbon reduction [40]. These efforts placed Taiwan 8th in the 2021 World Digital Competitiveness Ranking, showcasing its innovation and digital capacity [41]. Taiwan's initiatives in sustainability and alignment with international trends make it an ideal case study for this research. The rest of this article is organized as follows. First, the theory and hypotheses are put forward. Subsequently, the research design is described. Then, the hypotheses are tested by conducting hierarchical regression analysis and bootstrapping. Finally, the research findings, implications, and limitations are discussed.

2 Theoretical Underpinning and Hypotheses

2.1 Organizational Information Processing Perspectives

The understanding of how organizations process market information can be advanced substantially on the basis of principles derived from information-processing theories and organizational learning. Research on knowledge flows derives to some degree from earlier research on information processing and organization design. According to information-processing theories [19], information flows are an organizational response to task uncertainty. Task uncertainty is defined as the difference between the amount of information needed to perform a task and the amount of information already present in an organization [19]. The basic model is that organizations face uncertainty, that resolving uncertainty requires processing information, and that organizations can cope with these information-processing requirements through a range of design strategies. To cope with the high level of uncertainties arising from executing complex tasks, organizations must systematically structure and efficiently utilize information [42]. OIPT was developed to explain how the management of information helps organizations address various forms of uncertainty [19].

In contrast to information-processing theories, organizational learning theories provide rich perspectives on the processes that generate and change organizational knowledge. Organizational learning was addressed by Cyert and March [31] as a process by which organizations as collectives learn through interaction with their environments. Learning theories describe how organizations change their knowledge or their behavior in response to experiences [30]. This information processing perspective assumes that "an entity learns if, through its processing of information, the range of its potential behaviors is changed" [30, p. 89]. Huber [30] builds one of the most foundational models of organizational learning by breaking the process into four interrelated subsystems:

acquisition, distribution, interpretation, and memory. His model emphasizes that learning is not merely the act of absorbing new information, but a systemic organizational capability that must be designed, supported, and evolved over time. Learning occurs not when behavior changes immediately, but when the range of possible behaviors expands through enhanced cognition, understanding, and memory. Sinkula [20] argued that market information processing, then, is a function of what the organization has learned in terms of both facts about its relevant markets and its particular way of acquiring, distributing, interpreting, and storing information.

It is surprising that two closely related research streams—information processing and organizational learning—have remained largely unintegrated. Organizations, as information-processing entities, must manage uncertainty and complexity in their environments [19]. To handle complex, interdependent tasks, they need strong information processing capabilities [22]. Schulz [21] emphasized that organizational success relies not on choosing between these paradigms but on integrating them effectively. From the OIPT perspective [19, 42], reducing information lead times and improving information accuracy require robust infrastructure and advanced digital technologies. These tools enable rapid data acquisition and analysis, enhancing information processing capacity and reducing uncertainty in volatile environments [24, 28]. The information-processing view of organizational learning builds on well-established concepts like information acquisition, distribution, and interpretation [20, 23]. Market learning is thus seen as part of a firm's information-processing activities within the new product development process [43]. Following this logic, organizations that innovate based on thorough intelligence gathering and decision-making are more likely to achieve superior performance ,44]. Building on the above discussion, this study connects OIPT [19 [42] and the information-processing perspective of organizational learning [20, 30, 43] to investigate the relationship between digital transformation, OGLO, and ambidextrous green innovation.

2.2 Digital Transformation (DT) and Adoption of Digital Technologies (ADT)

Digital transformation (DT) has become a strategic priority for organizations, involving the integration and application of digital technologies to reshape business models and enhance value-creation strategies. It is increasingly recognized as essential for firms striving to attain and sustain competitive advantages in today's volatile, uncertain, complex, and ambiguous (VUCA) environments [1]. Despite still being in its developmental phase [45], DT has been conceptualized in diverse ways by scholars [28]. One comprehensive definition describes DT as a fundamental change process, driven by the innovative deployment of digital technologies and the strategic utilization of organizational resources and capabilities, with the aim of radically enhancing the organization and redefining its value proposition to stakeholders [46]. As such, DT has increasingly become embedded within modern business models and strategic practices [47]. DT involves digitizing processes to reshape core operations and value creation through emerging technologies [46]. Although no single definition is universally accepted [1], scholars agree on its substantial impact on business processes, capabilities, and market reach [28]. Building on prior work, this study defines DT as a fundamental change process driven by digital technologies to improve operations, customer relationships, and performance [1, 46].

OIPT explains how effective information management helps firms navigate uncertainty by enhancing their information processing capacity [22]. Digital transformation (DT) has become a priority, reshaping business models and value creation through technologies like AI, big data, cloud computing, mobile platforms, and social media [24]. These technologies, integrating information, computing, communication, and connectivity functions, are now embedded in many business processes [35]. While they enhance resource utilization and information systems, they also disrupt existing business models and structures [48]. From the OIPT viewpoint, adopting digital technologies strengthens organizational infrastructure, helping firms reduce information lead times and improve information currency [24].

2.3 Organizational Green Learning Orientation (OGLO)

Organizational learning (OL)—the creation and use of knowledge—is essential for refining insights and adapting to changing environments [49]. When learning is embedded across members, a learning orientation emerges, enabling organizational learning [50]. From a strategic perspective, organizational learning is seen as a key driver of strategic renewal, helping organizations balance continuity and change through a strong learning orientation [32]. Learning orientation refers to an organization-wide commitment to creating and applying knowledge to strengthen competitive advantage [33]. This involves acquiring and sharing information on customer needs, market dynamics, and competitor actions, as well as developing new technologies to produce superior products [15, 16]. Some scholars further define organizational learning orientation as a market knowledge process encompassing knowledge acquisition, dissemination, shared interpretation, and responsive action [17, 20, 33].

Organizational Green Learning (OGL), which involves preserving critical environmental functions for future generations, has become essential for sustaining global competitiveness [51]. This study examines OGL as a strategic approach to internalizing environmental sustainability through organizational change [52]. However, many organizations face challenges in adopting environmental strategies due to entrenched profit-oriented business models [52]. Organizational Green Learning Orientation (OGLO) reflects a firm's commitment to valuing the creation and application of green knowledge [7, 14]. Guided by green values, firms consciously engage in green learning to promote environmentally driven innovation. OGLO represents a firm's learning orientation that drives the sourcing, reuse, and generation of green knowledge [53]. Therefore, OGLO plays a critical role in enabling firms to identify, acquire, and generate green knowledge, which in turn supports both exploitative and exploratory innovations [54]. Building on the work of Sheng and Chien [33], this study defines OGLO as a market-oriented green knowledge process comprising four key dimensions: acquisition, dissemination, shared interpretation, and responsive action.

2.4 Ambidextrous Green Innovation

The aim of green innovation is to realize a win-win solution for reducing the conflicts between economic development and environmental protection [55]. On the one hand, green innovation is conducive to firms' sustainable development, realizing environmental benefits, such as saving energy, reducing carbon emissions, saving water, and facilitating product recovery. On the other hand, green innovation is beneficial to improve firms' green images, obtain customer recognition, and exploit international market [56], and then economic benefits, such as increasing green production and operation capability, can be achieved. Therefore, green innovation becomes an important way for firms to respond to the call for environmental protection and obtain sustainable competitive advantages. The core focus of GI includes technological advancements, pollution reduction, recycling, eco-product design, and ecological management [6]. Broadly, GIs involve the development, application, or introduction of new ideas, behaviors, products, and processes that reduce environmental burdens or meet sustainability targets [3]. Most definitions emphasize products, processes, and management or technical practices that minimize environmental impacts [11]. Although terms like green innovation, environmental innovation, and eco-innovation are often used interchangeably, this study uses "green innovation (GI)." GI is crucial for organizations to create value, gain a competitive edge, and enhance performance. It is essential for sustainable development (SD) to help firms improve environmental performance to meet protection standards. GI typically includes innovations in energy savings, pollution prevention, waste recycling, green product design, and environmental management technologies [5, 11, 58].

To meet environmental challenges, firms should adopt both exploitative and exploratory GI simultaneously, known as Ambidextrous GI [7, 28, 58]. Ambidextrous GI helps balance profitability with sustainability goals, resolving tensions in GI processes and enhancing corporate sustainability performance [59]. This study categorizes ambidextrous GI into exploratory and exploitative GI. Drawing on prior research [7, 12, 28], exploratory GI involves experimenting with new green alternatives and knowledge, as well as developing entirely new green skills and practices to generate future business opportunities. In contrast, exploitative GI leverages existing green knowledge to enhance current products and markets, making them more environmentally friendly without radical transformation.

2.5 Research Model

To address the research gap, this study uses a multi-theoretical approach. It first classifies ambidextrous GI into exploratory and exploitative types. Based on prior research [7, 12, 28], exploratory GI involves experimenting with new green alternatives and skills to create future opportunities, while exploitative GI applies existing green knowledge to enhance current products and markets with minimal change. Furthermore, this study applies OIPT [19, 42] and the information-processing perspective of organizational learning [20, 30, 43] to investigate the relationship among adoption of digital technology, OGLO, and ambidextrous GI. The proposed research framework is presented in Fig. 1.

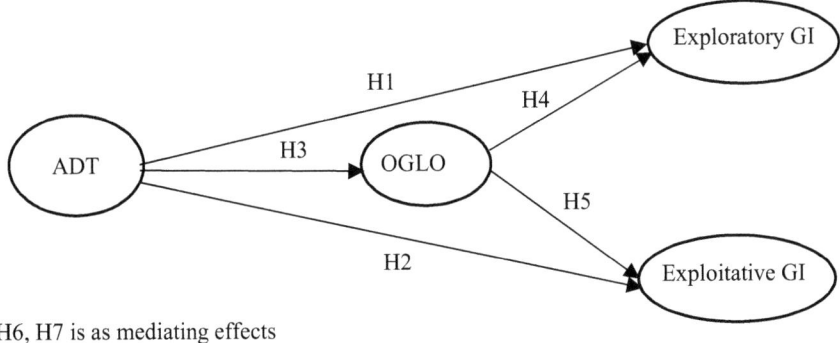

H6, H7 is as mediating effects

Note: ADT: Adoption of digital technologies; OGLO: Organizational green learning orientation; GI: Green innovation.

Fig. 1. Conceptual framework

2.6 Hypotheses Development

Based on the discussion above, this study proposes seven hypotheses and then examines these relationships within Taiwan's manufacturing industries.

H1. ADT has a positive effect on exploratory GI.
H2. ADT has a positive effect on exploitative GI.
H3. ADT has a positive effect on OGLO.
H4. OGLO has a positive effect on exploratory GI.
H5. OGLO has a positive effect on exploitative GI.
H6. An indirect positive relationship exists between the ADT and exploratory GI through OGLO.
H7. An indirect positive relationship exists between the ADT and exploitative GI through OGLO.

3 Methods

3.1 Measures

The development of the scales followed the procedures of Churchill [60]. This included defining the domains of the constructs and conducting a literature search to identify relevant scales. Existing scales were either adopted or modified, and new measures were developed as needed. A five-point Likert scale was utilized to measure all constructs except for demographic statistics. On this scale, a rating of 1 denoted "strongly disagree," while a rating of 5 indicated "strongly agree".

3.2 Questionnaire Development and Analysis of Common Method Biases

3.2.1 Questionnaire Development

To test our research hypotheses, we collected data from digitally transformed organizations using a survey method, which has been endorsed by several scholars for its ability to establish causal relationships between constructs and facilitate generalizable findings. The questionnaire development involved several steps. Survey items were sourced from literature and adapted to fit the research focus. Using Van de Vijver and Leung's [61] translation/back-translation technique, consistency between the English and Chinese versions was ensured. A panel of four DT experts and five sustainability researchers reviewed the content validity, leading to revisions. Feedback from four industry practitioners further improved clarity. The questionnaire was then piloted with 50 part-time MBA students in manufacturing, whose feedback enhanced readability and overall clarity.

The Foundation of Taiwan Industry Service, a non-profit focused on environmental protection and safety, manages a database of 1,320 ISO 14001-certified manufacturing firms in Taiwan (as of June 2024). For this study, we prioritized firms actively undergoing DT, selecting 800 firms (60.6% of the total) randomly. The survey targeted operational managers in areas such as production and manufacturing, technology, marketing and sales, and purchasing. After piloting the survey, questionnaires were distributed with a cover letter explaining the study's objectives. The response deadline was July 1, 2024, with two follow-up emails sent to boost participation. Between July and August 2024, we received 320 responses, with 10 incomplete ones discarded, resulting in 310 valid questionnaires and a response rate of 38.8%. This exceeds response rates reported in prior research.

3.2.2 Nonresponse Bias and Common Method Bias

To assess nonresponse bias, we used the extrapolation method [62] and t-tests [2] to compare study and demographic variables across survey waves. No significant differences ($p > 0.05$) were found between respondents from the first and second waves. Following Lambert and Harrington's [63] approach, we sent a condensed questionnaire with the top 30 items to 160 non-respondents. Of 140 surveys distributed, 45 were completed. Analysis showed no significant differences in early and late responses ($p > .05$), indicating no nonresponse bias. Harman's one-factor test [64] confirmed that common method variance was not a significant concern.

4 Results

We examined the interrelationships among four constructs—ADT, OGLO, exploratory, and exploitative GI—through seven hypotheses using Structural Equation Modeling (SEM). Following Fornell and Larcker[65] and Bagozzi and Yi [66], SEM was used to test measurement, predictive, and causal hypotheses, leveraging Maximum Likelihood Estimation (MLE) to account for measurement errors [67]. IBM SPSS v26 was used for model evaluation, and Confirmatory Factor Analysis (CFA) verified construct reliability and discriminant validity before analyzing inter-construct.

4.1 Reliability and Discriminant Validity

4.1.1 Reliability

Using Anderson and Gerbing's [68] two-step approach, we assessed convergent and discriminant validity to confirm construct accuracy and distinctiveness. Reliability was evaluated via CFA, composite reliability (CR), and average variance extracted (AVE). High standardized factor loadings ($p < 0.001$) confirmed convergent validity [67]. Convergent and discriminant validity were evaluated using Cronbach's alpha, composite reliability (CR), and average variance extracted (AVE) [67]. Cronbach's alpha values exceeded 0.7, CR surpassed 0.7, and AVE was above 0.5, indicating high reliability, internal consistency, and strong convergent validity for the constructs [69].

4.1.2 Discriminant Validity

To assess discriminant validity, we followed Fornell and Larcker's [100] method, comparing the square root of each construct's AVE with its correlations. A higher AVE indicates reliable measurement of the construct. In Table 1, diagonal elements (square roots of AVEs) exceed off-diagonal correlations, confirming the constructs' distinctiveness and supporting the measurement model's discriminant validity [65, 67]. Table 1 presents the means, standard deviations, and correlations among the study variables. ADT constructs show mean values above 4.32, indicating satisfactory levels in the sampled Taiwanese industries. Standard deviations highlight variability among firms, and positive, significant correlations between constructs support the research model, offering insights for firms aiming to enhance GI by targeting key constructs.

Table 1. Descriptive statistics, correlations and discriminant validity

	1	2	3	4
1. ADT	0.81			
2. OGLO	.61**	0.77		
3. ERGI	.53**	.56**	0.82	
4. EIGI	.47**	.49**	.76**	0.81
Mean	4.32	4.28	4.13	4.17
SD	.55	.57	.78	.85

Notes: ADT: Adoption of Digital Technology; OGLO: Organizational Green Learning Orientation; ERGI: Exploratory Green Innovation; EIGI: Exploitative Green Innovation. $n = 310$, * $p<0.05$; ** $p<0.01$; Values on the diagonal are the square root of average variance extracted.

4.2 Hypothesis Testing

4.2.1 Testing for Direct Effect

In the second analysis phase, we assessed the structural equation model using fit indices recommended by Bagozzi and Yi [66], including χ^2/df, GFI, AGFI, NFI, SRMR,

RMSEA, and CFI. The model showed acceptable fit: $\chi^2/df = 2.58$, GFI = 0.93, AGFI = 0.91, NFI = 0.91, CFI = 0.92, SRMR = 0.045, and RMSEA = 0.08, meeting criteria from Schermelleh-Engel et al. [70].

For hypothesis testing, covariance-based SEM was employed. The outcomes of the path analysis and hypothesis verification are detailed in Table 2. Hypothesis 1 (H1) stated that ADT positively influences ERGI, with the analysis supporting this through a significant path coefficient ($\beta = 0.35$, $p < 0.001$), thereby confirming H1. Hypothesis 2 (H2) argued that ADT enhances EIGI, which was supported by a significant path coefficient ($\beta = 0.31$, $p < 0.001$), affirming H2. Hypothesis 3 (H3) suggested a positive effect of ADT on OGLO, which was verified with a path coefficient ($\beta = 0.63$, $p < 0.000$), supporting H3. Furthermore, Hypothesis 4 (H4) posited that OGLO positively impacts ERGI; this was corroborated by a significant path coefficient ($\beta = 0.74$, $p < 0.000$), validating H4. Lastly, Hypothesis 5 (H5) proposed that OGLO positively influences EIGI, and this was substantiated by a significant path coefficient ($\beta = 0.71$, $p < 0.000$), confirming H5.

Table 2. Path coefficient and hypothesis verification results

Hypotheses	Path coefficient	p-value	Result
H1. ADT → ERGI	0.35	0.001	Supported
H2. ADT → EIGI	0.31	0.001	Supported
H3. ADT → OGLO	0.63	0.000	Supported
H4. OGLO → ERGI	0.74	0.000	Supported
H5. OGLO → EIGI	0.71	0.000	Supported

Note: 1. ADT: Adoption of Digital Technology; OGLO: Organizational Green Learning Orientation; ERGI: Exploratory Green Innovation; EIGI: Exploitative Green Innovation
2. n=310; The coefficients in the table are the standardized coefficients.

4.2.2 Testing for Mediating Effects

We used the PROCESS procedure in SPSS Version 3.5 to examine the mediating effects of ambidextrous learning. Following Hayes' [71] mediation methodology, we analyzed the roles of exploratory and exploitative EOL between EDTC and EI using bootstrapping to estimate bias-corrected 95% confidence intervals (CIs). The results from the bootstrap analysis are detailed in Table 3. The analysis revealed a significant indirect effect of ADT on ERGI through OGLO (indirect effect = 0.47, 95% CI: 0.34–0.61), supporting Hypothesis 6 (H6). Additionally, the analysis showed a significant indirect effect of ADT on EIGI via OGLO (indirect effect = 0.45, 95% CI: 0.32–0.60), validating Hypothesis 7 (H7).

Table 3. Hypothesis testing: Bootstrap result (n = 2000)

Hyp.	Paths	Bootstrapping		95% CI	
		Estimate	SE	Lower	Upper
H6	ADT → OGLO → ERGI	.47	.06	.34	.61
H7	ADT → OGLO → EIGI	.45	.07	.32	.60

ADT: Adoption of Digital Technology; OGLO: Organizational Green Learning Orientation; ERGI: Exploratory Green Innovation; EIGI: Exploitative Green Innovation.

5 Conclusion

5.1 Conclusions

Climate change and environmental degradation have prompted countries and firms to adopt GI. This study analyzes how ADT enhance OGLO and both exploratory and exploitative GI, using data from 310 Taiwanese manufacturers. Results confirm that ADT positively impacts both GI types, supporting Hypotheses 1 and 2, and align with prior studies [1, 4, 27, 28]. Secondly, the findings confirm that ADT significantly enhances OGLO, supporting Hypothesis 3 and aligning with previous studies. Nandi et al. [35] emphasized that digital technologies enable timely information, transparency, and visibility for effective information sharing. Similarly, Zhang et al. [72] noted that firms with strong learning capabilities can better integrate environmental knowledge through digital tools. Thirdly, the study finds that OGLO positively influences both exploratory and exploitative GI, supporting Hypotheses 4 and 5. This aligns with prior research emphasizing OGLO's role in enhancing both types of GI [9]. Fourthly, OGLO mediates the link between ADT and both exploratory and exploitative GI, supporting Hypotheses 6 and 7. This aligns with research showing that digital capabilities are key to sustainable innovation [4] and have reshaped how firms acquire, store, and use knowledge in their innovation processes [29].

5.2 Theoretical Implications

First, GI focuses on technological innovation, pollution reduction, recycling, eco-design, and ecological management [6]. Broadly, it involves new ideas, products, or processes that reduce environmental impact or support sustainability [3]. This study contribution is to adopt an ambidextrous view, classifying GI into exploratory and exploitative types, an area previously underexplored. Second, organizations must manage environmental uncertainty and complexity by strengthening their information processing capabilities [19, 22]. Despite their relevance, information processing and organizational learning have rarely been connected. This study applies OIPT, suggesting that effective information management enhances firms' ability to handle uncertainty. As DT reshapes business models through technologies like AI, big data, and cloud platforms [24], this study contribution is to explore how DT supports ambidextrous GI from the lens of OIPT.

The information-processing view of organizational learning sees learning as changes in behavior potential through information use [30]. While learning orientation has been studied broadly, limited research has addressed OGLO in the GI context [7]. This study contributes by examining how digital transformation (DT) enhances OGLO using OIPT, and how OGLO, in turn, drives exploratory and exploitative GI through the len of information-processing view of organizational learning. Fourth, this study contributes to academic research by integrating OIPT with the information-processing perspective of organizational learning, thereby developing a DT-OGLO framework that supports ambidextrous green innovation.

5.3 Managerial Implications

First, the findings show that ADT significantly boosts ambidextrous GI, with DT manufacturing improving energy and material efficiency. However, many firms still struggle to fully leverage DT for innovation [25]. Taiwanese companies are encouraged to adopt EDTC to strengthen environmental innovation and social responsibility. Second, our research shows that ADT significantly boosts OGLO, offering practical guidance for managers to enhance organizational learning. As an export-driven economy, Taiwanese firms should track global trends like the European Green Deal [2]. Advances in DT have made digital tools vital for learning and achieving environmental goals. Third, our findings show that OGLO significantly boosts both exploratory and exploitative GI. Firms with strong learning capabilities are better positioned in dynamic markets. Managers should prioritize green learning to enhance resource access and GI expertise. Fourth, our research shows that OGLO mediates the link between ADT and both types of GI, highlighting DT's role in promoting green practices. However, high digital infrastructure costs may hinder innovation [1] GI depends on OGLO and a collaborative environment for sharing environmental knowledge. Firms with strong OGLO achieve better GI, so we recommend investing in OGLO, supported by ADT, to improve GI outcomes.

5.4 Limitations and Scope for Future Research

This study has limitations, including its reliance on subjective measures and cross-sectional design. Future research should use objective metrics and longitudinal methods to improve reliability. Qualitative approaches like interviews and case studies can further explore the dynamics between ADT, OGLO, and GI. As the study focuses on Taiwan, broader geographic research could reveal diverse views on sustainability transitions and natural capital constraints. Future studies could also examine open innovation, knowledge management dimensions [72], and explore both the antecedents and outcomes of digital transformation.

Funding Information. The authors would like to thank the National Science and Technology Council, Taiwan, R.O.C. for providing financial support for this research via grant number MOST 112-2410-H-992-030 -SS3.

6 References

1. Han, Y., et al.: Unraveling the impact of digital transformation on green innovation through microdata and machine learning. J. Environ. Manage. **354**, 120271 (2024). https://doi.org/10.1016/j.jenvman.2024.120271
2. Huang, C.H., Huang, Y.C.: Exploring the linkages among green digital transformation capability, ambidextrous green learning and sustainability performance: a case study of manufacturing firms in Taiwan. J. Manuf. Technol. Manag. **35**(5), 1103–1123 (2024)
3. Arfi, B.W., Hikkerova, L., Sahut, J.M.: External knowledge sources, green innovation and performance. Technol. Forecast. Soc. Chang. **129**, 210–220 (2018)
4. Xu, J., Yu, Y., Min Zhang, M., Zhang, J.Z.: Impacts of digital transformation on eco-innovation and sustainable performance: evidence from Chinese manufacturing companies. J. Clean. Prod. **393**, 136278 (2023)
5. Huang, Y.C., Huang, C.H.: Exploring institutional pressure, the top management team's response, green innovation adoption, and firm performance: evidence from Taiwan's electrical and electronics industry. Eur. J. Innov. Manag. **27**, 800–824 (2024)
6. Shahzad, M., Qu, Y., Zafar, A.U., Rehman, S.U., Islam, T.: Exploring the influence of knowledge management process on corporate sustainable performance through green innovation. J. Knowl. Manag. **24**(9), 2079–2106 (2020). https://doi.org/10.1108/JKM-11-2019-0624
7. Wang, J., Xue, Y., Sun, X., Yang, J.: Green learning orientation, green knowledge acquisition and ambidextrous green innovation. J. Clean. Prod. **250**, 119475 (2020)
8. Saunila, M., Ukko, J., Rantala, T.: Sustainability as a driver of green innovation investment and exploitation. J. Clean. Prod. **179**, 631–641 (2018)
9. Lee, S.U., Park, G., Kang, J.: The double-edged effects of the corporate venture capital unit's structural autonomy on corporate investors' explorative and exploitative innovation. J. Bus. Res. **88**, 141–149 (2018)
10. Lin, L.H., Ho, Y.L.: Institutional pressures and environmental performance in the global automotive industry: the mediating role of organizational ambidexterity. Long Range Plan. **49**(6), 764–775 (2016)
11. Zhang, F., Zhu, L.: Enhancing corporate sustainable development: Stakeholder pressures, organizational learning, and green innovation. Bus. Strateg. Environ. **28**(6), 1012–1026 (2019). https://doi.org/10.1002/bse.2298
12. Huang, Y.C., Chen, C.T.: Exploring institutional pressures, firm green slack, green product innovation and green new product success: evidence from Taiwan's high-tech industries. Technol. Forecast. Soc. Chang. **174**, 121196 (2022). https://doi.org/10.1016/j.techfore.2021.121196
13. Rui, Z., Lu, Y.: Stakeholder pressure, corporate environmental ethics and green innovation. Asian J. Technol. Innov. **29**(1), 70–86 (2021)
14. Baker, W.E., Sinkula, J.M.: The synergistic effect of market orientation and learning orientation on organizational performance. J. Acad. Mark. Sci. **27**(4), 411–427 (1999)
15. Calantone, R.J., Cavusgil, S.T., Zhao, Y.: Learning orientation, firm innovation capability, and firm performance. Ind. Mark. Manage. **31**, 515–524 (2002)
16. Hurley, R., Hult, T.M.: Innovation, market orientation, and organizational learning: an integration and empirical examination. J. Mark. **62**(3), 42–54 (1998)
17. Slater, S.F., Narver, J.C.: Market orientation and the learning organization. J. Mark. **59**, 63–74 (1995)
18. Brockman, B.K., Morgan, R.M.: The role of existing knowledge in new product innovativeness and performance. Decis. Sci. **34**(2), 385–419 (2003)
19. Galbraith, J.R.: Organization design: an information processing view. Interfaces **4**(3), 28–36 (1974)

20. Sinkula, J.M.: Market information processing and organizational learning. J. Mark. **58**, 35–45 (1994)
21. Schulz, M.: The uncertain relevance of newness: organizational learning and knowledges flows. Acad. Manag. J. **44**(4), 661–681 (2001)
22. Yu, W., Zhao, G., Liu, Q., Song, Y.: Role of big data analytics capability in developing integrated hospital supply chains and operational flexibility: An organizational information processing theory perspective. Technol. Forecast. Soc. Change **163**, 120417 (2021). https://doi.org/10.1016/j.techfore.2020.120417
23. Deshpande, R., Zaltman, G.: Factors affecting the use of market research information: a path analysis. J. Mark. Res. **19**(1), 14–31 (1982)
24. Xie, X., Wu, Y., Palacios-Marques, D., Ribeiro-Navarrete, S.: Business networks and organizational resilience capacity in the digital age during COVID-19: a perspective utilizing organizational information processing theory. Technol. Forecast. Soc. Chang. **177**, 121548 (2022)
25. Appio, F.P., Frattini, F., Petruzzelli, A.M., Neirotti, P.: Digital transformation and innovation management: a synthesis of existing research and an agenda for future studies. J. Prod. Innov. Manag. **38**, 4–20 (2021)
26. Appio, F.P., Cacciatore, E., Cesaroni, F., Crupi, A., Marozzo, V.: Open innovation at the digital frontier: unraveling the paradoxes and roadmaps for SMEs' successful digital transformation. Eur. J. Innov. Manag. **27**(9), 223–247 (2024)
27. Fernandez, S., Torrecillas, C., Labra, R.E.: Drivers of eco-innovation in developing countries: the case of Chilean firms. Technol. Forecast. Soc. Chang. **170**, 120902 (2021)
28. Lu, H.T., Li, X., Yuen, K.F.: Digital transformation as an enabler of sustainability innovation and performance – information processing and innovation ambidexterity perspectives. Technol. Forecast. Soc. Change **196**, 122860 (2023). https://doi.org/10.1016/j.techfore.2023.122860
29. Wang, K., Sun, K., Li, Y., Qian Xi, Q.: Can digital technology break the financing dilemma of innovative SMEs? Technol. Forecast. Soc. Chang. **214**, 124030 (2025)
30. Huber, G.P.: Organizational learning: the contributing process and literatures. Organ. Sci. **2**(1), 88–115 (1991)
31. Cyert, R., March, J.G.: A Behavioral Theory of the Firm. Prentice-Hall, Englewood Cliffs (1963)
32. Senge, P.: The Fifth Discipline: The Art and Practice of the Learning Organization. Doubleday/Currency, New York (1990)
33. Sheng, M.L., Chien, I.: Rethinking organizational learning orientation on radical and incremental innovation in high-tech firms. J. Bus. Res. **69**(6), 2302–2308 (2016)
34. Osburg, T., Lohrmann, C.: Sustainability in a digital world. New Opportunities Though New Technologies. Springer, Cham (2017)
35. Nadkarni, S., Prügl, R.: Digital transformation: a review, synthesis and opportunities for future research. Manage. Rev. Q. **71**(2), 233–341 (2021). https://doi.org/10.1007/s11301-020-00185-7
36. Tortorella, G.L., Cawley Vergara, A.M., Garza-Reyes, J.A., Sawhney, R.: Organizational learning paths based upon industry 4.0 adoption: an empirical study with Brazilian manufacturers. Int. J. Prod. Econ. **219**, 84–294 (2020)
37. Shin, H., Lee, J.-N., Kim, D., Rhim, H.: Strategic agility of Korean small and medium enterprises and its influence on operational and firm performance. Int. J. Prod. Econ. **168**, 181–196 (2015)
38. D'Angelo, A., Presutti, M.: SMEs international growth: the moderating role of experience on entrepreneurial and learning orientations. Int. Bus. Rev. **28**(3), 613–624 (2019)

39. National Development Council. Taiwan's Pathway to Net-Zero Emissions in 2050. National Development Council, Taiwan (2022). https://www.ndc.gov.tw/en/Content_List.aspx?n=B154724D802DC488
40. Small and Medium Enterprise Administration (SMEA). Green and Digital Transformation Joint Achievement Expo, Digital Transformation x Sustainability for a Better Taiwan (2022). https://www.moea.gov.tw/MNS/english/news/News.aspx?kind=6&menu_id=176&news_id=103745. Accessed 28 Apr 2022
41. Ho, J.L.: 2021 Taiwan SME digital transformation survey. PwC Taiwan (2021). https://www.pwc.tw/en/publications/2021-taiwan-sme-digital-transformation-survey.html. Accessed 16 Apr 2021
42. Srinivasan, R., Swink, M.: An investigation of visibility and flexibility as complements to supply chain analytics: an organizational information processing theory perspective. Prod. Oper. Manag. **27**, 1849–1867 (2018). https://doi.org/10.1111/poms.12746
43. Pentina, I., Strutton, D.: Information processing and new product success: a meta-analysis. Eur. J. Innov. Manag. **10**(2), 149–175 (2007). https://doi.org/10.1108/14601060710745233
44. Zaltman, G., Duncan, R., Holbeck, J.: Innovations and Organization. Wiley, New York (1973)
45. Guandalini, I.: Sustainability through digital transformation: a systematic literature review for research guidance. J. Bus. Res. **148**, 456–471 (2022)
46. Gong, C., Ribiere, V.: Developing a unified definition of digital transformation. Technovation **102**, 102217 (2021)
47. Ranjan, P.: Unraveling the mystery of the link between digital orientation and innovation performance: the interplay of digital business capability and environmental dynamism. Technovation **131**, 102966 (2024)
48. Fang, M., Liu, F., Xiao, S., Park, K.: Hedging the bet on digital transformation in strategic supply chain management: a theoretical integration and an empirical test. Int. J. Phys. Distrib. Logist. Manag. **53**(4), 512–531 (2023). https://doi.org/10.1108/IJPDLM-12-2021-0545
49. Abbas, J., Zhang, Q., Hussain, I., Akram, S., Afaq, A., Shad, M.A.: Sustainable innovation in small medium enterprises: the impact of knowledge management on organizational innovation through a mediation analysis by using SEM approach. Sustainability **12**, 2407 (2020)
50. Argyris, C.: Prologue: Toward A Comprehensive Theory of Management. Sage Publications, London (1996)
51. Graham, S.: Antecedents to environmental supply chain strategies: the role of internal integration and environmental learning. Int. J. Prod. Econ. **197**, 283–296 (2018)
52. Bianchi, G., Testa, F., Boiral, O., Iraldo, F.: Organizational learning for environmental sustainability: internalizing lifecycle management. Organ. Environ. **35**(1), 103–129 (2022)
53. Khedhaouria, A., Jamal, A.: Sourcing knowledge for innovation: knowledge reuse and creation in project teams. J. Knowl. Manage. **19**(5), 932–948 (2015)
54. Bierly, P.E., Damanpour, F., Santoro, M.D.: The application of external knowledge: organizational conditions for exploration and exploitation. J. Manage. Stud. **46**(3), 481–509 (2009)
55. Leal-Rodríguez, A.L., Ariza-Montes, A.J., Morales-Fern_andez, E., Albort-Moranta, G.: Green innovation, indeed a cornerstone in linking market requests and business performance. Evidence from the Spanish automotive components industry. Technol. Forecast. Soc. Change. **129**, 185–193 (2018)
56. Albort-Morant, G., Leal-Rodríguez, A.L., De Marchi, V.: Absorptive capacity and relationship learning mechanisms as complementary drivers of green innovation performance. J. Knowl. Manag. **22**, 432–452 (2018). https://doi.org/10.1108/JKM-07-2017-0310
57. Opazo-Basaeza, M., Monroy-Osorio, J.C., Maric, J.: Evaluating the effect of green technological innovations on organizational and environmental performance: a treble innovation approach. Technovation **129**, 102885 (2024). https://doi.org/10.1016/j.technovation.2023.102885

58. Rehman, S.U., Kraus, S., Shah, S.A., Khanin, D., Mahto, R.V.: Analyzing the relationship between green innovation and environmental performance in large manufacturing firms. Technol. Forecast. Soc. Chang. **163**, 120481 (2021)
59. Gomes, P.J., Silva, G.M., Sarkis, J.: Exploring the relationship between quality ambidexterity and sustainable production. Int. J. Prod. Econ. **224**, 107560 (2020). https://doi.org/10.1016/j.ijpe.2019.107560
60. Churchill, G.A.: A paradigm for developing better measures of marketing constructs. J. Mark. Res. **16**(1), 64–73 (1979). https://doi.org/10.1177/002224377901600110
61. Van de Vijver, F., Leung, K.: Methods and data analysis of comparative research. Handb. Res. Cult. Cross-Cult. Psychol. **1**, 257–300 (1997)
62. Larson, P.D., Poist, R.F.: Improving response rates to mail surveys: a research note. Transp. J. **43**(4), 67–74 (2004)
63. Lambert, D.M., Harrington, T.C.: Measuring nonresponse bias in customer service mail surveys. J. Bus. Logist. **11**(2), 5–25 (1990)
64. Podsakoff, P.M., MacKenzie, S.B., Podsakoff, N.P.: Sources of method bias in social science research and recommendations on how to control it. Annu. Rev. Psychol. **63**(1), 539–569 (2012)
65. Fornell, C., Larcker, D.F.: Structural equation models with unobservable variables and measurement error: Algebra and statistics. J. Mark. Res. **18**(3), 382–388 (1981). https://doi.org/10.1177/002224378101800313
66. Bagozzi, R.P., Yi, Y.: On the evaluation of structural equation models. J. Acad. Mark. Sci. **16**(1), 74–94 (1988)
67. Hair Jr, J., Page, M., Brunsveld, N.: Essentials of Business Research Methods, Routledge (2019)
68. Anderson, J.C., Gerbing, D.W.: Structural equation modeling in practice: a review and recommended two-step approach. Psychol. Bull. **103**(3), 411–423 (1988)
69. Cronbach, L.J.: Coefficient alpha and the internal structure of tests. Psychometrika **16**(3), 297–334 (1951)
70. Schermelleh-Engel, K., Moosbrugger, H., Müller, H.: Evaluating the fit of structural equation models: tests of significance and descriptive goodness-of-fit measures. Methods Psychol. Res. Online **8**(2), 23–74 (2003)
71. Hayes, A.F.: Partial, conditional, and moderated moderated mediation: quantification, inference, and interpretation. Commun. Monogr. **85**, 4–40 (2018). https://doi.org/10.1080/03637751.2017.1352100
72. Zhang, X., Chu, Z., Ren, L., Xing, J.: Open innovation and sustainable competitive advantage: the role of organizational learning. Technol. Forecast. Soc. Chang. **186**, 122114 (2023)

Consumer Attitude Toward QR Code Payments with Smartphone in Japan: Satisfaction and Dissatisfaction

S. Nagayoshi(✉) and K. C. Abisikha

Shizuoka University, 3-5-1, Johoku, Chuo-ku, Hamamatsu, Shizuoka, Japan
nagayoshi@inf.shizuoka.ac.jp, k.c.abisikha.23@shizuoka.ac.jp

Abstract. The rapid adoption of QR code payments using smartphones in Japan has led to significant changes in the country's payment systems. This study aimed to clarify the mechanism behind the continued use of QR code payments services by focusing on consumer satisfaction and dissatisfaction. A hypothetical model based on the Expectation Confirmation Theory and Technology Acceptance Model was developed and tested using data collected through questionnaires from active users of QR code payments. Path analysis revealed that the fulfillment of expectations positively influenced perceived usefulness, satisfaction, and continuity intention while negatively affecting dissatisfaction. However, perceived usefulness did not significantly influence satisfaction or continuity intention, and, surprisingly, it positively affected dissatisfaction. The results suggest that users continue to use QR code payments not because they are satisfied with the service but because they are not dissatisfied with it, considering it as one of the various payment methods available in Japan. These findings provide valuable insights for QR code payment providers to develop their services. Further improvements in the model and questionnaire design are necessary to enhance the reliability of the collected data and the goodness of fit of the model.

Keywords: QR code payments · Continued use · Satisfaction · Dissatisfaction

1 Introduction

In Japan, QR code payments using smartphones have become popular since October 1, 2019, when the consumption tax increased from 8% to 10% [1]. In China, QR code payments using smartphones have been rapidly gaining popularity because credit cards were not widely used as a means of cashless payment before this and because they could be introduced at a relatively low cost [1].

According to the Payments Japan Association [2], the value of store spending and the number of store transactions in Japan expanded rapidly from 165 billion yen and 52 million cases in 2018 to 18.7 trillion yen and 1.15 billion cases in 2024, respectively. In addition, the number of active users expanded from 3.55 million users at the end of December 2018 to 87 million users by the end of December 2024, indicating a dramatic expansion in terms of the number of users.

There was an increase in the number of active users from 2018 to the end of December 2024; the number of active users peaked at 17.8 million at the end of December 2020 and has been on a downward trend since then. Considering that Japan's growing population is around 124 million, it is safe to assume that about 70% of the Japanese population is users and that QR code payments have penetrated the Japanese domestic market.

Rogers [3], who proposed a diffusion model of innovation, found that the number of adopters of new ideas, behaviors, objects, and innovations approximates a normal distribution with very few adopters in the early stages of market growth, followed by a gradually increasing growth rate, and finally slowing down again. Innovation adopters are divided into five stages, defined as Innovators, Early Adopters, Early Majority, Late Majority, and Laggards, from the initial stage, and the percentage of adopters in those markets is generally 2.5%, 13.5%, 34%, 34%, and 16%, respectively. Considering these five stages, QR code payment services using smartphones have already crossed the casm [4] between early adopters and the early majority, the next stage after innovators.

According to the Japan Payment Association [2], the use of QR code payments has consistently increased from 2018 to 2024 in terms of both the amount of money spent at stores and the number of cases of use at stores. As mentioned earlier, we believe that growth reached its peak in terms of the number of users, but we infer that further growth can be expected in terms of the amount spent and the number of cases of use in the future.

Considering an individual's experience with QR code payments, it is thought that there is diversity in the number of people who find QR code payment services convenient, useful, satisfied, and unsatisfied with the service. As the number of users, amount of money spent, and number of transactions have been increasing, it is assumed that, in general, once people start using QR code payment services in Japan, they may continue to use them.

In Japan, QR code payments services using smartphones were introduced as a new service in 2018, and both the amount of money spent at stores and the number of store transactions increased by 2024. Many people continue to use these services, which can be considered to have brought about changes in the payment mechanism in the Japanese society. The innovation of QR payment services using smartphones has spread [3].

Fujiki [5] states that the reduction in the cost of using cash due to the widespread use of cashless payments is overestimated and that the use of cashless payments is unlikely to reduce the frequency of cash payments.

Why do consumers accept QR code payments using smartphones?

Ajzen [6] proposed the theory of Planned Behavior/Reasoned Action was proposed by Ajzen [6] and suggested that behavior is determined by intentions, attitudes, and subjective norms [7]. Davis [8] proposed the Technology Acceptance Model (TAM). It is a model that predicts the acceptance and use of a new technology or system by its users, and is believed to explain their intended use primarily by how useful they perceive the technology to be and how easy it is to use.? TAM is continuously studied and is used as a model of how users perceive a technology to be Technology Acceptance Model 2 (TAM2) [9, 10] and the Unified Theory of Acceptance and Use of Technology [11], and has also been discussed in the context of electronic commerce [12, 13]. In addition, studies have been conducted on the continuous use of information systems have been conducted [8,

11, 14–17]. In addition to technology acceptance studies, studies on the continued use of information systems have also adopted Expectation Confirmation Theory [18], which incorporates the concepts of expectations for information system use and satisfaction with use.

The number of QR code payment users is due to the introduction and use of QR code payments in Japan triggered by the consumption tax hike in 2019. Many introduction and use promotion campaigns have been conducted by the Japanese government and others, which were designed to encourage consumers to use QR code payments with their smartphones as an incentive for consumers to use QR code payments using smartphones [19]. It can be inferred that the campaign led consumers to accept QR code payments by trying out QR code payments using smartphones and recognizing their usefulness and ease of use.

Recently, it has been said that an increasing number of businesses are no longer welcoming QR code payments from consumers because of the loss of payment fee exemptions [19]. In Japan, consumers continue to use QR code payments, as the amount of money spent in stores and the number of store transactions continue to increase, despite the disappearance of the campaign to promote the use of QR code payments. However, the reason for this continued consumer use is unclear.

Therefore, the purpose of this study is to clarify why QR code payment services using smartphones continue to be used in Japan; that is, the mechanism of continued use of QR code payment services.

2 Research Methodology

2.1 Research Progress

Abisikha and Nagayoshi [20], while employing previous studies [5–11] related to the continuous use of information systems, studied the relationship among "Confirmation of Expectation", "Perceived Usefulness", "Satisfaction", and "Continuity Intention." The model in the study [20] consists of "Confirmation of Expectation", "Perceived Usefulness", "Satisfaction", and "Continuity Intention". However, the goodness of fit of the model to the collected data was low and has not yet been validated. While Abisikha and Nagayoshi [20] confirmed that expectation and satisfaction are the influencing factors for continued use, they treated satisfaction and dissatisfaction together and treated the inverted data of the dissatisfaction questionnaire composed of a Likert scale as satisfaction.

2.2 Research Question

As Herzberg's [21] two-factor theory suggests, satisfaction and dissatisfaction should be analyzed separately; in this study, we corrected the fact that the authors' previous study [20] treated satisfaction and dissatisfaction together and made a clear distinction between the two.

The general research question in this study is: Why are QR code payments services that use smartphones being used continuously in Japan? In this study, we define the following individual Research Questions 1–4, focusing on satisfaction and dissatisfaction.

RQ1. How is satisfaction with QR code payments formed?
RQ2. How is dissatisfaction with QR code payments formed?
RQ3. How does satisfaction with QR code payments affect their continued use?
RQ4. How does dissatisfaction with QR code payments affect their continued use?

2.3 Research Method

This research is conducted as a hypothesis-verification study, in which hypotheses are constructed and tested with the aid of previous studies. To test the hypotheses, the quantitative data collected through questionnaires is statistically analyzed.

2.4 Hypothesis

Based on studies on the continuous use of information systems [8, 11, 14–17], the fulfillment of expectations about information systems positively influences the perception of the usefulness of information systems, satisfaction, and intention to continue using information systems. It is also expected to have a negative impact on dissatisfaction. Based on this, we derive the following H1-H4 for QR code payments.

H1: Confirmation of Expectation (C) has a positive influence on Perceived Usefulness (PU).
H2: Confirmation of Expectation (C) has a positive influence Satisfaction (S).
H3: Confirmation of Expectation (C) has a negative influence on Dissatisfaction (DS).
H4: Confirmation of Expectation (C) has a positive influence on Continuity Intention (CI).

Similarly, based on information system continued use studies [8, 11, 14–17], the perceived usefulness of an information system has a positive impact on satisfaction and the continued use of information systems, and a negative impact on dissatisfaction. Based on these findings, we derive the following H5-H7 for QR code payments.

H5: Perceived Usefulness (PU) has a positive influence on Satisfaction (S).
H6: Perceived Usefulness (PU) has a negative influence on Dissatisfaction (DS).
H7: Perceived Usefulness (PU) has a positive influence on Continuity Intention (CI).

Satisfaction is also considered to have a positive impact on the continued use of information systems [8, 11, 14–17], which is also true for QR code payments. Logically, dissatisfaction has a negative impact on continued use of QR code payments. Based on these considerations, we derive the following H8–H9.

H8: Satisfaction(S) has a positive influence on Continuity Intention (CI).
H9: Dissatisfaction (DS) has a negative influence on Continuity Intention (CI).

These H1–H9 hypotheses are illustrated in Fig. 1.

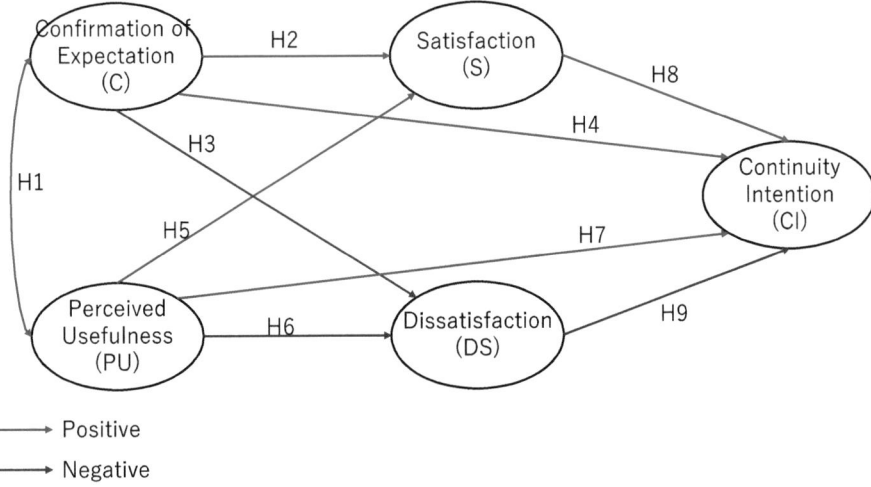

Fig. 1. Hypothetical Model

3 Data Collection and Analysis

3.1 Questionnaire Survey

Data Collection. The survey was conducted by iBridge Corporation, a Japanese survey implementation company. The company has 13 million members and has the largest number of monitors in Japan. The questionnaire was created using a survey platform called Freeasy owned by the company, and the survey was sent to the company's monitor members through the platform.

The questionnaire questions were partially adapted by the authors to be consistent with the purpose of this study, while taking great care not to alter the meaning of the questions using previous research as a guide.

The survey was conducted in two stages: the first survey asked whether the questionnaire participants used QR code payments using smartphones and was conducted on January 28, 2025; the second survey was conducted on January 29, 2025.

In the first survey, responses were received from 2,000 members of the company monitors. As a result, 1,260 respondents indicated that they used QR code payments using smartphones, while 740 respondents did not use QR code payments using smartphones.

The second survey asked the 1,260 respondents who were using QR code payments about their intention to continue using QR code payments. A total of 501 responses were received from the survey company. The survey company provided an AI-based fraudulent response elimination service as an option. According to the company, this service would collect 120% of the target sample size and eliminate short- and straight-lining responses as fraudulent. However, it should be noted that if the number of fraudulent responses exceeds 20%, it will not be possible to eliminate all fraudulent responses.

Data Cleaning. As mentioned, for the second questionnaire survey targeting active users of QR code payments using smartphones, 1260 people were asked to respond, and data from 501 responses were obtained. We then visually checked the obtained survey data and found that it contained many straight-lining responses, so we decided to exclude from the data for analysis those responses in which more than 80% of the respondents selected the same options, considering them to be highly suspicious of fraudulent responses. Finally, the survey response data of 461 respondents were used for analysis.

A summary of these procedures is presented in Table 1.

Data Reliability. Cronbach's alpha is a way of assessing reliability by comparing the amount of shared variance or covariance among the items making up an instrument to the overall variance [22]. Score of 0.9 or higher is considered excellent, that of 0.8 to 0.9 is adequate, that of 0.7 to 0.8 is marginal, that 0.6 to 0.7 is seriously suspect and that of less than 0.6 is totally unacceptable. For the analysis of Cronbach's alpha, we used the Bell Curve for Excel Version 4.08, provided by Social Survey Research Information Co.

Table 1 presents the results of the reliability test.

Table 1. Result of Data Reliability Test

		Active User	
Screening Survey		January 28, 2025	
Questionnaire Survey		January 29, 2025	
Statistical Population		1,260	
Originally Collected Data Sample		501	
AI Cleaning		YES	
Manual Cleaning		YES	
Sample to be analyzed		461	
Latent Variable		Cronbach Alfa	Evaluation
C	Confirmation of Expectation	.830	Adequate
PU	Perceived Usefulness	.894	Adequate
S	Satisfaction	.865	Adequate
DS	Dissatisfaction	.867	Adequate
CI	Continuity Intention	.693	Seriously Suspect

3.2 Path Analysis

Next, we tested the hypothetical model by analyzing quantitative data collected through a questionnaire using a covariance structure analysis. For the statistical analysis, we used IBM SPSS AMOS30 software for statistical analysis. The results of the analysis of survey data collected from active users are shown in Fig. 2. Table 2 lists the results of the analysis and contrasts them with those of the hypotheses.

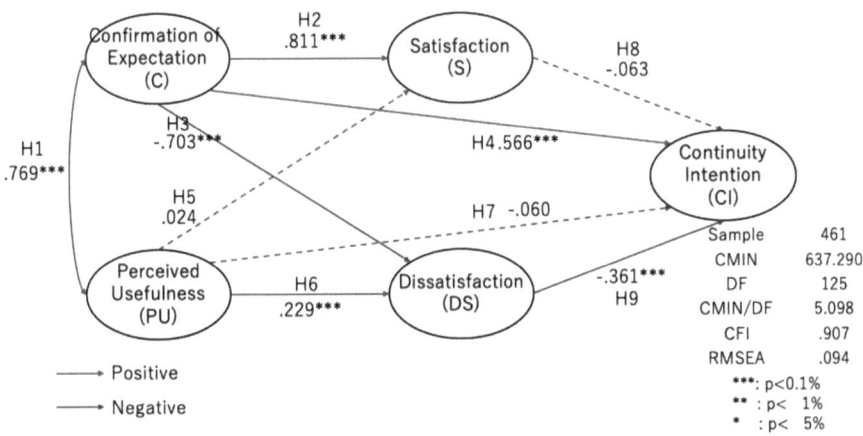

Fig. 2. Analysis Result of Active User

Table 2. Result of Analysis

			Value		
	Participant		Active User		
	Sample		461		
	CMIN		637.290		
	DF		125		
	CMIN/DF		5.098		
	CFI		.907		
	RMSEA		.094		
	Model Evaluation		Borderline		
	Hypothesis	Path	Value	Sig.	Support
H1	Confirmation of Expectation (C) has a positive influence on Perceived Usefulness (PU).	C → PU	.769	***	YES
H2	Confirmation of Expectation (C) has a positive influence on Satisfaction (S).	C → S	.811	***	YES
H3	Confirmation of Expectation (C) has a negative influence on Dissatisfaction (DS).	C → DS	-.703	***	YES
H4	Confirmation of Expectation (C) has positive influence on Continuity Intention (CI).	C → CI	.566	***	YES
H5	Perceived Usefulness (PU) has a positive influence on Satisfaction (S).	PU → S	.024		NO
H6	Perceived Usefulness (PU) has a negative influence on Dissatisfaction (DS).	PS → DS	.229	**	REJECT
H7	Perceived Usefulness (PU) has a positive influence on Continuity Intention (CI).	PU → CI	-.060		NO
H8	Satisfaction(S) has a positive influence on Continuity Intention (CI).	S → CI	-.063		NO
H9	Dissatisfaction (DS) has a negative influence on Continuity Intention (CI).	DS → CI	-.361	***	YES

4 Discussion

4.1 Model Assessment

CMIN/DF is the chi-square (CMIN) divided by Degree of Freedom (DF) and is a model fit index commonly used in structural equation modeling (SEM). Marsh and Hocevar [23] recommend using ratios as low as two or as high as five to indicate a reasonable fit. The CMIN/DF value of the model shown in Fig. 2 and Table 2 is slightly above 5, so the hypothetical model is not a good fit for the data collected.

Comparative Fit Index (CFI) is also a widely used statistical measure for assessing model fitness to the observed data. CFI values ranged from 0 to 1. Traditionally, CFI value of 0.90 or higher is considered acceptable, and 0.95 or higher is considered indicative of a very good fit. Traditionally, a CFI value of 0.90 or higher is considered acceptable, and 0.95 or higher is considered indicative of a very good fit. The CFI value of the model shown in Fig. 2 and Table 2 is .907, which is within the acceptable range.

Root Mean Square Error of Approximation (RMSEA) is widely used as a fit index in SEM. RMSEA = 0.05: Excellent fit, $0.05 <$ RMSEA = 0.08: Acceptable fit, $0.08 <$ RMSEA = 0.10: Borderline fit, and RMSEA > 0.10: Poor fit [24]. The RMSEA value of the model shown in Fig. 2 and Table 2 is .094; therefore, it is considered a borderline fit.

Based on the results of these analyses, we believe that the hypothetical model has a borderline fit. In this study, we recognize the need for model improvement. We also assumed that since the data collected by the questionnaire contained many fraudulent responses, it was not always possible to eliminate all fraudulent response data, and that the data analyzed contained fraudulent responses. Therefore, we recognize the need to improve the questionnaire questions to make them easier to understand and answer, as well as the need to improve the method of fraudulent response detection. While recognizing these problems, we assume that the hypothetical model is sound enough to examine the reasons for the continued use of QR code payments using smartphones. The results are discussed in the following sections.

4.2 Discussion

Hypothesis testing and discussion are conducted based on the results of analyzing data collected from users of QR code payments using smartphones.

[H1: Confirmation of Expectation (C) has a positive influence on Perceived Usefulness (PU).] is supported by the fact that there is a positive and significant relationship between (C) and (PU). Hence, if expectations for QR code payments are fulfilled, users consider its usefulness to be perceived.

[H2: Confirmation of Expectation (C) positively influences satisfaction (S).] is supported by the fact that there is a positive and significant relationship between (C) and (S). Hence, if the expectation of QR code payments is fulfilled, users are satisfied with the QR code payments.

[H3: Confirmation of Expectation (C) has a negative influence on Dissatisfaction (DS).] is supported by the fact that there is a negative and significant relationship between

(C) and (DS). Hence, if expectations for QR code payments are fulfilled, users are expected to be less dissatisfied with QR code payments.

[H4: Confirmation of Expectation (C) has a positive influence on Continuity Intention (CI).] is supported by the fact that there is a positive and significant relationship between (C) and (CI). Hence, if expectations for QR code payments are fulfilled, users are likely to continue using QR code payments. As this analysis is based on the data of continued users of QR code payments, it can be inferred that these active users fulfilled their expectations for QR code payments.

[H5: Perceived Usefulness (PU) has a positively influences satisfaction (S).] is not supported because there is no significant relationship between (PU) and (S). Hence, we consider that even if users recognize the usefulness of QR code payments, they are not necessarily satisfied with QR code payments.

[H6: Perceived Usefulness (PU) has a negative influence on Dissatisfaction (DS).] is rejected because there is a positive and significant relationship between (PU) and (DS). Hence, the results confirm that the perceived usefulness of QR code payments increases user dissatisfaction with QR code payments. Currently, we do not have a rational interpretation of the result.

[H7: Perceived Usefulness (PU) has a positive influence on Continuity Intention (CI).] is not supported because there is no significant relationship between (PU) and (CI) with respect to (PU). Hence, we believe that even if users recognize the usefulness of QR code payments, they do not necessarily use QR code payments continuously. In Japan, cash is still the main payment method in addition to QR payments using smartphones, and credit card payments are widespread [1]. Even if QR code payment users are aware of the usefulness of QR code payments, we can assume that QR code payments are just one of the various payment methods and that they are easily shifting to other payment methods or using them in combination.

[H8: Satisfaction(S) has a positive influence on Continuity Intention (CI).] is not supported because there is no significant relationship between (S) and (CI). Hence, we believe that even if users are satisfied with QR code payments, they do not necessarily continue to use QR code payments. What we mentioned in the discussion of H7 can also be applied to that of H8. Since cash is still the main payment method in Japan, besides QR code payments using smartphones, credit card payments are also widespread [1], even if users of QR code payments are satisfied with QR code payments. QR code payments are just one of the various payment methods, and they can easily move on to other payment methods or use them in combination.

[H9: Dissatisfaction (DS) has a negative influence on Continuity Intention (CI).] is supported by the fact that there is a negative and significant relationship between (DS) and (CI). Hence, if a user is not dissatisfied with QR code payments, the user is likely to continue using QR code payments. Active users of QR code payments using smartphones are considered to use QR code payments only because they are not dissatisfied with it.

To summarize these considerations, users of QR code payments using smartphones are aware of their usefulness. On the other hand, since QR code payments are just one of the various payment methods in Japan, it can be inferred that they are using it not because they are satisfied with it but because they are not dissatisfied with it.

5 Conclusion

This study clarified the relationship between consumer satisfaction and dissatisfaction with QR code payments using smartphones in Japan and their willingness to continue using them by conducting a quantitative analysis of data collected through a questionnaire survey. Through this study, it is inferred that users of QR code payments using smartphones recognize their usefulness and that they are continuously using QR code payments not because they are satisfied with it, but because they are not dissatisfied with it.

In Japan, the introduction and use of QR code payments were promoted by the consumption tax hike in 2019, and many introduction and use promotion campaigns were conducted by the Japanese government and others. Recently, an increasing number of businesses have been unwilling to welcome QR code payments from consumers because of loss of payment fee exemptions. However, the mechanism by which consumers continue to use QR payments is unclear. In this study, we approached the mechanism of continued use of QR code payments by focusing on consumer satisfaction and dissatisfaction with QR code payments, using the model of continued use of information technology as an aid. We believe that this is a novel approach.

The Findings of this study also indicate how consumers feel about the continued use of QR code payments and thus provide prerequisite knowledge for QR code payment providers to develop their services.

In the analysis of active users, H6: Perceived Usefulness (PU) has a negative influence on Dissatisfaction (DS). Although H6 was rejected because there was a positive and significant relationship between (PU) and (DS), a rational interpretation of this result has not been reached. Continued consideration of a reasonable interpretation is needed.

The model constructed in this study was built as a hypothetical model by employing the models proven in previous studies. We then attempted to validate the model based on the data collected through the questionnaire but were unable to ensure a reliable model fit. Therefore, it is necessary to improve this model through continuous exploration.

Additionally, the collected data contained many fraudulent responses. The survey company collected 120% of the target sample size and used the AI functionality provided by the company to eliminate fraudulent data to obtain data for the target sample size. However, it cannot be said that all fraudulent responses have been eliminated.

In future studies, we will improve the model constructed in this study and aim to construct a model with a higher degree of fit. In doing so, we will not only review the latent variables but also focus on the observed variables that constitute the latent variables to improve the model.

Finally, as part of increasing the accuracy of the observed variables that comprise the latent variables, a review of the questionnaire design will also be conducted to improve the reliability of the collected data.

Acknowledgments. This study was partially supported by a Grant-in-Aid (Grant #2318) sponsored by the Zengin Foundation for Studies on Economics and Finance.

Disclosure of Interests. The authors have no competing interests to declare relevant to the content of this article.

Generative AI Usage. Generative AI has been used to perform english editing to write this paper.

References

1. Shimizu, M.: Will QR code payments become popular in Japan in the first year of the cashless society? Comparison from cases in China and the U.S., Regional and Analytical Reports, Japan External Trade Organization (JETRO) (2019). (in Japanese). https://www.jetro.go.jp/biz/areareports/2019/1619a2493a52b0a4.html. Accessed 28 Apr 2025
2. Payments Japan Association: Survey of Code Payment Usage Trends, Published March 14, 2025 (2025). (in Japanese). https://paymentsjapan.or.jp/code-payments/202410-12/. Accessed 28 Apr 2025
3. Rogers, M.E.: Diffusion of Innovations, 5th edn. Free Press (2003)
4. Geoffrey, A.M.: Crossing the Chasm. Perfect Bound (1991)
5. Fujiki, H.: Will the widespread use of cashless payments reduce the frequency of the use of cash payments? Rev. Econ. Anal. **14**(1), 89–120 (2022) https://doi.org/10.15353/rea.v13i3.4522
6. Ajzen, I.: From intentions to actions: a theory of planned behavior. In: Kuhl, J., Beckmann, J. (eds.) Springer, Heidelberg (1985)
7. Dawn W.F., Nicole, F.: Chapter 33 - Peer influences on addiction. In: Principles of Addiction, pp. 323–331. Academic Press (2013). https://doi.org/10.1016/B978-0-12-398336-7.00033-4
8. Davis, F.D.: Perceived usefulness, perceived ease of use, and user acceptance of information technology. MIS Q. **13**(3), 319–340 (1989). https://doi.org/10.2307/249008
9. Venkatesh, V., Davis, F.D.: A theoretical extension of the technology acceptance model: four longitudinal field studies. Manage. Sci. **46**(2), 186–204 (2000) https://doi.org/10.1287/mnsc.46.2.186.11926
10. Venkatesh, V.: Determinants of perceived ease of use: integrating control, intrinsic motivation, and emotion into the technology acceptance model. Inf. Syst. Res. **11**, 342–365 (2000)
11. Venkatesh, V., Morris, M.G., Davis, G.B., Davis, F.D.: User acceptance of information technology: toward a unified view. MIS Q. **27**(3), 425–478 (2003) https://doi.org/10.2307/30036540
12. Venkatesh, V., Bala, H., Technology acceptance model 3 and a research agenda on interventions. Decis. Sci. **39**(2), 273–315 (2008). https://doi.org/10.1111/j.1540-5915.2008.00192
13. Gefen, D., Karahanna, E., Straub, D.W.: Trust and TAM in online shopping: an integrated model. MIS Q. **27**(1), 51–90 (2003). https://doi.org/10.2307/30036519
14. Bhattacherjee, A.: Understanding information systems continuance: an expectation-confirmation model. MIS Q. **25**(3), 351–370 (2001). https://doi.org/10.2307/3250921
15. Venkatesh, V., Thong, J.Y.L., Xu, X.: Consumer acceptance and use of information technology: extending the unified theory of acceptance and use of technology. MIS Q. **36**(1), 157–178 (2012). https://doi.org/10.2307/41410412
16. Venkatesh, V., Morris, M.G.: Why don't men ever stop to ask for directions? Gender, social influence, and their role in technology acceptance and usage behavior. MIS Q. **24**(1), 115–139 (2000). https://doi.org/10.2307/3250981
17. Venkatesh, V., Brown, S.A., Maruping, L.M., Bala, H.: Predicting different conceptualizations of system use: the competing roles of behavioral intention, facilitating conditions, and behavioral expectation. MIS Q. **32**(3), 483–502 (2008). https://doi.org/10.2307/25148853
18. Shukla, A., Mishra, A., Dwivedi, Y.K.: Expectation Confirmation Theory: A review. In: Papagiannidis, S. (ed.) TheoryHub Book (2024). https://open.ncl.ac.uk

19. Cashless Promotion Office, Commerce and Services Group, Ministry of Economy, Trade and Industry: Summary of the Study Group on Environmental Improvement for Further Promotion of Cashless Payments to Small and Medium-Sized Stores. (2023). https://www.jetro.go.jp/biz/areareports/2019/1619a2493a52b0a4.html. 28 Apr 2025
20. Abisikha, K.C., Nagayoshi, S.: Analyzing factors influencing continued QR code payments use: a study of Japan's post-government campaigns. in the presentation material of the international conference on business. Economy and Information Technology 2025 (ICBEIT 2025), 22–23 March 2025, Cebu, Philippines (2025)
21. Herzberg, F., Mausner, B., Snyderman, B.B.: The Motivation to Work, 2nd edn. John Wiley, New York (1959)
22. Collins, L.M.: Research Design and Methods. Encyclopedia of Gerontology, 2nd edn., pp. 433–442. Elsevier (2007) https://doi.org/10.1016/B0-12-370870-2/00162-1
23. Marsh, H.W., Hocevar, D.: Application of confirmatory factor analysis to the study of self-concept: first- and higher-order factor models and their invariance across groups. Psychol. Bull. **97**, 562–582 (1985)
24. Statistische, B.L.M.: How to interpret SEM model fit results in AMOS (2025). https://www.StatistischeDatenAnalyse.de/images/services/How_to_interpret_SEM_model_fit_results_in_AMOS.pdf. Accessed 1 May 2025

Feasibility Analysis on Green Business Plan from Biomass

Danai Tanamee and Suppanunta Romprasert[✉]

Faculty of Economics, Srinakharinwirot University, Bangkok, Thailand
{danait,suppanunta}@g.swu.ac.th

Abstract. The development of alternative energy from the residual resources within the community, such as soybean meal, coconut fiber, corn, sugar cane and rice straw, can create a source of energy called "fuels" for the development of alternative energy in the community sustainable self-reliance. The objectives are 1) to study the factors influencing the decision to use alternative energy as biodiesel in addition to the cost and return factors. 2) to study the trend of community demand for alternative energy using biodiesel and community energy management as alternative energy. The community may have people, who want or do not want to use alternative energy distributed in the community. This research will use the Accidental Sampling method to survey all 400 community members. The best case is a decrease in the cost of biodiesel equipment. The cost of construction and renovation of biodiesel plants has decreased to 12%. For the worst case, there is an increase in the cost of biodiesel equipment. The cost of construction and renovation of biodiesel plants has increased to 12%. In the best case, as a result of the decrease in cost of biodiesel. The cost of construction and renovation of biodiesel plants has decreased. As a result, the project cost is reduced to only Baht 162,400 when analyzing the project's cost and benefits. The project will have a net present value (NPV) of Baht162,029.86, representing a cost to income ratio. The Benefit-to-Cost Ratio is 2 times, and the IRR is 47.47% per annum.

Keywords: Biomass · Green Energy · Cost · Benefit · Community

1 Introduction

Energy plays a crucial role in almost every activity on our planet today, especially in relation to electricity consumption. It is an essential part of life, influencing personal activities from the moment an individual wakes up in the morning until the evening—whether cooking meals, relaxing with air conditioning, or using an elevator in a high-rise building at work. If energy supply diminishes to a crisis level, it results in a shortage, preventing people from carrying out their daily activities effectively.

Gasoline is considered a key factor in driving the economies of many countries worldwide, as reflected in gasoline price trends from 2024 to 2025. In various countries, including the United States, the United Kingdom, Belgium, Sweden, France, Australia, and Germany, biodiesel has been in use for a long time. Biodiesel production involves blending specific proportions of biodiesel and diesel fuel, such as a 2:98 ratio in refineries or a 5:95 blend in diesel fuel production. Additionally, in France, biodiesel has been incorporated into public transportation, with a 40% biodiesel composition. (Energy Policy and Planning Office, Ministry of Energy, 2017).

In Thailand, the demand for cars continues to grow each year, and every vehicle requires gasoline to operate. If the country's gasoline supply fails to meet the ever-increasing national demand, the proportion of imported gasoline will become significantly high. (Energy Policy and Planning Office, Ministry of Energy, 2017) (Tables 1 and 2).

Table 1. Gasoline Prices in 2024.

2024	Hi-Premium-Diesel S	Hi-Diesel S	Gasohol E85	Gasohol E20 S	Gasohol 91	Gasohol 95
20/12	47.14	32.94	33.89	34.14	35.88	36.25
26/11	47.14	32.94	33.99	34.24	35.98	36.35
31/10	47.14	32.94	32.89	33.14	34.88	35.25
28/09	47.14	32.94	32.89	33.14	34.88	35.25
23/08	47.14	32.94	33.99	34.24	35.98	36.35
25/07	47.14	32.94	35.99	36.24	37.98	38.35
29/06	47.14	32.94	36.39	36.64	38.38	38.75
31/05	47.14	32.94	35.79	36.04	37.78	38.15
27/04	45.14	30.94	37.99	38.24	39.28	40.35
26/03	43.64	29.94	36.79	37.04	37.68	39.15
20/02	43.64	29.94	36.09	36.34	36.68	38.45
31/01	43.64	29.94	35.59	35.44	35.78	37.55

Source: Bangchak.co.th (Unit: Baht/Kilogram)

Table 2. Gasoline Prices in 2025.

2025	Hi-Premium-Diesel S	Hi-Diesel S	Gasohol E85	Gasohol E20 S	Gasohol 91	Gasohol 95
11/1	47.14	32.94	33.89	34.14	35.88	36.25

Source: Bangchak.co.th (Unit: Baht/Kilogram)

When considering the national agenda for the policy on renewable energy, the government has been trying to support the production and the promotion of renewable energy. It is evident that efforts to develop biofuels and biomass encourage the use of renewable energy at all levels, ranging from households to villages also toward to the national level- through appropriate incentives. Additionally, there has been a push to integrate natural gas systems into the transportation sector, alongside ongoing research into various forms of alternative energy.

If one reflects carefully, it becomes clear that another form of energy, often overlooked, is just as vital as electricity, coal, and natural gas. This natural energy has existed for hundreds of millions of years and remains an immense and inexhaustible resource—unless the Earth were to become completely cold. Dr. Somporn, Deputy Director of the Academic Resources Center at Prince of Songkla University, has spearheaded a solar energy project based on his recognition that Thailand will likely face electricity and oil shortages in the near future. Solar energy is considered as a clean alternative energy used in the activities of the service. The production and installation of electrical equipment have been used to illuminate the facade of service centers instead of electricity. It is considered as a model for dissemination of solar energy. (Department of Educational Technology, Resource Center Prince of Songkla University, 2017). But the promotion of solar power generation for distribution to the state is still of high value today, especially in terms of funding needed to install solar panels. Therefore, the distribution of benefits may be concentrated among select groups (Chanovit, 2014).

However, alternative clean energy that is used to replace oil in countries around the world is "biodiesel", which is most used in Germany followed by the United States and France. Biodiesel consumption in the world is on the rise each year. In the near future, the volume is nearly half of the production and consumption of global fuel. Thailand's "biodiesel" must comply with the government regulations in 2007, which established two main formulas: Biodiesel B5, used for cars and biodiesel B100-commonly known as "Biodiesel Community", colored with purple- used for agricultural engines. PTT Energy Technology Research Institute (2007) at Wang Noi district Ayutthaya Province found that biodiesel is considered as a very environmentally friendly energy source compared to diesel in terms of sulfur dioxide toxicity, ozone depletion, small particulates, carbon monoxide, hydrocarbon and the exhaust from the engine. His Majesty King Bhumibol Adulyadej has advocated the use of biodiesel. The use of animal fat as a fuel source for cars has been experimented with. This alternative fuel would replace the import of oil from abroad in large numbers each year. Biodiesel is a source of energy that can be easily found in Thailand because plants used as agricultural raw materials are available in the country. This also creates opportunities for farmers to earn supplementary income (Office of Energy Policy and Planning, 2017).

Therefore, the development of alternative energy from the residual resources within the community, such as soybean meal, coconut fiber, corn, sugar cane and rice straw, can create a source of energy called "fuels" for the development to be alternative energy in local communities, contributing sustainable self-reliance.

The objectives of this paper are (1) to study the trend of community demand for alternative energy using biodiesel and community energy management as alternative energy and (2) to analyze the costs and benefits of using biodiesel in the community.

2 Literature Review

Energy is available everywhere on this planet, but the energy that makes people comfortable and allows factories to run their machines to produce goods is electrical energy. But there are also other forms of energy, such as oil, natural gas, coal, and biodiesel, which are considered the most modern form of alternative energy today because biodiesel can be developed to replace diesel. In terms of policy, especially renewable energy, there is government support for production and use, as well as for the development of alternative energy.

In terms of the measurement of project's economic value, it is considered an important index for investment decision making. When analyzing many projects, it can show the sequence of projects in investment, which is generally measured through 3 indicators as follows: (1) Net Present Value (NPV) is a calculation of the difference between the present value of income expected to be received each year throughout the life of the project and the present value of the expenses paid (Hardacer, Hurnie, Anderson, and Lien, 2004); (2) Benefit-Cost Ratio (B/C Ratio) is an indicator for a decision to invest when the value of the benefit-cost ratio is greater than or equal to one. If the number is less than one, the investment in that project should not be decided. It is calculated by the present value of income throughout the project's life divided by the present value of costs throughout the project's life; and (3) the project's internal rate of return (IRR), which is commonly used for project feasibility analysis. It is a tool for Long-Term Capital Budgeting because the obtained value reflects the rate of return on investment at the break-even point. It is used to compare with the interest rate or financial cost or project cost to determine whether it is more or less than the interest rate or financial cost (Lee, Boehlje, Nelson, and Murray, 1980; Warren, 1982; Illes, 2002; Ross, Westerfield, and Jaffe, 2005).

Discount rate There is a need to adjust future costs and benefits to their current worth, called "present value". In calculating the present value (Present Worth), it is necessary to choose the discount rate, which can be divided into two rates:

a) Cutoff rate. If considered for financial analysis, it is calculated from the final cost of money to the business or the rate that the enterprise can borrow. If economic analysis is used, then the opportunity cost of capital is considered to be a reflection of social choice.

b) Borrowing rate refers to the economic analysis of cost that countries have to pay for projects that the country expects to borrow for investment, and the rate of social preference applies to the future return to society. Typically, it is lower than the discount rate per person, because social projects, such as public projects, has a longer duration than a private project.

Sensitivity Measurement is one of the simplest and most widely used methods for analyzing uncertainty. It determines, if there is a change in assumptions, what will be the effect on the suitability of the project? (Cheung & Chappell, 2002).

Positive and Negative Impact is considered to be a cost and part of the project, either public or private, such as the environment, society, community, etc. (Coase, 1960).

3 Methodology

Questionnaires are used to collect data. The community may have people who want or do not want to use alternative energy distributed in the community. This research will use the Accidental Sampling method to survey all 400 community members. In this case, communities with less than one year of alternative energy use will be excluded from the sample.

Data collection was conducted through in-depth interviews to further research by community representatives who decided to use biodiesel alternatives.

Research tools use data analysis from the questionnaires to analyze data relationships by testing statistical data, including data analysis from in-depth questionnaires. The results are then analyzed for the cost and return of alternative energy, biodiesel, using the payback period, yield, and cost-benefit ratio.

In this research, the assumption is made that analyzing the cost and benefits to the family for using normal energy such as electricity, cooking gas, and conventional fuel, is to be compared with biodiesel from waste oil in the community, using a discount rate of 12% (Discount Rate).

4 Discussion

We conduct a survey to develop a community biodiesel energy development plan by distributing questionnaires. The questionnaires are divided into 2 types as follows: Type 1 is a questionnaire for members. The content includes general information of respondents, such as gender, occupation, income, biodiesel test, opinions about current energy community awareness, and future energy needs, including the acceptance of alternative fuels like biodiesel from the community. An open-ended questionnaire was used to provide feedback and suggestions for community biodiesel management. Type 2 is a questionnaire for community leaders, consisting of questions about the community energy situation, community energy management, and community participation in community energy management. It is used as primary data in terms of overall picture, including cost data from community representatives in making biodiesel for use in the community. Data collected from related sources such as the Department of Alternative Energy and Energy Conservation, Ministry of Energy, academic reports and related journals, the Internet, and other research sources are considered secondary data.

Descriptive and quantitative analysis To show the background and to see the appropriateness of investment in community biodiesel projects using an economic method called cost-benefit analysis by comparing the cost of biodiesel production and the benefits with time adjustment. It emphasizes on financial considerations. On the other hand, it is an economic analysis using community resources by combining the value of resources used in biodiesel production that are saved and the value reduced from the discharge of used plants in the community, which is an indirect benefit (Tables 3 and 4).

Type 1 questionnaire is as follows:

1: Perception of alternative energy education
1.1 Gender consists of

    1.1.1 male
    1.1.2 female

1.2 Age specific the year

Table 3. Section 1.

Month	%Change in Q_B	%Change in P_B	Elasticity of Benzene
January	−0.0254	0.0220	−1.1545
February	0.0219	0.0170	1.2906
March	−0.0351	0.0190	−1.8474
April	0.0185	0.0170	1.0853
May	−0.0216	−0.0080	2.7000
June	0.0257	−0.0080	−3.2113
July	−0.0246	0.0090	−2.7333
August	0.0166	−0.0470	−0.3536
September	−0.0095	−0.0360	0.2639
October	−0.0387	−0.0080	4.8375
November	−0.0011	−0.0050	0.2200
December	0.0682	−0.0100	-6.8200
		Average Elastic	-0.4769

Table 4. GasolineE20.

Month	%Change in Q_E	%Change in P_E	Elasticity of E20
January	0.0042	0.0170	0.2465
February	0.0616	0.0190	3.2405
March	-0.0018	0.0180	-0.1000
April	0.0613	0.0190	3.2242
May	-0.0079	-0.0100	0.7900
June	0.0125	-0.0100	-1.2520
July	0.0152	0.0090	1.6911
August	0.0045	-0.0280	-0.1600
September	-0.0220	-0.0080	2.7500
October	-0.0742	-0.0060	12.3667
November	-0.0571	-0.0060	9.5167
December	0.1653	-0.0110	-15.0236
		Average Elastic	1.4408

1.3 Educational area consists of

   1.3.1 Vocational Management (Primary)

1.3.2 Vocational Management (Master)
1.3.3 Educational Technology in Vocational and Technical Education
1.3.4 Curriculum and Vocational Education
1.3.5 Science Education
1.3.6 Electrical Engineering
1.3.7 Electronic
1.3.8 Architecture
1.3.9 Agricultural education
1.3.10 Industrial Management
1.3.11 Industrial Design Technology/Industrial Products Technology
1.3.12 Applied Linguistics/English for Science and technology

1.4 Career consists of

    1.4.1 Government services
    1.4.2 Officers in the State Regulatory Agency
    1.4.3 State enterprise
    1.4.4 Private company
    1.4.5 Private business
    1.4.6 Studying
    1.4.7 Others (specific)

1.5 Monthly income is stated in baht.
2. The questionnaire about the information received from the renewable energy media illustrated the sequence of media forwarding the information to the respondents about renewable energy.

    2.1 Newspaper, Weekly magazine, Weekly magazine
    2.2 Television
    2.3 Radio
    2.4 Publications of different agencies
    2.5 Exhibitions organized by public or private organizations
    2.6 Electronic media (website) of various agencies
    2.7 Seminars or training provided by government organizations; private

3. A questionnaire about awareness of renewable energy, indicating the level of feedback on renewable energy awareness, contains 32 questions and divided into 5 levels

    3.1 Score 5 Most Commented
    3.2 Score 4 Very Comments
    3.3 Score 3 Moderate level
    3.4 Score 2 Less comment level
    3.5 Score 1 Minimum level of feedback

4. Questionnaire for comments and suggestions. It is an open-ended questionnaire that asks for more ideas on why people want to gain knowledge about alternative energy.

4.1 You know about other types of renewable energy. In addition to the above-mentioned questionnaire.
4.2 What do you think is the main reason you want to know about renewable energy?
4.3 Other suggestions.

As above tables the price elasticity of demand for Benzene and Diesel are inelastic at 0.4768 and 0.3925. It shows the consumers will not change their consumption much when price changes. On the other hand, the demand of gasoholE20 is elastic. For our discussions, the car that use Benzene or gasoline91 are high engine and high quality. Its price is usually higher than normal cars. The owners will take a good care; moreover, they can afford it and also higher price of gasoline. So, the demand is not elastic. Similarly, the consumers of Diesel are mostly business sector. So regardless of how much the price increases, they have to use it. Conversely, gasolineE20 is the alternative fuel introduced for reduce using crude oil which also decrease cost. Most people change to use gasolineE20 as its price is lower that others. Therefore, its demand is sensitive to price change.

Regarding the cost analysis and financial benefits of the project, the case study of the community is a small community consisting of about 150 households. In terms of cost, there will be costs of construction and improvement of biodiesel, and expenses for equipment used in making biodiesel. It is worth about 180,000 baht. The cost of consuming raw materials from households falls by about 50 baht per household, which costs about 7500 baht per month in terms of reducing the cost of community energy. The project will save Baht 15 per liter of biodiesel and a biodiesel capacity of around 1000 L per month. The monthly benefit is 15,000 baht, minus the monthly cost, which is a net benefit of 7500 baht per month or 90,000 baht per year. If assuming a 5-year project life and a discount rate of 12%, then the net present value (NPV) is 144,429.86 baht. The Benefit-to-Cost Ratio is 1.8 times, and the IRR is 41.04 percent per year (Table 5).

Regarding the cost analysis and economic benefits of the project, it is found that the cost of buying raw materials from households falls by about 50 baht/household for about 7500 baht per month, even though it is the cost of the project. The households in the community have increased income at 50 baht per month. It also reduces the amount of waste discharged from used plants in the community. Therefore, if this benefit is used to offset the cost of raw materials, the net benefit of the project will be increased to Baht15,000. With the assumption that the project is 5 years old and 12% discounted, the project will have a net present value (NPV) is Baht 468,859.72, representing a cost to income ratio. The Benefit-to-Cost Ratio was 3.6 times, and the project's return on investment (IRR) was 96.59 percent per annum. The researchers collected insights from one household in the community. The average monthly household income is 15,000 baht, from the salary of household members and the income from repairing trucks. The monthly cost of about 12,000 baht is for food, electricity, water, equipment, miscellaneous costs, and energy costs of about 3000 baht. After the project, the energy cost reduced to only 600 baht per month, saving up to 2400 baht per month. In addition, the income from plants used as raw material for biodiesel production is about 50 baht per month. This household can save up to 29,400 baht per year (Tables 6, 7 and 8).

Table 5. Diesel.

Month	%Change in Q_D	%Change in P_D	Elasticity of Diesel
January	0.0123	0.0248	0.4960
February	0.0087	0.0000	undefined
March	0.0156	−0.0005	−31.2800
April	0.0026	0.0005	5.1600
May	−0.0079	0.0000	undefined
June	−0.0079	0.0000	undefined
July	−0.0439	0.0000	undefined
August	0.0051	−0.0219	−0.2347
September	0.0045	−0.0116	−0.3914
October	−0.0165	−0.0055	3.0000
November	0.0402	0.0184	2.1848
December	0.0484	0.0020	24.2050
		Average Elastic	0.3925

Table 6. Analysis of Costs and Benefits of Bio-Diesel Project on Community Projects.

Cost of Biodiesel Project (Direct Cost)	Benefit of Biodiesel Project (Direct Benefit)
The cost of purchasing raw materials from raw materials. household	Reduce community energy costs.
Buildings and improvements to biodiesel.	
Expenditure on equipment used for making Biodiesel	

After that, the sensitivity to change is analyzed to determine whether the change in situation will affect the feasibility of the biodiesel project in the community. The analysis is conducted under extreme assumptions. Key variables relevant to select biodiesel projects are identified, as they have the potential to impact the project's net present value. These values are then replaced with the highest and lowest possible values to recalculate the net present value. The results before and after the analysis are compared, as shown in the table.

The best case is a decrease in biodiesel equipment cost. The cost of construction and renovation of biodiesel plants has decreased to 12 percent. For the worst case, there is an increase in the cost of biodiesel equipment. The cost of construction and renovation of biodiesel plants has increased to 12 percent.

In the best case, as a result of the decrease in cost of biodiesel, the cost of construction and renovation of biodiesel plants has decreased. As a result, the project cost is reduced

Table 7. Analysis of the costs and benefits of the biodiesel project in economics. When there are projects in the community.

Cost of Biodiesel Project (Direct Cost)	Benefit of Biodiesel Project (Indirect Benefit)
The cost of purchasing raw materials from raw materials household.	The cost of biodiesel production cost savings.
Buildings and improvements to biodiesel.	The value is based on the amount of waste discharged from the community.
Expenditure on equipment used for making Biodiesel	

Table 8. Compare before and after analysis.

Factor	Best Case	Worst Case
The cost of biodiesel equipment	−7%	+7%
Construction and improvement of biodiesel plant systems	−12%	+12%

to only 162,400 baht when analyzing the project's cost and benefits. The project will have a net present value (NPV) of Baht162,029.86, representing a cost to income ratio. The Benefit-to-Cost Ratio is 2 times, and the IRR is 47.47% per annum.

While in the worst case when there is an increase in cost of biodiesel equipment, the cost of construction and renovation of biodiesel plants has increased 12%. This will increase the investment cost of the project to 197,600 baht. When analyzing the cost and benefits of the project.

The project will have a net present value (NPV) of Baht128,829.86. The Benefit-to-Cost Ratio is 1.64 times, and the IRR is 35.62% per annum. The project is still worth the money.

5 Conclusion and Policy Implication

Academic aspects Able to apply economic analysis and financial analysis principles to create a shared social concept based on the community.

Policy aspects Used as a tool to help decision-makers and increase social welfare for communities to see the most efficient use of resources in the community, including creating fairness in the joint use of local resources.

Economic/Commercial aspects It is an important factor in deciding on investment and the value of projects in the community.

Social and community aspects Used as a reflection of social choices in the community. Agencies that can use the research results.

Department of Energy and Provincial Office aspects To be used as a guideline for drafting development plans for other communities leading to the selection of communities that are ready to be established as prototype communities, groups of village headmen and villagers in the community to see the opinions and criticisms of the ideas of people in the community.

References

Bangchak Petroleum Public Company Limited: Oil prices back. Bangchak.co.th. (11 February 2018) (2018)

Bodie, Z., Kane, A., & Marcus, A.J.: Investment, 4th edn. McGraw-Hill (1999)

Chanovit, W.: "Analysis of Cost and Return of Solar Power Project Installed on Residential Roofs in Different Areas of Thailand. Term Paper Business Economics Program. National Institute of Development Administration. Educational Technology Department, Academic Resources Center Prince of Songkla University. techno.oas.psu.ac.th (February 5, 2017) (2014)

Cheung, Y.K., Chappell, R.: A Simple Technique to Evaluate Model Sensitivity in the Continual Reassessment Method. Biometrics **58**, 671–674 (2002)

Coase, R.H.: The Problem of Social Cost. J. Law Econ. **3**, 1–44 (1960)

Energy Policy and Planning Office, Ministry of Energy. (2017). www.eppo.go.th/index.php/en/ (5 February 2017)

Hardacer, J.B., Hurnie, R.B.M., Anderson, J. R., & Lien, G.: Coping with Risk in Agriculture, Second Edition, CABI Publishing, pp. 234–244 (2004)

Illes, M.: Vezetoi gazdasagtan. Bovitett masodik kiadas, Kossuth Kiado, pp. 85–175 (2002)

Lee, W.F., Boehlje, M.D., Nelson, A.G., & Murray, W.G.: Agricultural Finance, Seventh Edition, the Iowa State University Press, Ames, pp. 59–80 (1980)

Office of Energy Policy and Planning. Renewable Energy Knowledge. Energy Conservation Promotion Fund Public Relations "Total Power Division 2". www.eppo.go.th (5th February 2017)

PTT Research and Technology Institute. Form 56-1, Year 2007. www.pttplc.com/th/About/rti/Pages/default.aspx (5 February 2017)

Ross, S.A., Westerfield, R.W., & Jaffe, J.: Corporate Finance, International Edition, McGraw-Hill (2005)

Warren, M.F.: Financial Management for Farmers, the Basic Techniques of 'Money Farming', Third Edition. Stanly Thornes Ltd, pp. 240–260 (1982)

Author Index

A
Abisikha, K. C.   515
Afriansyah, Aidil   131
Alonso-González, Miguel   118

B
Bello, Daisy Ipatzi   369
Benítez-Morejón, Yahaira   247
Bunno, Teruyuki   217, 226

C
Cai, Yuanyuan   455
Candás, Juan Luis Carús   354
Chang, Yong-Yi   70
Chang, Yu-Ching   3
Chao, Chian-Hsueng   311
Chen, Chun-Hao   298
Chicaiza, Janneth   247
Chiu, Miao-Hui   498
Chiu, Yen-Chih   59
Cui, Yu   142

E
El Hassani, Samir   396
Espada, Jordán Pascual   354

F
Firmansyah, Hafiz Budi   131
Fu, Zhen-Xin   156

G
Gao, Yanzheng   431, 443
García-Díaz, Vicente   118
Goto, Kentaro   178
Guo, Zhiwei   419, 455

H
Hernández, Diana Patricia Eljach   263, 279
Hernández, Helmer Muñoz   263, 279
Herraiz, José Javier Martínez   192, 206

Hong, Tzung-Pei   298
Hou, Jian-Ren   3
Hsiao, Han-Wei   106
Hsu, Mei-Yun   16
Hsu, Pei-Chen   311
Hu, Xia   419
Huang, Haiyan   431, 443
Huang, Yi-Chun   498

I
Idota, Hiroki   142, 217, 226

J
Julio, Yamid Fabian Hernández   263, 279

K
Khulyati, Laila Diana   81
KOGA, Hiroshi   39
Kuo, Ying-Feng   3

L
Lee, Chung-Hong   156
Lee, Yen-Chan   94
Lee, Yun-Zhen   106
Liao, Yuxin   419
Liberona, Dario   324, 338
Lin, Jia-Chi   70
Lin, Yi-Hsun   3
Liu, Bo   455
Liu, Li   431
Liu, Yun-Hsiu   16, 59
López Pérez, Benjamín   118
Luengo, Alberto Larena   206

M
Mejía, Alex   167
Merodio, José Amelio Medina   192, 206
Minetaki, Kazunori   16, 59, 217, 226, 238
Montenegro, Laura   354

N
Nagayoshi, S.   47, 515
Nagayoshi, Sanetake   81
Nakamura, J.   47

O
Oliva, Mikel Ferrer   192, 206
Ota, Masaharu   142

P
Paladini, Stefania   369
Pan, Xueni   431, 443
Perez, Charles   396
Pertuz, Leonardo Antonio Díaz   263, 279
Prioló, Kavir Ala Oviedo   263, 279

R
Rodriguez, Carlos Cilleruelo   192
Romprasert, Suppanunta   481, 527
Romprasert, Tanapat   481
Rother, Marcel   338

S
Sano, Natsuki   408
See, Tian-Yi   311
Shen, Yu   431, 443
Sokolova, Karina   385
Sun, Jun-Teng   156

T
Tahir, Sabeen   369
Tanamee, Danai   481, 527
Tao, Yejun   408
Ting, I-Hsien   16, 59, 226, 238
Tsai, Yu-Chuan   298
Tu, Yuping   419

V
Valdiviezo-Diaz, Priscila   167, 247

W
Wang, Kai   70, 94
Wang, Quyuan   419
Wang, Ying   467
Wei, Xiu-Hua   311
Weng, Tzu-Ting   156

X
Xiao, Ruiwen   443

Y
Yang, Jing-Chi   298
Yang, Junchao   431, 443
Yang, Min-Li   498
Yeh, Chih-Hsuan   94
Yen, Chia-Sung   30, 59
Yu, Rongkui   467

Z
Zeng, Run   419
Zhao, Han   455
Zhou, Jiachun   431

MIX
Papier aus verantwortungsvollen Quellen
Paper from responsible sources
FSC® C105338

If you have any concerns about our products,
you can contact us on
ProductSafety@springernature.com

In case Publisher is established outside the EU,
the EU authorized representative is:
**Springer Nature Customer Service Center GmbH
Europaplatz 3, 69115 Heidelberg, Germany**

Printed by Libri Plureos GmbH
in Hamburg, Germany